ORGANIC MICROPOLLUTANTS
IN THE AQUATIC ENVIRONMENT

Proceedings of a symposium organized by the Commission of the European Communities, Directorate-General Science, Research and Development, Environment Research Programme, in cooperation with
– Bundesministerium für Gesundheit und Umweltschutz, Wien
  and
– Der Oesterreichische Wasserwirtschaftsverband
held in Vienna, Austria, 22–24 October 1985

Commission of the European Communities

ORGANIC MICROPOLLUTANTS
IN THE
AQUATIC ENVIRONMENT

Proceedings of the Fourth European Symposium
held in Vienna, Austria, October 22–24, 1985

Edited by

A. BJØRSETH
SCATEC, Slependen, Norway

and

G. ANGELETTI
Directorate-General for Science, Research and Development,
Commission of the European Communities, Brussels, Belgium

D. REIDEL PUBLISHING COMPANY

A MEMBER OF THE KLUWER  ACADEMIC PUBLISHERS GROUP

DORDRECHT / BOSTON / LANCASTER / TOKYO

Library of Congress Cataloging in Publication Data

Organic micropollutants in the aquatic environment.

At head of title: Commission of the European Communities, Bundesministerium für Gesundheit und Umweltschutz, Wien, Der Österreichische Wasserwirtschaftsverband.

Based on papers given at the Fourth European Symposium on Organic Micropollutants in the Aquatic Environment, held Oct. 22–24, 1985, in Vienna, Austria.

Includes index.

1. Organic water pollutants–Environmental aspects–Congresses. 2. Aquatic ecology–Congresses. 3. Water–Analysis–Congresses. 4. Water chemistry–Congresses. I. Bjorseth, Alf. II. Angeletti, G., 1943- . III. Commission of the European Communities. IV. Austria, Bundesministerium für Gesundheit und Umweltschutz. V. Österreichischer Wasserwirtschaftsverband. VI. European Symposium on Organic Micropollutants in the Aquatic Environment (4th : 1985 : Vienna, Austria)

TD427.O7O75   1986   628.1'68   86–3897

ISBN-13: 978-94-010-8571-7   e-ISBN-13: 978-94-009-4660-6

DOI: 10.1007/978-94-009-4660-6

Publication arrangements by
Commission of the European Communities
Directorate-General Information Market and Innovation, Luxembourg

EUR 10388
© 1986 ECSC, EEC, EAEC, Brussels and Luxembourg
Softcover reprint of the hardcover 1st edition 1986

LEGAL NOTICE
Neither the Commission of the European Communities nor any person acting on behalf of the Commission is responsible for the use which might be made of the following information.

Published by D. Reidel Publishing Company
P.O. Box 17, 3300 AA Dordrecht, Holland

Sold and distributed in the U.S.A. and Canada
by Kluwer Academic Publishers,
190 Old Derby Street, Hingham, MA 02043, U.S.A.

In all other countries, sold and distributed
by Kluwer Academic Publishers Group,
P.O. Box 322, 3300 AH Dordrecht, Holland

FOREWORD

The "Fourth European Symposium on Organic Micropollutants in the Aquatic Environment" was held in Vienna (Austria) from 22 to 24 October 1985. The Symposium was organized within the framework of the Concerted Action COST 641* which is included in the Third R & D Programme on the Environment of the Commission of the European Communities.

The aim of the Symposium was to review recent scientific and technical progress in the area of organic micropollutants in the aquatic environment and to present relevant research papers related to analytical methodologies, transformation reactions and transport of organic micropollutants in water, and water treatment processes. A special session was devoted to theoretical aspects and future activities. Furthermore, special poster sessions were organized where original contributions were presented.

This book presents the Proceedings of the Symposium including all review papers, presentations of research papers and extended versions of all posters. We believe that these Proceedings provide a good overview of the activities in this field in Europe. We are confident that it will constitute a valuable contribution to the understanding and solution of the problems posed by organic micropollutants in the aquatic environment.

The Commission of the European Communities whishes to express its gratitude to the co-organizers of the Symposium, Bundesministerium für Gesundheit und Umweltschutz, Wien, and Der Osterreichische Wasserwirtschaftsverband.

Brussels/Oslo, November 1985

G. ANGELETTI   A. BJØRSETH

* COST 641 :   Scientific and Technical Cooperation among
                                European Community Member Countries and
                                the Non-Member Countries Finland, Norway,
                                Sweden and Switzerland in the field of
                                "Organic Micropollutants in the Aquatic
                                Environment"

CONTENTS

TRANSPORT OF ORGANIC MICROPOLLUTANTS IN THE AQUATIC ENVIRONMENT

TRANSFORMATION IN THE AQUATIC ENVIRONMENT

WATER TREATMENT PROCESSES

ANALYTICAL METHODOLOGIES

Low dispersion liquid chromatography : Application of fast and narrow bore liquid chromatography to the analysis of micropollutants

Identification of surfactants and study of their degradation in surface water by mass spectrometry

Recent LC/MS methods for water analysis

The status of the development of standard procedures for the analysis of organic priority pollutants in water samples in the Federal Republic of Germany

Determination of organometallic compounds in the aquatic environment

HPLC determination of priority list aromatic hydrocarbons

FAB - Mass spectrometric study of the non volatile organic fraction in raw and drinking water extracts

Pollution conditions of the aquifer beneath a plant producing alkyd resins

Analyse de micropolluants organiques des eaux par couplage CG/SM - Collection de spectres de masse de référence

Improvement in estimation of polyaromatic hydrocarbons in seawater, marine sediments and organisms by spectro-fluorometry by using standard additions method

A comparison of procedures for trace enrichment of organophosphorus pesticide residues from water

The Kovats indices of some organic micropollutants on an SE54 capillary column

A dual detector gas chromatographic system for the routine determination of retention indices

Methods for the determination of EEC priority pollutants

Evaluation of sample preparation procedures for the determination of organic micropollutants in river sediments

Tracing a source of pollution by determination of specific pollutants in surface- and groundwater

Determination of extractable organohalogens in living tissue

Determination of acidic and neutral residues of alkyl-phenol polyethoxylate surfactants using GC/MS analysis of their TMS derivates

Determination of PCB-congeners in sediment samples using a retention index system

LOW DISPERSION LIQUID CHROMATOGRAPHY: APPLICATION OF FAST AND NARROW BORE LIQUID CHROMATOGRAPHY TO THE ANALYSIS OF MICROPOLLUTANTS.

C. Dewaele
Laboratory of Organic Chemistry, State University of Gent
Krijgslaan 281 (S4) 9000 Gent Belgium

Summary

The importance of dispersion in High Performance Liquid Chromatography (HPLC) is emphasised. The state of the art of Micro HPLC techniques such as high speed HPLC and narrow bore HPLC are reviewed.

1. Introduction

In the analysis of organic micropollutants by HPLC one is often confronted with two problems:
-the complexity of the sample and
-the detectibility of the pollutants in the sample.

All steps in the analytical procedure which improve this situation are benificial for both the qualitative elucidation and the quantitative determination :

1. A selective sample preparation can decrease the complexity of the sample and a preconcentration of the sample can increase the sensitivity.

2. Selective detection systems which show excellent sensitivity are also nessesary.

3. The column itself is the third important part in the chromatographic setup which can handle complexity and increase sensitivity. How this, from a point of view of column technology, can be achieved, is the topic of this paper.

The solution to the problem is often low dilution or low "dispersion". Low dispersion should be strived for. Dispersion is equivalent to band broadening and can therefore be discussed as follows:

$$N = 16(Vr/B)^2 \qquad Vr = Vo(k+1) \qquad Vo = 2\pi^2 L\emptyset$$

$$\text{thus } B = 4Vr/\sqrt{N} = 4\pi^2 L\emptyset(k+1)/\sqrt{N}$$

$$\text{and } D = 4Vo/\sqrt{N} \quad \emptyset \ r^2 L/\sqrt{N} \tag{1}$$

with N the plate number, Vr the retention volume, B the band width on the baseline, r the column radius, L the column length, Vo the column void volume, k the capacity factor, Ø the column porosity and D the dispersion.

The dispersion D as defined above is the band width of a non retarded peak (k = 0).

The dispersion can be decreased through a miniaturisation of three important characeristics of the column : the internal diameter (ID) , the column length (L) and the particle diameter of the packing material. HPLC systems where one or more of these parameters are miniaturised are called Micro HPLC systems. At the moment two Micro HPLC techniques, e.g. Fast or High Speed Liquid Chromatography (HSLC) and Narrow or Microbore HPLC are of interest in the practical analytical HPLC of micropollutants.

2.FAST HPLC : Micro HPLC through a miniaturisation of column length and particle diameter

Basic aspects

Over the past decade much effort has been directed towards increasing speed and efficiency of HPLC. At present, most HPLC analyses are carried out on 10 to 30 cm long columns (ID 0.4-0.5 cm) packed with 5 to 10 μm particles. One clear trend in column technology is the use of smaller particles packed in shorter columns. Small particles of 3 μm size are commercially available and their use is slowly being accepted by the chromatographic community(1). We demonstrated the use of 1 and 2 μm particles in short columns(2,3). From a historical point of view the use of small particles and short columns is not new. In 1930 Zechmeister (4) already mentioned the use of 1.2 to 3 μm aluminum oxide, clay or calcium carbonate particles while Tswett used 2 to 3 cm long columns for preparative separations. However at that time nor the instrumentation nor the theoretical background existed to explore the full potential of the small particle-small columns approach.

The following basic equations relate the analysis time to the most important HPLC parameters.

$$Tr = To\,(1+k) = L/u\,(1+k) \quad (2) \qquad Rs = \sqrt{N}/4\,(\alpha - 1/\alpha)(k/k+1) \quad (3)$$

$$Tr = H/u\,(\alpha/\alpha - 1)^2\,(k+1/k)^2\,(4Rs)^2\,(k+1) \qquad (4)$$

$$Tr = speed \times selectivity \times retention$$

where Tr the retention time, L the column length, k the capacity ratio of the last eluting peak of the mixture, k the capacity factor and α the selectivity factor for the least separated peaks, H the plate height and u the linear velocity of the mobile phase in the column and Rs the resolution required for the two least separated peaks. As can be derived from equation 2 , reducing the columns length has a direct impact on speed. Columns of 3 to 5 cm length can be commonly used. Even 0.5 to 1 cm long columns have their practical possibilities. The factors retention and selectivity in equation 4 are very specific to the sample and to the phase system. The H/u relation for different particles is of more general significance and is given in Figure 1.

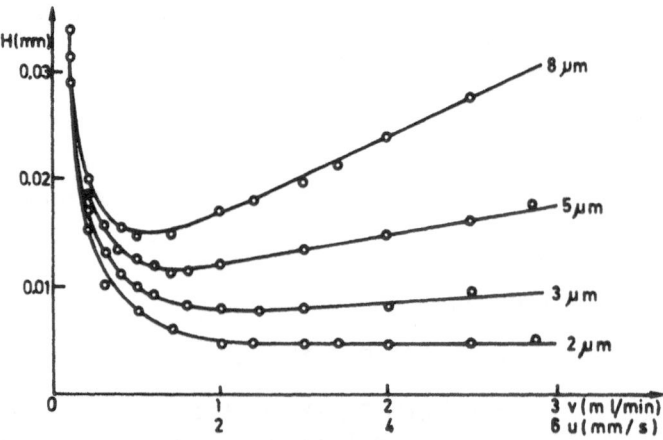

Figure 1 : van Deemter curves for different particle sizes
Column : 25 x 0.46 cm for 8 and 5 μm ROSil C18-D, 15 x 0.46 cm for 3 μm ROSil C18-D and 4 x 0.46 cm for 2 μm , mobile phase : 75:25 acetonitrile/water, sample : pyrene(k~6)

This figure shows that smaller particles :1) produce a lower plate height and thus more plates per unit length, 2) have a higher optimum flow rate and 3) have a gentle slope of the van Deemter plot (C term). Especially the 2 μm particles have a nearly flat H/u curve indicating that the efficiency and the sensitivity is nearly independant of the flow rate.
These three factors illustrate that small particles are better for fast analyses.

<u>Advantages of reducing column length and particle size</u>
  -Speed : Using short 1 to 5 cm long columns filled with 2 to 3 μm particles at moderate eluent speed leads to analysis times of 0.5 to 2 min and thus makes them very suitable for routine analysis. An example is given in Figure 2a where 10 PAH's are separated in less than 1 min.. Fig 2b shows the separation of phthalates in less than 1 minute. HSLC is also very interesting for research work, particulary in the development of chromatographic methods. The time for determining the optimum solvent composition can be reduced to minutes. At present, optimum isocratic conditions for an analysis are often deduced by running a gradient from low to high eluting power. Possible isocratic conditions are then often deduced from the solvent compositon at which elution of peaks of interest occurs. With HSLC a reverse approach can be used. Isocratic elution at 100, 90, 80, 70 % methanol or acetonitrile in water establishes the optimal solvent composition in a matter of minutes. This is not only due to fast chromatography but also to the speed with which small particle reversed phase systems equilibrate. The combination of automatic method development and HSLC has also been demonstrated and looks very promising(5). Another possibility of HSLC is the use of very high k values(6). Because of the low void volume and column lengths, k values of 50 to 200 can be obtained in an acceptable time. To a certain extent gradient elution can be avoided in this way.

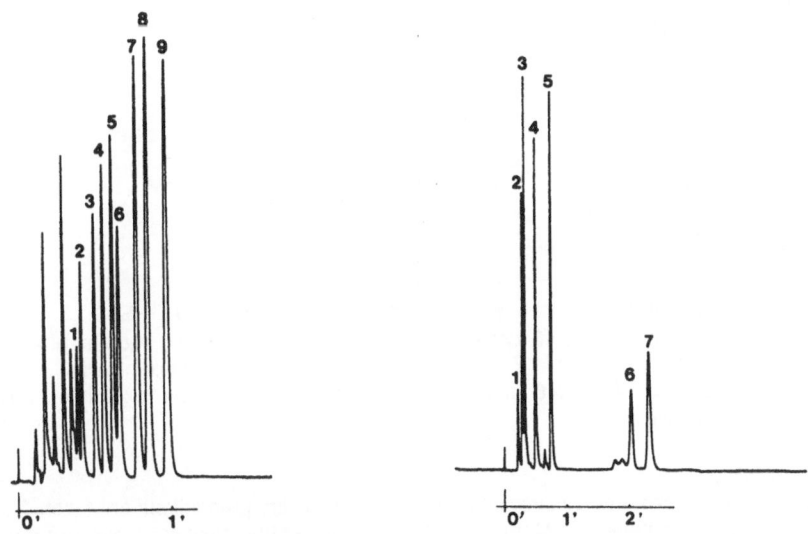

Figure 2a: HSLC separation of PAH mixture(left)
        Column : 4x 0.46 cm filled with 2 μm ROSil C18-D, mobile phase : 75:25 acetonitrile/water, 2 ml/min, P = 200 atm, detection : 254 nm, sample : 1. toluene, 2. naphthalene, 3 . diphenyl, 4. fluorene, 5. phenantrene, 6. anthracene, 7. fluoranthene, 8. pyrene and 9. triphenylene

Figure 2b: HSLC separation of phthalates(right)
        Column : 4 x 0.46 cm filled with 2 μm ROSil C18-D, mobile phase : 90:10 acetonitrile/ water, 1 ml/min P = 90 atm, detection = 254 nm, sample : 1. phthalic acid, 2. dimethylphthalate, 3. diethylphthalate, 4. dibutylphthalate, 5. dicyclohexylphthalate, 6. dioctylphthalate and 7. dinonylphthalate

-System Sensitivity : Sensitivity is closely related to dispersion. Equation 1 shows the influence of a reduction of L and dp on dispersion. Figure 3 illustrates the comparison of mass sensitivity for a standard 25 cm column filled with 10 μm particles and a 4 cm long column filled with 2 μm particles. Both analyses are run under identical conditions and both columns have nearly the same plate number. According to equation 1 the dispersion and sensitivity are 4 to 5 times higher for the HSLC column. The sample size required can also be reduced by a factor of 5 in comparison with conventional columns.

-Economical advantages : One of the direct advantages is a reduced solvent consumption, mostly a decrease of one order of magnitude. In Figure 2a more than 10 compounds are separated in less than one minute and with a solvent consumption of only 2 ml. In general instrument productivity increases also by one order of magnitude.

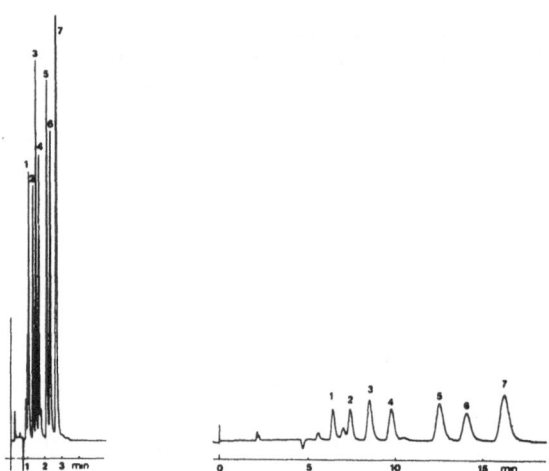

Figure 3 : Comparison of mass sensitivity between conventional HPLC and   HSLC.
    Left : Column : 4 x 0.46 cm filled with 2 μm ROSiI C18-D, mobile phase : 75:25 acetonitrile/water, 1 ml/min P = 120 atm, detection : 254 nm, sample : PAH's
    Right : Column : 25 x 0.46 cm filled with 10 μm RSil C18-D, other experimental conditions same as above

Practical limitations
    In practice HSLC can only be succesfull when the necessary precautions are taken. Table1 lists some column and instrumental problems and solutions. One of the major concems with regard to the use of HSLC columns was its reliability and lifetime. If the necessary precautions are taken is was demonstrated that these columns could undergo several thousands of injections(6,7). Because long valve switching times cause flow stoppage, injection can give a pressure surge between the valve and the column responsible for rapid column deterioration. Two possible solutions have been proposed : quick valve actuation (<0.1 s)(8) and injection valves with a bypass(9). Plugging of the small particle columns is always stressed as a potential problem(6). In general, this is not due to the small particles but to the small connecting tubes and small pore frits. In practice the installation of small prepacked columns between the pump and the injector can prevent these problems. Installing a Rheodyne in-line filter with very low extra bandbroadening is also proposed. Additionally this column can be used as a presaturation column to prevent silica dissolution, one of the mean reasons for

chemical instability(10). Because dissolution increases exponentionally with dp, this factor is most important for the small 1-3 µm particles(3). One of the unresolved problems is the influence of temperature on column lifetime and bandbroadening. From a theoretical point of view the influence of viscous heat dissipation on band broadening and peak tailing have been discussed(11,12,13).

Table 1 : Column and instrumental problems for HSLC

Problems	Causal factors	Solution	Resolved?
Column lifetime	Injection pulses	use injectors with bypass or quickly actuating valve	+
	Plugging	use filtered solvents, install guard column between pump and injector	+
	Silica dissolution	use silica saturator columns	+
	Temperature	keep column temperature around room temperature, keep flows under 10 mm/s	-
Compatible instrumentation	Injector	use low bandwith injector with small volume (<10 µl) (with bypass or quickly actuated)	+
	Connecting tubes	short length, small id (0.005-0.007 inch)	+
	Detector	detector with low bandwith (<10 µl) and fast response(<200 ms)	+
	Data systems	fast data acquisition	+

From our experience (3) it appears that under normal operating conditions (v<3-4 ml/min or u< 6-8 mm/s), viscous heating is not the major contribution to band broadening. This is also demonstrated in Figure 1. Recently we found that for 1-3 µm particles at high linear velocities(> 6 mm/s) columns were irreversibly damaged, yielding double peaks. This is due to a different thermal expansion of the metal column tube and the packing itself, resulting in increased permeability near the column wall and subsequent peak doubling(14). The use of small ID columns is favourable in this respect. Instrumental considerations are also summarised in Table 1. Instrumental dispersion can be reduced by miniaturisation of the instrument components and by improved flow characteristics. The contribution of each component has been studied extensively(15). In general the detector is the most important contributor to extra band broadening. At the moment several instruments are available specially designed for HSLC (Perkin Elmer, Kontron, Varian...)

Applications to the analysis of micropollutants

In general HSLC can be used for all types of micropollutant analysis, performed on a conventional column having a plate number of 3000 to 10000, with the advantage of decreased analysis time, increased instrument productivity and improved signal to noise ratio. In practice not more than 5 to 10 peaks can be separated in an isocratic run. In the literature not very much is found about the HSLC of micropollutants. Di Cesare et al. demonstrated the HSLC separation of phenolic priority pollutants within the analysis time of 2 minutes and with a detection limit of 150 to 900 pg.(16) . Sixteen PAH priority pollutants are separated in 1 min. by a gradient run on a 3 x 3 C18 column(6). The identification and quantification of phthalates in river water was performed in 3 minutes(17).
HSLC is also interesting for rapid screening of water samples. For very complex samples however these columns have to be used in combination with selective detection(18). An example is given in

Figure 4 where an ethyl acetate sublation water extract of a waste water treatement works is analysed by a very fast acetonitrile-water gradient on a one cm long column filled with 3 μm particles.

High Speed Liquid Chromatography is particulary useful when a restricted amount of compounds is to be monitored. An example is the monitoring of the herbicide diclorobenyl (dichloro benzonitrile) in water for controlled release experiments. One ppb of diclorobenyl can be determined in less than one minute(19). Sub ppb levels can be detected with on-line preconcentration.

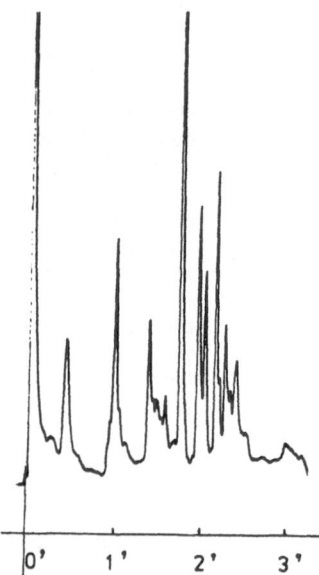

Figure 4 : HSLC of water extract
       Column : 1 x 0.46 cm column filled with 3 μm ROSil C18-D, mobile phase : gradient from water/acetonitrile 95/5 to acetonitrile 100 in 2 min., detection : 254 nm, sample : ethyl acetate sublation water extract (from waste water treatement works)

3. NARROWBORE HPLC : Micro-HPLC through a miniaturisation of the column diameter

Introduction

Narrow-or Microbore HPLC is that kind of Micro HPLC technique where the column diameter is decreased to 0.5-2 mm. Microbore HPLC has become an accepted instrumental technique since its renaissance nearly one decade ago, due to the work of Scott and Kucera(20). 0.5 to 2 mm ID columns where already used by Horvath and Lipsky(21), Kirkland(22), Snyder(23) and others in the late 1960's. At that time the low dispersion instrumentation for this type of columns did not exist. Therefore shorter conventional 4 to 4.6 mm ID columns filled with porous 5 to 10 μm particles were introduced. These columns will remain the "workhorses" in HPLC for many years. At the moment columns and instrumentation are available for microbore HPLC. However, microbore HPLC although used frequently for research, is not accepted for routine analyses.

Advantages of reducing the internal diameter

-Economical advantages : If a separation is performed under identical conditions (same linear velocity) on columns with different ID the solvent consumption is inversibily proportional to ID^2. The solvent consumption can be reduced by one order of magnitude when going from 4.6 to 1 mm ID. However , it must be considered that compared to a conventional 4.6 mm column, already 85 % solvent saving is obtained when using a 2 mm ID column with which it is easier to use conventional instrumentation with reasonable results. Going to 1 mm ID column results only in an additional saving of 10 % which is not significant no lower. At the moment an intermediate 2 mm column would satisfy the needs of many people and will probably remain the choice for practical HPLC for several years to come.

Another advantage is the use of very rare and expensive mobile and stationary phases as are e.g. used for optical isomer separation(24). However this is not relevant for micropollutant analysis.

-System sensitivity : It is now generally recognised that the claim of higher sensitivity for microbore columns is not justified. This can be derived from equation 5 relating dispersion and sensitivity :

$$C_{max} = \frac{V_{inj} \cdot C_{inj}}{4\, D\, (k+1)\, \sqrt{2\pi}} \qquad (5)$$

When the sample size is unlimited , the column diameter has no effect on the sensitivity as long as the sample size is adjusted to the diameter of the column. Whereas the sensitivity decreases with r^2 the allowable sample size increases with r^2. However when the sample size is limited, the dispersion will be smaller and the sensitivity better if narrow bore columns are used. In theory going from a 4.6 to a 1 mm ID column increases the sensitivity 20 times. In practice however this is not obvious because detectors for narrow bore HPLC have smaller cell volumes and are concequently less sensitive.

-LC/MS coupling : Probably the most important advantage of microbore HPLC is the easy coupling with a mass spectrometer. Whether direct introduction of the eluent into the source or an interface (nebulisation or moving belt) is used, the vaporisation and disposal of the vapors is made easier through the use of narrow bore columns. This subject will be discussed in extenso elsewere(25). Recent contributions on the subject are given in reference 26,27,28.

-Potential gain in efficiency : Although in practice most separations are carried out on columns generating 2000 to 10000 plates, for some problems more plates are needed. This is the case for complex mixtures like micropollutant samples. The simplest way to increase the efficiency is by lengthening the column. This can be done by coupling columns or filling longer columns. It is often demonstrated that this can be done easily with small bore columns. The latest contribution to this field was by Rosset et al(29) generating one million plates in 30 hours on a 22m x 1mm column filled with 7-8 μm particles. However if only 50000 to 150000 plates are needed this can be performed more easily on conventional columns in an acceptable time(Tr<1hour)(30,31)

Practical limitations

One of the disadvantages of small bore columns at the moment is the instrumentation. Describing instrumental details would lead us too far. Details can be found in ref. 32. At the moment, no complete chromatographs designed for small bore HPLC exclusively, are commercially available. They always have the possibility of handling conventional columns. Chromatographs for use with small bore columns are often made from various modules obtained from a range of suppliers to achieve the required specifications. Commercial low flowrate pumps are available from Beckman, Gilson, Jasco, HP, Perkin Elmer, Shimadzu, Varian , Waters, LKB ,LDC/Milton Roy, EM Science and Brownlee. Low volume injectors are available from Rheodyne and Valco. Low volume UV detectors are available from Kratos, Jasco, Perkin-Elmer, Schimadzu, LKB, Varian and Em Science. Low volume fluorescence detectors are available from Kratos and Perkin Elmer.

Applications to the analysis of micropollutants.

Two applications of narrow bore HPLC are relevant to micropollutant analysis: trace analysis of organic material and the analysis of complex mixtures. Trace analysis by sample concentration followed by separation on a narrow bore column lends itself to water analysis. A water sample can be pumped in almost any volume to a sample prepacked column, focusing the organic material as a sharp band, and then eluting the band with a gradient on a narrow bore column with their inherent high mass sensitivity(33,34). An example of the combination of trace enrichment and microbore column technology is given by Kok et. al.(35). He determined nitrobenzene in river water at the 2 ppb level using an electrochemical detector. An on line precolumn was fitted with a bypass and the separation was carried out on a 250 x 1.1 mm columns filled with 8 μm Lichrosorb RP 18. If other off-line sample preparation techniques are used, it is also easier to handle smaller amounts of water for use with microbore columns.

On reviewing the literature about micropollutant separations on narrow bore columns two thing are striking. First there is the enormous amount of PAH separations in the literature. PAH's are of significant environmental concern but they are also the favourite test sample for column performance (36,37,38).

Second most micropollutant separations are performed on smaller ID columns as will be discussed further. Polychlorinated biphenyls were separated by Takeuchi et al.(39) by reversed phase chromatography on a 1.5 m x 0.26 mm column. They also illustrated the separation of C1 to C9 dialkylphthalates in the GPC mode on a 143 x 0.35 mm column(40). Alkylated phenols were separated on a 50 x 0.5 mm column filled with 6 μm reversed phase packing by Slais et al(41). The detection limit for pyrocatechol with electrochemical detection was 4 pg. Most applications are found in LC/MS work wich will be treated elsewhere. Although in most cases the primary goal is to interface both techniques and not the optimisation of the chromatographic process. Several interesting separations are shown. PAH's, as well as nitro - and chlorophenols are separated on a 140 to 270 x 0.5 mm column filled with RP 18 material by Schafer et al.(42). LC/MS of organophosphorous pesticides(43)and triazine was also described(44) Phenylurea herbicides were separated on a 200 x 0.5 mm RP 18 column by Leusen et al.(45)

4. Future perspectives :
   High speed packed capillary HPLC (3): Micro-HPLC through a miniaturisation of column length, column diameter and particle size

Packed capillary columns of 50 to 200 cm used by Novotny (46) and Yang (36) show very high plate numbers, but require a very long analysis time. An example of the separation of micropollutants on this column type is given in figure 5. Remarkably these two meter long columns can be packed with 3 to 5 μm particles with an acceptable permeability. In conventional columns packed as tightly as possible the maximum column length for a backpressure of 250-300 atm for 3 and 5 μm would be 40 and 100 cm respectively . This high permeability can only be the result of a loose packing in capillary columns. Loose packing in conventional columns would be unstable. The reason for this not being the case for packed capillaries is the very small I.D. It seems acceptable that the packing structure stability is higher in a very small ID column than for a conventional column type.

With 1 μm spherical silicagel particles the longest traditional column that could be run at a linear velocity of 1 mm/s for a pressure of 250-300 Bar would be 4 cm long. With a reduced plate height of 3 that can be obtained on 1μm octadecylated particles this would be equivalent to 13500 plates. A separation demonstrating the high plate number per length unit on a conventional column is given in figure 6. We are evaluating now the use of 1 μm particles in fused silica columns of 320 μm ID. With the loose packing in such columns a 10 to 20 cm column does seem possible. This would produce 30000 to 70000 plates with analysis times around 15 min. Then Micro LC would have equalled capillary GC in efficiency and speed. Several attempts in this direction have been published in the literature(7, 47). An example of the separation of phenols is given in figure 7. Of course this would tremendously increase the possiblities of HPLC to micropollutant analysis. .

ACKNOWLEDGEMENTS
    We thank the "Nationaal Fonds voor Wetenschappelijk Onderzoek - NFWO", the "Instituut ter
Aanmoediging van het Wetenschappelijk Onderzoek in Nijverheid en Landbouw - IWONL" and the
"Ministerie voor Wetenschapsbeleid" for financial support to the laboratory.

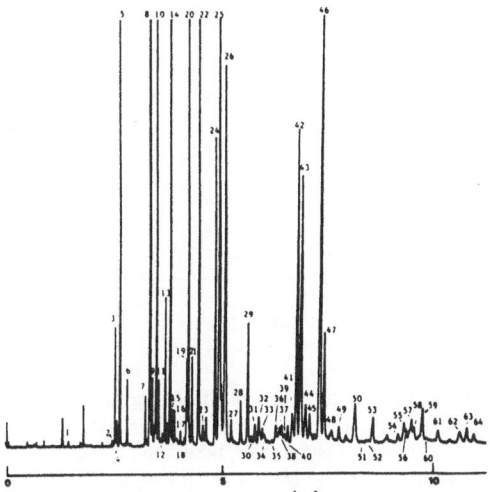

Figure 5 : Packed capillary chromatogram of large polyaromatic compounds extracted from carbon
        black
        Column : 1.8 meter x 250 µm fused silica capillary packed with 3 µm C18 spherical,
        detection : fluorescence, stepwise gradient. Reproduced with permission from ref. 46

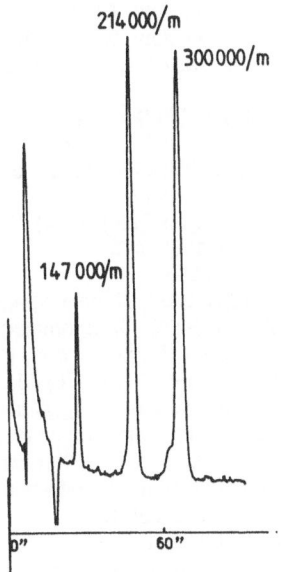

Figure 6 : Micro HPLC on 1 µm particles
        Column : 1.5 cm x 0.4 cm filled with 1 µm ROSil C18-D, mobile phase : 75:25
        acetonitrile/water, 0.8 ml/min, ΔP = 80 atm, detection : 254 nm (1.4 µl cell and 20 ms time
        constant), sample:naphthalene, anthracene, pyrene (k~6), HETP 3µm

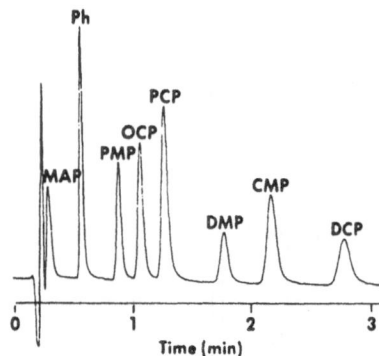

Figure 7: Example of a High Speed Packed Capillary chromatogram.
Column : 75 mm x 0.32 mm, 3 μm MicroPak SP C18, mobile phase :34:0.1:0.1:66 acetonitrile/acetic acid/triethylamine/water, flow rate : 20 μl/min (390 atm), detection : electrochemical detection, peaks : MAP = m. aminophenol, Ph = phenol, PMP = p. cresol, OCP = o. chlorophenol, DMP = 2.6 dimethylphenol, CMP = 4 chloro, 3 methylphenol, DCP = 2,4 dichlorophenol(reprinted with permission from ref.7)

REFERENCES
1. R. Majors, G. Barth and C. Lockmüller, Anal. Chem, 56 (1984) 300R
2. M. Verzele and C. Dewaele, J. Chromatogr. 282 (1983) 341
3. M. Verzele and C. Dewaele, in "Science of Chromatography" ed. F. Bruner, Elsevier Amsterdam, 1985, p. 435.
4. L. Zechmeister and L. Von Cholnoky, Die Chromatografische Adsorptionmethode. Wien, Springer Verlag, 1937
5. J.G. Gfeller and F. Erni, 15 th International Symposium on Chromatography, Nurnberg, 1-5 oct. 1984.
6. M. W. Dong and J.R. Gant , LC Magazine, 2(4) (1984) 294
7. S. J. Van Der Wal, LC Magazine,3 (1985) 488
8. F. Erni, J. Chromatogr., 282 (1983) 383.
9. J.L. Di Cesare, M.W. Dong and J.R. Gant, Chromatographia,15 (1982) 596.
10. J.G. Atwood, G.J. Schmidt and W. Slavin, J. Chromatogr., 171 (1979)109
11. I. Halàsz, R. Endele and J. Asshauer, J. Chromatogr., 112 (1975) 37
12. Cs. Horvath and H. Lin, J. Chromatogr., 149 (1978) 43.
13. H. Poppe, J. Kraak, J. Huber and J. van den Berg, Chromatographia, 14 (1981) 515.
14. Y. Yambo, C. Dewaele and M. Verzele, unpublished results.
15. J.L. Di Cesare, M.W. Dong and J.G. Atwood, J. Chromatogr. 217 (1981) 369.
16. J. L. Di Cesare, M.W. Dong and L.S. Ettre, "Introduction to High Speed Liquid chromatography" Perkin Elmer Corp., Norwalk, Connecticut, 1981
17. M.W. Dong and J.L. Di Cesare, J. Chromatogr. Sc., 20 (1982) 517-522.
18. J.C. Kraak, "Analysis of Organic Micropollutants in water", Proceedings of the Third European Symposium, Oslo 1983, p. 110.
19. J. Van Dichel, C. Dewaele and E. Schacht, J. Chromatogr. (in preparation)
20. R.P.W. Scott and P. Kucera, J. Chromatog., 169 (1979) 51
21. C. Horvath and S.R. Lipsky, Anal. Chem. 39 (1967) 1422
22. J.J. Kirkland, J. Chromatog. Sci., 9 (1971) 206
23. L.R. Snyder, J. Chromatog., 63 (1971) 15

24. V. Mc Guffin, LC Magazine, 2(4) (1984) 282
25. P. Arpino, 4 th European Symposium on Organic Micropollutants in the aquatic Environment, Vienna, 22-24 October 1985
26. J. Henion, Microcolumn High Performance Liquid Chromatography, ed P.Kucera, Elsevier, Amsterdam, 1984, p. 260
27. K. Ogan, Small Bore Liquid Chromatography Columns, ed. RPW Scott, J. Wiley & Sons, New York,1984, p. 104.
28. E. D. Lee and J.D. Henion,J. Chromatog. Sci. 23 (1985) 253
29. H.G. Menet, P.C. Gareil and R.H. Rosset, Anal Chem, 56 (1984) 1770
30. M. Verzele and C. Dewaele, J. HRC & CC, 5(1982)245
31 J.L. Di Cesare, M.W. Dong and F.L. Vandemark, Am. Lab. 13 (1981)52
32. Chapter 3 and 4 in "Small bore liquid chromatography columns" ed. RPW Scott, Wiley Interscience, New York 1984
33. R.P.W. Scott and P. Kucera, J. Chromatog., 284 (1982)
34. Sj. Van der Wal, poster at the 9th International Symposium on Column Liquid Chromatography, Edinburgh 1985
35. W.T. Kok, U.H.T. Brinkman, R.W. Frei, H.B. Hanekamp, F. Nooitgedacht and H. Poppe, J. Chromatog., 237 (1982) 367
36. F. Yang, J. Chromatogr., 236 (1982) 263
37. Sj. Van der Wal and F. Yang, J. HRC & CC,6 (1983)216
38. R.P.W. Scott, J. Chromatog. Science, 18 (1980) 49
39. D. Ishii and T. Takeuchi, J. Chromatog. 255 (1983) 409
40. T. Takeuchi and D. Ishii, J. Chromatog., 238 (1982) 409
41. K. Slais and D. Kourilova , J. Chromatog., 258(1983) 57
42. K.H. Schafer and K. Levsen, J. Chromatog., 206 (1981) 245
43. C.E. Parker, C.A. Haney, D.J. Harvan and J.R. Hass, J. Chromatog. 242 (1982) 77
44. C.E. Parker, C.A. Haney and J.R. Hass, J. Chromatog. 237 (1982) 233
45. K. Leusen, K.H. Schafer and J. Freudenthal, J. Chromatog., 271 (1983) 51
46. A. Hirose, D. Wiesler and M. Novotny, Chromatographia, 18 (1984) 239
47. T. Takeuchi and D. Ishii, J. HRC & CC, 6 (1983) 683

IDENTIFICATION OF SURFACTANTS AND STUDY OF THEIR DEGRADATION IN SURFACE WATER BY MASS SPECTROMETRY

E. Schneider and K. Levsen
Institute of Physical Chemistry, University of Bonn,
Wegelerstr. 12, D 5300 Bonn, F.R. Germany

Summary

Pure surfactants have been analyzed by newer mass spectrometric techniques, i.e. by field desorption (FD), desorption chemical ionization
(DCI) and fast atom bombardment (FAB). FD leads to spectra which
are dominated by the quasimolecular ions while fragments are missing.
On the contrary, both the DCI and the FAB method lead to spectra
which contain both quasimolecular ions and structure specific fragments. Using the FD technique in combination with collisionally activated decomposition (CAD) in a tandem mass spectrometer cationic,
nonionic and anionic surfactants have been identified in river water
without prior separation. Moreover, cationic surfactants have been
quantified in sewage water and surface water.
Finally, the biodegradation of a nonionic surfactant (nonylphenolethoxylate) and a cationic surfactant (distearyldimethylammoniumchlorid) in surface water has been monitored using the field desorption
technique. Biodegradation of the nonionic surfactant is a much faster
process than that of the cationic surfactant.

1. INTRODUCTION

Surfactants are organic compounds that when added to a liquid change the
interfacial properties of that liquid. They are used as wetting, foaming,
dispersing, emulsifying and penetrating agents and are the major organic
constituents of detergents. Surfactants are mostly classified by the nature
of their ionic charges. Thus one distinguishes between three types of
surfactants: (a) cationic (such as quaternary ammonium salts), (b) nonionic
(formed, e.g., by condensation reactions between ethyleneoxide and fatty
alcohols or alkylphenols) and (c) anionic, (e.g., alkylbenzenesulfonates
or alkylsulfonates). Surfactants are typically manufactured as a mixture
of homologous compounds, the constituents of which differ in the length
of the alkyl or ethoxylate chain. The release of surfactants into the
environment (in particular into surface water) has required the development
of sensitive analytical methods for their determination at trace levels.
The various methods developed and employed up to date have been reviewed
recently (1, 2). The most common procedure for the determination of cationic
and anionic surfactants is by colorimetry (3, 4). Although simple, these
methods are neither very specific nor very sensitive. Moreover, most of
the other analytical techniques proposed so far do not allow the determination of the exact constitution, in particular the alkyl moiety, nor do
they permit the determination of the distribution of chain lengths.
While organic compounds at trace levels in various sample matrices
are most readily identified by combined gas chromatography/mass spectro-

metry (GC/MS) this method can not be applied directly to the analysis
of ionic surfactants as these compounds are too polar to be amenable either
to gas chromatography or to conventional electron impact (EI) mass spectrometry. Im contrast, nonionic surfactants have been analyzed by GC (5)
and by EI mass spectrometry (6).

In recent years a variety of mass spectrometric ionization techniques
have been developed which also allow the analysis of strongly polar compounds (7). Among these techniques the field desorption (FD), fast atom
bombardment (FAB) and desorption chemical ionization (DCI) methods are
now commercially available. In addition it has been demonstrated that
tandem mass spectrometry now commonly referred to as mass spectrometry/mass
spectrometry (MS/MS) (8) can be used for a direct mass spectrometric mixture analysis. In this paper the use of these new mass spectrometric techniques for the analysis of pure surfactants as well as for the analysis
of surfactants in surface water will be reviewed and new results on the
primary biodegradation of surfactants in surface water reported.

2. PURE SURFACTANTS

Cationic Surfactants. The most common cationic surfactant are quarternary ammonium salts with one or several long alkyl chains as hydrophobic
moiety. Cationic surfactants have been analyzed by FD, FAB and DCI (9-11,20)
and by tandem mass spectrometry (13, 20). The FD spectra are dominated
by signals of the intact cations while fragments are almost completely
missing. Thus the spectra give molecular weight information and information
on the distribution of chain lengths while structure information is missing.
Such structure information is however readily available when the field
desorption technique is combined with tandem mass spectrometry using the
collisional activated decomposition (CAD) method (13, 20) (vide infra).

Both molecular weight and structural information is also available
if the FAB method is employed. Fig. 1 represents the FAB spectrum of a

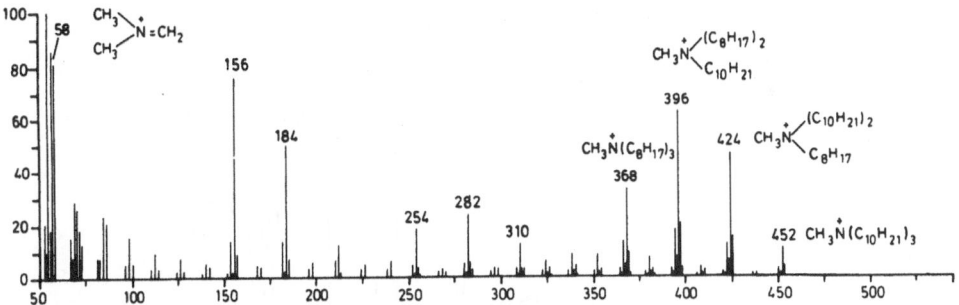

Fig. 1. FAB spectrum of a trialkylammonium compound

trialkylammonium compound . The upper mass range is dominated by signals
due to the intact cations at mz 368, 396, 424 and 452. The most abundant
fragments are highly structure specific. Thus the fragments at m/z 254,
282 and 310 are formed by loss of a substituent as alkane while the dominant fragments at m/z 156 and 184 are formed by loss of one substituent
followed by α-cleavage of a second substituent as shown in scheme I:

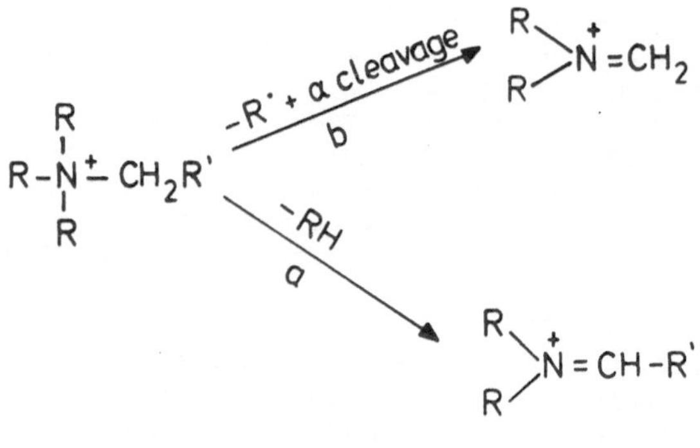

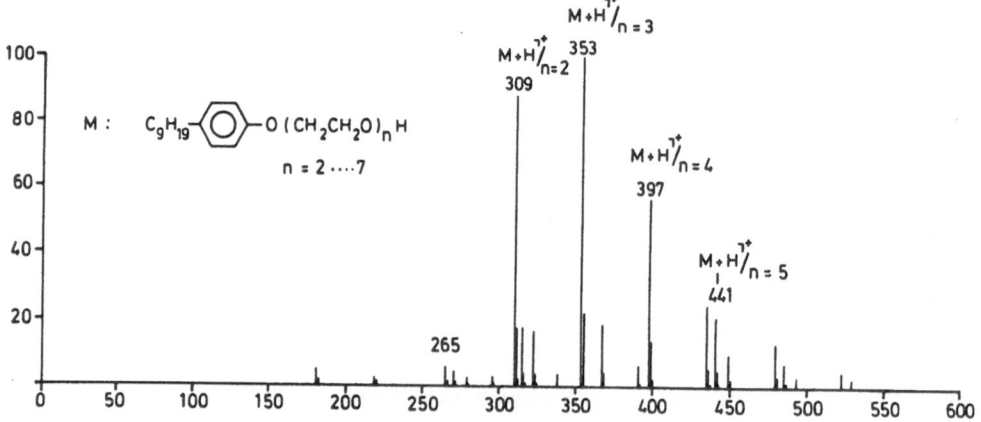

Scheme 1

Finally also DCI spectra of cationic surfactants have been obtained. Under DCI conditions partial thermal degradation of the sample occurs, but structure information can still be derived from such spectra.

Nonionic Surfactants. Pure nonionic surfactants have been analyzed by FD, FAB and DCI (10, 11, 13 - 16). FD spectra of nonionic surfactant are again very simple. They show almost exclusively $M^{+\cdot}$ or $(M + H)^{+}$ ions and thus give information on the molecular weight and the distribution of chain lengths. This distribution can, however, not be determined quantitatively as the relative abundance of the quasimolecular ions changes with the temperature of the field ion emitter. Fig. 2 shows the FD spectrum

Fig. 2. FD spectrum of a nonylphenolethoxylate

of a nonylphenolethoxylate with two to seven ethyleneoxide units as an example. The FAB method is less suited for the analysis of nonionic surfactants as this method gives quasimolecular ions of low abundance. In contrast, abundant molecular ions are observed if the DCI technique is employed. This is demonstrated for a nonylphenolethoxylate with 4 - 12 ethylenenoxide units in Fig. 3. In this spectrum also intensive fragments are found which , however, in most instances do not reflect the structure of the compound.

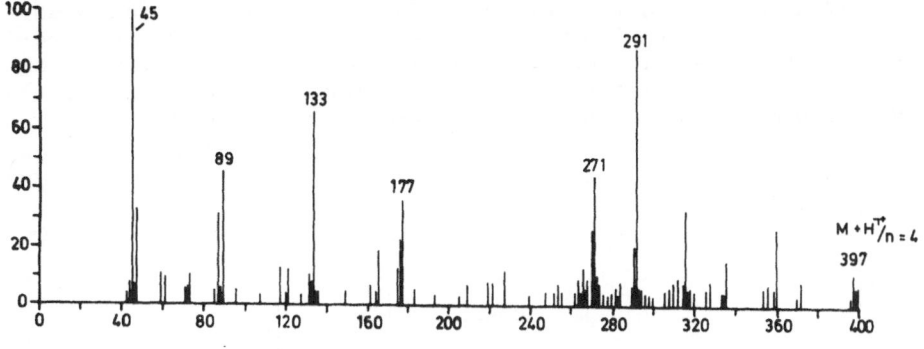

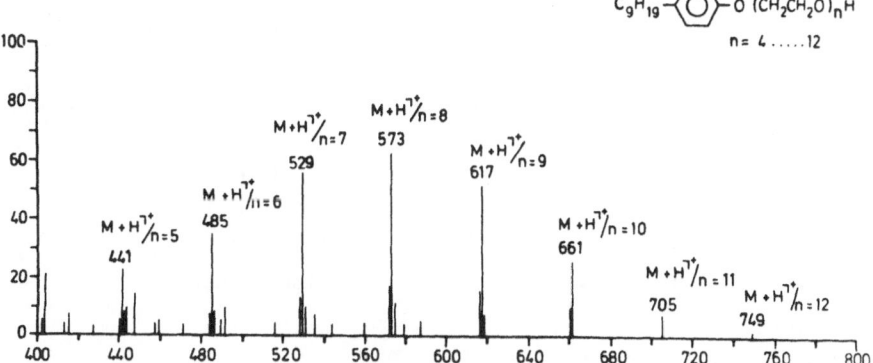

Fig. 3. DCI spectrum of a nonylphenolethoxylate

Anionic Surfactants. Anionic surfactants comprise about 80% of all
surfactants manufactured worldwide. Their intact ionization is again possible with modern ionization techniques (11 - 13, 17). This is illustrated

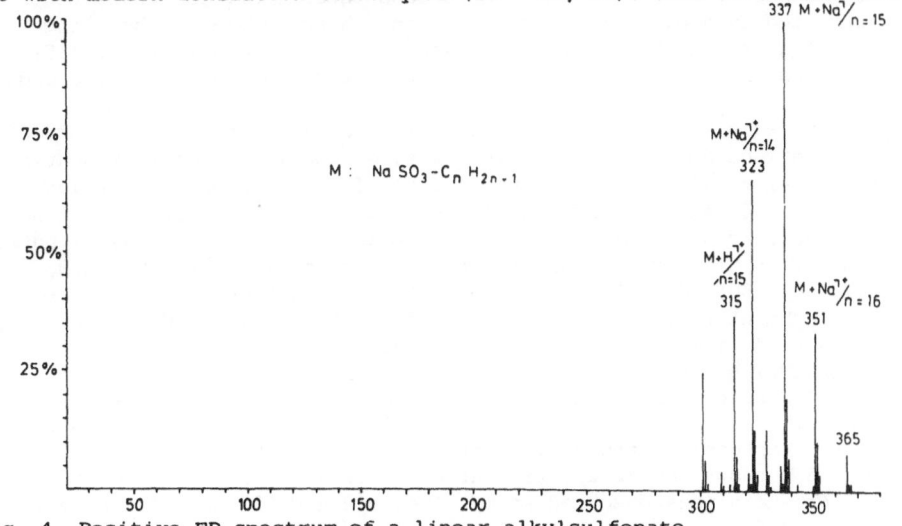

Fig. 4. Positive FD spectrum of a linear alkylsulfonate

in Fig. 4 and Fig. 5 which show the FD spectra obtained in the positive
ion and negative ion mode, respectively. Quasimolecular ions are formed
by alkali attachment and detachment, respectively. The spectra are free
from fragments. Quasimolecular ion information can also be obtained from
FAB and DCI spectra. Moreover, these spectra display abundant fragments,
which, however, do not reflect the structure of the surfactant.

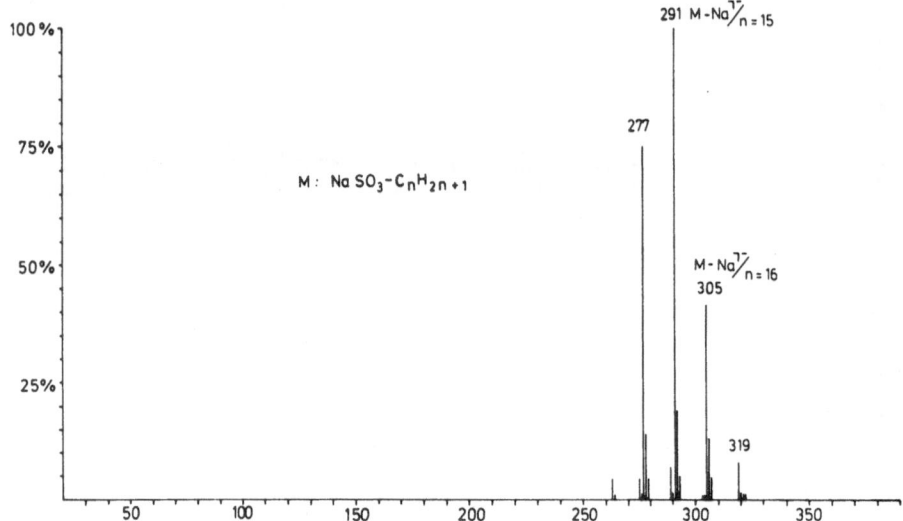

Fig. 5. Negative ion FD spectrum of a linear alkylsulphonate

3. SURFACTANTS IN SURFACE WATER

Qualitative Identification. Using pure surfactants it has been demonstrated previously that a combination of field desorption and collisionally activated decomposition (CAD) in a tandem mass spectrometer is parti-
cularily well suited to characterize this compound class (13) as this
allows a direct mixture analysis of the surfactant. This approach has
been used to identify surfactants in surface water (18). The surfactants
were extracted by adding ethylacetate to the water sample and bubbling
a stream of nitrogen through the solution as described by Wickbold (19).
By this means the surfactants are transfered from the aqueous to the organic layer.

Fig. 6 a-c displays the FD spectra of a Rhine River sample recorded
at three distinct emitter heating currents. At the lowest emitter heating
current (5 mA) a major and two minor series of ions are observed which
are spaced by 44 daltons (Fig. 6a). These series are characteristic of nonionic surfactants with polyethyleneoxide chains of different chain lengths
as discussed in the second chapter. The main ion series (m/z 385, 429,
473, 517, 561, 605, 649, 693, 781 and 825) can be assigned to arise from
sodium attachment to polyethyleneoxide dodecylether, $C_{12}H_{25}-O-(CH_2CH_2O)_nH$,
with n = 4 - 14. The two minor series can be assigned to arrive from
$C_{14}H_{29}O(CH_2CH_2)_nH$ with n = 4 - 14 and $C_9H_{19}-C_6H_4-O-(CH_2CH_2O)_nH$ with
n = 6-12. CAD spectra have been recorded from the main precursors. However,
these CAD spectra showed a very unspecific fragmentation and could not
be used to characterize the surfactant structure further.

Fig. 6 b shows the FD spectrum at a medium emitter heating current
(20 mA). The main ions observed at m/z 494, 522 and 550 have tentatively
been assigned to arise from cationic surfactants. This is corroborated

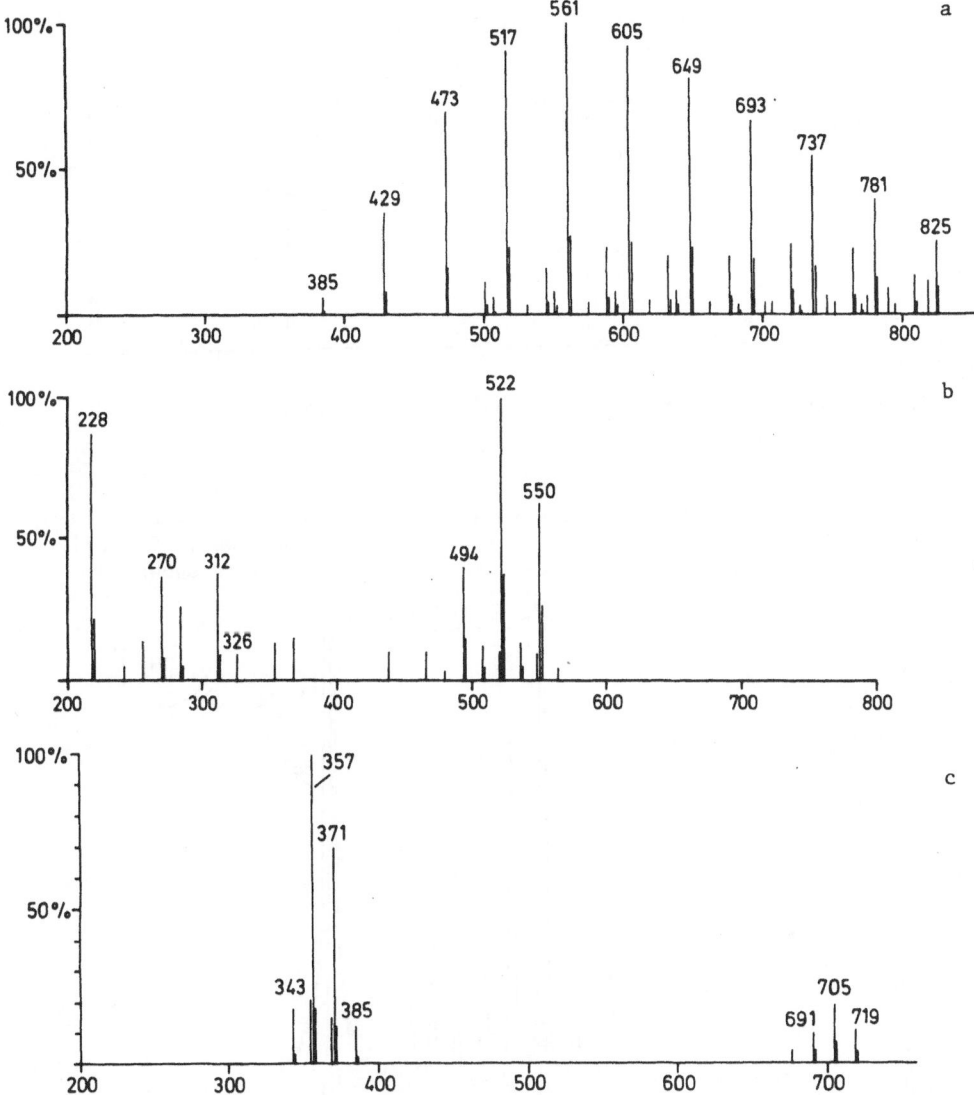

Fig. 6. FD spectra of a water sample from the Rhine River. Emitter heating current was (a) 5 mA, (b) 20 mA and (c) 23 mA.

by the CAD spectra (18). Fig. 7 shows the CAD spectrum of m/z 522 as example. Making use of the known fragmentation behavior of quaternary ammonium salts as shown in Scheme I in Chapter 1, the precursor can be assigned as $(CH_3)_2N^+(C_{16}H_{33})(C_{18}H_{37})$: The abundant fragments at m/z 268 and 296 arise by loss of one substituent as alkane while m/z 58 is formed by loss of one substituent as alkyl radical followed by α-cleavage of the second substituent.

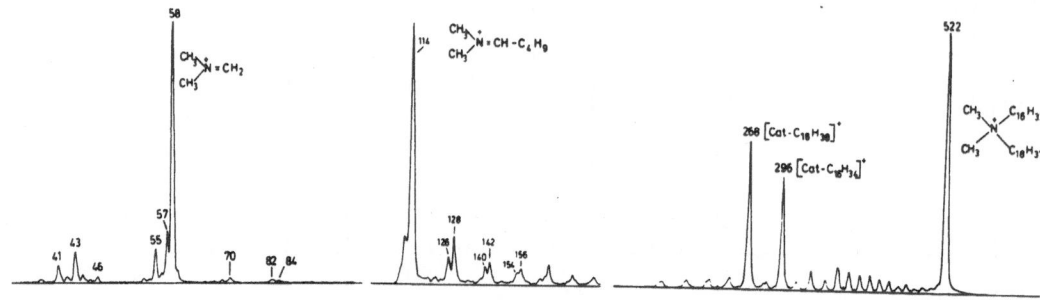

Fig. 7. Collisionally activated decomposition spectrum of m/z 522 in Fig. 6b.

At higher emitter heating currents (23 mA) signals due to the cationic surfactants disappear. Instead a series of ions at m/z 343, 357, 371
and 385 is observed (Fig. 6c). The CAD spectrum of m/z 357 is shown in
Fig. 8.

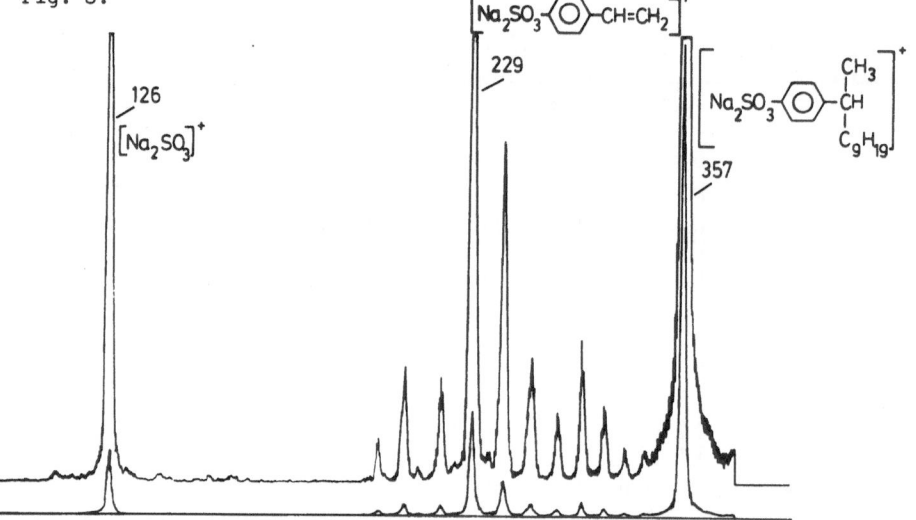

Fig. 8. Collisionally activated decomposition spectrum of m/z 357 in
Fig. 6 c.

This Spectrum unambiguously demonstrates that an alkylbenzenesulphonate
is present. The abundant fragment at m/z 126 reflects the functional group
while m/z 229 reveals how the alkyl chain is predominantly branched.

The results demonstrate that the FD technique is well suited for
the identification of surfactants in surface water. Is is of advantage
that the three types of surfactants desorb at distinct emitter heating
currents. Thus by proper choice of the emitter heating current a partial
separation of nonionic, cationic and anionic surfactants is achieved.

Quantization. Cationic surfactants in surface water have been quantified using the FD technique. To this end a deuterated standard, i.e.
$(CD_3)_2N(C_nH_{2n+1})_2Cl$ was synthesized which had a similar distribution

of chain lengths as those cationic surfactants   which are predominantly
present in surface water. Fig . 9 shows part of an FD spectrum of a sewage
water sample to which the standard was added. The peaks m/z 494, 522 and

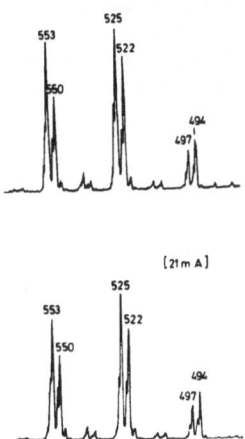

Fig. 9. Partial FD spectrum of a sewage water sample to which an internal
standard was added.

550 result from the sample, those of m/z 497, 525 and 553 from the standard.
Using this approach sewage water as well as surface water near Bonn has
been analyzed. In sewage water 350-480 µg/l, in surface water 6-12 µg/l
of distearyldimethylammonium salt was found.

The quantization of anionic surfactants turned out to be considerably
more difficult. The FD technique was not suited for this purpose. Thus
FAB method was employed to quantify sewage water samples. After extraction
of the sample a further clean up by thin layer chromatography was necessary.
As deuterated standards were not available homologues were used as standard.
It turned out that the ratio of standard to sample changed during the
desorption which leads to less reliable data. Thus using decylbenzenesul-
phonate as standard repeated measurements of the same sewage sample lead
to concentrations ranging from 18-24 µg/l.

4.PRIMARY BIOGRADATION OF SURFACTANTS IN SURFACE WATER
The primary biodegradation of nonionic and cationic surfactants in
surface water has been studied by field desorption mass spectrometry.
To this end a defined amount of the surfactant was added to 4 l of River
Rhine water and stirred continuously. 50 ml samples were taken in regular
intervals over a period of several weeks and analyzed after addition of
the standard.
The result is represented in Fig. 10 which shows the FD spectra of
a nonylphenolethoxylate. Peaks belonging to the nonylphenolethoxylate
are marked by numbers which represent the numbers of ethyleneoxide units
present, while unmarked peaks belong to the standard (octylphenolethoxy-
late). Fig. 10 a shows the result at the beginning of the experiment.
The FD spectrum is dominated by peaks which belong to nonylphenolethoxylate
homologues with 4-8 ethylene units while the homologue with 2 ethylene
units is of minor abundance. (Besides peaks of the sample and the standard

additional peaks are observed in the spectrum marked by an arrow. These peaks could be assigned as belonging to dinonylethoxylates which are obviously present in the sample as major inpurity).

Fig. 10b shows the spectrum after 7 days of biodegradation. The spectrum is now dominated by the homologue with two ethyleneoxide units, while the homologues with three and four ethyleneoxide units are of minor abundance and all higher homologues are missing (except the dinonylethoxylates which are uneffected). The results demonstrate that all higher homo-

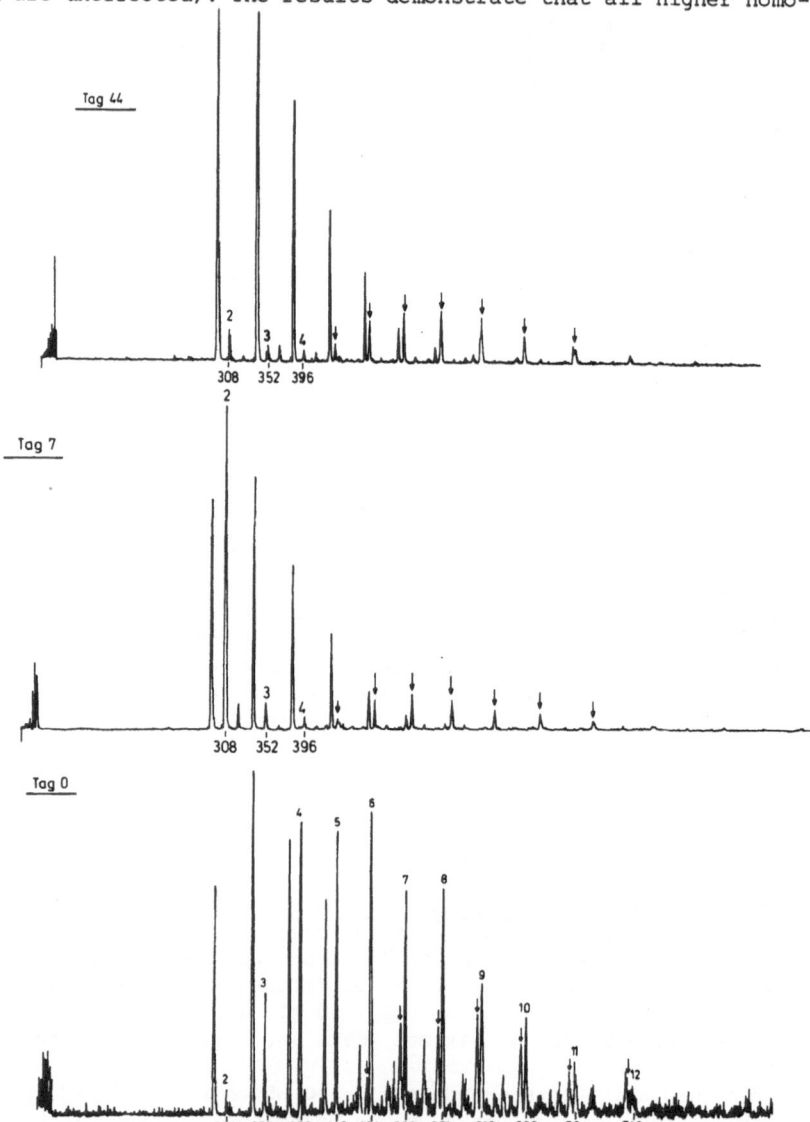

Fig. 10. Biodegradation of a nonylphenolethoxylate in surface water. Field desorption spectrum obtained after 0 (a), 7 (b) and 44 (c) days. (Peaks belonging to the nonylphenolethoxylate are marked by a number, which represents the number of ethyleneoxide units. Dinonylethoxylates are indicated by an arrow. Unmarked peaks belong to the standard)

logues are biodegraded mainly to the homologue with two ethylenoxide units within this time interval. Finally, the FD spectrum after 44 days is shown in Fig. 10c. Now also the homologue with two ethyleneoxide units is almost completely biodegraded while the dinonylethoxylates are persistent.

Fig. 11 shows the development of the homologue with two ethyleneoxide units over a period of 38 days. It is obvious that this homologue is formed after three days and passes through a maximum after 6 - 10 days. The results demonstrate that the primary biodegradation to the homologue with two ethyleneoxide units is a relatively fast process, while further biodegradation is slow. This is important as the homologue with two ethylene oxide units is more toxic than the precursors. Moreover this homologue is no longer detected by standard colorimetric methods.

The biodegradation of cationic surfactants has been studied by a similar approach. Fig. 12 shows the result for a distearyldimethylammoniumchlorid for two different primary concentrations. It is obvious that primary biodegradation starts after ∿ 14 days and is not complete after 50 days, i.e. biodegradation of cationic surfactants in surface water is considerably slower than that of nonylphenolethoxylates.

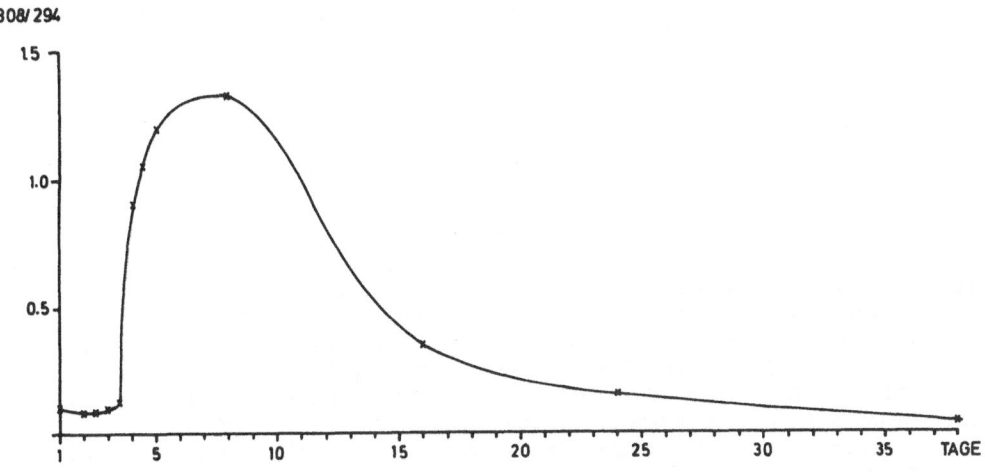

Fig. 11. Development of the homologue with two ethyleneoxide units, m/z 308, (relative to the standard, m/z 294).

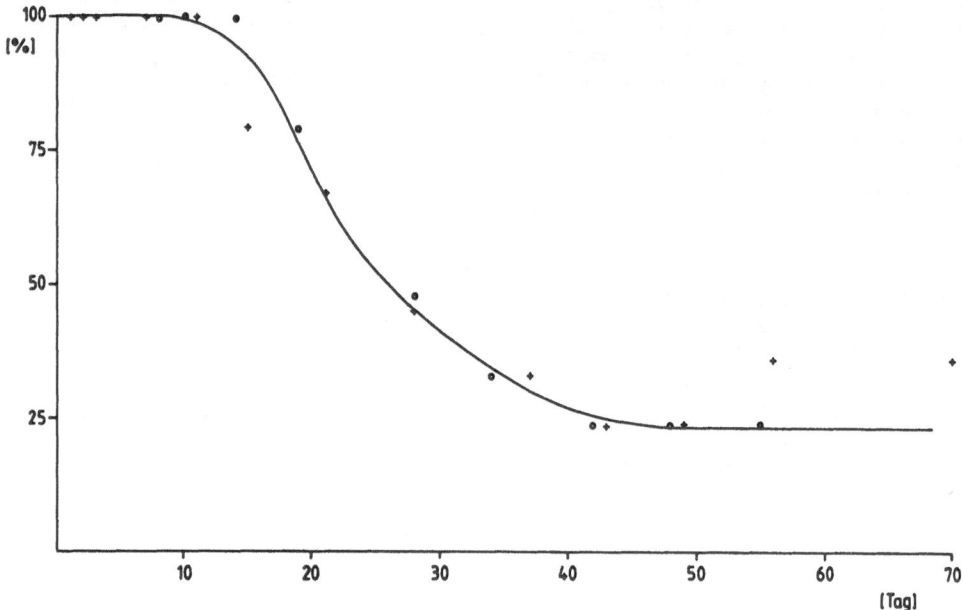

Fig. 12. Biodegradation of a distearyldimethylammoniumchlorid in surface
water: + = 33 mg in 4 l Rhine water, 0 = 2 mg in 4 l Rhine water.

REFERENCES

(1)   LLENADO, R.A. and JAMIESON, R.A. (1981), Anal. Chem. 53, 174 R.
(2)   LLENADO, R.A. and NEUBECKER, T.A. (1983), Anal. Chem. 55, 93 R.
(3)   LONGWELL, J. and MANIECE, W.D. (1955), Analyst (London), 80, 167.
(4)   MALAT, M. (1979), Fresenius Z. Anal. Chem. 297, 417.
(5)   JULIA-DANÉS, E. and CASANOVAS, A.M. (1979), Tenside Deterg. 16, 317.
(6)   SCHAFFNER, C., STEPHANOU, E. and GIGER, W. (1982) in:Bjørseth, A.,
      Angeletti, G. (eds), Analysis of organic micropollutants in water.
      D. Reidel, Dordrecht, Holland, p. 330.
(7)   MORRIS, H.R. (ed.) (1981), Soft ionization biological mass spectro-
      metry, Heyden, London.
(8)   MCLAFFERTY, F.W. (ed.) Tandem mass spectrometry, Wiley, New York, 1983.
(9)   COTTER, R.J., HANSEN, G. and JONES, T.R. (1982) Anal. Chim. Acta
      136, 135.
(10) SCHNEIDER, E.; LEVSEN, K.; DÄHLING, P. and RÖLLGEN, F.W. (1983),
      Fresenius Z. Anal. Chem. 316, 277.
(11) LEVSEN, K.; SCHNEIDER, E.; RÖLLGEN, F.W.; DÄHLING, P.; BOERBOOM,
      A.J.H.; KISTEMAKER, P.G.; MCLUCKEY, S.A. (1984) in: Bjørseth A.,
      Angeletti, G. (eds), "Analysis of Organic Micropollutants in Water",
      D. Reidel, Dordrecht, p. 132.
(12) LYON, P.A.; STEBBINGS, W.L.; CROW F.W.; TOMER, K.B.; LIPPSTREU, D.L.
      and GROSS, M.L. (1984) Anal. Chem. 56, 8.
(13) WEBER, R.; LEVSEN, K.; LOUTER, G.J.; BOERBOOM, A.J.H. and HAVERKAMP,
      J. (1982), Anal. Chem. 54, 1458.

(14) OTSUKI, A. and SHIRAISHI, H. (1979), Anal. Chem. 51, 2329.
(15) YASUHARA, A.; SHIRAISHI, H.; TSUJI, M. and OKUNO, T. (1981), Environ.
     Sci. Technol. 15, 571.
(16) SHIRAISHI, H.; OTSUKI, A. and FUWA, K. (1982), Bull. Chem. Soc. Jpn.
     55, 1410.
(17) SCHNEIDER, E.; LEVSEN, K.; DÄHLING, P. and RÖLLGEN, F.W. (1983),
     Fresenius Z. Anal. Chem. 316, 488.
(18) SCHNEIDER, E.; LEVSEN, K.; BOERBOOM, A.J.H.; KISTEMAKER, P.; MCLUCKEY,
     S.A. and PRZYBYLSKI, M. (1984), Anal. Chem. 56, 1987.
(19) WICKBOLD, R. (1972), Tenside Deterg. 9, 173.
(20) LYON, P.A.; CROW; F.W.; TOMER, K.B. and GROSS, M.L., (1984), Anal.
     Chem. 56, 2278.

RECENT LC/MS METHODS FOR WATER ANALYSIS

P.J. ARPINO

Department of Chemistry, Ecole Polytechnique, Palaiseau, France.

Summary

This review surveys publications describing methods for LC/MS that have appeared during the period 1983-1985. Emphasis is placed on commercial interfaces. In particular, the combination of Fast Atom Bombardment with a moving-belt interface has progressed significantly; the transfer of interest from direct liquid introduction to thermospray appears to have developed in several laboratories and instrument manufacturers. More reliable LC/MS systems are now available for the analysis of thermolabile substances in water.

1.INTRODUCTION

Combined liquid chromatography/mass spectrometry (LC/MS) has been the subject of intensive research during the last decade and is well documented. In particular, publications describing early LC/MS methods and applications to environmental studies were reviewed in depth by Dai Games et al. (1) in the proceedings of the Oslo symposium. The current review surveys methodology and applications that have appeared since this report. An emphasis has been placed on the methodology based on the "Thermospray" (TS), as this method has developed rapidly during recent years, while other LC/MS techniques have either not changed appreciably or have declined in importance (2).

Active research programs on LC/MS by the end of the 70s were soon followed by the introduction on the market of different types of LC/MS interfaces, adaptable to magnetic sector or quadrupole instruments. As a consequence, it was hoped that the utility of the technique in the environmental area would increase considerably. This has not yet been verified. Although the potential of combined LC/MS for water analyses is still very high, application to specific determinations of pollutants are few. A possible reason for this situation could be that early LC/MS interfaces, such as the original moving belt interface (3,4), or first models of Direct Liquid Introduction interfaces (DLI), were rather expensive, with a limited range of application, and were sometime difficult to operate routinely in a service laboratory. They consisted of interfaces to standard mass spectrometers also used with other conventional inlets, i.e. a GC interface and a direct insertion probe.

Running LC/MS analyses often required long installation and tuning time, and results were sometimes unsatisfactory. Many laboratories put their LC/MS interface aside, preferring to use the mass spectrometer for more well established analysis methods such as GC/MS, or for off-line LC/MS techniques; for instance, the combination of preparative high performance liquid chromatography (HPLC) with field desorption (FD) (5,6) or Fast Atom Bombardment (FAB) (7,8) was found to be particularly useful. The application of off-line LC/MS to environmental studies has been reviewed previously (9) and both the advantages and the disadvantages are the same today. This review does not cover recent aspects of decoupled techniques.

No new revolutionary concept for LC/MS has appeared during the last two years, and in the following, the focus is placed on progress made to improve established direct coupling methods. Because reports on application of LC/MS to environmental studies are still rare, and also because the subject is not studied in this laboratory, this review addresses the question of LC/MS in general, with particular emphasis on commercial interfaces. In the conclusion, the potential of modern LC/MS methods for water analysis will be reexamined.

2.TRANSPORT INTERFACES

This approach to LC/MS aims at the complete removal of the solvent from the column eluent prior to introduction of solutes into the mass spectrometer. Solutions are fed continuously to a moving Kapton belt. Commercial systems are available from Finnigan-MAT and V.G. Analytical, and a comparison between different models was published recently (10). The influence of construction details, such as extention of the moving belt into the ionization chamber of the mass spectrometer, was emphasized. Consequently, the quality of results in recent reports varies depending on the instrument that was available to the analyst.

An old Finnigan moving belt was used by Krost (11) for the analysis of polynuclear aromatic mixtures, including anthracene, pyrene, chrysene and benzo(a)pyrene. As already reported in the past, the sensitivity was found to be a direct function of compound volatility, and the lower detection limit in the single-ion monitoring mode was approximately 10 ng.

Levsen et al. (12) used the new Finnigan-MAT interface and a quadrupole mass spectrometer for the analysis of non-ionic surfactants of the alkylphenolethoxylate and oxoalcoholethoxylate classes. Information on the distribution of homologues and the presence of impurities was obtained. Biodegradation of non-ionic surfactants in surface water was investigated. After adding 0.3 g of nonylphenolethoxylate to 4 l of river water and stirring the solution for 3 days, the LC/MS analysis revealed that biodegradation to the homologue with two ethylene oxide units had occurred.

In another report (13), it is noted that deactivation of the Kapton belt in the same interface with a 50-ppm Carbowax 20M solution reduced memory effects during recycles of the belt, and in general increased the detection limit for various compounds. For instance, a detection limit of 35 pg of bromazepam was obtained under CI mode. Various isomeric tocopherols and α-tocatrienol were identified in a maize germ oil.

These techniques have been applied to substances of environmental concern other than those listed in ref. 11 and 12; for example, in the analysis of linuron, diuron and monuron, which belong to two groups of thermally labile pesticides: carbamates and ureas (14). The method was satisfactory for regulatory analysis of crop residues at the ppm level. Application to other classes of substances may be of interest to the reader for evaluation of recent instrument performance. Such studies include the identification of antioxidant and ultraviolet light stabilizing additives in plastics (15); sequence analysis of derivatized peptides (16) and identification of C-methylation artifacts from the permethylation reaction of peptides (17). There are many applications to analyses of natural substances, such as a series of naturally occurring phospholipids (18) and sphingoid bases (19); gossypol and some of its derivatives (20); pepper and capsicum oleoresins (21); vanilla extracts (22); dexmethasone and betamethasone (23).

From compilation of the recent literature, three major trends in research on improvement of moving-belt performance can be delineated: improved methods for depositing the solution onto the belt, use of small-bore chromatographic columns, investigation of new ionization methods.

2.1. Spray deposition onto the belt

Special attention has been paid to the solution deposition onto the moving belt.

Initially, the solution was simply fed onto the belt in a continuous stream via a narrow tube located close to the belt. Liquid deposition was uneven especially for aqueous solutions. The liquid film used to break into beads that degraded the chromatographic resolution and caused vacuum instabilities inside the mass spectrometer envelope.

Vouros et al. (24,25) have shown that spray deposition from a nebulizer located 4-8 mm above the belt and at an angle of 60° with the belt, eliminated this problem. Fine spraying is obtained by means of a coaxial pneumatic nebulizer. In recent models, the gas is preheated, or the solution is directly heated electrically (26) in a manner similar to that used in a thermospray nebulizer (vide supra). The heating still improves the spray deposition and in addition, part of the solvent can be evaporated before the concentrated mist reaches the belt.

2.2 Micro-HPLC instrumentation and post-column chemistry

There has been some movement in the past couple of years toward greater use of small-bore liquid chromatography in conjunction with moving belt interfaces, since it dramatically reduces the solvent which must be removed by the interface (25,27). Instrumentation which is compatible with small-bore LC columns (i.e. 1 to 2 mm i.d. packed tubes and pumps delivering constant flow-rates at ca 50-200 µl/min) has become available and has been the subject of recent reviews (28,29). As a consequence, the infrared heater used in early belt interfaces, which was difficult to adjust and caused severe problems when running aqueous solutions, and splitting devices to remove solvent excess are no longer necessary with small-bore columns.

Karger and Vouros (27) believe that even smaller columns should be investigated for optimum chromatographic and mass spectrometric conditions, for instance, 200 µm fused-silica capillary columns packed with small particle (3µm) diameter supports. They have also reemphasized the utility in conjunction with moving-belt LC/MS of a post-column solvent/solvent extraction and on-line synthesis of volatile derivatives. Although they have pioneered post-column chemistry since 1979 and have demonstrated its large possibilities (30), they have been followed by only one other moving-belt user (110), probably because of the additional instrumental constraints posed by this approach. However, a group of DLI users is also following this approach (31).

2.3 Improved ionization methods

Conventional ionization methods following thermal desorption of the samples from the belt continue to be electron impact and ammonia or isobutane chemical ionization. However, investigation on the use of soft ionization techniques applied to samples still adsorbed onto the belt has continued during the last two years.

Several research programs on combination with Secondary Ion Mass Spectrometry (SIMS) appear to have been stopped (see 32 and 33 for early work). A same situation is observed in the case of laser desorption (LDMS) (34). Only one recent report describing combination of either of these two methods with a moving carbon steel belt and a quadrupole mass spectrometer has appeared in the literature (35). The authors compared the performance of the technique based on analyses of 14 amino acids and 4 nucleosides by the two ionization methods. They estimated that LDMS provided more reliable molecular weight information. However, since the laser energy absorbed by the sample varies uncontrollably, the mass spectra as well as the LC response areas were less reproducible. On the other hand, chromatographic peak shapes, mass spectral reproducibilities and resolution were much better in SIMS than with LDMS.

Future progress on the on-line combination with SIMS and LDMS appear uncertain. Another effective soft ionization technique, ^{252}Cf plasma desorption, has so far been exclusively applied in discontinuous mode. Sample solutions were sprayed onto a mechanized support and solutes were ionized after complete solvent evaporation (36). On the other hand, one trend to look for is the combination with Fast Atom Bombardment. Although research reports are still few, preliminary results using LC/FABMS appear encouraging. The experimental set-up is now commercially available from VG Analytical and Finnigan-MAT.

FAB implies bombardment by high kinetic energy atoms of samples in solution in liquid matrices such as glycerol of thioglycerol. Initially the matrix was spread separately onto the belt using a reservoir and an applicator (37), however, this method appears uncompatible with the maintenance of a high vacuum in the mass spectrometer. In a recent report (26), the matrix was dissolved into the HPLC solvent at very low concentration, typically 0.05% of the solvent employed. Possible influence of this additive on the chromatographic separation was not investigated. The clean-up heater used in conventional belt systems for regenerating the belt surface was replaced by a washbath containing methanol for removing nonionized samples and the liquid matrix.

The instrument was applied to molecular weight determination and sequencing of a series of peptides including the underivatized hexadeca- peptide antiamoebin I (M=1669 u) and a mixture of tetra- to nonapeptides obtained by partial hydrolysis of antiamoebin I. Total ion current profiles similar to the trace from a chromatographic UV detector were produced. One weakness of this method is the poor sensitivity; for instance, 125 ng of an N-acylpentapeptide gave an M+Na$^+$ ion with a signal-to-noise ratio of 3 (26).

In his 1983 review (1), Games concluded that conventional moving belt systems were limited in the range of compounds that can be handled. The situation appears to be the same today, thus explaining why the number of application reports has not increased significantly during the past two years (2). On the other hand, the combination with FAB could change this situation, although more examples of application must be obtained for the evaluation of its analytical potential, in particular for the application to water analysis.

3.DIRECT LIQUID INTRODUCTION

In this method, a portion of the eluent from the liquid chromatograph is fed directly into the ion source of a mass spectrometer configured for CI where solvent induced CI spectra are produced from the solute (1,38).

3.1. General instrumentation

DLI probes are now available from major mass spectrometer manufacturers and from accessory dealers, e.g. Vacumetrics and Hositrad, and can be fitted to any model of mass spectrometer. Quadrupoles are generally used but work using sector instruments has also been reported (39-41). A diaphragm is no longer the only method of vacuum nebulization. Other devices including fused-silica capillary tubes (42-45) and coaxial pneumatic nebulizers (46,47) have been described, some of which are commercial products. The particular model introduced by Extranuclear was based on the work done at the National Bureau of Standard in Washington D.C. (48) but it has subsequently been abandoned. The online concentrator of this interface was an interesting idea, but the liquid pressure of the concentrated solution in front of the DLI nebulizer was too low and could not be adjusted.

Classical DLI models comprise a metal diaphragm pierced by a 2-5 μm pinhole for vacuum nebulisation of chromatographic effluents. Modifications of a standard mass spectrometer for DLI/LCMS operations include the addition to the CI source block of a desolvation chamber for better droplet concentration (49) and a cryotrap for a better vacuum pumping (50). However, the trap imposes some practical constraints during routine work; it can be advantageously replaced by a rough pump directly attached to the ion source (43), in a manner first described for thermospray experiments.

DLI interfaces are simple to design and to install, rather easy to operate, and have often produced good analytical results when applied to difficult samples (49-53). Modification of standard mass spectrometers for DLI coupling, designs of desolvation chambers and laboratory-built DLI probes, source and interface parameters are well described (51,52). The life-time of the heated metallic ribbon, which emits the electron beam for inducing solvent CI reactions, can be very short if it is surrounded by a high pressure of solvent vapours; replacement of the filament assembly by a discharge tube has been proposed (44).

Nevertheless, all DLI interfaces suffer more or less from severe plugging problems and irregular responses, especially those models which utilize a small diaphragm (54). Another inherent limitation is the poor heat transfer to liquid droplets in a vacuum when attempting to vaporize the solvent, even when the droplets pass through a very hot desolvation chamber (55). The situation can be improved by preheating the solution while it flows through the DLI probe, a method sometime called "hot DLI" (43). The difference from a normal thermospray is that a straight jet is still produced in the air from the nebulizer, although the difference between some devices referred to as DLI (42-44) and a microthermospray is probably very small. Another method, pneumatic nebulization, has also been claimed to improve nebulization and desolvation conditions and has been incorporated into some DLI models (46,47).

3.2. Micro-HPLC instrumentation

Because the liquid flow rate sampled by the DLI interface into the ion source is usually very small, in the 10-80 μl/min interval, combination with small-bore columns has become increasingly popular among DLI users(56-62). Splitting the column effluent, an operation which negatively affects the limit of detection and the sensitivity, is avoided when using small-bore columns; the full column effluent is fed into the ion source, through the DLI nebulizer.

Small-bore packed column with 0.5 to 2 mm I.D. are usually selected Efforts have been made to reduce dead volumes between the column outlet and the nebulizer in the DLI probe, and various designs of improved DLI interfaces, specially optimized for coupling with small-bore columns, have been published (42,44,56,61).

Preliminary work using 5 and 10 μm I.D. fused-silica open-tubular columns in combination with a 4 μm diaphragm nebulizer has been reported by Niessen and Poppe (63). Although the detection limits of the system look promising, being in the low picogram range, no application to difficult samples was shown. The potential for routine work remains unproven.

With respect to moving-belt interfaces, the use of post-column exchange of polar solvents, which are difficult to vaporize, for apolar solvents, by means of on-line liquid/liquid extraction has been advocated for DLI systems (31,64).

3.3. Recent application of DLI interfaces

Concerning substances of interest for environmental studies, I found no recent application on water extracts. However, the literature of the last couple of years contains several investigations of thermally labile herbicides and pesticides, including the active ingredient of two herbicides, chlosulfuron and sulfometuron (65); 25 carbamates, chlorinated carboxylic acids and methylureas (52,66,67); and several organophosphorus pesticides (68). Of related interest is the analysis of two mycotoxins, nivalenol and deoxynivalenol (69).

Another group of application reports address the determination of biological substances, including ranitidine and its metabolites in urine (70); underivatized sialic acids (71); several optically active drugs (72, 73); corticosteroids in equine biological fluids (74), - the same compounds were also analysed by others using a moving-belt interface (23), allowing a direct comparison of performance-; quantitation of dethiobiotin and biotin in biological fluids (75); sequencing of underivatized peptides (53); liposoluble vitamins (76); glycerolipids and glycerophospholids including diacylglycerol moieties of natural phosphatidylcholines (77,78).

Explosives certainly belong to a class of thermally sensitive compounds and DLI instrumentation was found particularly useful for their determination (39,79). However, reduction of nitro- groups to amino groups may occur. When studying nitropolycyclic aromatic hydrocarbons by several analytical methods, Quilliam and co-workers (80) observed that the reduction was most extensive during LC/MS with a DLI interface. They estimated that both qualitative and quantitative analyses at trace levels can be severely impaired by the irreproducibility of the mass spectra.

Although the number of reports on applications of DLI systems is still high, it has decreased somewhat compared to previous years (2). Many users have shifted to thermospray (TS) because the change from a DLI set-up requires little modification. There are several direct comparison studies on both techniques (65,67,81). Although the sensitivities for given compounds were often comparable, and molecular information sometime better in DLI than in TS, the later was found to be easier to use routinely and less prone to -although not completely free of- plugging problems. A remaining interest in DLI system is the combination with small-bore columns. For the moment, TS appears to accommodate a low liquid flow-rate input with some difficulties, unless the column effluent is diluted to 1 ml/min before entering the interface.

4. THERMOSPRAY

4.1. Principles of nebulization and solute ionization by thermospray

Thermospray has evolved from large crossed beam machines assembled in Salt Lake City by Marvin Vestal and co-workers in the early 70s (see 82-84 for recent reviews). It consists of a mechanically simple method which simultaneously achieves two different processes: first, the nebulization and the nearly complete desolvation of the HPLC solution, prior to entering the vacuum in the mass spectrometer, and second the soft ionization of nonvolatile and thermally labile solute samples.

Typically, a solution of the sample in a solvent containing ca 0.1 M of a volatile salt, e.g. ammonium acetate, is heated by an appropriate source of heat. Part of the solvent evaporates and the vapors rapidly expand to sonic velocities. The vapors cause the pneumatic nebulization of the remaining liquid into a fine mist. The resulting droplets are electrically charged as a result of liquid film ruptures and random repartition of positive and negative moities of the salt molecules between individual liquid droplets. One assumes that the high electrical field between droplets of opposite charges induces the desorption, from the liquid solution to the gas phase, of preformed solute ions (i.e. ions derived from the solute molecule by protonation, cationization or addition of solvent cluster ions (84)).

4.2 Specific advantages

Three facts have contributed to the rapid recent development of thermospray:

(i) The design of the interface was continuously simplified in order to achieve a simple device applicable to different models of mass spectrometer. For example, nebulization was achieved successively by a high powered laser, an oxy-hydrogen torch, several electrical cartridges (85), and finally by a tube directly heated by an electric current (86). The original cost was reduced by at least 4 orders of magnitude. Direct electrical heating is now used in most commercial versions of this method, for instance those of Vestec, Finnigan, Nermag, V.G. Analytical, Kratos and Extranuclear. In general they are adapted to quadrupole mass spectrometers, but the coupling to magnetic sector instruments is also possible and shows equal performance (87-89).

(ii) The experimental observation of solute ions produced by simply spraying electrolytic solutions in the absence of other ionization devices (85), the so-called filament-off mode, has received considerable attention, and is still one of the most spectacular features of the method. In reality, today it has been established that this effect applies only to polar compounds and preformed ions. Nevertheless, these molecules are often separated exclusively by liquid chromatography and are difficult or impossible to analyse by conventional MS. For nonpolar molecules and a large number of medium polarity samples, electrons from a heated filament are necessary to induce the series of chemical ionization reactions that produce sample derived ions. This "abnormal" mode is referred to as the filament-on mode or solvent chemical ionization. It is equivalent to the ionization conditions which prevail in DLI. Which of these ionization modes should be applied to an unknown molecule cannot be predicted. An increasing number of molecules appear to need the filament-on conditions.

Nevertheless, filament-off operation is an interesting simplification when applicable. It also leads to interesting theoretical considerations of ionization mechanisms under the conditions of TS, and also under those of other MS techniques such as field desorption (90), FAB (91) and SIMS (92); direct comparison studies have been conducted (90-92).

(iii) TS interfaces are more reliable with respect to plugging than previous DLI interfaces, as long as the heated capillary tube has a diameter larger than 100 μm. Many DLI users have turned to thermospray for this reason.

4.3. Optimization of interface and source parameters

As for DLI, sensitivity is compound dependent. The detection limit is comparable or even lower with TS (67,71,81). There is a direct relationship between the diameter of the heated tube and the optimum liquid flow-rate. For 100-150 μm I.D. tubes, the flow-rate is 1-1.5 ml/min. There are several consequences:

1) The mass spectrometer must be customized to high vapor throughputs by connecting a rough pump directly to the ion source (85,86).

2) A special ion source block must be used. It is not suitable for GC/MS, or solid probe analyses. Changing instrument mode of operation requires the exchange of source blocks, i.e. a half-day work for a skilled operator.

3) The salt required for filament-off conditions may alter the chromatographic retention parameters. It is advisable to add the solution of the salt at the HPLC column outlet (89,93).

4) Small-bore columns cannot be connected directly. Attempts to use narrow capillary tubes, ca 10-25 μm, into the thermospray probe have failed because of a rapid plugging. The column effluent must be diluted up to 1 ml/min before entering the interface. Note however that post-column addition also solves the problem raised in the previous paragraph.

5) Thermal vaporization of a constant fraction of the HPLC solvent is required for stable operating conditions, even if this fraction is close to 99% (85,86). In particular, and as in DLI, the liquid flow-rate entering the interface must be rigourously maintained. Flow pulsations produced by HPLC pumps using one or several reciprocating pistons induce oscillations in the recorded ion currents. A pulseless solvent delivering system, such as a syringe-type pump (e.g. the Brownlee lab pump, or equivalent models) avoids this particular problem.

Finally, the critical influence of the temperature and the voltage at different locations in the thermospray probe and the ion source block has been observed (85,94). Small changes in these parameters often produce strong variations in the intensities of ion currents and modify the recorded mass spectra. No round robin test has yet been conducted to evaluate the reproducibility of mass spectra produced by TS instruments, and it remains to be done if, for instance, official analytical methods for environmental analyses are to be established.

4.4 Recent applications

As for the previous LC/MS methods, applications of thermospray during the past two years generally concern analyses of biological substances. The important difference with the previous years is that a larger number of studies describe real case problems and analyses of a priori unknown mixtures.

Peptide sequencing has been achieved by combination of sequential enzymic hydrolysis on a column of immobilized enzymes, followed by TS-LC/MS (95-97). Thermospray LC/MS has also permitted the study of patients suffering carnitine deficiency; carnitine and its derivatives (98), and a drug metabolite, valproylcarnitine, in patient urine (99) were determined at trace levels. In addition, several other examples of analyses of drugs and their metabolites in biological fluids were reported (67,89,100). In many cases, a rapid clean-up extraction from an aqueous solution was followed by TS-LC/MS investigation; the operating protocol could be adapted to aqueous solutions from environmental studies.

In other field of application, one finds the investigation of diquaternary ammonium salts (100); and constituents of two synthetic dyes, Basic Red 14 and Basic Orange 21 (102).

4.5 Future prospects

Thermospray is at a stage of exponential development. Most instrument manufacturers who offered DLI accessories have now replaced them by TS probes. More and more application examples from new users are reported, but some of them also contain criticisms. Optimization of experimental conditions and studies of samples containing unknowns can present difficulties; the molecular weight information can be obscured by the presence of multiply-charged ions, solvent clusters and cationated molecules, the intensities of which may vary according to changes in the experimental conditions.

It remains that thermospray, which has proven its capability to produce good mass spectral information from difficult samples, attracts to LC/MS new teams of scientists who were previously concerned by the practical difficulties involved with existing commercial interfaces. Environmental chemical analysts coud do well to investigate the utility of this LC/MS coupling method in solving their difficult analytical problems.

5. MISCELLANEOUS LC/MS COUPLING METHODS

None of the previous LC/MS methods is perfect. Work in several academic laboratories is still under way to find new concepts that would permit improved instrument design.

5.1.Ion evaporation and electrospray

The two names in fact cover the same physical process. The HPLC effluent is nebulized into a spray of small droplets which are charged electrically by direct contact with, or induction from, metallic surfaces held at a high voltage. Stripping the solvent in a nitrogen bath at atmospheric pressure sets free the solute ions, which are subsequently sampled into the vacuum of a mass spectrometer and analysed. The method has features similar to thermospray, however, it differs in that nebulization and ionization are achieved at nearly ambient temperature and at atmospheric pressure, and that the initial electrical charge carried by the droplets is probably higher in electrospray than in TS.

Ion evaporation was the name given by Thomson and co-workers (103). An instrument has been available from Sciex (Canada) since 1980; however, it seems that no application report has been published during the period covered in this review. Dole et al. (104) had called the method electrospray in the late 60s. Application for LC/MS purpose is currently reinvestigated by Fenn and co-workers (105,106). They have reported mass spectra from a series of nonvolatile substances and have estimated that, from the standpoint of flexibility, convenience, sensitivity, cleanliness of the mass spectrometer, and ease of maintenance, the method may comprise an effective and practical LC/MS interface. However, chromatograms and extracted ion current profiles from electrospray or ion evaporation experiments are seldom shown; thus, the possibility of an application to the on-line monitoring of effluents from an HPLC column has not yet been proven.

5.2. Magic LC/MS

Magic is the acronym for Monodisperse Aerosol Generation Interface for Combining Liquid Chromatography with Mass Spectroscopy (107). It employs a sophisticated cross-flow pneumatic nebulizer for producing a monodisperse aerosol from the HPLC solution. The solvent is removed at ambient temperature and pressure. The sample forms a cloud of dust particles and is directed toward the ion souce in which it is ionized by in-beam electron impact. Although this elegant method gives detection limits in the nanogram range for several substances, it appears to apply only to moderately volatile molecules.

5.3. Pervaporation through semi-permeable membranes

One side of the membrane is in contact with the liquid solution, the other side, under a vacuum, is connected directly to the MS ion source. Many volatile nonpolar solutes in polar solvents, e.g. water, can selectively cross the membrane and be mass analysed. Eustache and Histi have already shown an interesting application to the analysis of chloro-alkanes in water (108).

A new version of this method, using a semipermeable silicon capillary tubing and a triple quadrupole MS, has recently been described (108). No chromatographic equipment was used. The solution from a flask was continuously circulated through the capillary tube. A number of mono- and diaromatic hydrocarbons gave detection limits in the range $10^{-3}M - 10^{-6}M$, depending on the sample. The authors suggest that the membrane device be used in conjunction with a robot-controlled organic reactor or as an industrial process monitor.

6. CONCLUSION

It should be reemphasized that today, there is still no universal LC/MS system. However, the potential of modern LC/MS methods for organic analyses, and in particular environmental chemical studies, is very high. This is especially true for thermally sensitive compounds with masses of 300-500 u. The new methods extend the range of molecules amenable to the combination of a chromatographic separation and a mass spectrometric investigation to marginally volatile molecules, present for instance in river and drinking water, which are difficult or impossible to elute from a gas chromatographic column. For these substances, the recent LC/MS systems can reconstitute the chromatographic signal and produce mass spectral information at trace levels without sacrificing chromatographic separation. The results obtained in the case of several pesticides are good examples of the present possibilities.

On the other hand, it must be stated that for higher molecular weight substances, e.g. oligosaccharides, oligopeptides, porphyrins, polyaromatic compounds, etc., with masses of 500-2000 daltons, the molecular weight information can often be ascertained by employing recent LC/MS systems, but at the expense of some more or less serious damages to the chromatographic parameters. In the LC/MS literature, there are several cases of misuse and abuse of the term LC/MS for work in which only the mass spectral data were reported. Even in thermospray, nonvolatile samples that will show a mass spectrum may become adsorbed on metallic surfaces while they pass through the ionizer. The reconstructed chromatographic profiles then exhibit prominent tailing and sensitivity is strongly reduced. Note however that for these substances, direct water analysis can be employed obviating a separation step, but with liquid solutions sampled in a flow-injection mode directly into the MS detector. The method could be valuable for solving practical problems in a routine analytical laboratory.

7. ACKNOWLEDGMENTS

I thank Dr. G.R.B. Webster (University of Manitoba, Winnipeg, Manitoba, Canada) for his assistance during the writing of this review.

REFERENCES

1. Games, D.E.; McDowall, M.A.; Glenys Foster, M.; Merez O. "Analysis of Organic Micropollutants in water"; Angeletti, G., Bjørseth, A., Eds.; D. Reidel Publishing Co: Dordrecht, 1984,68.
2. Arpino, P.J. J. Chromatogr. 1985,323,3.
3. McFadden, W.H.; Schartz, H.L.; Evans, S. J. Chromatogr. 1976,122,389.
4. Millington, D.S.; Yorke, D.A.; Burns P. Adv. Mass Spectrom. 1980, 8B, 1819.
5. Schulten, H.R. J. Chromatogr. 1982, 251, 105.
6. Schulten, H.R.; Bahr, U.; Monkhouse, P.B. J. Biochem. Biophys. Methods 1983,8,239.
7. Brueckner, H.; Przybylski, M. J. Chromatogr. 1984,296,263.
8. Crawthorne, B.; Fielding, M.; Steel, C.P.; Watts, C.D. Environ. Sci. Technol. 1984,18,797.
9. Watts, C.D.; Crawthorne, B.; Fielding, M.; Steel, C.P. "Analysis of Organic Micropollutants in water", Angeletti, G., Bjørseth, A., Eds.; D. Reidel Publishing Co: Dordrecht, 1984,120.
10. Games, D.E.; McDowall, M.A.; Levsen, K.; Schafer, K.H.; Dobberstein, P.; Gower, J.L. Biomed. Mass Spectrom. 1984,11,87.
11. Krost, K.J. Anal. Chem. 1985,57,763.
12. Levsen, K.; Wagner-Redeker, W.; Schäfer, K.H.; Dobberstein, P. J. Chromatogr. 1985,323,135.
13. Van der Greef, J.; Tas, A.C.; Ten Noever de Brauw, M.C.; Höhn, G.; Meijerhoff, G.; Rapp, U. J. Chromatogr. 1985,323,81.
14. Cairns, T.; Siegmund, E.G.; Doose, G.M. Biomed. Mass Spectrom. 1983,10,24.
15. Vargo, J.D.; Olson, K. Anal. Chem. 1985,57,672.
16. Yu, T.J.; Schwartz, H.A.; Cohen, S.A.; Vouros, P.; Karger, B.L. J. Chromatogr. 1984,301,425.
17. Yu, T.J.; Karger, B.L.; Vouros, P. Biomed. Mass Spectrom. 1983,10,633.
18. Jungalwala, F.B.; Evans J.E.; McCluer, R.H. J. Lipid Res. 1984,25,738.
19. Jungalwala, F.B.; Evans, J.E.; Kadowaki,H.; McCluer, R.H.. J. Lipid Res. 1984,25,209.
20. Martin, S.A.; Zhou, R.H.; Games, D.E.; Jones, A.; Ramsey, E.D. HRC CC, J. High Resolut. Chromatogr. Chromatogr. Commun. 1984,7,196.
21. Games, D.E.; Alcock, N.J.; Van der Greef, J.; Nyssen, L.M.; Maarse, H.; Ten Noever de Braw, M.C. J. Chromatogr. 1984,294,269.
22. Fraisse, D.; Maquin, F.; Stahl, D.; Suon, K.; Tabet, J.C. Analusis, 1984,12,63.
23. Cairns, T.; Siegmund, E.G.; Stamp, J.; Skelly, J.P. Biomed. Mass Spectrom. 1983,10,203.
24. Hayes,M.J.; Lankmayer, E.P.; Vouros, P.; Karger, B.L.; McGuire, J.M. Anal. Chem. 1983,55,1745.
25. Hayes J.M.; Schwartz, H.E.; Vouros, P.; Karger, B.L. Thruston A.D.,Jr.; McGuire J.M. Anal. Chem. 1984,56,1229.
26. Stroh, J.G.; Cook, J.C.; Milberg, R.M.; Brayton, L.; Kihara, T.; Huang, Z.; Rinehart K.L.,Jr.; Lewis, I.A.S. Anal. Chem. 1985, 57,985.
27. Karger, B.L.; Vouros, P. J. Chromatogr. 1985,323,13.
28. Scott, R.P.W. J. Chromatogr. Sci. 1985,23,233.
29. Rabel, F.M. J. Chromatogr. Sci. 1985,23,247.

30. Vouros, P.; Lankmayr, E.P.; Hayes, M.J.; Karger, B.L.; McGuire, J.M. J. Chromatogr. 1982, 251,175.
31. Maris, F.A.; Geerdink, R.B.; Frei, R.W.; Brinkman, U.A.Th. J. Chromatogr. 1985,323,113.
32. Benninghoven, A.; Eicke, A.; Junack, M.; Sichtermann, W.; Krizek, J.; Peters, H. Org. Mass Spectrom. 1980,15,459.
33. Smith, R.D.; Burger, J.E.; Johnson, A.L. Anal. Chem. 1981,53,1120.
34. Huber, J.F.K.; Dzido, T.; Heresch, F. J. Chromatogr. 1983,271,27.
35. Fan, T.P.; Hardin, E.D.; Vestal, M.L. Anal. Chem. 1984,56,1870.
36. Jungclas, H.; Danigel, H.; Schmidt, L. Int. J. Mass Spectrom. Ion Phys. 1983,46,197.
37. Dobberstein, P.; Korte, E.; Meyerhoff, G.; Pesch, R. Int. J. Mass Spectrom. Ion Phys. 1983,46,185.
38. P.J.Arpino. "Liquid Chromatography Detectors", Vickrey, T.M. Ed.; Marcel Dekker: New York, N.Y. 1983,243.
39. Yinon, J.; Hwang, D.G. J. Chromatogr. 1983, 268,45.
40. Chapman, J.R.; Hardin, E.H.; Evans, S.; Moore, L.E. Int. J. Mass Spectrom. Ion Phys. 1983,46,201.
41. Alborn. H.; Stenhagen, G. J. Chromatogr. 1985,323,47.
42. Bruins, A.P.; Drenth, B.F.H. J. Chromatogr. 1983,271,71.
43. Arpino, P.J.; Beaugrand, C. Int. J. Mass Spectrom. Ion Proc. 1985,64,275.
44. Hirter, P.; Walther, H.J.; Dätwyler, P. J. Chromatogr. 1985,323,89.
45. Tiebach, R.; Blaas, W.; Kellert, M. J. Chromatogr. 1985,323,121.
46. Apffel, J.; Brinkman, U.A.Th.; Frei, R.W.; Evers, E.A.I.M. Anal. Chem. 1983,55,2280.
47. Yoshida, H.; Matsumoto, K.;Itoh, K.; Tsuge, S.;Hirata, Y.; Mochizuki, K.; Kokubun, N.; Yoshida, Y. Fresenius' Z. Anal. Chem. 1982,311,673.
48. Christensen, R.G.; Hertz, H.S.; Meiselman, S.; White, E. Anal. Chem. 1981,53,171.
49. Dedieu, M.; Juin, C.; Arpino, P.J.; Guiochon, G. Anal. Chem. 1982,54,2372.
50. Eckers, C.; Skrabalak, D.S.; Henion, J.D. Clin. Chem. 1982,28,1882.
51. Sugnaux, F.R.; Skrabalak, D.S.; Henion, J.D. J. Chromatogr. 1983,264,357.
52. Voyksner, R.D.; Bursey, J.T. Anal. Chem. 1984,56,1582.
53. Kenyon, C.N. Biomed. Mass Spectrom. 1983,10,535.
54. Mauchamp, B.; Krien, P. J. Chromatogr. 1982,236,17.
55. Arpino. P.J.; Bounine, J.P.; Dedieu, M.; Guiochon, G. J. Chromatogr. 1983,271,43.
56. Henion, J.D. J. Chromatogr. Libr. 1984,28,260.
57. Lee, D.E.; Henion, J.D. J. Chromatogr. Sci. 1985,23,253.
58. Tsuge, S. J. Chromatogr. Libr. 1985,30,217.
59. Westwood, S.A. Anal. Proc. 1984,21,418.
60. Henion, J.D. J. Chromatogr. Libr. 1985,30,243.
61. Esmans, E.L.; Geboes, P.; Luyten, Y.; Alderweireldt, F.C. Biomed. Mass Spectrom. 1985,12,241.
62. Tiebach, R.; Blaas, W.; Kellert, M.; Steinmeyer, S.; Weber, R. J. Chromatogr. 1985,318,103.
63. Niessen, W.M.A.; Poppe, H. J. Chromatogr. 1985,323,37.
64. Apffel, J.A.; Brinkman, U.A.Th.; Frei, R.W. J. Chromatogr. 1984,312,153.
65. Shalaby, L.M. Biomed. Mass Spectrom. 1985, 12,1985.
66. Voyksner, R.D.; Bursey, J.T.; Pellizari, E.D. J. Chromatogr. 1984,312,221.

67. Voyksner, R.D.; Bursey, J.T.; Hines, J.W.; Pellizzari, E.D. Biomed. Mass Spectrom. 1984,11,616.
68. Parker, C.E.; Yamaguchi, K.; Harwan, D.J.; Smith, R.W.; Hass, J.R. J. Chromatogr. 1985,319,273.
69. Tiebach, R.; Blaas, W.; Kellert, M.; Steinmeyer, S.; Weber, R. J. Chromatogr. 1985,318,103.
70. Lant, M.S.; Martin, L.E.; Oxford, J. J. Chromatogr. 1985,323,143.
71. Shukla, A.K.; Schauer, R.; Schade, U.; Moll, H.; Rietschel, E. Th. J. Chromatogr. 1985, 337,231.
72. Crowther, J.B.; Covey, T.R.; Dewey, E.A.; Henion, J.D. Anal. Chem. 1984,56,2921.
73. Eckers, C.; Henion, J.D. Chromatogr. Sci. 1985,32,115.
74. Skrabalak, D.S.; Covey, T.R.; Henion, J.D. J. Chromatogr. 1984,315,359.
75. Azoulay, H.; Desbene, P.L.; Frappier, F.; Georges, Y. J. Chromatogr. 1984,303,272.
76. Milon, H.; Bur, H. LC, Liq. Chromatogr. HPLC Mag. 1984,2,455.
77. Pind, S.; Kuksis, A.; Myher, J.J.; Marai, L. Can. J. Biochem. Cell Biol. 1984,62,301.
78. Kuksis, A.; Myher, J.J.; Marai, L. JAOCS, J. Am. Oil Chem. Soc. 1984,61,1582; 1985,62,762; 1985,62,767.
79. Yinon, J.; Hwang, D.G. J. Chromatogr. 1985,339,127.
80. Quilliam, M.A.; Messier, F.; D'Agostino, P.A.; McCarry, B.E.; Lant, M.S. Spectrosc.: Int. J. 1984,3,33.
81. Crowther, J.B.; Covey, T.R.; Silvestre, D.; Henion, J.D. LC Mag. 1985,3,240.
82. Vestal, M.L. Mass Spectrom. Rev. 1983,2,447
83. Vestal, M.L. Science, 1984,226,275.
84. Vestal, M.L. "Ion formation from organic solids"; Benninghoven, A, Ed.; Springer-Verlag: Heidelberg, 1983,246.
85. Blakley, C.R.; Vestal, M.L. Anal. Chem. 1983,55,750.
86. Garteiz, D.A.; Vestal, M.L. LC Mag. 1985,3,334.
87. Vestal, M.L. Anal. Chem. 1984,55,2591.
88. Chapman, J.R. J. Chromatogr. 1985,323,153.
89. Catlow, D.A. J. Chromatogr. 1985,323,163.
90. Schmelzeisen-Redeker, G.; Giessman, U.; Röllgen, F.W. J. Phys. Colloq. 1984,C9,297.
91. Fenselau, C.; Liberato, D.J.; Yergey, J.A.; Cotter, R.J.; Yergey, A.L. Anal. Chem. 1984,56,2759.
92. Schmelzeisen-Redeker, G.; Wong, S.S.; Giessman, U.; Röllgen, F.W. Z. Naturforsch.A: Phys. Phys. Chem. Kosmophys. 1985,40A,430.
93. Voyksner, R.D.; Bursey, J.T.; Pellizari, E.D. Anal. Chem. 1985,56,1507.
94. Schmelzeisen-Redeker, G.; McDowall, M.A.; Giessman, U.; Levsen, K.; Röllgen, F.W. J. Chromatogr. 1985,323,127.
95. Kim, H.Y.; Pilosof, D.; Dyckes, D.F.; Vestal, M.L. J. Am. Chem. Soc. 1984,106,7304.
96. Pilosof, D.; Kim, H.Y.; Dyckes, D.F.; Vestal, M.L. Anal. Chem. 1984,56,1236.
97. Pilosof, D.; Kim, H.Y.; Vestal, M.L.; Dyckes, D.F. Biomed. Mass Spectrom. 1984,11,403.
98. Yergey, A.L.; Liberato, D.J.; Millington, D.S. Anal. Biochem. 1984,139,278.
99. Millington, D.S.; Bohan, T.P.; Roe, C.R.; Yergey, A.L.; Liberato, D.J. Clin. Chem. Acta 1985,145,69.

100. Covey, T.; Crowther, J.; Dewey, E.A.; Henion, J.D. Anal. Chem. 1985,57,474.
101. Schmelzeisen-Redeker, G.; Giessman, U.; Röllgen, F.W. Angew. Chem. 1984,96,889.
102. Betowski, L.D.; Ballard, J.M. Anal. Chem. 1984,56,2604.
103. Thomson, B.A.; Iribarne, J.V.; Dziedic, P.J. Anal. Chem. 1982,54,2219.
104. Gieniec, J.; Mack, L.L.; Nakamae, K.; Gupta, C; Kumar, V.; Dole, M. Biomed. Mass Spectrom. 1984,11,259.
105. Yamashita, M.; Fenn, J.B. J. Phys. Chem. 1984,88,4451; 1984,88,4671.
106. Whitehouse, C.M.; Dreyer, R.N.; Yamashita, M.; Fenn, J.B. Anal. Chem. 1985,57,675.
107. Willoughby, R.C.; Browner, R.F. Anal. Chem. 1984,56,2625.
108. Eustache, H.; Histi, G. J. Membrane Sci. 1981,8,105.
109. Brodbelt, J.S.; Cooks, R.G. Anal. Chem. 1985,57,1155.
110. Bigley, F.; Grob, R.L.; Brenner, G. J. Chromatogr. 1984,288,293.

THE STATUS OF THE DEVELOPMENT OF STANDARD PROCEDURES FOR THE
ANALYSIS OF ORGANIC PRIORITY POLLUTANTS IN WATER SAMPLES IN
THE FEDERAL REPUBLIC OF GERMANY

P.G.LAUBEREAU
Hessische Landesanstalt fuer Umwelt
Wiesbaden, FRG

SUMMARY

An overview is given about the existing and planned standard methods for the analysis of organic priority pollutants in water samples in the Federal Republic of Germany.
The priority pollutants referred here are those of the
EEC-List of dangerous substances, issued in 1982.- Taking
a proposed standard for "Low Volatile Chlorinated Hydrocarbons" as an example, some of the difficulties and problems which arise in case a method has to be standardized,
are discussed.

1.   INTRODUCTION
       With this lecture I want to present to you a status report
about the existing and planned German standard procedures for
the determination of organic "priority pollutants". In addition
to this I want to discuss some of the problems which arise
during the practical realization of a standard. It is hoped
that my lecture and a discussion can give to you some sti-
mulation and some awareness about work which needs to be done.
       At the beginning please allow that I give some short infor-
mation about the organization which deals with the standardization of water analysis in Germany and that I explain which
kind of "priority pollutants" is discussed here.

1.1   THE ORGANIZATION FOR THE STANDARDISATION OF PROCEDURES
                FOR WATER ANALYSIS IN GERMANY
       The availability of standardized methods is of great im-
portance for the determination of organic micropollutants in
water. Only by good harmonization, by an appropriate description of a procedure and by an uniform practice of methods, or
at least by a harmonized correlation of the techniques and procedures of analysis to equal criteria of quality and analytical
objectives, it is possible to achieve the necessary comparability of results, with regard to parameters measured, matrices,
local and temporal necessities and other variables of an environmental investigation.
       As shown in figure 1, the standardization of methods for

water analysis in Germany is done by a joint operation of the
"Deutsches Institut fuer Normung (DIN)" and by the "Fachgruppe
Wasserchemie der Gesellschaft Deutscher Chemiker", supported
by the federal and the state governements, universities,
industry and other groups.

The work is done by the leading committee for the "German
standard methods (Deutsche Einheitsverfahren) for examination
of water, wastewater and sludge" , its sub-committees and its
related working groups.

A compilation of these standards is issued as "Deutsche
Einheitsverfahren zur Wasser-, Abwasser- und Schlamm-Unter-
suchung" (1), in a loose-bound manner and updated regularely.

1.2 THE WORKING GROUPS FOR THE STANDARDISATION OF METHODS FOR THE ANALYSIS OF ORGANIC COMPOUNDS IN WATER-SAMPLES

The existing working groups for organic compounds are
listed in figure 2. As you see the 9 groups cannot cover the
whole range of possible organic pollutants and their analysis.
It is urgent therefore to expand the number of groups in immediate future.

1.3 THE "PRIORITY POLLUTANTS"

The priority pollutants about which I am talking here are
those from the EEC-list of priority pollutants (2).

This is not the place to discuss the completeness and the
relevance of this selection of compounds and to make comparisons with other lists of priority pollutants. My task here
simply is to give to you a review about standards which are
available for a selected variety of highly different organic
compounds which are of certain importance to the aquatic environment.

For a general information I have tried to arrange the
EEC priority pollutants into groups a shown in figure 3.

2. EXISTING STANDARDS AND STANDARDS UNDER DEVELOPMENT

2.1 METHODS FOR THE DETERMINATION OF AROMATIC HYDROCARBONS AND POLYNUCLEAR AROMATIC HYDROCARBONS (PAH)

2.1.1 METHODS FOR THE DETERMINATION OF POLYNUCLEAR AROMATIC HYDROCARBONS IN DRINKINC WATER
(DIN 38 403 Part 13)

The German standard DIN 38 403 Part 13 includes three
procedures, all based on thin layer chromatography (TLC). Their
fields of application are:

      Procedure 1:   qualitative screening test
      Procedure 2:   semi-quantitative screening test
                       (to check if the concentration of PAH
                       exceeds 20% of the threshold value)
      Procedure 3:   quantitative determination

Figure 4 presents a flow-diagram of procedure 3.

The standard is very much oriented to the needs of the
German "Trinkwasser Verordnung (TVO)". As shown in figure 5
the procedure principally allows for the separate determina-

tion of 6 PAH. However for survey purposes with respect to the TVO, the sum of the concentrations of the PAH found, is reported as the result of the analytical determination

2.1.2 DETERMINATION OF POLYNUCLEAR AROMATIC HYDROCARBONS WITH HPLC
(Proposal for a DIN 38 407 Part 8)

Since several years efforts are made at the DIN as well as at the ISO (International Standard Organization) to issue a standard for the determination of PAH by high performance liquid chromatography (HPLC). As shown in figure 6 a german proposal includes two HPLC-based procedures, one using an enrichment by extraction and the other an enrichment by adsorption. The proposed standard is very much like the standard proposed in an ISO-Draft (ISO DP 43 of the respective working group (TC 147 WG 19)).

The performance characteristics of this proposal is summarized in figure 7.

At the level of the international standard organization a procedure for the determination of 22 PAH by the HPLC-method is under discussion. This procedure is determined to measure the contents of PAH in all kinds of water, in sediment and in fish. It is an extensive multi-step method using gel-permeation chromatography and alumina column chromatography for clean-up purposes. On demand it would be possible to show to you figures about the procedure and its performance characteristics.

2.1.3 DETERMINATION OF BENZENE AND HOMOLOGUES
(Proposal for a DIN 38 407 Part 7)

The determination of benzene and its homologues, as it is proposed for a standard DIN 38 407 Part 7, is a straightforward gas chromatographic method. The flow diagram is shown in figure 8. The detection occurs via flame ionization detector (FID).

And as shown in figure 9 the method is applicable for the investigation of sewage and effluents.

2.2 PROCEDURES FOR THE DETERMINATION OF HALOGENATED HYDROCARBONS AND ORGANOCHLORINE PESTICIDES

2.2.1 THE DETERMINATION OF VOLATILE HALOGENATED HYDROCARBONS IN WATER SAMPLES
(Proposal for a standard DIN 38 407 Part 4)

A reliable standard for the determination of volatile halogenated hydrocarbons is of special importance for environmental pollution control. It took several years and many discussions to harmonize a procedure. The method which now is in its final stage of elaboration is shown in figure 10. The basic concept is the gas chromatographic analysis with electron capture detector (GC-ECD) after a step of separation and enrichment by extraction. For some of the compounds FID-detection is necessary or recommended.

The performance characteristics of the procedure is summarized in figure 11. As you see, the standard is applicable for all kinds of ·halogenated hydrocarbons with 1 to 6 carbon atoms. The standard gives rise for a series of questions with respect to the substances determined and the type of sample. These questions I want to postpone for the discussion at the end of my talk and for the comments I will make afterwards with respect to problems which arise if you design a standard. At the moment I want to point on the following :

(-) The necessary, statistically proven, data about the performance of the method (accuracy, yields, precision, limits of detection etc.) have yet to be determined by inter-laboratory comparison. A first round robin test, performed 2 years ago, produced data which could not be interpreted.

(-) The method does not use additional clean-up and enrichment procedures besides the step of extraction.

(-) Concerning a head-space GC-ECD method, the discussion is still going on, if it is necessary and possible for volatile halogenated hydrocarbons to standardize this kind of method in addition to the extraction method, beyond the case of chloroethene which is described in section 2.2.2.

2.2.2 DETERMINATION OF CHLOROETHENE (VINYL CHLORIDE) BY HEAD-SPACE GAS CHROMATOGRAPHY
(Draft DIN 38 413 Part 2)

Because of the properties of chloroethene it was decided to issue a separate standard for this compound based on the head-space technique. The flow diagram of this method is shown in figure 12 and the performance characteristics is summarized in figure 13. The method has yet to be tested by an inter-comparison run.

2.2.3 GAS CHROMATOGRAPHIC DETERMINATION OF LOW VOLATILE HALOGENATED HYDROCARBONS AND ORGANOCHLORINE PESTICIDES
(Proposal/"Blaudruck" for a DIN 38 407 Part 2)

For the determination of low volatile halogenated hydrocarbons and organochlorine pesticides a so called "Blaudruck" (blueprint), issued under the responsibility of the working group, is ready now for beeing tested in the public before becoming a definite standard. As I shall talk in a further part of the lecture a bit more about basic problems occuring with the development of standards, using this "Blaudruck"as an example, in the section here I only want to present to you a general information about the procedure, as shown in the figures 14 (flow-diagram), 15 (performance characteristics) and 16 (list of selected compounds).

2.3   PHENOLS

2.3.1 DETERMINATION OF SELECTED PHENOLS BY GAS CHROMATOGRAPHY
(Proposal for DIN 38 407 Part 10)

At this stage of the standardization the basic concept for the determination of phenols from the EEC-list of priority pollutants is the analysis by GC. For some of them like : 2-amino-4-chloro-phenol, 2-chloro-phenol, 3-chloro-phenol and 4-chloro-phenol, their analysis by HPLC is discussed and under consideration for a standard.

The analysis for the determination of phenols by GC according to the proposed standard follows the scheme shown in figures 17 and 17a, the performance characteristics is listed in figure 18 and the compounds wich can be analyzed are shown in fig. 19.

3. PRIORITY POLLUTANTS FOR WHICH, UNTIL NOW, NO STANDARD
PROCEDURE IS PROPOSED

3.1 POLAR HALOGENATED ORGANIC COMPOUNDS
This group includes the compounds :
(-) Chloral hydrate
(-) Chloro acetic acid
(-) 2-Chloro ethanol
(-) Epichlorohydrin
(-) 1,3-Dichloropropane-2-ol
First ideas for the analytical concept within a standard are:
(-) the determination of chloro acetic acid by GC via derivatives,
(-) the determination of epichlorohydrin by GC-FID after extraction with chloroform
and
(-) the determination of 1,3-dichlorpropane-2-ol by HPLC
Concerning the epichlorohydrin one should mention that one of the procedures under consideration originally is working with packed columns (stationary phase 1,2,3-tris(2-cyanethoxy)-propane on chromosorb P/AW) and does not give a good limit of detection (about 100 ug/l!). Another proposed procedure is using capillary columns with carbowax as the stationary phase. The main problems are not the GC separation and the detection but the separation and the enrichment from the water phase. Therefore some work must be done to complete and to improve the method with respect to these points.
For chloral hydrate and for 2-chloro ethanol until now no procedure is proposed.

3.2 AROMATIC AND ALIPHATIC AMINES
This group includes the compounds listed in Figure 20. A possible way for the determination seems to be by application of HPLC. First discussions are under way to check if such a procedure would be sufficient for the needs of enviromental pollution control.

3.3 ORGANONITROGEN AND ORGANOPHOSPHOROUS PESTICIDES,
AND PHENOXYALKYL ACID HERBICIDES,
INCLUSIVE EEC-COMPOUND NO.114 : TRIBUTYLPHOSPHATE
The present situation with respect to a standard procedure concerning the organonitrogen and organophosphorous pesticides as well as with the phenoxyalkyl acid herbicides is similar to the situation with the search for an appropriate procedure for the aromatic and aliphatic amines. A definite proposal for a standard does not exist. There is a certain tendency now to consider the HPLC as one possible method for the determination.

The compounds under consideration here are listed in figure 21. For another group, listed in Figure 22, a standard based on GC-analysis may be developed one day.

Concerning the phenoxyalkyl acid herbicides an old and very preliminary proposal, alternative to the HPLC-concept, with a gas chromatographic procedure, already exists, as follows:

The phenoxyalkyl acid pesticides are extracted from the water sample with acetic acid ethyl ester. After derivatization under nitrating conditions and esterification with methanol/ sulfuric acid, and after a clean-up with an alumina column, the derivatives are analyzed with GC/ECD. The limit of detection is in the range of 0.5 ug/l, applying packed column techniques.

3.4   ORGANOTIN COMPOUNDS

For the group of those eight positions of the EEC-list of priority pollutants which concern the organotin compounds it is planned to develop a standard with a HPLC-method for their determination. Naturally, at this stage of the discussion one also is evaluating the advantages of a gas chromatographic procedure as alternative.

4.   PROBLEMS ARISING WITH THE STANDARDISATION OF PROCEDURES FOR THE DETERMINATION OF ORGANIC MICROPOLLUTANTS

4.1   GENERAL REMARKS

The first question is : Do we need standard procedures?

If we affirm this question, the second question arises : How does an ideal standard for organic micropollutants look like?

The working group for the development of a standard for the "determination of low volatile chlorinated hydrocarbons and organochlorine pesticides" has had many discussions about these questions and in the following I want to report about some aspects and some of the results of this group in which I am working (3). We think that we have made some progress in answering these questions, although we are far from having solved every problem.

To our opinion the first question can be definitely answered with : YES, and the reason for this can be given in a short way as follows :

A modern concept of analytical quality assurance, as it is practised for instance by the WHO global environmental monitoring program and as it was originally proposed by a group of the Water Research Centre in Medmenham/England (4), includes the following parts :

   1.   establish working group
   2.   define objects
   3.   chose analytical methods
>> 4.   ensure unambigious description of methods <<
   5.   estimate within laboratory precision
   6.   set up control charts
   7.   compare calibration standard solutions
   8.   check inter-laboratory bias

As you see the unambigious description of the analytical

method is essential part of the analytical quality assurance. <u>We think the adequate described standard procedure, which in addition to this must be thoroughly practically proven</u>, is the proper means of fulfilling this requirement.

On the one hand it cannot be the task of every single laboratory to develop and to describe its own procedure, on the other hand it is clear to everyone working in the field of environmental routine analysis of organic micropollutants, that even if the basic and essential quality data of a laboratory method, - like : accuracy, reproducibility or yields - , are known, the equivalency and comparability of results coming from different laboratories <u>can be improved</u> if methods as uniform as possible are applied.

In this respect I have to hint especially on the following points and problems :
- The many steps and the complexicity of the methods involved!
- The question : which steps are included in the method and in the evaluation of the quality data?
- The individual and often changing composition of environmental samples!

This leads to the discussion and the answering of the second question.

4.2 DISCUSSION OF SELECTED PROBLEMS AND REQUIREMENTS IN CONNECTION WITH THE DEVELOPMENT OF A STANDARD METHOD
(Shown by example of the proposed standard for the determination of low volatile halogenated hydrocarbons and organochlorine pesticides (5))

On principle there exists always a great manyfold of requirements, objectives and limitations which must be considered if one develops a standard procedure. The available time for this lecture is not enough to present to you a systematic and exhausting overview with respect to this. It might be worth trying it one day at another opportunity because the experience shows that gaining of these necessary informations often is neglected.

Today I want to stress some selected points, especially those requirements which gave rise for problems during the development of a standard for the determiniation of low volatile halogenated hydrocarbons and organochlorine pesticides. From this you will see, among others, that there is or may be a distinct difference between a procedure which a single laboratory is developing in a straightforward manner for its very special own needs and the standard which has to satisfy a (possibly) broad range of needs and applications.

I am emphasizing this very much because there is often a lack of comprehension between scientists and legislators if it takes some while to issue a harmonized standard procedure for the determination of certain parameters despite the fact that some analysts and some laboratories are already able to detect and to quantify these parameters in some of the samples under consideration.

The standard to which I am refering now was already discribed in its outline in part 2.2.3 of this lecture.Therefore

I can step directly into the discussion of the problems :

Problem No.1: Which substances shall be determined?

The group of low volatile chlorinated hydrocarbons and organochlorine pesticides includes a nearly innumerable quantity of compounds. From this it is not feasable to publish a separate standard for each compound (like it is the case for chloroethene) or even for a greater number of subgroups of compounds. On the other hand it seems very risky to issue a standard which, without restrictions, allows the determination of all the respective substances.

Therefore it was decided that the standard discussed here can be applied on principle for the whole group of compounds under consideration. However by the editing working group the standard is tested explicitly only for about 25 - 30 representative compounds, which are of : special interest, of special importance and of different types of chemical structure, physical behaviour and analytical detectability. For this group of compounds the standard assumes a particular "analytical responsibility".

If an analyst has the task of analyzing for other compounds of the group of low volatile halogenated hydrocarbons and organochlorine pesticides, beyond the list of 25 - 30 substances,

        (-) he is obliged to apply this standard, with
            its instructions, with the respective instrumen-
            tation, with the criteria, etc., mentioned therein,
however
        (-) he has to proof the feasability of the procedure
            with respect to the new compounds, in a thorough
            and responsible manner, before analyzing samples.

A special problem was the treatment of the" PCB-question". About this I will report in a separate section of my lecture.

(Remark : Concerning the complementary standard for the volatile halogenated hydrocarbons, as described before, to my opinion these borderlines are not drawn so clear!)

Problem No. 2: Which type of sample is allowed?

The question arose if different standard methods should be developed for different types of samples, like drinking water, surface water, or sewage? The result of extensive dis_ cussions was, that in many practical cases and from the analytical point of view it is difficult to make a sharp and unequivocal distinction between the different kinds of samples.

Therefore ist was decided to cover with this standard all possible types (matrices) of samples and simply to distinguish between:

Group 1:   water with no or low pollution
              (normally drinking water, ground water, certain
              surface waters)
Group 2:   water with moderate pollution
              (certain surface waters)
Group 3:   heavily polluted water
              (sewage and certain effluents; certain ground
              water)

in connection with the effort necessary for using clean up procedures and for the confirmation of the results.

It remains to the responsibily of the analyst to adjust his analytical efforts in a proper way to the sample which has to be analyzed and to prove the feasability of the method by checking carefully a set of necessary characteristic data for the procedure (e.g. separation quality of columns, yields, limits of detection, repeatability and so on).

Problem No. 3: The Identification and the confirmation of the qualitative result

Prerequisite for the determination of a compound is the unequivocal identification. Many thoughts had to be devoted to the problem of adequate ways of identification, because there were no rules available from other sources. Without going into details the following are the outlines of our concept:

1. The analytical effort has to be adjusted to the problem to be solved.
2. The quality of separation efficiency of the columns used must be checked and proven to fulfill certain (minimum) needs (Remark : in most cases capillary column techniques has to be applied).
3. For an adequate confirmation of a result the gas chromatographic evidence of a compound, obtained with at least two columns of different polarity, is mandatory.
4. Different stages of the degree of identification and confirmation are formulated:
   (-) not identified (if no peak appears in case of analyzing only with one GC-column)
   (-) identity possible (if a peak with the expected retention time appears in case of analyzing with one GC-column)
   (-) identity very likely (if corresponding peaks appear at gas chromatograms of two columns of different polarity)
   (-) identity confirmed/ensured (in case of group 1 and group 2 samples identity confirmed if corresponding peaks appear at gas chromatograms of two columns of different polarity; in case of group 3 samples further confirmation, e.g. 3rd column or GC/MS identification may be necessary, depending in the problem to be solved.)
5. Together with the quantitative results, the analyst has to report about the degree of identification/confirmation (including an information about the quality of separation and clean-up).

Problem No. 4: The ways of calibration and of calculation of the result

There may be many reasons for getting wrong results. However one source for doubtful or incomperable results, which very often is neglected, is an inadequate or obscured way for calibration and calculation of the results.

With respect to the different objectives for an analysis and to the considerable analytical effort which possibly is necessary, it was decided to offer within the standard the possibility for a problem-oriented choice of options for calibration and calculation of results. Within this frame the aspects of :

(-) single-point-calibration
(-) multiple-point-calibration (at least 5 points)
(-) standard addition
(-) internal standardisation
(-) external standardisation
(-) calibration including the total procedure
(-) calibration not including the total procedure
(-) correction of the result by typical yields and by background
(-) calibration with linear and with non-liniar calibration functions

are treated. The different possibilities are described separately and the analyst can use them according to his needs. However he is obliged to cite the way of his procedure together with the results.

Problem No. 5: The general concept, the structure and the
            volume of the standard

From the discussion of problems 1 trough 4 you probably feel that the question about a general concept, the structure and the extent of the contents of the standard had to be solved. As a result of many considerations, especially with respect to

(-) the optimization of the analytical effort

and to

(-) an optimized quality of the analytical result

the over-all concept is the following:

The standard provides and describes all necessary steps for the identification and determination of the compounds in different kinds of samples. In addition to this it provides, if necessary, minimum requirements and instructions for the check and the control of the procedure.

This leads to the structure of a standard with "analytical modules" which can be used according to the needs. For instance concerning the clean-up the following methods are options :

(-) Separation with acetonitrile
(-) column chromatography with Florisil
(-) column chromatography with alumina
(-) column chromatography with silica

Concerning gas chromatographic separation capillary column techniques is very much recommended, although, for certain purposes, packed column techniques is considered as an option.

It was decided to describe the "analytical modules" as detailed as possible and in a way that they can be cited by the analyst. The analyst on the other hand if he uses the standard is obliged to report with the result in a detailed manner about :

(-) the steps used with the procedure,
(-) the characteristic data of his selected procedure and/ or about the verification of typical performance data (including ways of calibration, calculation, identification as mentioned before)

Problem No 6: The confirmation of the practical feasability of the new standard

The reason that the proposed standard is at the moment at the so called stage of a "Blue print (Blaudruck)" is based in the fact that one felt that it is necessary to carefully check the practical applicability of a standard with a broad field of application, concerning compounds, samples, concentrations, objectives of investigation and so on. To test the proposed standard an extensive 3 year program on the base of inter-laboratority tests was started .

In a first round of experiments the following was checked:
(-) The multiple point calibration of the GC-step with own and independent standard solutions (compounds: 1,2,3,4-tetrachlorobenzene, HCB, gamma-HCH, PCB-180, beta-endosulfane, endrine, p,p'-TDE),
(-) the determination of compounds in a sample of pure water to which the compounds mentioned before were added,
(-) the constancy of the retention time in connection with the degree of identification.

First results of the determination of HCB (52.0 ng/l), beta-endosulfane (204.8 ng/l) and PCB-180 (117.5 ng/l), obtained with non-polar and polar columns, with standard solutions of own or independent origin, show that :
(-) a recovery between 93 and 104 % can be obtained,
(-) the inter-laboratory coefficient of variation can be assumed to be better then 15 %,
(-) tested with HCB the inter-laboratory coefficient of variation is slightly better in case of multiple point calibration compared to a single point evaluation.

Problem No. 7: The "PCB Problem"

The question of an unequivocal, accurate and comparable determination of polychlorinated biphenyls (PCB) and the imple-mentation of a procedure into a standard was, and still is, object of many discussions of the working group. The status now is the following :
1.) Among the following possible alternatives for a PCB-analysis :
a.) determination of the total contents of PCB after per-chlorination,
b.) determination of the total contents of PCB after de-chlorination,
c.) determination by packed-column GC,
d.) determination of, or via single compounds by capillary column GC,
only the case d.) will be part of a standard.
2.) As there are different objectives for the determination of PCB, like :
a.) to get an information about products (e.g. concentration of Aroclor 1260 etc.)
b.) to get an information about the sum of PCB (e.g. total contents of PCB)

c.) to get an information about single components
(e.g. concentration of 2,2'-dichlorobiphenyl etc.)
it seems necessary to provide separate procedures.
3.) It was decided finally to offer the following three
options and analytical procedures, either in a separate
standard or in an appendix to the standard for low
volatile halogenated compounds :
a.) The determination of 6 (to 10?) "indicator PCB".
(The procedure is more or less the procedure al
ready given by the proposed standard for low-volatile halogenated compounds. It follows the concept
of "indicator PCB" which is under discussion at
other groups (6)(7)(8)(9). The indicators are the
PCB 28, 52, 101, 138, 153, 180 (according to the
systematic numbering of Ballschmiter (10))).
b.) The identification, determination and comparison
of patterns of products.
No final decision is made until now with respect
to the pattern recognition procedure although
there are some models under tentative consideration.
c.) The unequivocal identification and determination
of a single PCB in special (industrially polluted)
samples
This method uses the mass-spectrometry for identification, confirmation and determination.

5. Final remarks
With this lecture I hope that by directly and indirectly
pointing on deficits I gave some stimulations especially to
those who have the knowledge and the possibility for the development of suitable analytical methods.
To my opinion one has to promote and to improve the
following :
1.) The formation of new working groups.
With this respect in Germany working groups for the
(-) determination of polar, water soluble organic
compounds (e.g. chloroethanol, chloral hydrate,
epichlorohydine),
(-) determination of polymer, water soluble organic
compounds (e.g. polyamides, polysaccharides),
(-) GC determination of organohalogenic compounds
which cannot be determined by ECD detection
(e.g. chlorobenzene, chlorinated ethers, dichloromethane, 1,2-dichloroethane) however by
e.g. Hall detectors or MS,
(-) GC determination of organophosphorous and organonitrogen compounds,
(-) determination of adsorbable organic sulfur compounds,
are planned.
2.) The participation of a greater number of experienced
analysts in the production of standards.
3.) Better and more precise objectives for the projects of
standardisation.

4.) Production of standards which are true laboratory pro-
    cedures. This means complete methods according to
    which analytical work can be done. Procedures which
    consider and treat all the relevant problems and cases
    of application, and last but not least procedures
    with suitable information about characteristic data.
    In this respect it is necessary to reach agreement in
    many important details, for instace about the questions
    (-) What shall be summarized under the term "analytical
        procedure"?
    or
    (-) what is the definition of the "limit of detection"
        of an analytical procedure?
    and so on....
    With these questions let me finish, thanking you very
much for your attention.

REFERENCES
(1)   DEUTSCHE EINHEITSVERFAHREN ZUR WASSER-, ABWASSER- UND
        SCHLAMM-UNTERSUCHUNG, Herausgegeben von der Fachgruppe
        Wasserchemie in der Gesellschaft Deutscher Chemiker,
        VCH Verlags GmbH, Weinheim
(2)   Mitteilung der Kommission an den Rat ueber die ge-
        faehrlichen Stoffe im Sinne der Liste 1, Amtsblatt der
        Europaeischen Gemeinschaften Nr.C 176/3 vom 14.7.82
(3)   P.G.LAUBEREAU, Zum Stand der Normung eines Verfahrens
        fuer die Bestimmung von schwerfluechtigen Halogenkohlen-
        wasserstoffen und Organochlorpestiziden in Wasser,
        Korrespondenz Abwasser, 31, 499 - 508 (1984)
(4)   UNEP/WHO/UNESCO/WMO Project on Global Water Quality
        Monitoring, GEMS/WATER Operational Guide, Charpter IV,
        Analytical Quality Control (Prepared in cooperation
        with the Water Research Centra Medmenham Laboratory,
        Medmenham-Marlow, Buckinghamshire, England)
        World Health Organization Document ETS/78,9
(5)   P.G.LAUBEREAU et al., Vorschlag fuer ein Deutsches
        Einheitsverfahren zur Wasser-, Abwasser- und Schlamm-
        untersuchung, gemeinsam erfassbare Stoffe (Gruppe F),
        Gas-chromatographische Bestimmung von schwerfluechtigen
        Halogenkohlenwasserstoffen und Organochlorpestiziden
        in Wasser (F 2), Deutsche Einheitsverfahren (DEV),
        14. Lieferung 1985, VCH Verlag GmbH, Weinheim
(6)   H.BECK und W.MATHAR, Analysenverfahren zur Bestimmung
        von ausgewaehlten PCB-Einzelkomponenten in Lebens-
        mitteln, Bundesgesundheitsblatt, 28, 1 (1985)
(7)   M.CARL, Hrsg., Rahmenkonzept fuer die Routineanalytik
        von polychlorierten Biphenylen (PCB), Manuskript erar-
        beitet von der Arbeitsgruppe "Organische Spurenanalytik",
        VDLUFA-Schriftenreihe, Heft 12, VDLUFA-Verlag, Darmstadt,
        1985, ISBN 3-922712-11-6
(8)   L.G.M.TH.TUINSTRA, A.H.ROOS and G.A.WERDMUELLER, BCR-
        Ringtest of individual chlorobiphenyls (2/1983) - Sum-
        mary of results (A concept), EC-Report 84.27, Buereau
        of Reference, Brussels, (1984)

(9)   COST 681, Working Party 2, Laboratories Intercomparison
      on PCBs in Sewage Sludges (1983 - 1984), Meeting, Langen
      (FRG), 20-21 march 1985
(10) K.BALLSCHMITER und H.J.ZELL, Analysis of Polychlorinated
      Biphenyls (PCB) by Glas Capillary Gas Chromatography,
      Fres. Z. Anal. Chem. 302, 20 (1980)

Figure 2

THE EXISTING WORKING GROUPS (WG) FOR THE STANDARDIZATION OF PROCEDURES FOR THE ANALYSIS OF ORGANIC COMPOUNDS

IN WATER SAMPLES

WG 1   Polyaromatic hydrocarbons (PAH)

WG 2   Volatile chlorinated hydrocarbons (VCH)

WG 3   Low volatile chlorinated hydrocarbons (LVCH)

WG 4   Phenolic compounds analyzed with GC (Ph-GC)

WG 5   Organic compounds analyzed with HPLC (OC-HPLC)

WG 6   Benzene (Bz)

WG 7   Chloroethene / Vinylchloride (VC)

WG 8   Nitrilo-tri-aceticacid (NTA)

WG 9   Carbondisulphide (CS2)

Figure 1

THE ORGANIZATION FOR THE STANDARDIZATION OF PROCEDURES

FOR WATER ANALYSIS IN GERMANY

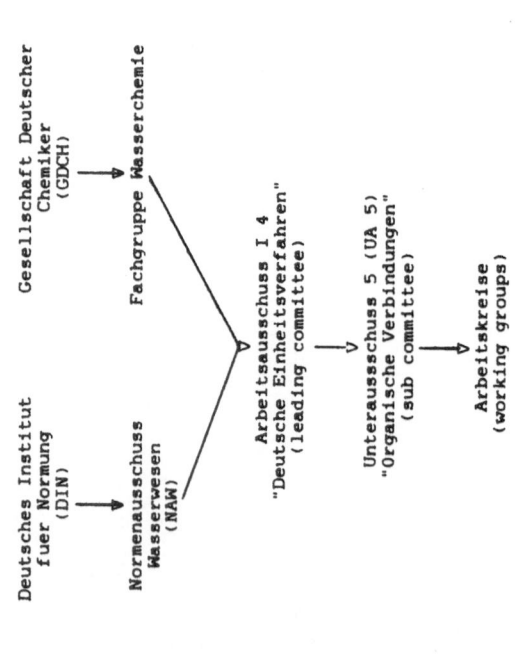

Deutsches Institut fuer Normung (DIN)

Gesellschaft Deutscher Chemiker (GDCH)

Normenausschuss Wasserwesen (NAW)

Fachgruppe Wasserchemie

Arbeitsausschuss I 4 "Deutsche Einheitsverfahren" (leading committee)

Unterausschuss 5 (UA 5) "Organische Verbindungen" (sub committee)

Arbeitskreise (working groups)

Figure 3

ATTEMPT OF A GROUP-ARRANGEMENT OF THE ORGANIC

"PRIORIY POLLUTANTS" OF THE EEC-LIST

1. Aromatic hydrocarbons, including polyaromatic hydrocarbons (PAH)

2. Halogenated hydrocarbons and organochlorine pesticides

   (-) volatile halogenated hydrocarbons
   (-) low-volatile halogenated hydrocarbons
        (including nitrated compounds)
   (-) organochlorine pesticides

3. Phenols

4. Polar halogenated organic compounds

5. Aromatic and aliphatic amines

6. Organonitrogen and organophosphorous pesticides, inclusive phenoxyalkyl acid herbicides and EEC-No. 114 tributylphosphate

7. Organometallic compounds of tin

Figure 4

DETERMINATION OF PAH IN DRINKING WATER ACCORDING

TO DIN 38 409 PART 3

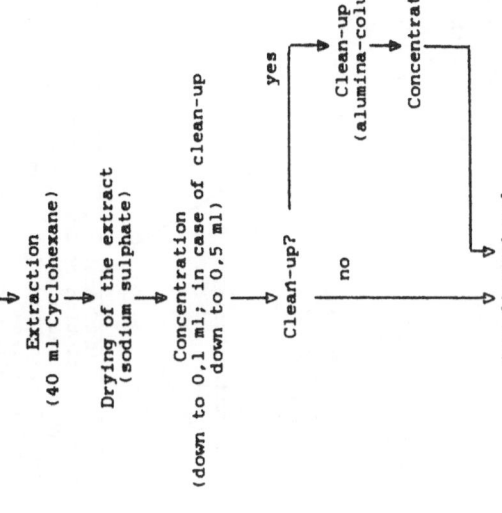

Sample
(about 2 l)

Extraction
(40 ml Cyclohexane)

Drying of the extract
(sodium sulphate)

Concentration
(down to 0,1 ml; in case of clean-up
down to 0,5 ml)

Clean-up?   → yes →   Clean-up
                        (alumina-column)

             no   Concentration

two-dimensional
thin layer chromatography
(Aluminum oxide / cellulose 40% acetylated)
TLC solvents:
1. n-hexane/toluene
2. methanol/diethylether

Measurement/Evaluation
(by scanner)

- 55 -

Figure 6

DETERMINATION OF PAH BY HPLC

(According to the proposol for a DIN 38 407 Part 8)

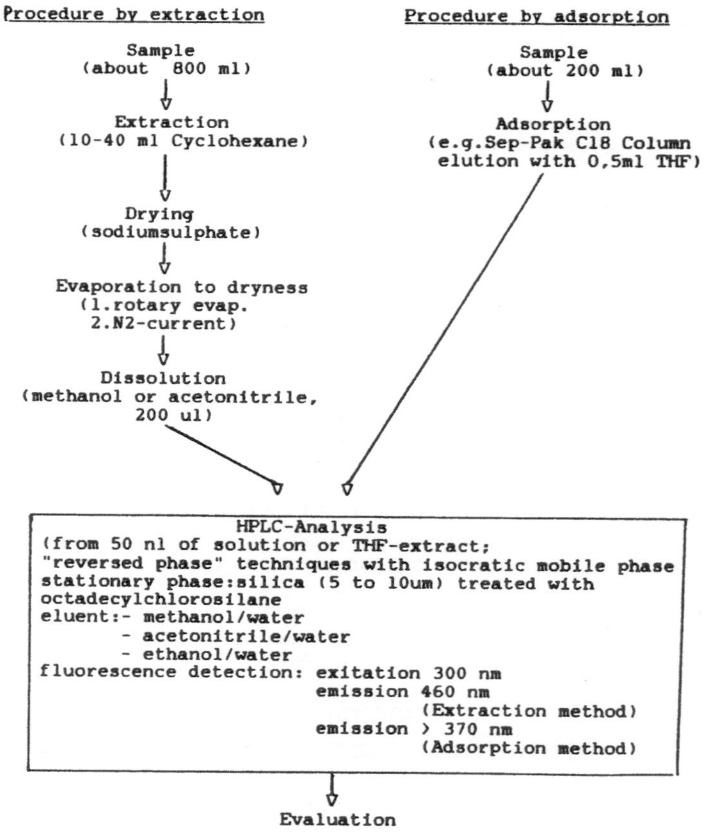

Procedure by extraction

Sample
(about 800 ml)
↓
Extraction
(10-40 ml Cyclohexane)
↓
Drying
(sodiumsulphate)
↓
Evaporation to dryness
(1.rotary evap.
2.N2-current)
↓
Dissolution
(methanol or acetonitrile,
200 ul)

Procedure by adsorption

Sample
(about 200 ml)
↓
Adsorption
(e.g.Sep-Pak C18 Column
elution with 0,5ml THF)

HPLC-Analysis
(from 50 nl of solution or THF-extract;
"reversed phase" techniques with isocratic mobile phase
stationary phase:silica (5 to 10um) treated with
octadecylchlorosilane
eluent:- methanol/water
       - acetonitrile/water
       - ethanol/water
fluorescence detection: exitation 300 nm
                        emission 460 nm   (Extraction method)
                        emission > 370 nm (Adsorption method)

Evaluation

Figure 5

PERFORMANCE CHARACTERISTICS OF THE STANDARD DIN 38 409 PART 3

(DETERMINATION OF PAH IN DRINKING WATER)

SUBSTANCES DETERMINED / RANGE OF APPLICATION

Substance	Range of concentration (ng/l; sample 2 l)
Fluoranthene	5 - 500
Benzo(b)fluoranthene	1 - 250
Benzo(k)fluoranthene	1 - 250
Benzo(a)pyrene	1 - 250
Benzo(ghi)perylene	1 - 250
Indeno(1,2,3-cd)pyrene	2,5- 250

TYPE OF SAMPLE

Drinking water or water with low degree of pollution

CALIBRATION

(-) "2-point-calibration"/external standard
(-) calibration not including the total procedure

IDENTIFICATION

"fixed"/not dependant on the type of sample

Figure 7

PERFORMANCE CHARACTERISTICS OF THE PROPOSAL FOR A
DIN 38 407 PART 8 (HPLC-PROCEDURE)

SUBSTANCES DETERMINED/RANGE OF APPLIKATION:

Substance	Range of concentration	
	Extraction	Adsorption
Fluoranthene	100 ug/l	500 ug/l
Benzo(b)fluoranthene	1-40 ug/l	100 ug/l
Benzo(k)fluoranthene	1-40 ug/l	100 ug/l
Benzo(a)pyrene	20 ug/l	100 ug/l
Benzo(g,h,i)perylene	20 ug/l	100 ug/l
Ideno(1,2,3-c,d)pyrene	20 ug/l	100 ug/l

TYPE OF SAMPLE

Drinking water, ground water, surface water moderately
polluted (samples analyzed without filtration)

CALIBRATION

(-) "6-point-calibration"/external standard
(-) calibration not including the total procedure

FURTHER CHARACTERISTIC DATA

yet to be determined by inter-laboratory comparison

Figure 8

THE DETERMINATION OF BENZENE AND HOMOLOGUES
(According to the proposal for DIN 38 407 Part 7)

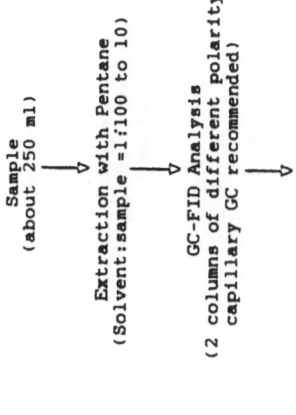

Sample
(about 250 ml)

Extraction with Pentane
(Solvent:sample =1:100 to 10)

GC-FID Analysis
(2 columns of different polarity;
capillary GC recommended)

Evaluation

Figure 10

THE DETERMINATION OF VOLATILE HALOGENATED
HYDROCARBONS IN WATER SAMPLES
(According to draft DIN 38 407 Part 4)

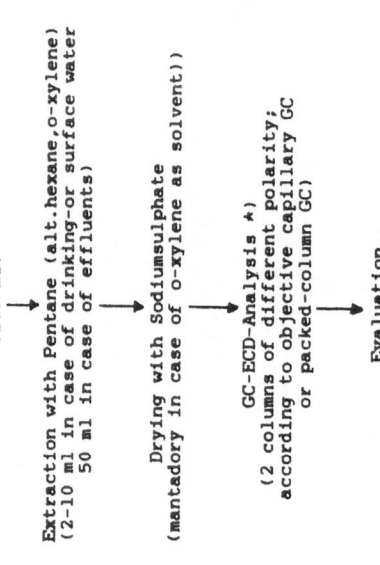

Sample
(200 ml)

↓

Extraction with Pentane (alt.hexane,o-xylene)
(2-10 ml in case of drinking-or surface water
50 ml in case of effluents)

↓

Drying with Sodiumsulphate
(mandatory in case of o-xylene as solvent))

↓

GC-ECD-Analysis *)
(2 columns of different polarity;
according to objective capillary GC
or packed-column GC)

↓

Evaluation

Remarks: *) dichloro methane, aliphatic mono-chlorinated
hydrocarbons and difluoro hydrocarbons with
GC-FID

Figure 9

PERFORMANCE CHARACTERISTICS OF THE PROPOSED-PROCEDURE
FOR THE DETERMINATION OF BENZENE AND HOMOLOGUES
(Proposal for a DIN 38 407 Part 7)

SUBSTANCES DETERMINED / RANGE OF APPLICATION:

Substance	Range of concentration (sample size 250 ml)
Benzene	>
Toluene	>
Xylene	> 1 ng/l to 10 mg/l
Ethylbenzene	>
(Isopropyl benzene)	>
(Biphenyl)	>

TYPE OF SAMPLE

Water of all kinds, including sewage and effluents

CALIBRATION

(-) multiple point calibration/external standard
(-) Calibration not including the total procedure

RECOVERY

controlled by internal standard (toluene d8?)

IDENTIFICATION

(-) Philosophy: 2 columns of different polarity
(-) if necessary special detector like photo-ionization
    detector

Figure 11

Performance Characteristics of the Procedure for the Determination of High-Volatile Halogenated Hydrocarbons

·(according to the proposal for the standard DIN 38 407 Part 4)

SUBSTANCES DETERMINED :

- fluorinated, chlorinated, bromated and iodated, mainly non-aromatic hydrocarbons with 1 to 6 carbon atoms

- halogenated hydrocarbons with boiling points from 20 to 180 C

- about 24 compounds from the EC-list of priority pollutants

TYPE OF SAMPLE

- groundwater, drinking water, surface water, sewage , effluents

RANGE OF APPLICATION :

- not definitively defined

- depending on compound and sample from about 1 to 100 ug/l

CALIBRATION :

- 6-point calibration with the options :
  -- calibration including or not including the total procedure
  -- calibration with external standard or calibration with additional internal standard

STANDARD DEVIATION :

- inter-laboratory precision VR : 20 - 50 % in synthetic sample, depending on conc. and compound

RECOVERY :

- 40 - 90 % after the first extraction, dep. on compound

IDENTIFICATION / CONFIRMATION :

- Philosophy : two columns with different polarity, under certain circumstances conductivity detector or coulometric detector

---

Figure 12

DETERMINATION OF CHLOROETHENE (VINYL CHLORIDE)

BY HEADSPACE GAS CHROMATOGRAPHY

(Draft DIN 38 413 Part 2)

WATER SAMPLE
↓
TRANSFER INTO HEADSPACE FLASK
↓
SEALING OF THE FLASK WITH SEPTUM
↓
HEATING TO 80 Deg. C
↓
EQUILIBRATION AT 80 DEG. C FOR 1 HOUR
↓

HEADSPACE GC-FID-ANALYSIS

FID
packed Columns 2 m Steal or Glass
Stationary Phase : Carbopack C80 + 0.19% Picric Acid
Temperature : 50 Grad C, isothermal

↓
Evaluation

Figure 14

Gas-chromatographic Determination of Low-volatile Chlorinated Hydrocarbons and Organochlorine Pesticides

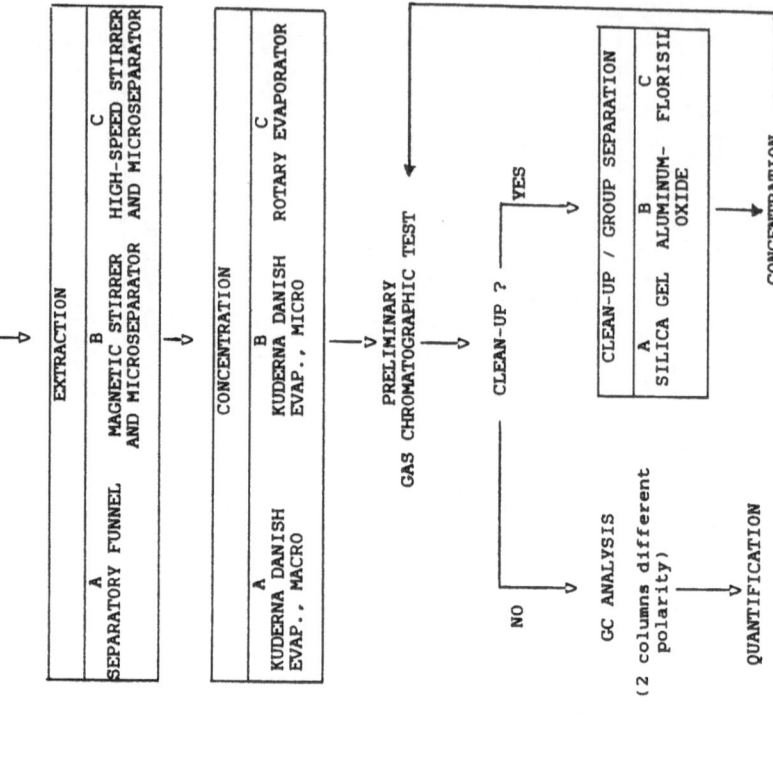

Figure 13

PERFORMANCE CHARACTERISTICS OF THE PROCEDURE FOR THE

DETERMINATION OF CHLOROETHENE-(VINYLCHLORIDE)-BY

HEADSPACE GAS CHROMATOGRAPHY

(Draft DIN 38413 Part 2)

SUBSTANCES DETERMINED

(-) CHLOROOETHENE (VINYL CHLORIDE)

TYPE OF SAMPLE

(-) all kinds of water samples

RANGE OF APPLICATION

(-) for concentrations of 5 to 500 ng/l

CALIBRATION

Two options:

(-) external 5-point calibration
(-) calibration by standard addition to the sample

STANDARD DEVIATION / RECOVERY

(-) Data to be determined by intercomparison runs

IDENTIFICATION/CONFIRMATION

(-) method as described delivers sufficiently
    unequivocal results

Figure 15
Part 2

PERFORMANCE CHARCTERISTICS OF THE PROCEDURE FOR THE GAS-

CHROMATOGRAPHIC DETERMINATION OF LOW -VOLATILE CHLORINATED

HYDROCARBONS AND ORGANOCHLORINE PESTICIDES

(Proposal "Blaudruck" for a DIN 38 407 Part 2)

TYPE OF SAMPLE

(-) all kinds of water including ground water,
surface water, effluents und sewage

RANGE OF APPLICATION

(-) not definitely defined
(-) depending on compound and sample, from
about 1 or 10 to 1000 ng/l

CALIBRATION:

(-) various options :

  (- -) "single point calibration"
  (- -) multiple point calibration
  (- -) calibration by standard addition
  (- -) calibration including or not including
     the total procedure
  (- -) calibration with external standard or
  (- -) calibration with additional internal standard

STANDARE DEVIATION / RECOVERY

(-) data to be determined by intercomparison runs

IDENTIFICATION / CONFIRMATION

Philosophy: (-) analysis with two columns of different
        polarity is minimum requirement for
        identification

      (-) the degree of the identification of a
        compound must be reported with the result

      (-) in certain cases confirmation of the
        identification with GC/MS

Figure 15
Part 1

PERFORMANCE CHARACTERISTICS OF THE PROCEDURE FOR THE

GAS -CHROMATOGRAPHIC DETERMINATION OF LOW-VOLATILE

CHLORINATED HYDROCARBONS AND ORGANOCHLORINE PESTICIDES

(Proposal "Blaudruck" for a DIN 38 407 Part 2)

SUBSTANCES DETERMINED:

(-) acyclic or carbocyclic hydrocarbons with more than
5 carbon atoms; hydrogen atoms substituted by halogens
(the NO2-group may be present); vapor pressure at
20 deg. C 10 exp-3 mbar; boiling point < 180 deg. C;

(-) Organochlorine pesticides like DDT and structural
related compounds; or compounds derived from hexachloro-
cyclopentadien (e.g. cyclodien-insecticides;
including oxygen and sulfur contaning compounds of this
group of pesticides

(-) 27 compounds named, for which the validity of the
procedure is especially checked (among these 13 post-
tions/compounds of the ECC-list of priority-list of
priority pollutants)

(-) however in principle applicable for about 24 compounds /
"positions" from the ECC-list of priority pollutants

(-) basic procedure for about 7 "PCB-indicator compounds"

Figure 16

SELECTED COMPOUNDS FOR WHICH THE VALIDITY OF THE PROCEDURE : "PROPOSAL DIN 38 407 PART 2", WILL BE CHECKED

EEC-No.	Substance	EEC-No.	Substance
1	Aldrin	–	Methoxychlor
46	P,p'-DDD (TDE)	–	Pentachlorobenzene
46	P,p'-DDE	–	Pentachloronitrobenzene (PCNB, Quintocene)
46	o,p'-DDT	101	2,4,4'-Trichlorobiphenyl (PCB No.18)
46	P,p'-DDT	101	2,2',5,5'-Tetrachlorobiphenyl (PCB No.52)
71	Dieldrin	101	2,2',4,5,5'-Pentachlorobiphenyl (PCB No.101)
76	alpha-Endosulfan (alpha-Thiodan)	101	2,2',3,4,4',5-Hexachlorobiphenyl (PCB No.138)
76	beta-Endosulfan (beta-Thiodan)	101	2,2',4,4',5,5'-Hexachlorobiphenyl (PCB No.153)
77	Endrin	101	2,2',3,4,4',5,5'-Heptachlorobiphenyl (PCB No.180)
82	Heptachlor	101	2,2',3,3',4,4',5,5'-Octachlorobiphenyl (PCB No.194)
82	Heptachloreopoxide	109	1,2,4,5-Tetrachlorobenzene
83	Hexachlorobenzene (HCB)	118	1,2,4-Trichlorobenzene
85	alpha-Hexachloro-cyclohexane (alpha-HCH)		
85	beta-Hexachloro-cyclohexane (gamma-HCH)		
85	gamma-Hexachloro-cyclohexane (gamma-HCH) (Lindan)		

Figure 17

GAS -GROMATOGRAPHIC DETERMINATION OF PHENOLS

(According to a proposal for a standard DIN 38 407 Part 10)

WATER SAMPLE

↓

FILTRATION IF NECESSARY

↓

ENRICHMENT AND CLEAN-UP

A	B	C
BY EXTRACTION	BY SORPTION	BY DERIVATIZATION
(e.g. Diethyl-ether) and	(e.g. XAD)	– either by aceticanhydride
COLUMN CHROMATO-GRAPHY		– or by penta-fluorobezoyl-chloride

↓

GC -Analysis with FID *)

↓

Evalution

Remarks: *) ECD or MS in certain cases

- 62 -

Figure 17a

Figure 18

GAS CHROMATOGRAPHIC DETERMINATION OF PHENOLS

(here : SCHEME OF THE STEP "ENRICHMENT BY EXTRACTION AND CLEAN-UP",Proposal DIN 38 407 Part 10)

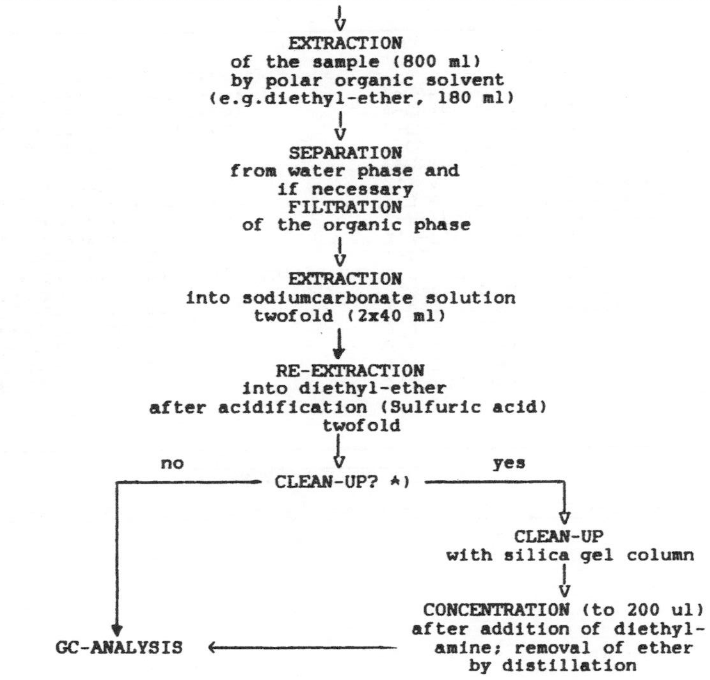

```
        EXTRACTION
     of the sample (800 ml)
     by polar organic solvent
     (e.g.diethyl-ether, 180 ml)
            ↓
        SEPARATION
     from water phase and
       if necessary
        FILTRATION
     of the organic phase
            ↓
        EXTRACTION
  into sodiumcarbonate solution
       twofold (2x40 ml)
            ↓
      RE-EXTRACTION
   into diethyl-ether
after acidification (Sulfuric acid)
        twofold
                         yes
  no ── CLEAN-UP? *) ────────┐
                             ↓
                          CLEAN-UP
                    with silica gel column
                             ↓
                    CONCENTRATION (to 200 ul)
                    after addition of diethyl-
                    amine; removal of ether
                         by distillation
GC-ANALYSIS ←────────────────┘
```

Remarks : *) necessary in case of heavy polluted samples

PERFORMANCE CHARACTERISTICS OF THE PROCEDURE FOR THE

GAS CHROMATOGRAPHIC DETERMINATION OF PHENOLS

(According to a proposal for a standard DIN 38407 part 10)

SUBSTANCES DETERMINED

(-) PHENOLS (about 36 compounds listed in a table)
(-) all the phenolic compounds of the EC-list of priority polutants (7 positions of the list = 9 compounds) with exeption of 2-amino-4-chlorophenol

TYPE OF SAMPLE

(-) all kinds of water samples, including ground water, surface water, effluents and sewage

RANGE OF APPLICATION

(-) determinations at concentrations above 0.1 ug/l possible (depending on the degree of enrichment; factors of 4000-10000 possible)

CALIBRATION

(-) various options:

(- -) calibration including or not including the total procedure
(- -) calibration with external standard or calibration with additional internal standard

STANDARD DEVIATION / RECOVERY

(-) data to be determined by intercomparison runs

IDENTIFICATION / CONFIRMATION

(-) Philosophy : gas chromatographic analysis with two capillary columns of different polarity

(-) in certain cases confirmation of the identification with selective detectors (like ECD or MS) recommended

Figure 19

Figure 20

PHENOLIC COMPOUNDS WHICH CAN BE ANALYZED WITH THE PROPOSAL FOR A STANDARD DIN 38 407 PART 10

Phenol
2-Methylphenol
3-Methylphenol
4-Methylphenol
2,4-Dimethylphenol
4-Ethylphenol
2,6-Di-tert.-butyl-4-methylphenol
2-Phenylphenol
2-Benzylphenol
4-Benzylphenol
2-Benzyl-4-methylphenol
2-Chlorophenol
3-Chlorophenol
6-Chlorthymol
4-Chlor-2-methylphenol
4-Chlor-3-methylphenol
6-Chlor-3-methylphenol
2,4-Dichloro-3,5-dimethylphenol

2,3-Dichlorophenol
2,4-Dichlorophenol
2,6-Dichlorophenol
2,4,6-Trichlorophenol
2,3,5-Trichlorophenol
2,4,5-Trichlorophenol
2,3,6-Trichlorophenol
2-Nitrophenol
2,4-Dinitrophenol
2-Methyl-4,6-dinitrophenol
2,3,4,5-Tetrachlorophenol
2,3,4,6-Tetrachlorophenol
Pentachlorophenol
6-Chlorthymol
2-Chlor-4-tert.-butylphenol
4-Chlor-2-benzylphenol
2-Cyclopentyl-4-chlorophenol

AROMATIC AND ALIPHATIC AMINES OF THE EEC LIST OF PRIORITY POLLUTANTS

EEC-No.	COMPOUND	POSSIBLE WAY OF DETERMINATION BY :
2	2-Amino-4-chloro-phenol	HPLC
8	Benzidine	HPLC
17	2-Chloroaniline	HPLC
18	3-Chloroaniline	HPLC
19	4-Chloroaniline	HPLC
27	4-Chloro-2-nitroaniline	HPLC
41	2-Chloro-p-toluidine	HPLC
42	Chlorotoluidine (others than 2-Chloro-p-toluidine)	
52	Dichloroaniline	HPLC
56	Dichlorobenzidine	HPLC
72	Diethylamine	HPLC
74	Dimethylamine	HPLC

Figure 22

ORGANONITROGEN AND ORGANOPHOSPHOROUS-PESTICIDES.

AND PHENOXYALKYL ACID HERBICIDES

OF THE EEC-LIST OF PRIORITY POLLUTANTS

PART 2

(inclusive EEC-No. 114 :TRIBUTYLPHOSPHATE)

EEC-No.	COMPOUND	POSSIBLE WAY OF DETERMINATION BY :
43	COUMAPHOS	GC
44	CYANURCHLORIDE (2,4,6,-Trichloro-1,3,5-triazin)	GC
47	DEMETON (Iclusive Demeton-o,De-meton-s,Demeton-s-methyl, Demeton-s-methyl-sulphon)	GC
70	DICHLORVOS	GC
73	DIMETHOAT	GC-NFID
75	DISULFOTON	GC-NFID
80	FENITROTHION	GC
81	FENTHION	GC-PFID
89	MALATHION	GC
93	METHAMIDOPHOS	GC
94	MEVINPHOS	GC
97	OMETHOATE	GC
98	OXYDEMETON-METHYL	GC
103	PHOXIM	GC
104	PROPANIL	GC
105	PYRAZON	GC
106	SIMAZIN	GC
113	TRIAZOPHOS	GC
116	TRICHLORFON	GC
124	TRIFLURALIN	GC
114	TRIBUTYLPHOSPHATE	GC

Figure 21

ORGANONITROGEN AND ORGANOPHOSPHOROUS PESTICIDES,

AND PHENOXYALKYL ACID HERBICIDES

OF THE EEC-LIST OF PRIORITY POLLUTANTS

PART 1

EEC-No.	COMPOUND	POSSIBLE WAY OF DETERMINATION BY :
5	Azinphos-ethyl	HPLC
6	Azinphos-methyl	HPLC
45	2,4-D (including salts and esters)	HPLC
69	Dichlorprop	HPLC
80	Fenitrothion	HPLC
81	Fenthion	HPLC
88	Linuron	HPLC
90	MCPA	HPLC
91	Mecoprop	HPLC
95	Monolinuron	HPLC
107	2,4,5-T (including salts and esters)	HPLC

DETERMINATION OF ORGANOMETALLIC COMPOUNDS IN THE AQUATIC ENVIRONMENT

Markus D. Müller
Federal Research Institute
CH-8820 Wädenswil, Switzerland

Summary

Organometallic compounds of several elements occur in the environment. Among them, the most important are tin, lead and mercury, which are of anthropogenic origin. Methyl-compounds of tin and gallium have been detected in natural waters in the ppt-range and are likely of natural origin. The analysis of these compounds at trace levels involves sample enrichment and chromatographic separation techniques followed by specific detection. The environmental fate of the most important anthropogenic organometallic species is summarized together with the procedures necessary for their detection. Derivatization steps and limits of detection are discussed.

1. INTRODUCTION

Many metals form various organometallic compounds, however only a few of them have found widespread technical use. These are the organometallic compounds of tin, lead and mercury: methyl-and ethyl lead compounds are used in large amounts (estimated 400 000 t/y worldwide) as gasoline additives, organomercurials are in decreasing use as seed dressing agents and disinfectants and organotins have found widespread applications as stabilizers for PVC, catalysts, agrochemicals and biocides with an estimated use of 33 000 t/y. These synthetic compounds have a profound impact on the environment due to their toxicity, stability and widespread use. In addition, another group of organometallic compounds has received increasing interest: the methyl compounds of the elements Ga and Sn were detected in natural waters in the low ppt levels and their occurrence is ascribed to formation in the environment.

2. ENVIRONMENTAL BEHAVIOUR AND TRACE LEVEL DETECTION OF ORGANOMETALS

The physico-chemical properties of the organometallic compounds of a certain metal are strongly dependent on the number and type of ligands present in the compound. Tin and lead (group IVa elements) form stable nonpolar compounds with four organic ligands, mercury (group IIb element) with two ligands. Cleavage of organic side chains leads to more or less polar and charged species, depending on the polarity of the inorganic ligand. There are two major processes determining

the environmental fate of organometallic compounds:
degradation by cleavage of side chains and methylation. The
first process leads to complete mineralization of organometals
through a series of intermediates of increasing polarity by
successive cleavage of side chains and may occur abiotic (e.g.
photolysis and hydrolysis of tributyltin in water [1]) as well
as in organisms [2]. Methylation, on the other hand, likely to
be associated with biotic processes [3], leads to a mobilization of the metals. As organometals containing methylgroups
generally exhibit high toxicities together with good
solubility in water and stability, this process may become
very important. It has been thoroughly investigated in the
case of mercury after serious intoxications (Minamata-desease
[4]).

Trace analysis of organometallic compounds has to cope
with a broad spectrum of different compounds which may be
closely related but may have widely different properties. As
an example, the fate of tetraethyllead from incompletely
combusted gasoline can be described as follows: the volatile
and nonpolar compound is converted in the free atmosphere
after some time to a triethyllead cation (e.g. as sulfate),
which is now a polar, nonvolatile compound adsorbed at
particles. The particles are washed out by rain, the
triethyllead is either adsorbed on soil or deposited in lake
sediments. Anaerobic or aerobic degradation both lead to
inorganic Pb-II by successive cleavage of ethyl groups. The
analysis of sediment samples may thus reveal a series of
dealkylated species containing one to three ethylgroups.

Trace level analysis procedures for organometals can be
divided in to two subgroups: one uses a direct assay (e.g.
colorimetry or an electrochemical method) the other combines a
separation step, where defined organometallic species are
separated with a sufficient sensitive and specific detection
method. In the present paper, only procedures of the latter
type are discussed because non-chromatographic methods are
less suited for the analysis of individual compounds of
complex patterns of organometallic species in the environment.

As the absolute sensitivity of most detector systems is
in the of pg or ng-range, environmental samples containing
organometallic compounds at the ppt- or ppb levels are too
diluted to be analysed directly. Sample enrichment is
therefore necessary prior to chromatographic analysis and
detection. A procedure for trace anlysis of these compounds
will therefore include at least a part of the steps as shown
in Table 1.

Table 1: Scheme for Trace Level Analysis of Organometals

Sample - Raw Extract - Derivat. - Separation - Detection
e.g. liq./liq.-Extraction - Methylation - GLC - AAS

The scheme resembles the well known procedure for trace
analysis of organic compounds. But there are some differences:
as described above, organometallic compounds are present in

the environment in a semiionic form which prevents their extraction into a suitable organic solvent (solvation of the ions in aqueous solutions, adsorption at ion-exchanging matrices). The ions have to be converted into a form suitable for extraction. This may be achieved by complexing agents in buffered solution rendering the organometallic cations sufficiently lipophilic for extraction. Dithizone and tropolone may be used for this purpose in the case of organolead- and organotin compounds, respectively [5, 6]. Monoorganomercurials as chlorides are sufficiently lipophilic to be extracted and analysed directly. As the ionic lead- and tin compounds are thermally labile and involatile, their analysis is either done by HPLC [7], or they are converted into the corresponding volatile tetraalkyllead-, tetraalkyltin- or alkyltinhydrides. This makes them amenable to gas chromatography, which may often be preferred because of the excellent separation capability of (capillary) gas chromatography and the wide choice of detection systems which can easily be coupled with the gas chromatograph.

The alkylation can be carried out using a Grignard reagent, the hydrides are generated in an aqueous solution of $NaBH_4$. There is a choice of alkylgroups for derivatization, but the decision is mainly governed by the volatility of the compounds to be analysed. The hydride generation procedure may be applied for organotin and other hydride-forming organometals (Se, Ge, Ga and Se) and combines derivatization with enrichment by a purge and trap-cycle of the volatile hydrides. Analysis is carried out by controlled evaporation into a suitable detection system (e.g. AAS or flame photometric detection). The sensitivity is excellent for methyl-organometals (sub -ppt-levels) but decreases rapidly with longer side chains.

In the case of organotin compounds, the preparation of alkylderivatives is often preferred over hydride formation in cases of longer alkylchains because the derivatives are more stable than the hydrides. Extracts can be purified and unknown organometallic compounds can be identified by means of GC/MS. Table 2 shows a summary of analysis procedures for the most important organometallic compounds.

Table 2: Analysis Scheme for Selected Organometallic Compounds

Compounds of interest[1]	MeHg	AlkylPb	MeSn	BuSn	PhSn
Step					
Enrichment	Extr. Toluene/ HCl	Benzene/ Dithizone	--	Pentane HCl	CH_2Cl_2 Acetate[2]
Derivat.	--	Phenyl.	$NaBH_4$	Methyl.	Methyl.
Separation	GLC	GLC	Purge/ Trap	GLC	GLC
Detection	ECD	ECD	AAS	FPD	FPD
Typ. lim. of detection	1 ppb	ppt	0.2 ppt	1 ppt	--
Levels found	20 ppb (fish)	--	5 ppt (water)	10 ppt (water)	--
References	[8]	[9]	[10]	[11]	[12]

[1] charges of ionic species omitted

Abbreviations: Me methyl-; Bu n-butyl; Ph phenyl-; GLC gas chromatography; ECD electron capture detection; FPD flame photometric detection.

The choice of references is somewhat arbitrary, however it is not the intention to give a full review on all methods published. When looking through the procedures, one realizes that up to now no universal method for trace analysis exists. This is due to the fact that a method has to be tailor-made to the problem it should cover. The following general considerations should be taken into account when a procedure for determination of a single compound or a group of compounds has to be worked out:

a) extraction procedures should not alter distribution and ratios of the various alkylated specis to be determined;
b) when derivatization is used, those groups introduced should be clearly distuingishible from those which are present before derivatization;
c) volatility and stability of the derivatives, and possible interferences from reagents have to be considered;

As an example, triphenyltin is sensitive to strong acids [13] resulting in disproportionation into diphenyltinoxide and tetraphenyltin. Phenyltins are therefore extracted as acetates, thus avoiding artifacts. Derivatisation can be done by methylation or ethylation, yielding sufficiently volatile derivatives for gas chromatographic analysis. The derivatization of methyl- or ethyllead compounds is preferably done by phenylation instead methylation or ethylation. The phenylderivatives are sufficiently involatile for clean up and the ratios of the various alkylated species can be determined.

4. CONCLUSIONS

The analysis of organometallic compounds in environmental samples requires detailed knowledge, both of the environmental behaviour of the compound and of their chemical properties because sample enrichment and analysis include critical steps. There is a lack of data concerning residue levels of individual organometallic compounds in the environment although a wealth of data on environmental pollution determined as inorganic lead or mercury is available. As the impact on human health and on the environment is strongly dependent on the form a metal is present, the trace analysis of organometallic compounds in environmental samples is a powerful tool for establishing risk assesments connected with the use and disposal of these compounds.

5. REFERENCES

1 J.R. Maguire, J.H. Carey and E.J. Hale, J. Agric. Food Chem. 31 (1983) 1060 .
2 E.C. Kimmel, R.H. Fish and J.E. Casida, J. Agric. Food Chem. 25 (1977) 1 .
3 L.E. Hallas, J.C. Means and J.J. Cooney, Science 215 (1982) 1505 .
4 S. Skerfving and J. Vostal, in "Mercury in the Environment" p.101, (L.T. Fridberg and J.J. Vostal Eds.) The CRC Press, Cleveland 1972 .
5 D.S. Forsyth and W.D. Marshall, Anal. Chem. 55 (1983) 2132 .
6 H.A. Meinema, T. Burger-Wiersma, G. Versluis- de Haan and E.C. Gevers, Env. Sci. Technol. 12 (1978) 288 .
7 T.-H. Yu and Y. Arakawa, J.Chrom. 258 (1983) 189 .
8 G.H. Alvarez, S.C. Hight and S.G. Capar, J. Assoc. Official Anal. Chem. 67 (1984) 715 .
9 D. Chakraborti, W.R.A. De Jonge, W.E. van Mol, R.J.A. van Cleuvenbergen and F.C. Adams, Anal. Chem. 56 (1984) 2692 .
10 M.O. Andreae and J.T. Byrd, Anal. Chim. Acta 156 (1984) 147.
11 M.D. Müller, Z. Anal. Chemie 317 (1984) 32 .
12 B.W. Wright, ML. Lee and G.M. Booth, J. High Res. Chrom. Chrom. Comm. 2 (1979) 189 .
13 J. Basters, A. Martijn, Th.van der Molen, P. Pasma, B. Rabenort, B. van Rossum and J. Smink, J. Ass. Official Anal. Chem. 61 (1978) 1507 .

HPLC DETERMINATION OF PRIORITY LIST AROMATIC HYDROCARBONS

S.Galassi

Istituto di Ricerca sulle Acque,CNR.Brugherio,Milano,Italy

SUMMARY

Aromatic hydrocarbons are usually determined by gas-
chromatography.This technique is generally very sensitive and selective.Unluckily, surface water samples,
industrial effluents and test solutions for toxicity
studies on aquatic organisms can not be introduced
into a gas-chromatograph as water is a bad solvent
for GLC analysis.Therefore,stripping devices have been
introduced before the GLC column,but this technique
needs tedious calibration procedures.Solvent extraction,
on the other hand,is time-consuming and requires very
expensive purified solvents. Reverse phase liquid chromatography has been used in our laboratory to analyse
aromatic hydrocarbons in aqueous solutions.Rapid direct
analyses of benzene,toluene,ethylbenzene,xylenes,
isopropylbenzene,naphthalene,anthracene have been
performed with an UV detector,at sensitivity levels
higher than FID-GLC.Advantages and limitations of this
technique in respect to GLC are discussed.

1. INTRODUCTION

Six aromatic hydrocarbons(benzene,toluene,ethylbenzene,
isopropylbenzene,naphthalene,anthracene) and two groups of
aromatic hydrocarbons(xylenes,PAH) have been included in
the List I of the CEC Directive (76/464),concerning water
pollution by dangerous substances discharged into the
water environment.An Italian law (76/319) lays down a limit
value of 0.2 mg/l for aromatic solvents to be discharged
into surface waters. In order to verify the protectivity of
this limit for aquatic life,ecotoxicological studies have
been undertaken in our laboratory on the aromatic hydrocarbons within the List I,adapting the standard testing procedures for these volatile compounds(1) and following the
test concentrations by HPLC analysis.HPLC has been found
to present some considerable advantages in respect to GLC,
generally employed for hydrocarbon analysis.The advantages
and limitations of this technique will be discussed,specially
in the case of aqueous solutions.

2. EXPERIMENTAL

2.1 Liquid chromatography

A Varian liquid chromatograph (Model 5000),with an

ultraviolet absorbance (UV 100) and a fluorescence detector
(Fluorichrom TM) was used. A C_{18}-bonded phase column (Micropak MCH 5-N-CAP,Varian,150x4mm,I.D.) with a guard column
(Hibar Lichrosorb RP-18,Merck,2 cm) was employed.
C_6-C_{10} hydrocarbons and commercial solvents were analysed
in the following conditions:mobil phase:CH_3OH-H_2O (75:25),
flow rate: 1 ml/min,injection volume: 50 μl,detector λ :200
nm. For dilute aqueous solutions (below 0.1 mg/l) a sample preconcentration was carried out as follows:the guard
column was disconnected,just before the chromatographic
column.The dilute aromatic hydrocarbon solution was put
into the reservoir C of the HPLC apparatus. 20 ml of sample were passed through the guard column at a flow rate
 of 2 ml/min and aromatic hydrocarbons were then eluted
with CH_3OH (reservoir B),collecting three eluates of 2 ml
each. Methanol eluates were analysed as reported before and
a complete recovery of aromatic hydrocarbons was obtained
in the third eluate.
PAH were analysed with CH_3CN-H_2O (75:25),as mobil phase,at
a flow rate of 2 ml/min with the fluorescence detector
using a tungsten-halogen source and an excitation filter
with a bandpass of 300-380 nm and an emission filter above
400 nm.In this case the injection volume was 10 μl.
2.2 Gas-chromatography
     A C.Erba (Model 4200) gas-chromatograph,adapted for
working with capillary columns,was used. A fused silica
column (25mx0.32mm,I.D.),coated with 2‰PS-255 stationary
phase,was employed.The gas-chromatographic conditions were
as follows:hydrogen carrier gas:1.8ml/min;splitting ratio:
7;injector temperature:250°C;oven temperature: 5 min at 60°
and then up to 200°C,4°C/min;FID detector temperature:250°C.
2.3 Chemicals
     Benzene,toluene and xylene were RP grade C.Erba reagents.
Ethylbenzene,o-,m-,p-xylene and isopropylbenzene were C.Erba
GC standards. Quantitative standards were prepared in CH_3OH
or water for HPLC analysis and in n-pentane for GLC analysis
adding liquid chemicals by volume and solid ones by weight.
Formamide was used as t₀ HPLC peak to calculate capacity
factors. PAH were EPA standards.

3.RESULTS AND DISCUSSION
3.1 C_6-C_{10} aromatic hydrocarbons
     The capacity factors (K') and the detection limits of
C_6-C_{10} aromatic hydrocarbons in CH_3OH-H_2O (75:25) are reported in table I and an example of calibration line for
benzene analysis is shown in fig.1. Ten times lower detection limits could be obtained with the preconcentration
procedure reported in the experimental part.Analogue,but

completely automated techniques are now available,allowing
to preconcentrate and analyse many samples in a short time.
However,it is interesting to point out that HPLC sensitivity,
without sample preconcentration,is comparable to that one
reported for GLC analysis,after a liquid-liquid sample extraction (2).
In fig.2 HPLC chromatograms of aromatic hydrocarbon mixtures
are shown.Chromatographic peaks are completely resolved with
the exception of xylenes and ethylbenzene. Single components
in such mixtures could be resolved with a polar GLC capillary column (2).

3.2 Commercial solvents

In order to determine the aromatic hydrocarbon content
in commercial mixtures,gasoline and "naphtha solvent" have
been considered. In spite of the complexity of such mixtures,very simple HPLC chromatograms have been obtained(fig.3)
and volatile aromatic hydrocarbons could be determined in
a short analysis time,owing also to inability of UV absorbance detector to detect saturated hydrocarbons. "Naphtha
solvent" was analysed also by HR-GLC (fig.4),confirming the
HPLC quantitative result for volatile aromatic hydrocarbons.
Benzene peak in the GLC chromatogram was a solvent impurity.

3.3 PAH

The use of high performance liquid chromatography for
the analysis of PAH mixtures has been widely reported and
the advantages of this technique over GLC recently discus-
sed(3). In table 2 the capacity factors and the detection
limits with fluorescence detector for PAH within the List I
are reported.Besides offering the advantage of a high
sensitivity,the selectivity of the fluorescence detector
allows to determine PAH from complex biological matrices.
We could determine PAH in environmental samples(water,sedi-
ment) at ppt level by extracting them with CH_2Cl_2 and inject-
ing concentrated extracts into the LC column,without any
clean-up procedure.

4. CONCLUSIONS

The HPLC technique allows to analyse the aromatic hydrocarbons considered in the present communication by means of
a single not too expensive apparatus. It should be able,
however,to determine other aromatic compounds,such as chlorobenzenes,nitrobenzenes,chlorophenols,nitrophenols,within
the CEC priority list. Although the HPLC technique does not
offer at the moment as efficient resolving power as capillary GLC,the choice of selective detection conditions can
allow to analyse priority compounds in complex mixtures,even
in aqueous medium. Sensivity can be improved by means of
on line preconcentration systems. Furthermore,if a spectro-

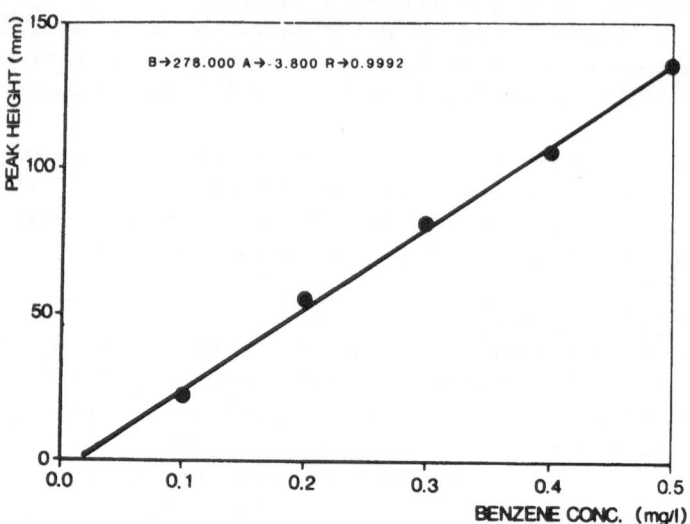

Figure 1. Calibration line for benzene HPLC determination
in water samples.

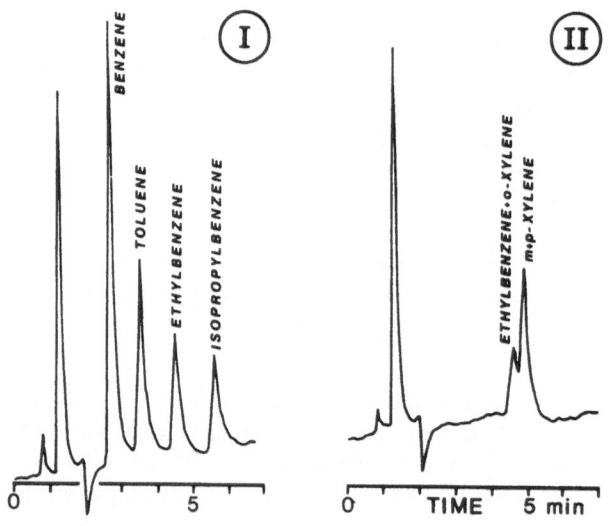

Figure 2.(I)Completely resolved solvent mixture by HPLC
(II)Incomplete resolution of C_8 aromatic
hydrocarbons

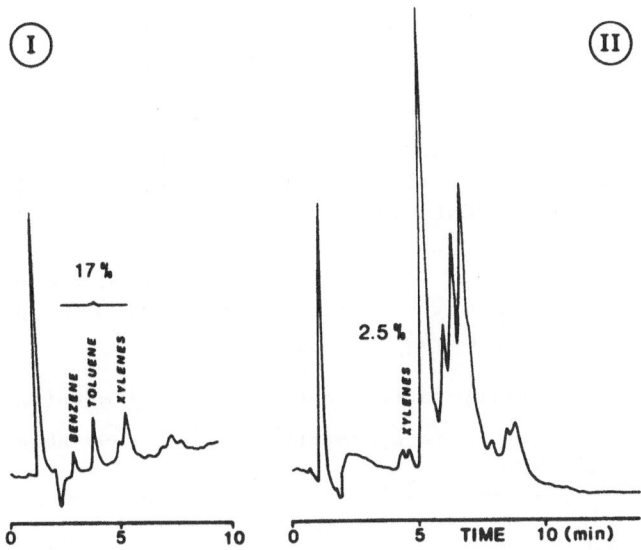

Figure 3. Volatile aromatic hydrocarbon determination in gasoline (I) and in "naphtha solvent (II) by HPLC

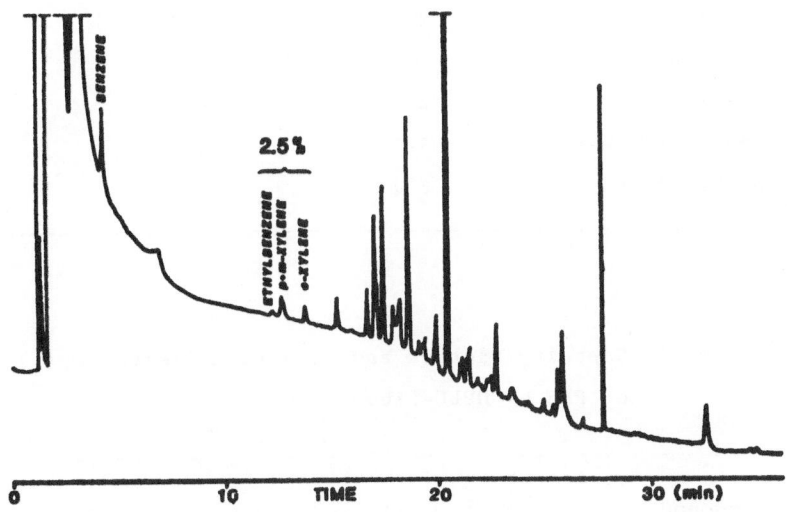

Figure 4. Gas-chromatogram of "naphtha solvent" on a 2%₀ PS-255 capillary column

photometric detector is available,the absorbance ratioing
technique can provide a simple tool for qualitative purposes,
confirming peak identity and purity.

5.REFERENCES

1. GALASSI,S. and VIGHI,M.(1981).Testing toxicity of volatile substances with algae.Chemosphere.10,1123-1126
2. ISTITUTO DI RICERCA SULLE ACQUE (1984).Solventi organici
   aromatici.E-016.Quaderni,11
3. McINTYRE,A.E. and LESTER,J.N.(1983).Organic contaminants
   in the aquatic environment.IV.Analytical techniques.
   Sci.Total Environ.27,201-230

Table I. Capacity factors and detection limits(mg/l)

of aromatic hydrocarbons by HPLC-UV absorption

Compound	K'	detection limit
Benzene	1.06	0.1
Toluene	1.69	0.1
Ethylbenzene	2.43	0.1
o-Xylene	2.46	0.1
m-Xylene	2.64	0.1
p-Xylene	2.65	0.1
Isopropylbenzene	3.28	0.2
Naphthalene	2.44	0.1

Table II. Capacity factors and detection limits (mg/l)
of PAH by HPLC-fluorescence

Compound	K'	detection limit
Anthracene	2.34	0.01
3,4 Benzopyrene	7.56	0.02
3,4 Benzofluoranthene	6.38	0.02

FAB MASS SPECTROMETRIC STUDY OF THE NON VOLATILE ORGANIC FRACTION IN

RAW AND DRINKING WATER EXTRACTS

J. Rivera , F. Ventura , J. Caixach and A. Figueras .
D. Fraisse and V. Blondot

Institut Química Bio-Orgànica, C.S.I.C. J. Girona Salgado, 18-26.
  08034-Barcelona (Spain).
Sociedad General de Aguas de Barcelona. Passeig de Sant Joan, 39.
  08009-Barcelona (Spain).
Centre de Spectrometrie de Masse, S.C.A. C.N.R.S. BP-22 69390-Vernai-
son (France).

Summary

Analysis of the non volatile organic extracts of granular activated
carbon filter installed at Barcelona's water works plant, raw and
drinking water were performed using HPLC fractionation, FAB mass spe-
trometry and combined FAB-CID-MIKE. Also LC/MS has been evaluated.
Among identified non volatile organics present in water, some of
them not reported previously, there are a wide range of nonionic
and anionic surfactants and polyglycols. The identification of the
components has been carried out by comparison of FAB and FAB-CAD-
MIKE spectra with those obtained from different HPLC fractions.

INTRODUCTION
     Gas chromatography/mass spectrometry has succesfully been applied
for identification and quantitation of a large number of "volatile"
organic pollutants in water (1,2). Nevertheless, these compounds
represent only 10-20% of the total organic material (3), the remaining
part referred as the "non volatile" organic fraction, is to be identified
at the compound level.
     Although the composition of the non volatile fraction is largely
unknown, the largest fraction consists of humic material and the
remainder, high molecular weight compounds, polar, ionic, thermolabile
and other complex mixtures.
     HPLC is a particularly well suited technique for analytical
fractionation of the non volatile organic fraction (4-7). The
identification of the components of this fraction can be carried out by
combination of liquid chromatography and mass spectrometry (LC/MS). The
main associated problems, such as low efficiency of HPLC columns to
separate organic complex mixtures, lack of sensitivity of the system with
the usual interfaces (moving belt, thermospray) requires high levels of
each compound to be detected. That makes this technique seldom used with
real water samples (8). Recently LC-FAB-MS (9) has opened a promising
pathway to the identification of components in complex mixtures.
     Increasing use of new ionization techniques such as direct chemical
ionization (DCI), field desorption (FD) or fast atom bombardment (FAB),

and the combination of FD or FAB with collisionally induced decomposition (CID) has been succesfully used for the characterization of non volatile organics, such as polyglycols (10), surfactants (11-17), rubber additives (18), dyes (19,20) and so on, and further applied to the analysis of non volatile organics in surface water (4-7, 21-23).

The present paper reports the application of HPLC. FAB mass spectrometry and FAB-CID-MIKE to the analysis of the organic extract retained by a granular activated carbon filter, non-volatile organic fraction of raw and drinking water from Llobregat river, supplying Barcelona city. HPLC fractions are tested for mutagenic activity prior to mass spectrometric analysis.

2. EXPERIMENTAL

2.1 Isolation techniques: Two approaches have been used. a) 50 g of wet granular activated carbon from a Water's Works filter (Pittsbourg, F-200) were freeze dried, extracted with methanol (Merck, Darmstadt, HPLC grade) during fifteen minutes in an ultrasonic bath. The total extract was concentrated on a Kuderna-Danish evaporator to c.a 1 ml prior to analysis. b) Extracts of organic micropollutants from river and drinking water were obtained by passing c.a 2000 liters of water through a column of activated carbon of the same type as used in the water works plant. The organics were desorbed with a Soxhlet with dichloromethane during 48 h. The total extract was concentrated on a K-D evaporator and redissolved with ether (8 ml). The ether soluble fraction was separated in acids and base + neutrals fractions and analyzed by GC/MS. The ether insoluble fraction was redissolved in methanol.

2.2 Separation and identification techniques: a) HPLC procedure: Separation and fractionation of the extracts of GAC filters and ether insoluble fraction of raw and drinking water were carried out using reversed phase HPLC. Chemical separations were achieved on a 7.8 mm x 30 cm μ-Bondapak C-18 (Water Assoc.) with 10 um diameter particles for semipreparative scale. A 3,9 mm x 30 cm μ-Bondapak C-18(10 μm) column for analytical scale and a 1 mm x 22 cm (Zorbax ODS 7 μm) home made column for microbore analysis. Also a precolumn was used in the semipreparative separations (ODS 30 μm home made). The gradient elution scheme shown in Table I was used for LC separations. Acetonitrile from Merck was used. Water was either bidistilled in glass and pirolyzed or MilliQ grade.

Time (min)	% solvent A	% solvent B
0	90	10
10	70	30
15	70	30
25	40	60
30	40	60
35	10	90

Table I: A - Water/Methanol: 90:10.   B - Acetonitrile.

b) Mass spectrometry: LC/MS, FAB and FAB-CID-MIKE were evaluated.

LC/MS: Mass spectrometric analysis were achieved with a VG-MM 30F equipped with a moving belt LC/MS interface (VG Analytical). Tipically solutes were desorbed from the polyimide moving belt at 250ºC. EI spectra were recorded from m/z 130 to 450 at 3 s/scan and 70eV.

FAB and FAB-CID-MIKE spectra were obtained using a ZAB-HF

spectrometer by using a Saddle field atom gun (Ion Tech Ltd). Samples were mixed in monothioglycerol saturated with NaCl and bombarded with 8kV xenon atoms. CID-MIKE spectra were obtained by selecting the appropriate ion in the spectrometer and then introducing helium into the collision cell between the magnet and the electrostatic sector to result in a 33% reduction of the intensity of the selected ion.

RESULTS AND DISCUSSION

a) Compounds adsorbed on G.A.C. filters of Barcelona's water works plant. Figure 1 shows an HPLC chromatogram of the total extract of GAC filter and the preparative fractionation carried out. The main objectives were: a) to obtain lesser complex fractions, in order to evaluate the LC/MS coupling. Previous optimization of chromatographic separation was achieved with microbore columns more suitable for the moving belt interface. b) To analyze these fractions (off line) by direct introduction to the mass spectrometer using FAB ionization.

Figure 2 shows the TIC of fraction 2 obtained via preparative liquid chromatography. High background present made impossible the interpretation of spectra obtained. This can be explained by means of the low concentration of compounds present in the samples, poor liquid chromatographic resolution in complex mixtures and low reponse in the EI mode. This assumption is in agreement with the identified compounds by FAB ionization.

Figure 3 displays positive and negative FAB spectra of this same fraction 2 Three main groups can be observed: alkylbenzene sulfonates $NaSO_3C_6H_4C_nH_{2n+1}$ ranging from n=9-14, with quasimolecular ions formed by sodium attachment dominate the lower mass range (m/z= 329,343,357 and 375). Negative FAB spectrum shows the presence of $(M-Na)^-$ and $(2M-Na)^-$ ions as major peaks of alkylbenzene sulfonates corresponding to n=10 and 11. The other identified compounds in this fraction are polyethoxylated nonylphenols $C_9H_{19}C_6H_4O(CH_2CH_2O)_n H$ a widely used nonionic surfactant, that gives 44 daltons spaced ions series. These series are characteristic of nonionic surfactants with poly(ethylen oxide) of different chain length. The main series (m/z = 287,331,375,449,463,507,...) are assigned to arise from sodium attachment to the poly(ethylen oxide) nonylphenol with n=0-21. Other series of polyethoxylated alkylphenols as decyl and isoctylphenols were identified by their $(M+Na)^+$ ions. The last group of compounds identified in this fraction, shows at higher masses, a series of ions which are spaced 58 daltons. These series are characteristic of polypropylenglycols, but they are not in agreement with pure standards reported in the literature (10). These ions have been tentatively assigned to arise from block copolymers of polyethylen and polypropylen mixtures $H(OCH_2CH_2)n (OCH_2CHCH_3)_mOH$. the limited mass range used for mass measurement did not allow us to precise the range of oligomers.

Other identified compounds in GAC filters were polyethylen glycol $H(OCH_2CH_2)_nOH$ with n up to 25. The spectrum is dominated by oligomer quasimolecular ions formed by sodium attachment. The relative oligomer abundances are in agreement with the results obtained by Lattimer (10) with PEG 600. These compounds were identified in fraction 1, whereas fractions 3 and 4 showed the main presence of alkylbenzene sulfonates already identified in fraction 2.

b) Identification of compounds in raw water (Llobregat river) and
drinking water of Barcelona. Figure 4 shows an HPLC chromatogram of the
non volatile extract of raw water in the middle course of the river, raw
water entering to the water works plant, drinking water and blank.

Samples were separated in 10 fractions monitored with a UV-Diode
Array detector using the method described before. Figure 5 displays the
positive FAB spectra of two examples of HPLC fractions of raw water in
the middle course of the river and two examples of drinking water. The
complexity of the obtained spectra showing different class of compounds
spaced 44 or 58 daltons, which are characteristic of polyethoxylated or
polypropoxylated compounds, led us to study the FAB spectra of a wide
range of nonionic surfactants, such as polyethoxylated alkylphenols,
linear and oxo alcohols, unsaturated fatty acids and amines (tallow,
coconut) and so on.

Figure 6 shows the positive FAB spectra of industrial polyethoxylated
linear alcohols (C_{12}-C_{14}) and polyethoxylated unsaturated fatty acids.
The spectra of these compounds are dominated by their $(M+Na)^+$ ions
showing the different number of units. FAB-CID-MIKE spectra of two
relevant peaks of the FAB spectra of these compounds are shown in figure
7. The CID-MIKE spectrum of lauric alcohol using m/z= 429 and
corresponding to $(M+Na)^+$ n=5 shows the cleavage of the alkylchain, the
breakdown of the ethoxy units and other losses of ethoxy units from M-H_2O
as main fragments. The precursor of polyethoxylated unsaturated fatty
acids used $(M+Na)^+$ n=4 m/z=525 for oleic acid, indicates the cleavage of
the alkyl chain, the double bond position (24) and the different ethoxy
units. More detailed information of the nonionic surfactants is described
elsewhere (17).

The results obtained with different types of polyethoxy and
polypropoxylated compounds, allowed us to built up a table for a rapid
identification of these type of compounds. FAB-CID-MIKE yields detailed
structural information and provides univocal identification of unknown
compounds by comparison with standards. This technique is particularly
well suited when different type of compounds give the same $(M+Na)^+$ peaks.

The applied methodology described above allowed us to identify a
wide range of nonionic compounds and polymers, some of them firstly
reported to occur in water. Figure 8 shows the identification of
nonylphenols in an HPLC fraction of raw water by comparison of the
CID-MIKE spectrum of the sample with their corresponding standard.

The main identified compounds in raw and drinking water are given in
Table II. The presence of these compounds is due to the presence of
surfactants' industries near the river, other industrial activity and
domestic discharges.

CONCLUSIONS

The use of FAB mass spectrometry and combined FAB-CID-MIKE has
proved to be a useful technique for the identification of surfactants and
polyglycols. Both techniques and HPLC, have brought an aid to the
identification of non volatile organics adsorbed on granular activated
carbon filters installed at Barcelona's water works plant, raw water and
drinking water. The presence of these compounds changes easily due to
industrial activity and weak flow of the river.

Despite the use of FAB has grown markedly there is not many data
available in the literature about compounds liable to be found in the non
volatile fraction. Future work is being done in order to create our own
FAB library of both usual and occasional non volatile compounds

discharged into the river.

ACKNOWLEDGEMENT
We wish to thank CIRIT for a postdoctoral fellowship for a one of us (J.C.). This work has been carried out in the frame of the French-Spanish Integrated Action 151/84. We also are indebted to Molins-Kao S.A. for a gift of the different type of surfactants.

BIBLIOGRAPHY

1.- Garbarino, J.R.; Steinheimer, T.R.; Taylor, H.E. Anal.Chem. 1985, 57, 46 R - 88 R.
2.- Keith, L.H. "Advances in the Identification & Analysis of Organic Pollutants in Water", Vol. 2; Ann Arbor Sci. Publish. Michigan, 1981.
3.- National Academy of Sciences, (1977). "Drinking water and Heath", p. 489-856. Natl. Acad. Sci. Washington, D.C.
4.- Crathorne, B.; Watts, C.D.; Fielding, M. J. Chromatogr. 1979, 185, 671-690.
5.- Crathorne, B.; Fielding, M.; Steel, C.P.; Watts, C.D. Environ Sci. Technol. 1984, 18, 797-802.
6.- Crathorne, B.; Watts, C.D. "Analysis of Organic Micropollutants in Water"; D. Reidel Publish Dordrecht, 1982; p. 159-173.
7.- Watts, C.D.; Crathorne, B.; Fielding, M.; Steel, C.P. "Analysis of Organic micropollutants in Water"; D. Reidel: Dordrecht, 1984; p.120-131.
8.- Schauenburg, H.; Schill, H.; Knöppell, H. "Analysis of Organic Micropollutants in Water"; D. Reidel: Dordrecht, 1982; p. 193-198.
9.- Stroh, J.G.; Cook, J.C.; Milberg, R.M.; Brayton, L.; Tsuyoshikihara; Zhaogeng Huang; Rinehart, K.L., Jr. Anal. Chem. 1985, 57, 985-991.
10.- Lattimer, R.P. Int. J. Mass Spectrom. and Ion Processes 1983/84, 55 221-232.
11.- Cotter, R.A.; Hansen, G.; Jones, T.R. Anal. Chimica Acta 1982, 136, 135-142.
12.- Weber, R.; Levsen, K.; Louter, G.R.; Boerboom, J.H.; Hoverkamp, J. Anal. Chem. 1982. 54, 1458-1466.
13.- Schneider, E.; Levsen, K.; Dähling, P.; Röllgen, F.W. Fresenius 'Z Anal. Chem. 1983, 316, 277-285.
14.- Schneider, E.; Levsen, K.; Dähling, P.; Röllgen, F.W. Fresenius 'Z Anal. Chem. 1983, 488-492.
15.- Lyon, P.A.; Stebbings, N.L.; Crow, F.W.; Tomer, K.B.; Lippstreau,D.L.; Gross, M.L. Anal. Chem. 1984, 56, 8-13.
16.- Lyon, P.A.; Stebbings, N.L.; Crow, F.W.; Tomer, K.B.; Lippstreau,D.L.; Gross, M.L. Anal. Chem. 1984, 56, 2278-2284.
17.- Rivera, J.; Caixach, J.; Ventura, F.; Fraisse, D.; Dessalces, G.; 10th. International Mass Spectrometry Conference. Swansea (U.K.). September 1985.
18.- Lattimer, R.P.; Harris, R.E.; Ron, D.B.; Diem, H.E. Rubber Chem. Technol, 1984, 57, 1013-1022.
19.- Brown, R.M.; Creaser, C.S.; Wright, H.J. Organic Mass Spectrom. 1984, 19, 311-314.
20.- Monaghan, J.J.; Barber, M.; Bordoli, R.S.; Sedgwick, R.D.; Tyler, A.N. Organic Mass Spectrom. 1983, 18, 75-82.
21.- Otsuki, A.; Shiraishi, H. Anal. Chem. 1979, 51, 2329-2332.
22.- Yasuhara, A.; Shiraishi, H.; Tsugi, M.; Toshidide, O. Environ. Sci. & Techn. 1981, 15, 570-573.
23.- Shinohara, R.; Kidd, A.; Eto, S.; Hori, T.; Koga, K.; Akuyama, T. Environ. Int. 1980. 4, 31-37.

24.- Tomer, K.B.; Crow, F.W.; Gross, M.K.; J. Am. Chem. Soc., 1983, 105, 5487-5488.

Table II: Main identified compounds in raw and drinking water.

RAW WATER

- Polyethoxylated alkylphenols C_nH_{2n+1} ⟨benzene ring⟩ $-O(CH_2CH_2O)_mH$
  nonylphenols   m = 0-13
  isooctyl and decylphenols (traces)   m = 0 - 13
- Polyethoxylated alcohols C_nH_{2n+1} $(OCH_2CH_2)_mOH$
  lauric   m ≤ 10
  isotridecyl   m ≤ 12
  myristic   m ≤ 12
  pentadecyl   m ≤ 18
- Poly (ethoxypropoxy) alkylph. C_nH_{2n+1} ⟨benzene ring⟩ $(OCH_2CH_2)_{n'}(OCHCH_2)_mOH$
  nonylphenols n' = 1-8; m = 1-4 CH_3
- Polyethoxylated fatty acids
  castor oil (ricinoleic acid) n ≤ 12   (tentatively assigned)
- Alkyl ether sulfates C_nH_{2n+1} ⟨benzene ring⟩ $(OCH_2CH_2)_mOSO_3Na$
  nonylphenols m ≤ 10
  isooctylphenols m ≤ 6
- Polyethylen glycols $H(OCH_2CH_2)_nOH$
  PEG 200, PEG 400, PEG 600
- Poly(ethylenpropylen)glycols block copolymers
  with   n = 1-5; m = 3-4 $H(OCH_2CH_2)_n(OCH-CH_2)_mOH$
                                                            CH_3

DRINKING WATER

- Polyethoxylated alkylphenols
  chlorinated and/or brominated nonylphenols n=0-15 (tentatively
  isoctylphenols   m = 0-15   assigned)
  dodecylphenols (traces) m = 0-10
- Polyethoxylated alcohols
  myristic   m ≤ 7
- Poly(ethoxypropoxy) alkylphenols
  dichlorinated nonylphenols m ≤ 10 (tentatively assigned).
- Alkyl ether sulphates
  nonylphenols   m ≤ 10
- Polyethylenglycols
  PEG 200
- Poly (ethylenpropylen)glycols block copolymers
  n, m ≤ 12.

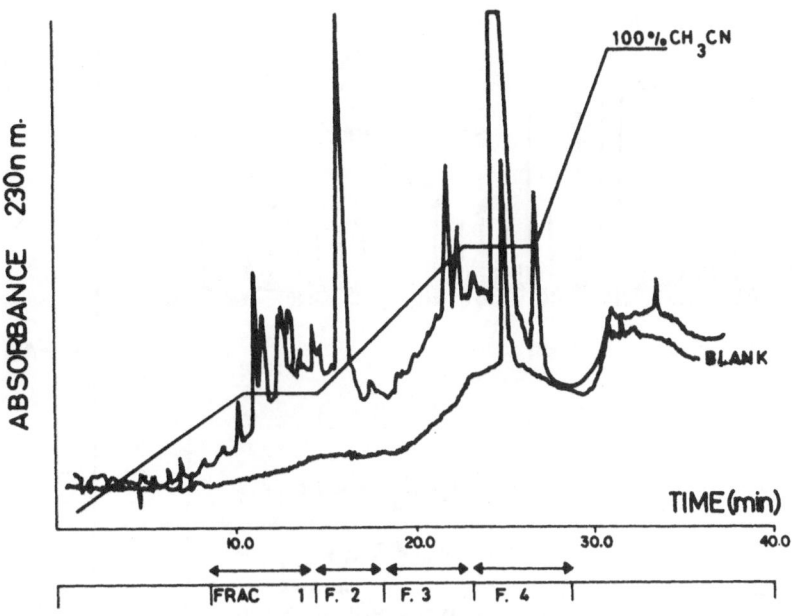

Fig 1: Granular activated carbon filter. HPLC profile of the total extract.

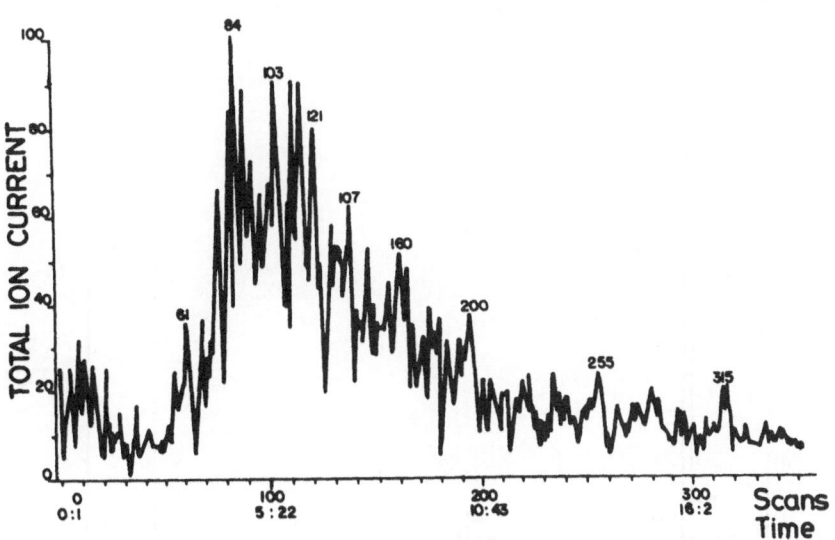

Fig 2: Total ion current LC/MS of HPLC fraction nº 2. Isocratic elution $H_2O/CH_3OH/AcH$: 50/48/2. C_{18} (1 mm x 220), flow rate 50 μl/min, belt speed 0.8 cm/s.

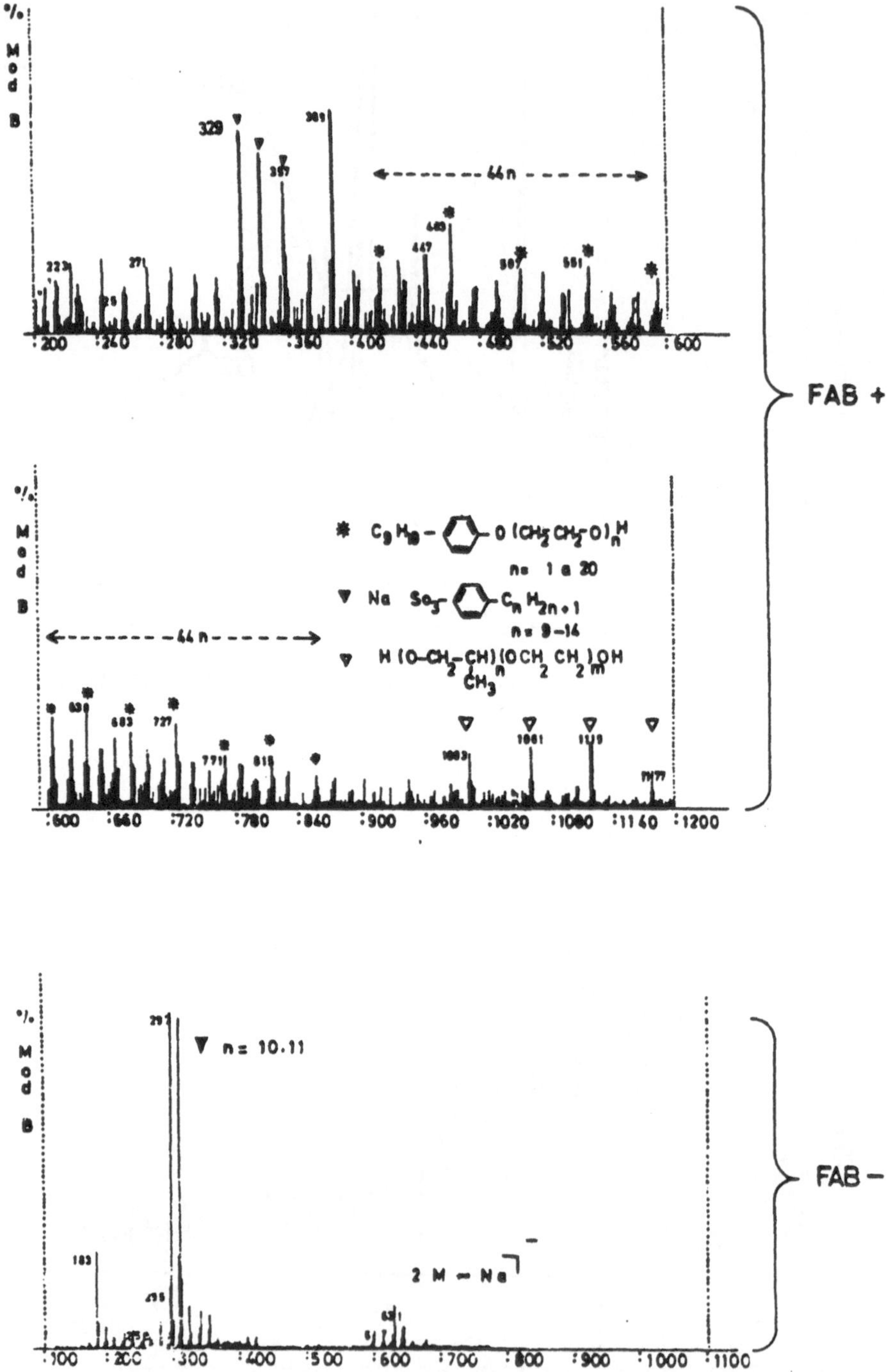

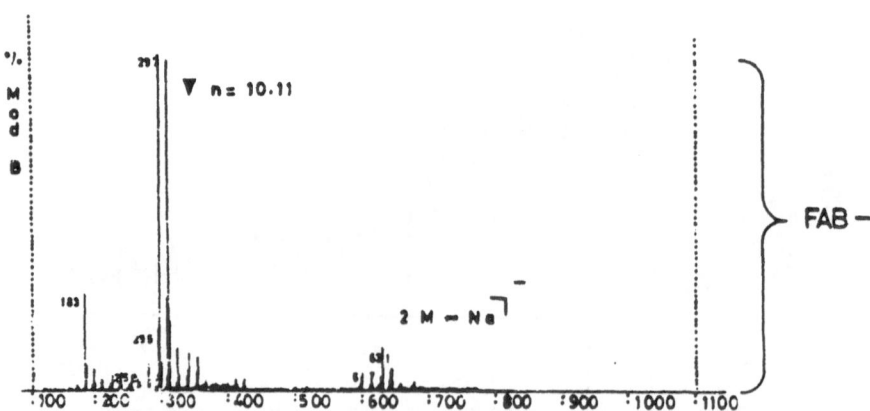

Fig 3: Top: FAB(+) spectrum and bottom: FAB (−) spectrum of HPLC
fraction nº 2.

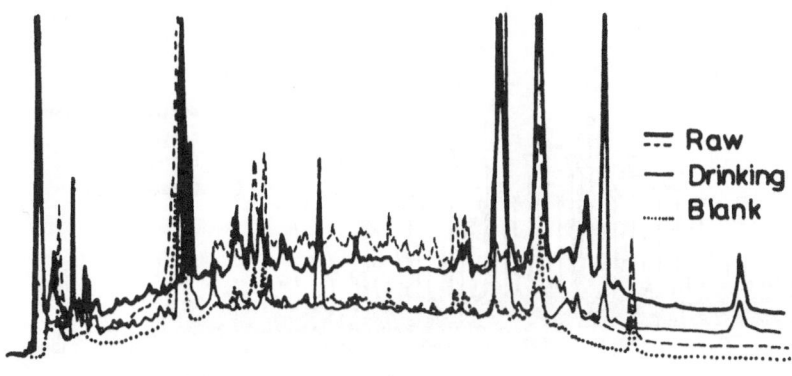

Fig 4: HPLC profile of raw water in the middle course of the river
(——), raw water entering the water works plant (----), drinking water and blank.

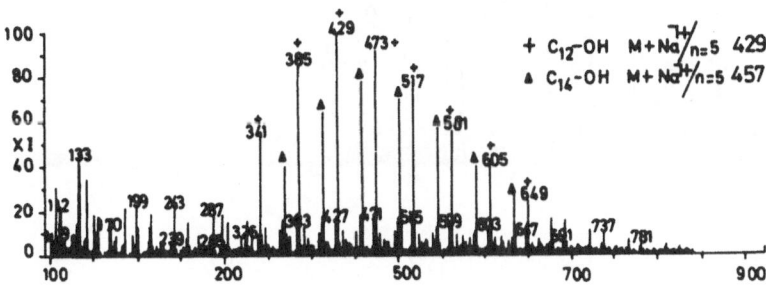

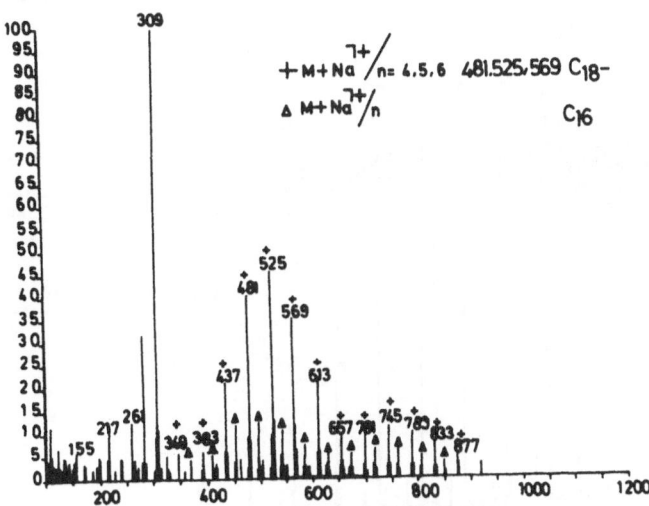

Fig 6: FAB spectra of standards.
    Top: polyethoxylated linear alcohols $(C_{12}-C_{14})$
    Bottom: polyethoxylated unsaturated fatty acids.

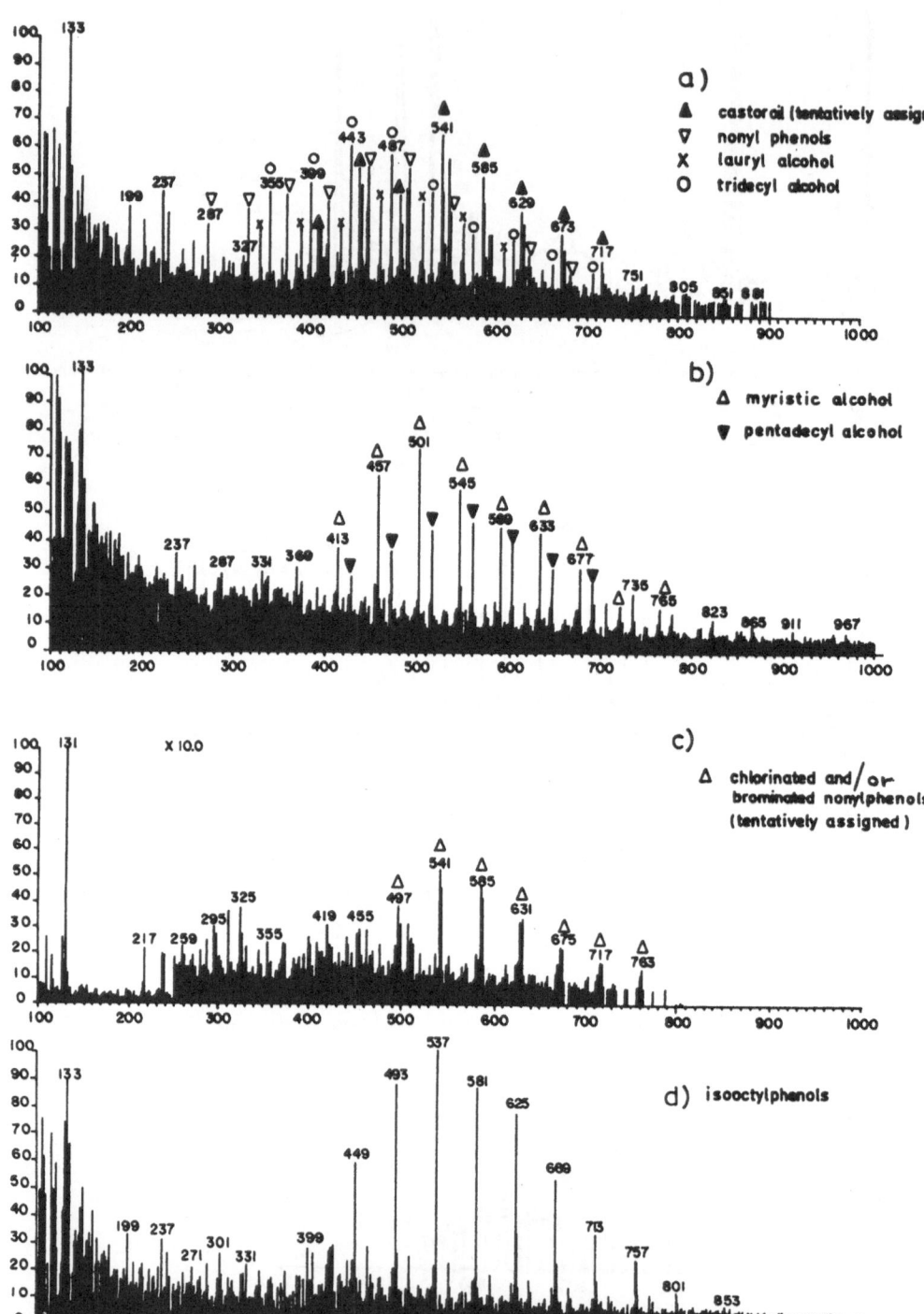

Fig 5: a,b) FAB(+) spectra of raw water. HPLC fractions 8 and 9.
c,d) FAB(+) spectra of drinking water. HPLC fractions 6 and 7

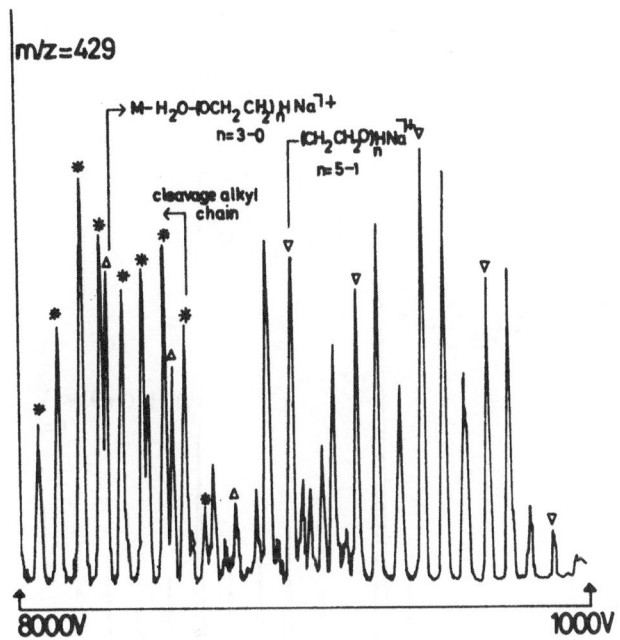

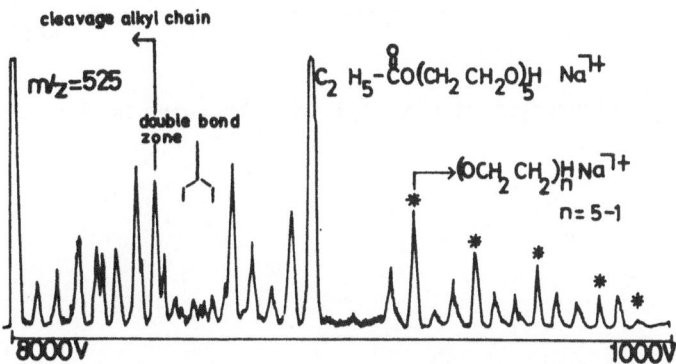

Fig 7: FAB–CID–MIKE spectra of standards from fig 7.
top: Lauric alcohol. m/z= 429 used as precursor.
bottom: Oleic acid. m/z= 525 used as precursor.

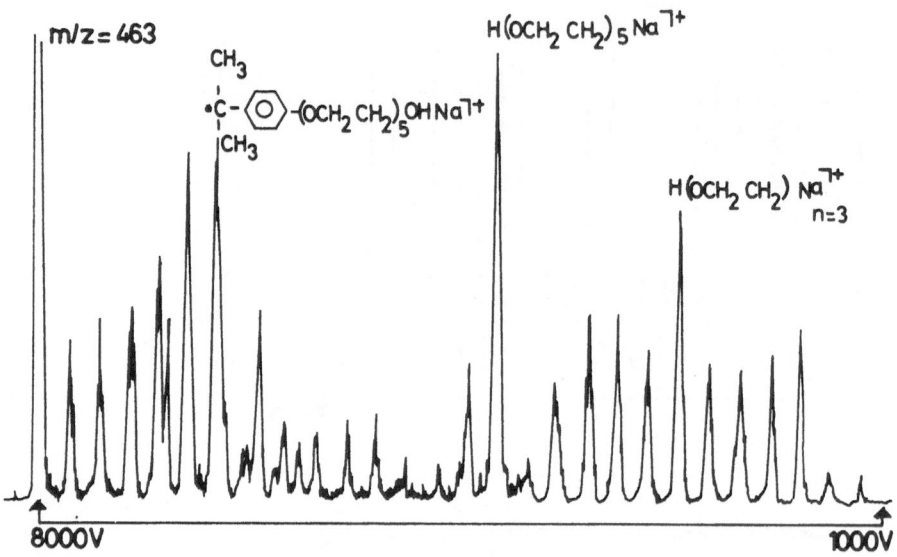

Fig 8: FAB-CID-MIKE spectrum of m/z= 463 from raw water HPLC fraction
nº 8.

POLLUTION CONDITIONS OF THE AQUIFER BENEATH A PLANT PRODUCING ALKYD RESINS

E. MANTICA, D. BOTTA, L. CASTELLANI PIRRI
Dept. of Industrial Chemistry and Chemical Engineering
Politecnico di Milano, Piazza Leonardo da Vinci, 32
20133 Milano (Italy)

Summary

Factories producing alkyd resins use many raw materials for the produc
tion of their numerous products. These are polybasic acids, monobasic
acids deriving from natural oils, polyhydric alcohols,   styrene, tolue
ne diisocyanate, dicyclopentadiene, glycol esters and ethers, and va
rious solvents. Furthermore, esterification reactions form volatile
products (aldehydes, ketones, heterocyclic oxygenated compounds) which
pass into the plant's gaseous and liquid waste products. Many of these
products constitute a danger for the quality of drinking water if acci
dental losses from piping or machinery reach the ground and the under
lying aquifer. The appearance of unpleasant odors and tastes in the
local water supply led in this case to an investigation of the level
of pollution of the soil and water beneath this factory. Analyses car
ried out by GC/MS/DS confirmed the presence of various organic compo
unds used as raw materials, or deriving from foreseeable secondary reactions or formed by oxidative transformations produced by microorga
nisms. The identity of some pollutants felt to be of special signifi
cance was confirmed by synthesizing these compounds and comparing their
mass spectra and retention times with those of the unknown contaminants present in water samples.

1. INTRODUCTION

The protection of ground waters against pollution by chemical compounds is becoming more and more important and even essential in the light of
increasing reports of contamination of aquifers supplying drinking water
for the population. The dispersion of toxic substances adversely affecting
human healt may come about in various ways (localized or widespread) and
may be due to incorrect or culpable practice. The most frequent causes of
water or soil contamination are: incorrect operation of some component of
apparatus or plants, leakages from pipes, tanks and industrial or municipal
sewers, improper application of chemical substances in agricolture (non
point sources), unsuitable methods of disposing liquid or solid wastes in
public or private impoundments or landfills. Dispersion initially affect
surface waters or surface strata of the soil, but later, after a variable
period of time, also involve ground waters thanks to the rinsing effect of
rain. Even more dangerous are direct immissions of liquid wastes into deep-
lying strata, due to the highly deprecable use of underground injection
wells, which bring the pollutants into direct contact with deep aquifers in
a short time.

Without underestimating the effects of municipal discharges and of chemicals used in agriculture or produced as animal wastes from commercial livestock and poultry facilities, it must be underlined that the greatest hazard for ground waters comes from the dispersion of the various chemicals used in industrial processes. Amongst such activities those of large chemical plants carrying out the synthesis and transformation of numerous products in quantities as great as hundreds or even thousands of tons per year are specially hazardous. The ground waters beneath a large chemical plant are particulary exposed to the risk of pollution, which is often intense and persistent. At such depths the self-purification processes, which are effective and sufficiently rapid in the case of surface waters, are extremely slow. Years or even decades may be required to reclaim an aquifer polluted by such a chemical plant.

The case considered in this paper involves a chemical plant of considerable size involved in highly diversified production. The main productive nucleus is a plant for the preparation of alkyd resins of various widely used types. This plant uses large quantities of many raw materials in the production of end-products having many applications, and eliminates gaseous, liquid and solid wastes.

The leading raw materials are polycarboxylic acids or their anhydrides (phthalic and maleic anhydride, adipic or fumaric acid), polyhydric alcohols (ethylene,diethylene, triethylene, propylene and dipropylene glycols, pentaerythritol, trimethylolpropane and its diallyl ether), fatty acids and oils (fatty acids from peanut or soybean oil, linseed, soybean or castor oil). Minor components used to modify the products from the various processes are dicyclopentadiene, which after depolymerization to cyclopentadiene, reacts, by Diels-Alder condensation, with maleic anhydride to produce endo-methylentetrahydrophthalic acid a component of unsaturated polyesters; styrene, which copolymerizes with ur aturated polyesters; toluendiisocyanate, which acts as a reticulating agent for polyesters containing free hydroxyl groups and is an intermediate in the production of polyurethane lacquers.

The end products are polyester resins, alkyd resins, resins for lacquers and paints having various properties according to the field in which they are to be used. All these materials are then diluted to various concentrations with solvents such as white spirit, xylene, styrene, etc.

Wastes from the plant requiring a proper disposal vary according to the process. Volatile wastes include water, methanol, acetic and propionic aldehydes, acetone, styrene and many other unidentified organic compounds. Liquid wastes contain water, unreacted organic compounds or by-products of the different reactions, whilst solids include tars and resinous materials.

This brief review of possible pollutants easily shows how the risk for the aquifer beneath the factory is very serious. It may indeed be expected that a part of the products dispersed will be found in the water either as such or as the result of more or less complex transformations, prevalently of the oxidative type, due to atmospheric oxygen or microorganisms in the soil or water.

2. EXPERIMENTAL

2.1 Sampling (fig. 1)

The sample of water was taken from a tap fed by a vertical immersion pump in a test well within the area occupied by the chemical plant in question. The well was 36 m deep, reaching a semi-permeable lenticular clay formation which isolates the phreatic aquifer from the artesian aquifer. The well casing is a 4" dia. PVC pipe with perforations between 22 and 34 m to permit sampling from the phreatic aquifer alone. The sample was maintained at low temperature until analysis.

2.2 Separation of Pollutants from the Aqueous Matrix and their Concentration

Isolation of pollutants from the aqueous matrix was achieved by extraction with methylene chloride at various pH values. Three fractions (neutral, acidic, basic) were thus obtained, as shown in fig. 2. All extracts were dried on anhydrous sodium sulphate and concentrated at 200 µl in a Kuderna-Danish distillator. The concentrated solutions were maintained at -20°C to avoid losses of volatile compounds.

2.3 Analysis of the Neutral Fraction

Analyses carried out in this early part of the study were limited to the neutral fraction isolated as explained above (2.2). Gas chromatographic analysis was carried out using a capillary glass column in a Carlo Erba Mod. 2150 gas chromatograph in the conditions listed in Table I. The complete chromatogram obtained is given in fig. 3, and the expansion of two areas ($t_R = 0 \rightarrow 8$ and $11 \rightarrow 33$ minutes) in figs. 4 and 5.

Successively the extract in CH_2Cl_2 was introduced into the split/splitless injector of a Hewlett-Packard Mod. 5985 B GC/MS/DS system in the experimental conditions reported in Table II. The identity of the pollutants was established by automatic comparison with the spectra present in the system library and, in part, by manual method by comparison with spectra available in literature (1),(2),(3). Table III lists some compounds identified by both methods. No reference spectra were found for numerous compounds, so that it was necessary to program a series of syntheses for certain classes of products, as explained in 2.4.

2.4 Syntheses of Reference Compounds

2.4.1 Dimethyldioxanes-1,4 (fig. 6)

A preliminary attempt to prepare dimethyldioxanes-1,4 was by the condensation of 1,2-propanediol in the presence of traces of a strong acid as catalyst. Glycol (50 g) and 1 g of a strong acid resin of the type Dowex-50-X2 (mesh 50/100) were introduced into a round bottom flask fitted with a magnetic stirrer. The flask was heated to 160°C in an oil bath. A fraction was distilled at b.p. 90-95°C which, after cooling, divided into two layers, which were separated in a separatory funnel. The lower aqueous phase was eliminated and the upper phase dried on anhydrous Na_2SO_4. The li quid obtained was injected into a OV-1 capillary column mounted in a Carlo Erba Mod. 4160 gas chromatograph, yielding the chromatogram displayed in fig. 7(a). Two largely prevalent peaks are found which, according to the order of increasing retention times and mass spectra, can be attributed to 2-ethyl-4-methyldioxolane-1,3 cis (peak 1) and to 2-ethyl-4-methyldioxolane-1,3 trans (peak 2). Only traces of the desired dimethyldioxanes-1,4 have been found.

A second preparation was attempted by condensing the dipropylene glycol (mixture of isomers) in the presence of small quantities of strong acid resin, operating in the same experimental conditions described above. A liquid was obtained which was found to be a mixture of compounds with a boiling range from 114 to 128°C. A 1 µl sample of a 1‰ solution in methylene chloride was introduced at split ratio 1:30 into the injector of the gas chromatograph and the chromatogram of fig. 7(b) was obtained. Peaks 1 and 2 correspond to the two above mentioned 2,4-dialkyldioxolanes-1,3, peak 3 is 2,6-dimethyldioxane-1,4 cis, peak 4 is 2,5-dimethyldioxane-1,4 trans, peak 5 is 2,6-dimethyldioxane-1,4 trans and peak 6 is 2,5-dimethyldioxane-1,4 cis. Attempts are now being made to fractionate this mixture by rectification, but results to date are hardly satisfactory since no component has been isolated at a sufficient degree of purity to permit its use in computing GC response factors.

2.4.2 2,4-Dialkyldioxolanes-1,3 (fig. 8)

The 2,4-dialkyldioxolanes-1,3 synthesized for this research were obtained by acid-catalyzed condensation of 1,2-propanediol with an aliphatic aldehyde (propionaldehyde, iso-butyraldehyde, n-butyraldehyde). The glycol and the aldehyde, in a ratio of 1:3 by weight after addition of a trace of an acid catalyst (Dowex-50-X2 Resin) were introduced into a closed reactor heated on a water bath for one hour. The liquid thus obtained was filtered to eliminate the solid resin and introduced into the injector of a gas chromatograph as described in 2.4.1. Chromatograms (a),(b),(c) of fig. 9 were thus obtained for the condensation products of the three aldehydes. Two largely prevalent peaks are evident, and may be attributed, on the basis of the retention parameters and mass spectra (base peak m/z 87), to the two diastereoisomers cis and trans. In this case too, rectification did not make it possible to separate the two components in the pure state: future attempts to do so will probably be made by preparative gas chromatography in the hope of obtaining better results.

2.4.3 Dicyclopentadiene Oxides (fig. 10)

Dicyclopentadiene (93% commercial product) was oxidized with perbenzoic acid, prepared by reaction of benzoyl peroxide with sodium methylate in a chloroform solution and further treated with H_2SO_4. The perbenzoic acid solution was mixed with a solution of dicyclopentadiene in chloroform and allowed to react at room temperature until iodometric controls indicated the disappearance of the peroxides. The crude product was composed of two oxides II and III, of a small quantity of dioxide IV, of unreacted dicyclopentadiene, of benzoic acid and its methyl ester. Acid was eliminated by washing the chloroform solution with an aqueous solution of sodium hydroxide. The identity of the various products was confirmed by GC (fig.11) and by GC/MS analysis.

2.4.4 Oxydation of Dicyclopentadiene with Potassium Permanganate

To increase the number of oxygenated derivatives of dicyclopentadiene useful as references to identify these compounds in the water sample under consideration, dicyclopentadiene was oxidized with potassium permanganate in alkaline solution. Fig.12 reports some possible oxidation products with molecular weights 146, 148, 162, 164, but numerous others may be hypothesized in relation to the various modes of oxidation (with oxygen, with inorganic oxidants, with various species of microorganisms).

Treatment of dicyclopentadiene with potassium permanganate in an aqueous solution of sodium carbonate at 100°C for about 1 hour yielded a brownish suspension of manganese dioxide. The precipitate was eliminated by filtration and the aqueous phase extracted 3 times with 30 ml of methylene chloride. The combined extracts were dried on anhydrous sodium sulphate, then analyzed by gas chromatography and GC/MS. The chromatogram in Fig.13(a) shows the results obtained and indicates the molecular weights of the compounds corresponding to some of the main peaks. Besides the unreacted dicyclopentadiene (m.w. 132, base peak 66) various compounds with molecular weights 146 ($C_{10}H_{10}O$), 148 ($C_{10}H_{12}O$) and 150 ($C_{10}H_{14}O$) were formed, which are ketones, oxides and alcohols produced by oxidation of the dicyclopentadiene. Some of these compounds have a base peak 66 (products of oxidation of the cyclopentadiene ring and in particular of allylic oxidation), others have base peak 81 or 82 (products of oxidation of the norbornene ring). The precipitate remaining on the filter, consisting prevalently of manganese dioxide, was treated with diluted sulphuric acid and the obtained suspension was filtered. The aqueous phase was extracted 3 times with 30 ml of methylene chloride, the combined extracts were dried on anhydrous sodium sulphate, a sample then being analyzed by gas chromatography and GC/MS. The chromatogram of fig. 13(b) shows the presence of small quantities of unreacted dicyclopentadiene, of a compound largely prevalent at m.w. 148 and base peak 82, together with a series of minor components at m.w. 164 ($C_{10}H_{12}O_2$) and 162 ($C_{10}H_{10}O_2$).

2.4.5 Oxidation of Dicyclopentadiene with Sodium Dichromate

Dicyclopentadiene was reacted for 2 h with sodium dichromate in a sulphuric acid solution. After completion of the reaction the liquid was neutralized with sodium carbonate and extracted 3 times with 20 ml of methylene chloride. The combined extracts were dried on anhydrous sodium sulphate and a sample was analyzed by gas chromatography and GC/MS. The chromatogram of fig. 13(c) shows the presence of unreacted didyclopentadiene, of a prevalent compound at m.w. 150 ($C_{10}H_{14}O$) and base peak 66, of appreciable quantities of three compounds at m.w. 162 ($C_{10}H_{10}O_2$) and a series of minor components at m.w. 146 ($C_{10}H_{10}O$), 148 ($C_{10}H_{12}O$) and 178 ($C_{10}H_{10}O_3$).

3. RESULTS AND DISCUSSION

An inspection of the chromatogram of the neutral fraction given in fig. 3 shows the high level of complexity of the mixture of pollutants present in the water sample under examination. The information obtainable both by automatic comparison of the mass spectra of the unknown products with those available in the HP 5985 B system library, and also by manual comparison with the various collections of spectral data in the literature (1), (2),(3) is very reduced. This is due to the limited number of reference spectra relating to the particular classes of products polluting the water. Furthermore, some of the information is incomplete, unreliable, or leads to compounds which do not correspond to the real production of the plant. To obtain adequate identity confirmation, associating the MS data with GC retention parameters, it was necessary to buy some compounds already available on the market and to synthesize others in our laboratory in an attempt to prepare molecules whose identity had been hypothesized on the basis of MS results. The task was by no means easy and has made it necessary to limit our studies for the time being to two groups of substances:
a- dioxane-1,4 and its dimethyl derivatives, and 2,4-dialkyldioxolanes-1,3;
b- dicyclopentadiene and its oxidation products (oxides, ketones, alcohols).

3.a Dioxane-1,4, dimethyldioxanes-1,4, 2-alkyl-4-methyl-dioxolanes-1,3

These are by-products of the main reaction of condensation between polyhydric alcohols and polycarboxylic acids which is the main object of the production process. They may derive from autocondensation processes on glycols (fig.6) (ethylene or diethylene glycol for dioxane-1,4, 1,2-propane diol or dipropylene glycol for dimethyldioxanes-1,4), or from the condensation of a glycol (1,2-propanediol) with an aldehyde (fig.8). Dioxane-1,4 is easily obtainable on the market as a pure reagent whilst the dimethyldioxanes-1,4 were synthesized by us as described in 2.4.1. Confirmation of the identity of the compounds prepared was obtained from MS data: dimethyldioxanes-1,4 yield reasonably intense peaks of the molecular ion 116, base peaks 42 and very intense peaks at mass 45. Distinction between the two isomers cis and trans was achieved on the basis of their GC retention times. Less-recent literature and, in part, mass spectroscopic data reveal some confusion because very frequently the 2-ethyl-4-methyldioxolanes-1,3 cis and trans are exclusively obtained in reactions addressed to obtain dimethyldioxanes-1,4, their spectra having a molecular ion 116 of insignificant intensity, the base peak at m/z 87, and a series of relatively intense peaks at m/z 31,41,57,59,72.

The identity of some compounds present in the water sample was established by comparison of the gas chromatographic retention times and the mass spectra: they are the dioxane-1,4 and the four dimethyldioxane-1,4 (2,5-isomer cis and trans; 2,6-isomer cis and trans). The persistence in the water and the toxicity of these compounds (4) led us to dedicate particular attention to this class and to consider their possible use as tracers even at a considerable distance from the source pollution.

Dialkyldioxolanes-1,3 are less persistent in the environment, being ace tals which are more easily hydrolized in water contaminated by acid inorganic or organic compounds. The products synthesized in our laboratory are given in fig. 8. Confirmation of the identity of the 2-alkyl-4-methyl-dioxolanes-1,3 (2-ethyl, 2-isopropyl, 2-n-propyl) was achieved on the basis of gas chromatographic retention times and peak intensities (distinction between cis and trans isomers) and of mass spectrometry data (base peak m/z 87, $C_4H_7O_2$ in all cases). The identity of the 2-ethyl-4-methyl-dioxolane-1,3 (cis and trans isomers), of the 2-i-propyl-4-methyl-dioxolane-1,3 (cis and trans isomers) and of the 2-n-propyl-4-methyl-dioxolane-1,3 (cis and trans isomers) in the water sample was confirmed by comparison of the GC retention times and MS data.

3.b Dicyclopentadiene and its Oxidation Products

Recognition of the presence in the neutral extract from the water sample of the exo- and endo-isomers of dicyclopentadiene (peaks 12 and 13 of the chromatogram in fig. 3 or fig. 5) was found to be very easy. Automatic library search suggested the identity of the two compounds with high probability (presence of molecular peak 132 and base peak 66 derived from a highly favored retro-Diels-Alder reaction). Confirmation of this identity was achieved by comparison of the gas chromatographic retention times and mass spectra with those of a authentic sample.

The same cannot be said of the compounds corresponding to the peaks 14-33 of the chromatogram. In these cases neither automatic library search nor manual comparison with spectra in the literature suggested reliable hypotheses because of the total absence of reference spectra. It was felt, thanks to the observation that many of the compounds which produce chromatographic peaks in this region have spectra with 66, 81 or 82 as base peaks, that these may at least in part be oxidation products of dicyclopentadiene. The reaction of the later compound with perbenzoic acid (see 2.4.3) made available retention times and mass spectra for the following references: 4,7-methano-1 H-indene-2,3-epoxide, 3a,4,7,7a-tetrahydro- (dicyclopentadiene-2,3-epoxide), 4,7-methano-1 H-indene-5,6-epoxide, 3a,4,7,7a-tetrahydro-(dicyclopentadiene-5,6-epoxide), 4,7-methano-1 H-indene-2,3,5,6-diepoxide, 3a,4,7,7a-tetrahydro- (dicyclopentadiene-2,3,5,6-diepoxide). This made possible a comparative identification of the compounds giving rise to peaks 15, 20 and 25 of figs. 3 or 5.

Oxidation of dicyclopentadiene witn inorganic oxidants, described in 2.4.4 and 2.4.5, was carried out in order to synthesize other products to be used as references. In these two reactions, however, the crude mixtures obtained are more complex, to such an extent that the isolation and structural characterization of the various components are still to be completed. An interesting observation based on the indications of the molecular weigths of the main components is that the oxidation products of m.w. 146,148, 150 and 162, 164 yield peaks in the same region of fig. 3 in which the presence of oxidized derivatives of dicyclopentadiene was hypothesized on the basis of the MS data. Significant analogies are to be observed in the mass spectra of the unknown compounds present in the neutral extract from the water sample and in the crude mixtures obtained in the two oxidations, al-

though as yet it has been possible to identify with certainty only 4,7-me
thano-1 H-indene-1-one, 3a,4,7, 7a-tetrahydro-(dicyclopentadiene-1-one).
The mass spectrum of this compound has molecular peak 146, base peak 66 (re
tro-Diels-Alder reaction); peaks 118 (M-28) and 117 (M-29) are quite intense and numerous other peaks of various intensities are found at m/z 39, 51,
65,77,81,91,103,131.

Other peaks of compounds at m.w. 146,148,162 and 164 are present in
the chromatogram of fig. 3, but none of these coincides exactly with those
of the products of the oxidation reactions of dicyclopentadiene. The mass
spectra too, though the m/z values of the various fragments are similar, show
intensity differences which prevent reliable identifications.   This fact
leads to the hypothesis that the mechanism by which the oxidized derivatives of dicyclopentadiene are formed is largely by microbial transformation
by various species of microorganisms. Preliminary in vitro experiments
with pure dicyclopentadiene have not as yet led to the identification of
strains capable of reproducing in the laboratory the processes taking place
in the soil, in conditions which involve the simultaneous presence of many
substrata and numerous species. This study is still under way, and findings
which may confirm or confute the above hypothesis will be the subject of a
forthcoming publication.

4. CONCLUSIONS

The presence in any given area of a large chemical plant always involves a considerable hazard for the phreatic aquifer, which may be reached by
considerable quantities of chemical compounds over periods of time which
may vary according to the permeability of the soil and the rainfall. The
substances dispersed into the environment may in their turn be modified by
chemical reactions or by transformation reactions by various types of micro
organisms in aerobic or anerobic conditions. This alters the organoleptic
properties of the water, which becomes unsuitable for domestic supply. The
risk is directly proportional to the persistence of the various chemical
compounds in the environment. In the case being studied certain pollutants
which are not easily degradable have been found downstream of the chemical
plant at distances of 10 km and even more, and the local authorities are
forced to purify the water on activated carbon. It is absolutely necessary
for any chemical plant to take all such measures as may be required to prevent spillage and leakage from sewers and underground tanks. The choice of
suitable methods for the elimination of wastes, whether gaseous, liquid or
solid, is fundamental if damage to a resource as precious as drinking water
is to be avoided.

ACKNOWLEDGEMENTS

We would like to thank Mr.Lucio Ogliani of our Department for his contribution in the preparation of capillary columns and in gas chromatographic analysis, and Mr. Alessio Rossini of Istituto di Chimica delle Macromo
lecole (CNR, National Research Council of Italy) for some of the GC/MS analyses.

BIBLIOGRAPHY

1 - S.R.Heller and G.W.A.Milne
    EPA/NIH Mass Spectral Data Base
    Vol. 1-4
    NSRDS-NBS 63
    U.S.Department of Commerce - NBS, Washington (1978)

2 - E.Stenhagen, S.Abrahamsson, F.W.McLafferty (editors)
    Registry of Mass Spectral Data
    Vol. 1-4
    J.Wiley and Sons, New York (1974)

3 - Eight Peak Index of Mass Spectra
    Vol. 1, Pt 1,2; 2 Pt 1,2; 3 Pt 1,2,3
    Third Edition 1983
    The Mass Spectrometry Data Centre - The Royal Society of Chemistry,
    Nottingham (1983)

4 - F.I.Dubrovskaya, M.Kh.Khachaturian, V.P.Levkin, E.V.Borodina, R.M.Ni
    kol'skaya, V.T.Smirnov
    Gig. Sanit. 1977, (4), 7-11 (Russ). C.A. 87, 28082y, 1977

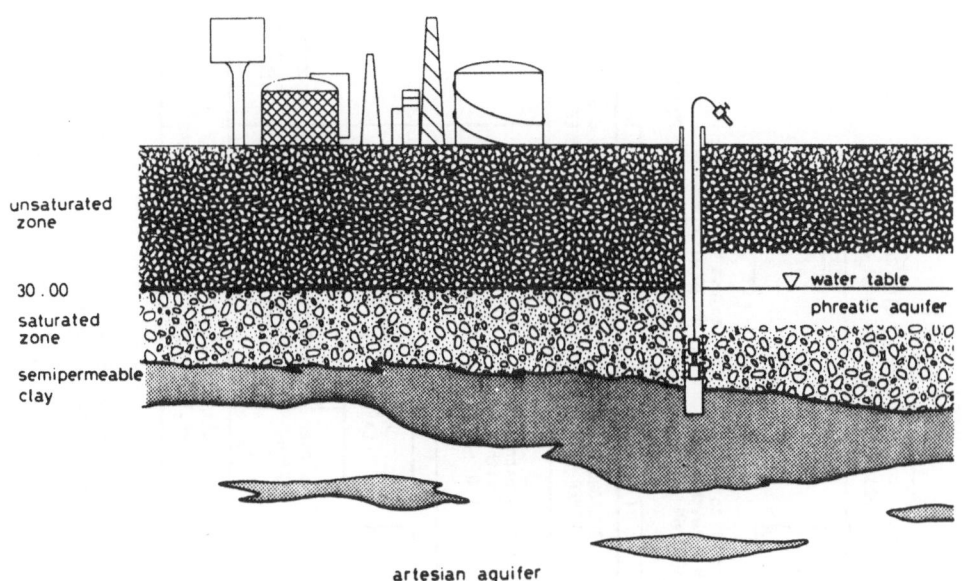

Fig. 1 - Phreatic aquifer and sampling conditions.

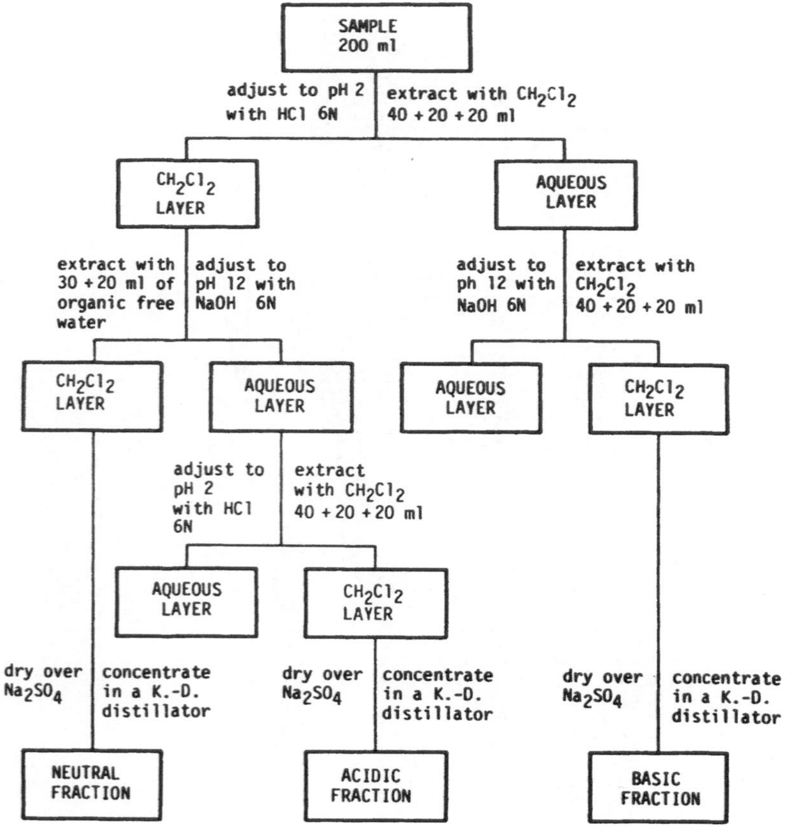

Table I

GC ANALYSIS CONDITIONS

Gas chromatograph Carlo Erba model 2150

Column : Pyrex glass capillary, persilanized, coated with SE - 30 silicone rubber. I.D. = 0,3 mm. L = 25m. Film thickness = 0,26 μm.

Injection : Splitless. T_{inj} = 250 °C. Sample injected : 0,5 μl of CH_2Cl_2 solution.

Carrier gas : Hydrogen (GC) at 2ml/min.

Oven : Room temperature at injection
to 40 °C ballistically after 45 s
40 °C isothermal for 2 minutes
4C ⟶ 250 °C at 3 °C/min.

Detector : Flame ionization (FID). T_{det} = 250 °C
H_2 Flow : 30 ml/min
Air flow : 300 ml/min

Recorder : Leeds and Northrup Italiana S.p.A. Model Speedomax XL 681 B
1 mV F.s.
Chart speed : 1 cm/min for fig. 3,11
2 cm/min for fig. 4,5,13.

Fig. 2 - Extraction of a 200 ml portion of the polluted water sample with methylene chloride at different pH values

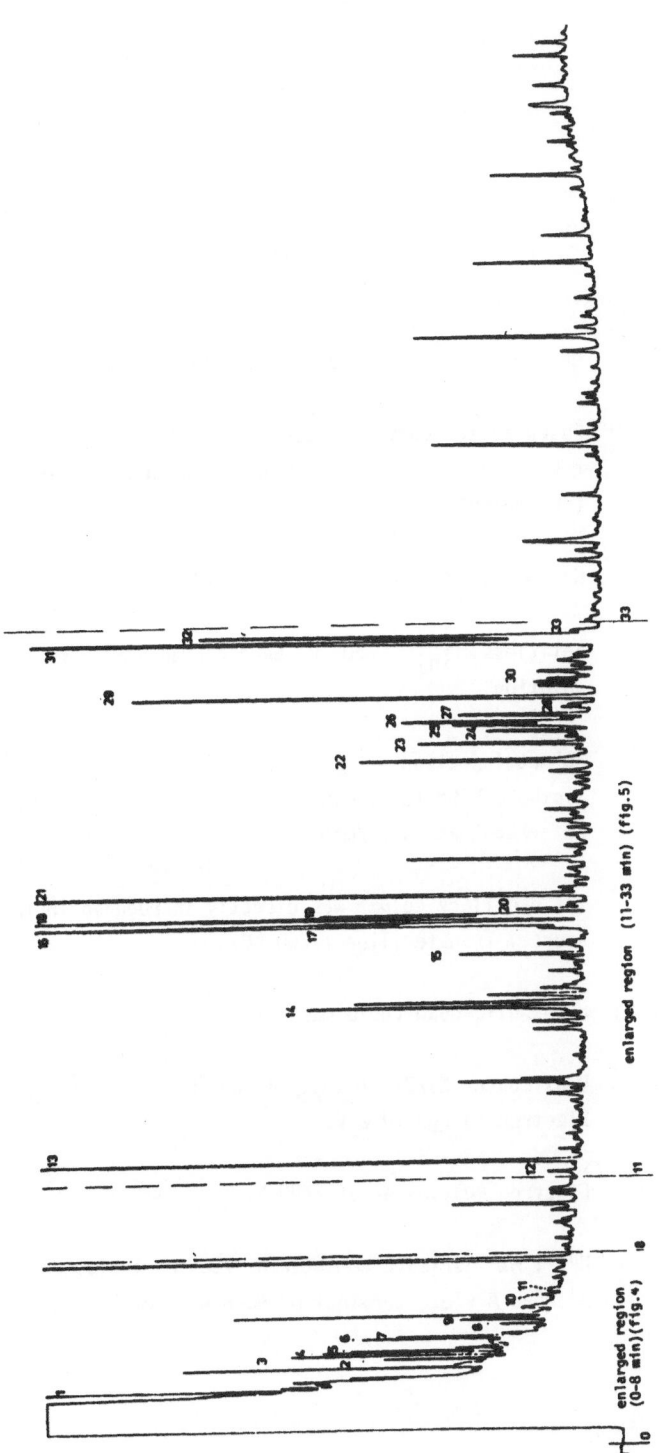

Fig.3 - Chromatogram of the neutral fraction

Table II

GC - MS analysis conditions

GC-MS-DS   Hewlett-Packard   Mod 5985 B

GC	:	HP Mod 5940 A modified for capillary column.
Column	:	Fused silica capillary column coated with SE 54 silicone rubber. I.D. 0.33 mm, L 50 m., film thickness 0.3 μm (Chrompack).
Carrier gas	:	Helium at 2ml/min.
Injection	:	Splitless. T_{inj} = 250 °C. Sample quantity: 1 μl of CH_2CL_2 solution.
Oven temperature	:	30° at injection Isothermal 30° for 2 min 30 ⟶ 260° at 4 °C/min.
Interface GC - MS	:	The capillary column was directly introduced in the ion source. Transfer line : 260 °C.
MS	:	Hyperbolic quadrupole mass filter.
Source	:	Dual source EI/CI. T_{source} = 200 °C Electron energy 70 e V.
Detector	:	Electron multiplier at 2200 V.
D.S.	:	HP 21 MX Series E computer. HP 7906 dual disc drive. HP 2648 A video terminal. HP 9876 A thermal printer.

Table III - List of compounds identified in the neutral fraction

Peak No.	Compounds or their molecular weights
1	Dioxane - 1,4 (88)
2	2 - Ethyl- 4 - methyldioxolane - 1,3 cis (116)
3	2 - Ethyl- 4 - methyldioxolane - 1,3 trans (116)
4	2,6 - Dimethyldioxane - 1,4 cis (116)
5	2,5 - Dimethyldioxane - 1,4 trans (116)
6	2,6 - Dimethyldioxane - 1,4 trans (116)
7	2,5 - Dimethyldioxane - 1,4 cis (116)
8	2-i - Propyl - 4 - methyldioxolane - 1,3 cis (130)
9	2-i - Propyl - 4 - methyldioxolane - 1,3 trans (130)
10	2-n - Propyl - 4 - methyldioxolane - 1,3 cis (130)
11	2-n - Propyl - 4 - methyldioxolane - 1,3 trans (130)
12	Dicyclopentadiene (exo isomer) (132)
13	Dicyclopentadiene (endo isomer) (132)
14	M.W. 148
15	Dicyclopentadiene 2,3 - oxide (148)
16	M.W. 148
17	Dicyclopentadien - 1 - one (146)
18	M.W. 148
19	M.W. 148
20	Dicyclopentadiene 5,6 - oxide (148)
21	M.W. 148
22	M.W. 162
23	M.W. 164
24	M.W. 164
25	Dicyclopentadiene 2,3 - 5,6 - dioxide (164)
26	M.W. 164
27	M.W. 164
28	M.W. 162
29	M.W. 164
30	M.W. 162
31	M.W. 164
32	M.W. 186
33	M.W. 164

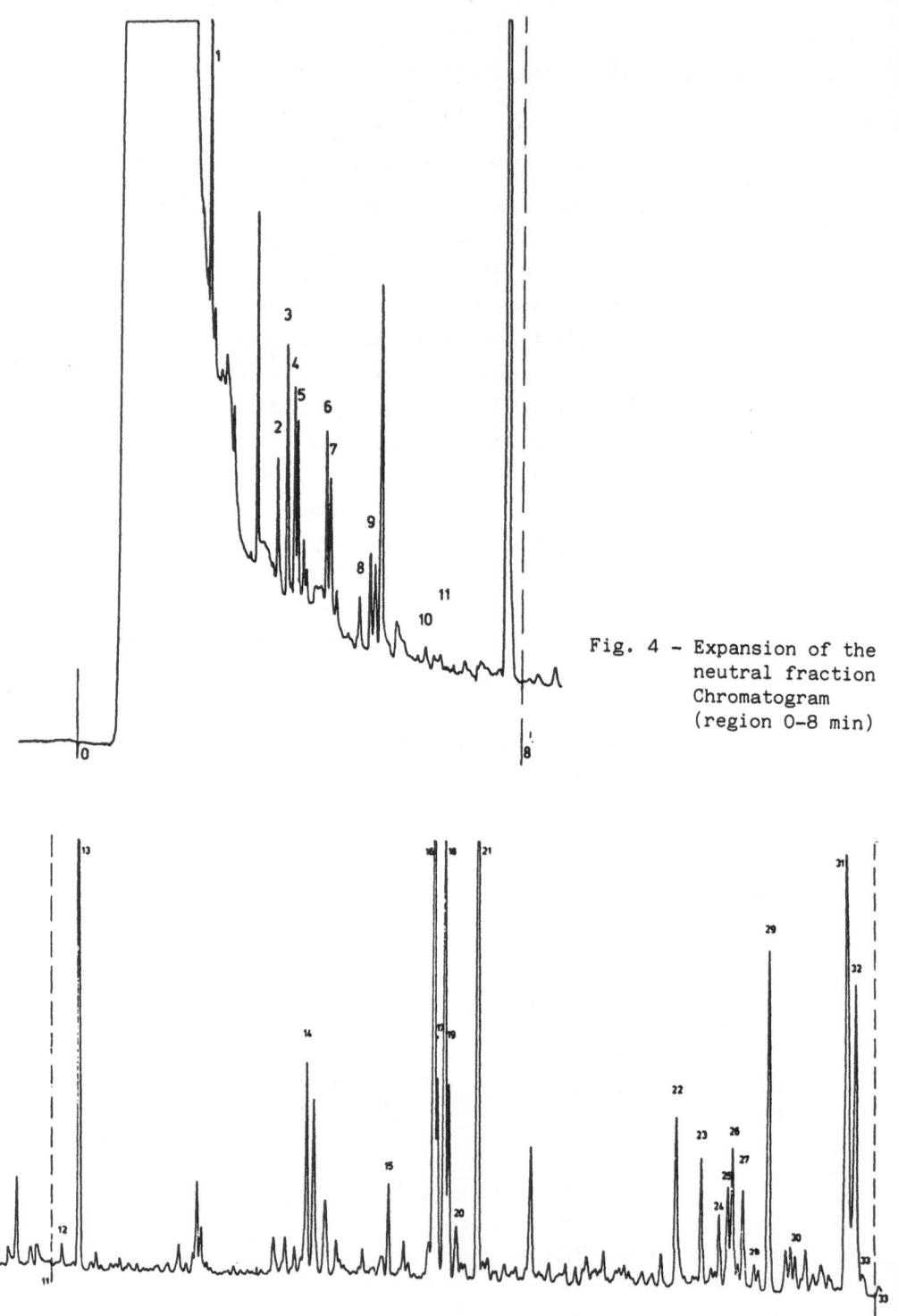

Fig. 4 – Expansion of the
neutral fraction
Chromatogram
(region 0-8 min)

Fig. 5 – Expansion of the neutral fraction chromatogram   (region 11-33 min).

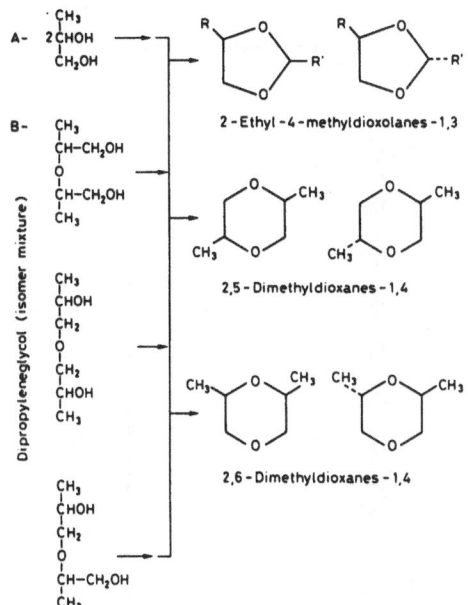

Fig. 6 – Syntheses of dimethyldioxanes – 1,4

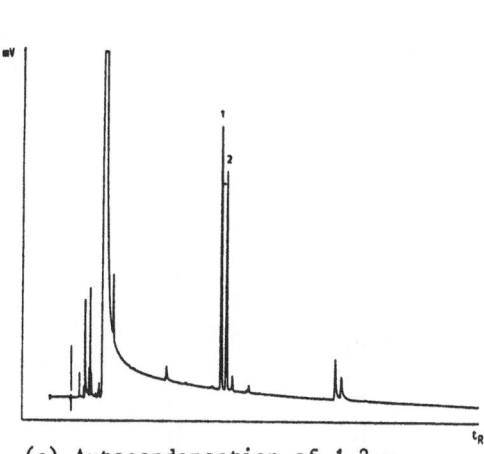

(a) Autocondensation of 1,2 –
    Propanediol

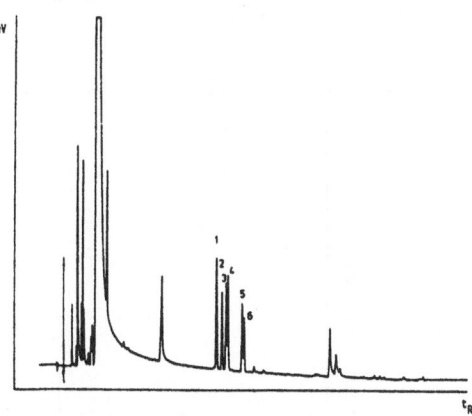

(b) Condensation of dipropylene
    glycol (Isomer Mixture)

Fig. 7 – Chromatograms of
Dimethyldioxanes – 1,4

Fig. 8 – Syntheses of 2,4 –
Diakyldioxolanes – 1,3

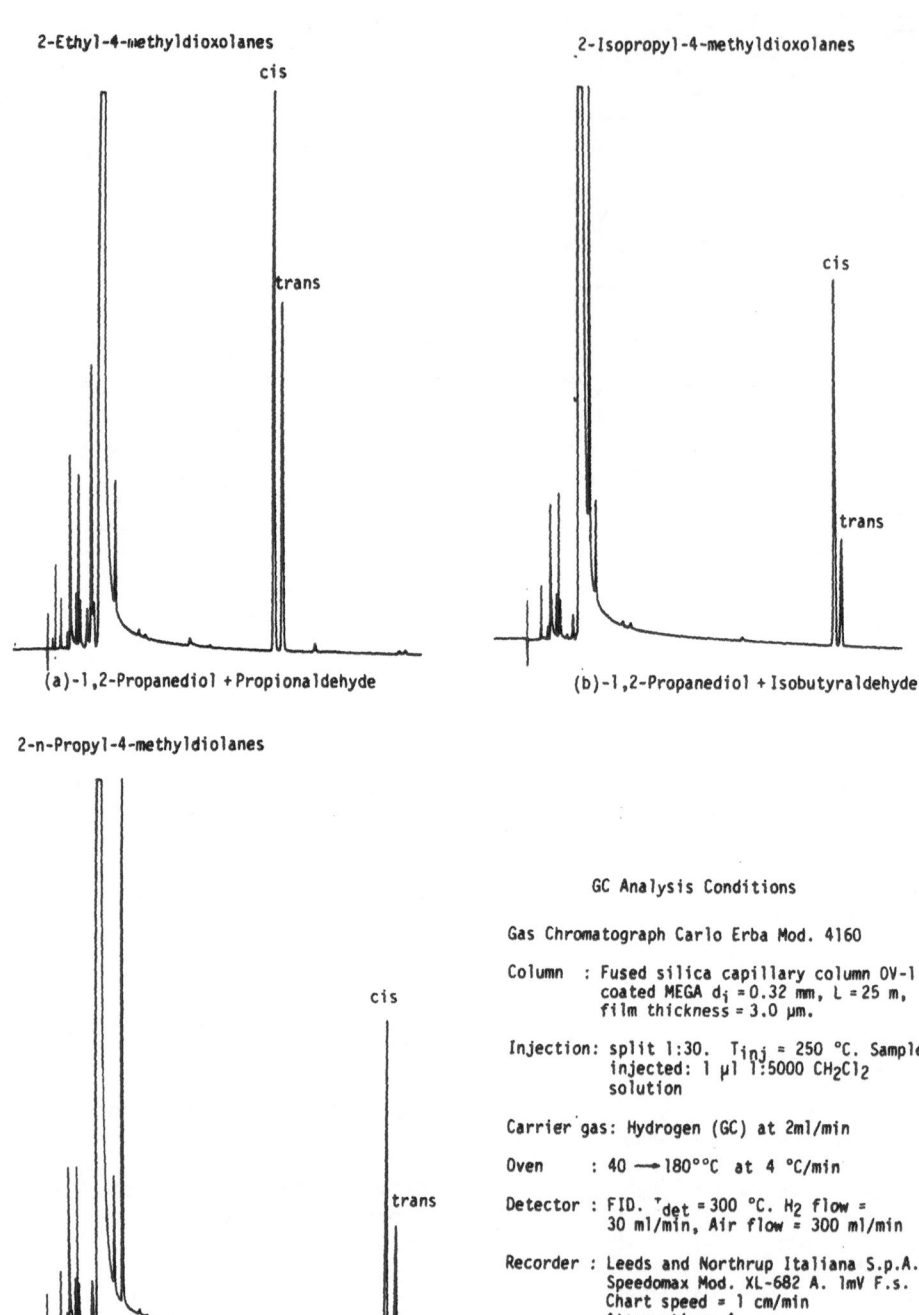

2-Ethyl-4-methyldioxolanes

cis

trans

(a)-1,2-Propanediol + Propionaldehyde

2-Isopropyl-4-methyldioxolanes

cis

trans

(b)-1,2-Propanediol + Isobutyraldehyde

2-n-Propyl-4-methyldiolanes

cis

trans

(c)-1,2-Propanediol + n-Butyraldehyde

GC Analysis Conditions

Gas Chromatograph Carlo Erba Mod. 4160

Column : Fused silica capillary column OV-1 coated MEGA $d_i = 0.32$ mm, L = 25 m, film thickness = 3.0 μm.

Injection: split 1:30. T_{inj} = 250 °C. Sample injected: 1 μl 1:5000 CH_2Cl_2 solution

Carrier gas: Hydrogen (GC) at 2ml/min

Oven : 40 ⟶ 180°°C at 4 °C/min

Detector : FID. T_{det} = 300 °C. H_2 flow = 30 ml/min, Air flow = 300 ml/min

Recorder : Leeds and Northrup Italiana S.p.A. Speedomax Mod. XL-682 A. 1mV F.s. Chart speed = 1 cm/min Attenuation: 4

Fig. 9 - Chromatograms of 2,4 - Dialkyldioxolanes - 1,3

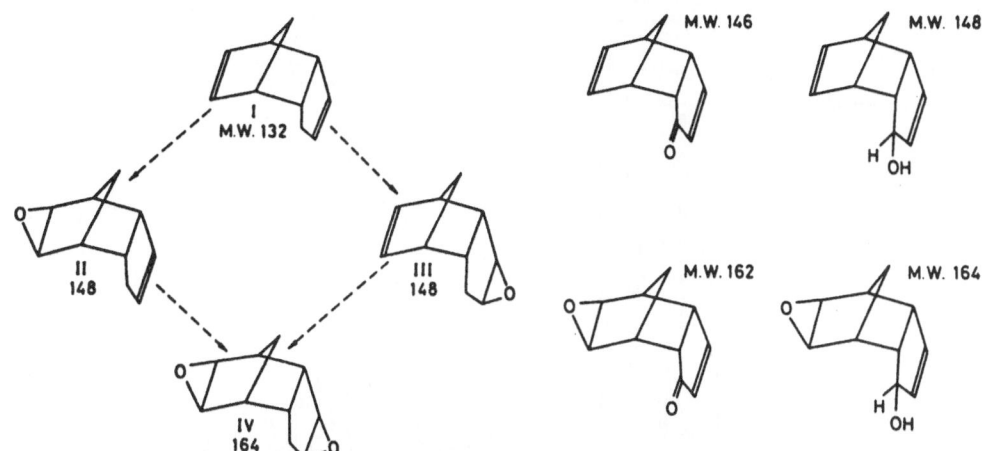

Fig. 10 - Oxidation of dicyclopentadiene
(endo isomer) with
perbenzoic acid.

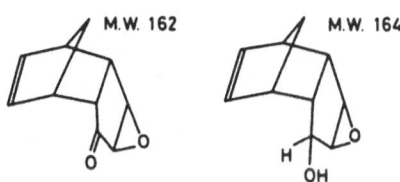

Fig. 12 - Other possible products of the
dicyclopentadiene oxidation
with molecular weights 146, 148,
162, 164.

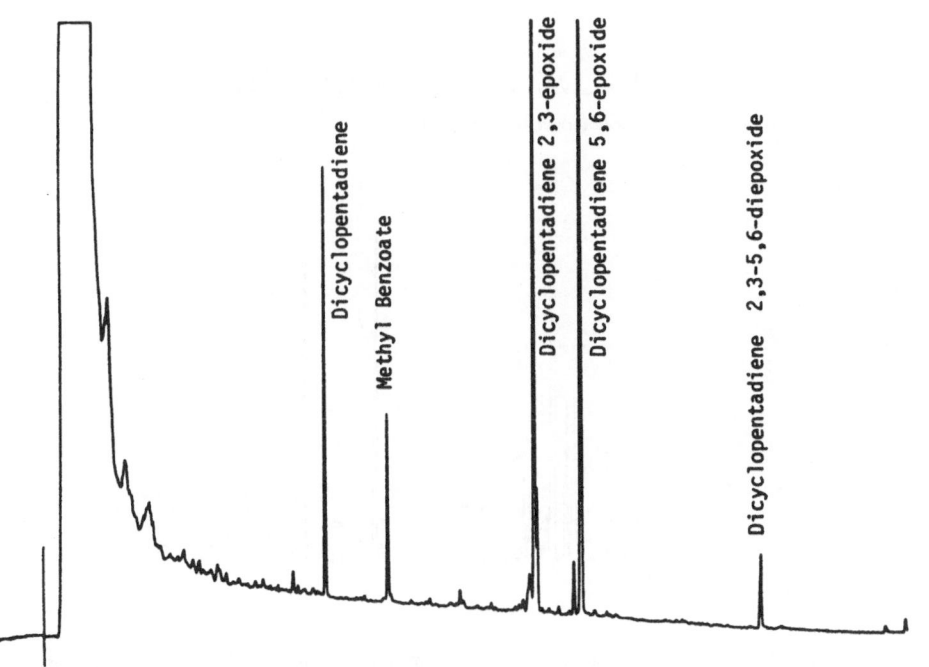

Fig. 11 - Chromatogram of dicyclopentadiene oxides.

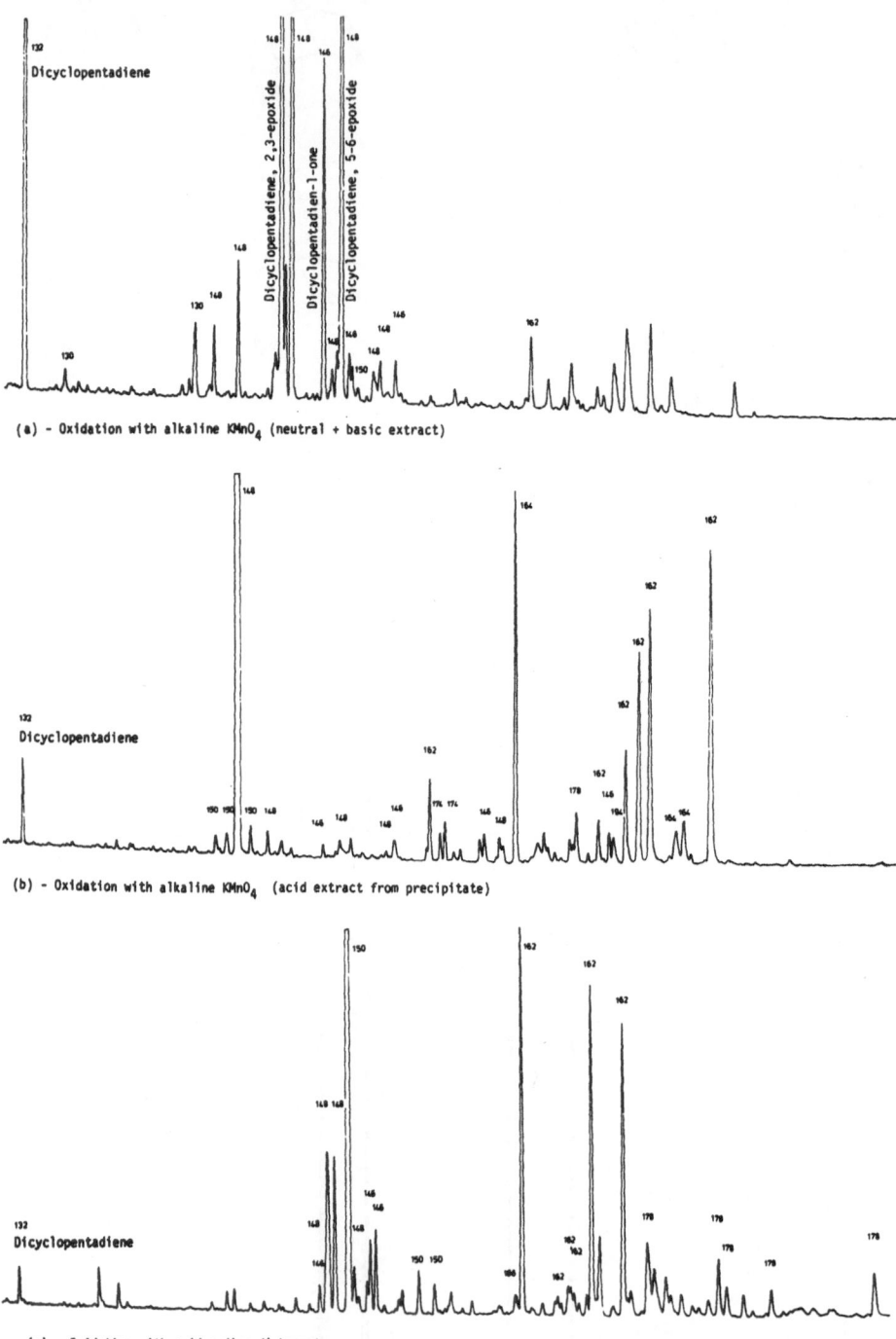

Fig. 13 – Chromatograms of dicyclopentadiene oxidation products
with permanganate and dichromate

ANALYSE DE MICROPOLLUANTS ORGANIQUES DES EAUX PAR COUPLAGE CG/SM

COLLECTION DE SPECTRES DE MASSE DE REFERENCE

Analysis of organic micropollutants in water by GC/MS
Mass spectral reference data

R. MASSOT et J.C. LECLERC
CEA-IRDI-SEA - Centre d'Etudes Nucléaires - 85X - 38041 Grenoble, France

Résumé

Le couplage CG/SM est l'une des méthodes les plus puissantes pour l'-
analyse des micropolluants organiques des eaux. L'identification s'effectue par comparaison des spectres obtenus à ceux d'une collection de
référence. Les collections commerciales qui rassemblent plusieurs dizaines de milliers de spectres et qui comportent de très nombreux produits hors du domaine d'intérêt sont lourdes d'emploi et d'une effica-
cité plutôt médiocre.Les problèmes d'identification sont considérablement simplifiés si l'on dispose d'une spectrothèque spécifique. L'un
des objectifs des comités COST successifs 64B, 64B bis et 641 a été
précisément de constituer une collection de spectres de masse de compo
sés organiques susceptibles d'être trouvés dans des eaux de rivière.
Les 2000 spectres qui la constituent actuellement viennent principalement des laboratoires participant à l'action COST.Après vérification
et corrections éventuelles ils sont mis en forme de tableaux normalisés et distribués par fascicules de 100 unités aux nations participan-
tes. Nous avons également réalisé un Index rassemblant les principales
caractéristiques des spectres. Cet Index à plusieurs entrées sert de
table de matières à la collection complète;mais, dans bien des cas, il
peut permettre, à lui seul, des identifications de polluants organiques.

Summary

GC/MS is one of the most widely used analytical techniques for identifying organic micropollutants in water. Identification is obtained by
comparing the produced spectra with those of reference collections.
Commercial collections which contain several tens of thousands of spec
tra and which include many products outside our area of interest are
quite difficult to use effectively. Identifying is greatly simplified
if a specific collection is used. One of the objectives of the COST
projects 64B,64B bis and 641 was to compile a collection of mass spectra of organic compounds likely to be found in water. The 2000 spectra
of the COST collection come mainly from laboratories working for the
COST projects. The mass spectra are collected, verified and corrected
if possible, and then converted to standardised tables, combined in
sections of 100 units and distributed to participating countries. We
have also compiled an Index which contains the main characteristics of
the mass spectral data. This Index, not only provides a thorough list
of the contents of the spectra collection, but it is also a means, in
many cases, of identifying the organic pollutants.

1 - INTRODUCTION

L'une des méthodes de choix pour l'étude des micropolluants organiques des eaux est le couplage chromatographie/spectrométrie de masse. L'identification des produits s'effectue par comparaison directe des spectres obtenus à ceux d'une collection de référence.

Pourquoi une collection COST?

On sait qu'il existe déjà des collections de spectres de masse de référence, commercialisées pour la plupart, certaines contenant plusieurs dizaines de milliers de spectres. Mais ces collections sont très générales: elles comportent de nombreux produits que l'on ne rencontre jamais ou de façon non significative dans les eaux de rivière (composés synthétisés dans des laboratoires de recherche par exemple) et, a contrario, elles peuvent présenter de graves lacunes dans le domaine concerné.

Il n'en est pas de même avec une spectrothèque spécifique qui contient en principe les composés cherchés et non les autres. Les critères pour l'identification peuvent alors être considérablement affinés et conduire, dans de nombreux cas, à la sortie d'une courte liste de produits possibles contenant bien le composé cherché.

L'un des objectifs des projets COST 64 successifs a été précisément l'élaboration d'une collection de spectres de masse de composés susceptibles d'être rencontrés dans les eaux de rivière.

2 - REALISATION DE LA COLLECTION

La réalisation de cette collection qui rassemble actuellement 2000 spectres de masse de référence comporte plusieurs phases consécutives :

Collecte des spectres disponibles dans les laboratoires participant à l'action COST (36 laboratoires de 12 pays différents) mais aussi dans la littérature technique et scientifique ou encore par échanges avec la collection générale française DARC-PLURIDATA.

Contrôle des spectres pour détecter certaines erreurs notamment par comparaison avec d'autres spectres du même composé.

Mise sous forme de tableaux normalisés comprenant quatre zones (Cf fig. 1) :
- la liste des pics du spectre et des intensités relatives,
- la liste des dix plus grands pics du spectre classés par ordre de taille,
- la formule développée,
- le nom du composé, sa formule brute et sa masse moléculaire ainsi que quelques conditions opératoires.

Diffusion des spectres par le canal du Secrétariat de l'action COST. Pour plus de commodité d'emploi ces spectres sont réunis par fascicules de 100 unités.

Réalisation d'un Index. Par ses entrées multiples ce document permet un accès facile à la collection complète; par les données qu'il comporte il peut également suffire à l'identification d'un composé par son spectre de masse. Il rassemble en effet, sous forme résumée, les principales données des spectres de masse (Cf fig. 2) :
- nom du composé,
- masse moléculaire
- formule brute,
- spectre de masse réduit à ses dix plus grands pics classés par ordre de taille décroissante et au pic moléculaire.

Il comporte trois entrées :
- un classement des spectres par masses moléculaires des composés. Par son utilisation on peut, connaissant la masse moléculaire et le spectre du produit inconnu, rechercher aisément les spectres analogues ou voisins et

aboutir, dans bien des cas, aux identifications.

- un classement par masses des pics les plus importants des spectres (Cf fig. 3). Si la masse du composé à identifier demeure inconnue l'identification sera tentée par comparaison directe des spectres de masse.

- un classement par formules brutes utile pour rechercher le spectre d'un composé donné.

Un premier Index portant sur 800 spectres a été réalisé en 1980. Un Index plus complet, concernant 2000 spectres est en cours de mise en forme.

3 - CONTENU DE LA COLLECTION

La collection COST 64 B bis comporte donc à ce jour 2000 spectres de masse de référence représentant 1817 composés différents et recouvrant dix familles chimiques (Cf tableau 4). Elle contient 160 des 188 polluants classés en priorité par l'E.P.A.. Les composés manquants seront ajoutés au fur et à mesure de leur obtention par les laboratoires concernés.

En principe la collection ne comporte qu'un spectre par composé. Cependant, dans quelques cas, il nous est apparu nécessaire de maintenir une certaine duplication, par exemple :

- lorsque les conditions expérimentales ne sont pas les mêmes (analyseur magnétique ou quadripolaire, énergie des électrons ionisants réglée à 70 eV ou à 24 eV, etc.),

- lorsque les spectres en notre possession sont assez différents et qu'il ne nous a pas été possible de trouver un critère d'élimination suffisamment sûr. C'est le cas, par exemple, de l'endrine (4 spectres) et du diazinon (3 spectres).

4 - CONCLUSION

Cette collection de spectres et l'Index qui la complète devrait faciliter les problèmes d'identification par couplage CG/SM des micropolluants organiques des eaux. Elle comporte certainement des lacunes et nous faisons appel aux laboratoires travaillant dans ce domaine pour tenter de les combler.

SPECTRE DE MASSE

BORDEREAU NO : 10701   REF.: FRO10005

M/E	I	M/E	I	M/E	I	M/E	I	M/E	I	M/E	I
35	25	105.5	46	174	98	319	142				
36	100	106	71	175	124	320	454				
38	35	110	60	176	395	321	66				
39	29	111	55	177	59	322	97				
50	100	122	100	184	24	324	9				
51	82	122.5	42	210	182						
61	28	123	153	211	52						
62	48	123.5	39	212	65						
63	53	124	64	233	31						
73	45	125	41	235	30						
74	126	126	20	245	35						
75	166	135	40	246	1000						
76	24	136	21	247	180						
84	21	137	35	248	698						
85	30	140	85	249	104						
86	50	141	84	250	115						
87	123	142	28	280	64						
87.5	41	149	52	281	89						
88	77	150	62	282	75						
92	27	160	40	283	87						
93	21	161	25	284	32						
97	26	163	31	285	29						
98	71	170	53	316	825						
99	102	172	40	317	113						
105	212	173	22	318	930						

FORMULE DEVELOPPEE : NO. 6839/5

DIX PLUS GRANDS PICS :

M/E	246	318	316	248	320
I	1000	930	825	698	454

M/E	176	105	210	247	75
I	395	212	182	180	166

NOM DU COMPOSE : 1,1-DICHLORO-2,2-DI-(4-CHLORO)PHENYL-ETHENE; D.D.E.

FORMULE MOLECULAIRE : C14H8CL4   MASSE MOLECULAIRE   : 316
TYPE DU SPECTROMETRE DE MASSE : JEOL JMS 07
MODE D'INTRODUCTION : COUPLAGE CG-SM   T. D'INTRODUCTION   :
TEMPERATURE DE SOURCE: 220°C   ENERGIE DES ELECTRONS: 70 EV
DATE : 13/3/73
LABORATOIRE: SEAPC-CENG-85X-38041 GRENOBLE CEDEX

/O 1/Q 2/P O/T M/

Fig. 1 - SPECTRE DE MASSE SOUS FORME DE TABLEAU NORMALISE

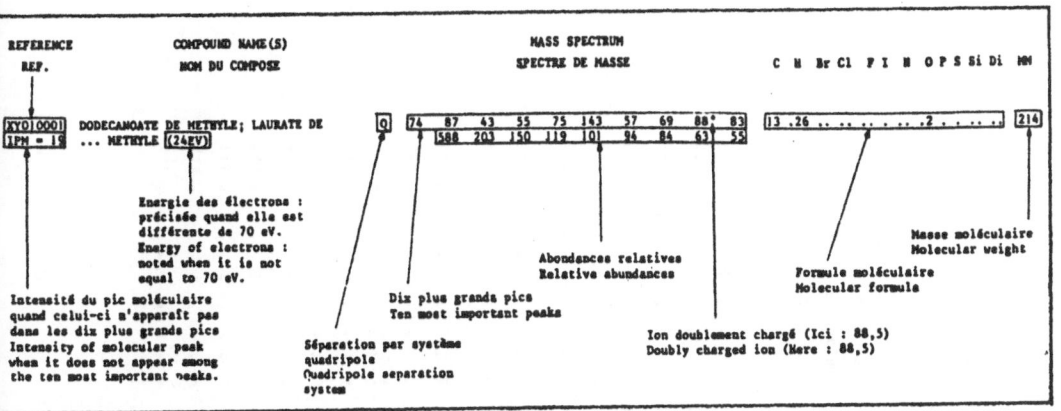

55

REF.	NOM DU COMPOSE	SPECTRE DE MASSE	C H Br Cl F I N O P S Si Di	MM
FR020214	DIMETHYL-BENZO(C)ACRIDINE	257 256 255 258 254 39 243 241 63 242 / 202 202 186 143 101 83 83 77 71	19 .15 .. ..1 1	257
ML010227	1-METHYL-1-METHOXY-3-(4-BROMO)PHENYL-UREE	61 46 258 91 260 172 170 63 60 90 / 216 141 136 130 100 98 93 89 73	.9 .11 .1 .. ..2 .2 ..	258
ML010026 IPM = 38	DEMETON	88 60 89 126 61 115 114 170 93 143 / 382 214 137 116 116 108 102 94 91	.8 .19 .. ..3 1 2 ..	258
ML010228 IPM = 20	0,0-DIETHYL-S-(2-ETHYLTHIOETHYL)-PHOSPHO- ... -ROTHIOATE	88 89 60 61 171 97 59 115 143 90 / 568 352 251 171 103 100 100 88 76	.8 .19 .. ..3 1 2 ..	258
ML010230 IPM = 49	0,0-DIETHYL-0-(2-ETHYLTHIOETHYL)-PHOSPHO- ... -ROTHIOATE	88 101 73 60 81 93 109 111 89 61 / 535 447 378 322 285 281 272 188 182	.8 .19 .. ..3 1 2 ..	258
FR020112 IPM = 14	1-METHYL-4-BUTYLCARBOXYLATE-5-CHLORO- ... -METHANOYL-1-CYCLOHEXENE	93 194 92 138 57 94 167 79 41 91 / 284 284 211 174 147 128 128 128 119	13 .19 ..1 .. .. .3 ..	258
FR050020 IPM = 40	1,2,2-TRIMETHYL-1-BENZOYL-3-ETHANOYL- ... -CYCLOPENTANE	105 43 147 77 160 153 41 109 152 136 / 900 400 320 200 180 160 150 110 100	17 .22 .. ..2	258
FR050021	1,2,2-TRIMETHYL-1-ETHANOYL-3-BENZOYL- ... -CYCLOPENTANE	105 77 43 133 106 215 41 198 216 258 / 220 130 110 80 80 70 40 40 40	17 .22 .. ..2	258
FR010257	5-BUTYL-6-HEXYL-INDANE	145 258 187 146 129 128 215 131 259 143 / 294 189 178 70 58 56 51 51 50	19 .30	258
FR010256	2-BUTYL-5-HEXYL-INDANE	187 258 117 43 131 188 129 41 115 259 / 471 190 163 150 144 125 93 89 86	19 .30	258
GB020099	5-METHYL-TETRAHYDROPYRANO(4,3-F)- ... -1,1,2,3,3-PENTAMETHYLINDANE	243 213 258 246 171 41 143 185 214 43 / 283 217 190 79 61 55 53 52 51	18 .26 .. .. .. .1 ..	258
ML010232	1-NITRO-2,3,5,6-TETRACHLORO-BENZENE	203 201 215 261 213 108 259 205 178 143 / 877 866 726 713 699 583 505 487 475	.6 ..1 .. .4 .. .. .1 .2 ..	259
FR020081	0,0-DIETHYL-S-(2-THIABUTYL)-PHOSPHORO- ... -DITHIOATE; THIMET	75 121 47 97 65 93 45 260 231 125 / 269 181 172 142 138 100 66 65 61	.7 .17 .. ..2 1 3 ..	260
ML010234	0,0-DIETHYL-S-(2-THIABUTYL)-PHOSPHORO- ... -DITHIOATE; THIMET	75 121 260 97 47 65 93 69 231 125 / 318 184 184 159 136 133 88 72 66	.7 .17 .. ..2 1 3 ..	260
FR010365 IPM = 43	5-HYDROXY-URIDINE	128 133 73 57 61 129 43 71 45 85 / 625 531 438 250 219 219 200 188 138	.9 .12 .. ..2 .7 ..	260
ML010103	2-(3,4-DICHLORO)PHENYL-4-METHYL-1,2,4- ... -OXADIAZOLIDINE-3,5-DIONE	159 124 -260 161 262 126 88 163 97 145 / 967 713 713 497 405 155 155 152 150	.9 ..6 .. .2 .. .2 .3	260
ML010233 IPM = 39	3-S-BUTYL-5-BROMO-6-METHYL-URACILE	205 207 40 39 70 206 190 162 188 164 / 939 421 202 194 170 165 152 147 141	.9 .13 .1 .. .. .2	260
DK020188 IPM = 110	TETRACHLORO-STANNANE	225 227 223 229 224 221 155 120 226 153 / 760 650 370 300 260 250 210 180 170	.. .. .. .4 1	260
FR010413	DIPHENYL-4-AMINOPHENYL-AMINE	260 261 259 130 77 167 182 78 166 51 / 208 75 75 46 44 42 42 36 35	18 .16 .. ..2	260
FR010202 IPM = 1	2,3-DIETHANOYL-TARTRATE DE DIMETHYLE	43 44 161 132 90 203 174 42 45 143 / 201 146 101 89 84 84 46 34 30	10 .14 .. ..8	262
ML010104 IPM = 0	0,0-DIMETHYL-S-(2-ETHYLSULFONYLETHYL)- ... -PHOSPHOROTHIOATE	169 109 125 168 79 110 142 170 171 59 / 631 444 122 115 113 108 69 62 59	.6 .15 .. ..5 1 2 ..	262
FR110008	2-(4-CHLOROPHENYL)-5-CHLORO-BENZOFURANE	262 264 199 40 263 266 265 163 131 81 / 640 230 230 180 130 120 120 100 100	14 ..8 .. .2 .. .. ..1 ..	262
FR110009	3-(4-CHLOROPHENYL)-7-CHLORO-BENZOFURANE	262 264 199 163 263 164 201 99 81° 266 / 680 380 230 180 160 130 130 120	14 ..8 .. .2 .. .. ..1 ..	262
FR110007	3-PHENYL-5,7-DICHLORO-BENZOFURANE	262 264 199 163 263 164 99 81° 201 256 / 660 400 250 170 150 150 150 140 125	14 ..8 .. .2 .. .. ..1 ..	262
FR020015	0,0-DIMETHYL-0-(4-NITROPHENYL)-PHOSPHO- ... -ROTHIOATE; METHYLPARATHION	109 125 263 79 63 47 93 264 62 64 / 840 683 262 183 165 155 79 75 73	.8 .10 .. ..1 .5 1 1 ..	263

Fig. 3 − CLASSEMENT PAR LES TROIS PICS LES PLUS IMPORTANTS

NOM DU COMPOSE — SPECTRE DE MASSE — C H O B C F I N O P S S D / R L I I

1er pic du spectre — m/e

Intensités relatives en °/..

2ème pic

3ème pic

NOM DU COMPOSE	1	2	3	4	5	6	7	8	9	10	C	H	O	B	C	F	I	N	O	P	S	S	D		
LINDANE (GAMMA 1.2.3.4.5.6 HEXACHLORO	111	51	109	73	30	75	77	239	38	39	6		6		6								111		
... ... CYCLOHEXANE)																									
N=E .013 M=288	967	936	697	669	487	474	467	421	395																
O.O DIETHYL S-((2 CYANOPROPYL) AMIDO-																									
... ... CARBOXYMETHYL)																									
... ... PHOSPHOROTHIOATE	111	138	81	109	82	93	97	68	65	170	10	19							2	4	1	1		111	
N=NL 1.121 M=296	993	714	664	567	483	470	435	416	398																
4.4 PRIME DICHLORO DIAZOBENZENE	111	139	113	250	75	252	141	50	112	76	12	8			2			2						111	
N=E 1.296 M=250	557	586	271	257	171	100	71	71	57																
4 CHLOROPHENYL 4 CHLOROBENZENE																									
... ... SULFONATE	111	175	75	99	113	177	73	63	302	127	12	8			2			3	1			111		
N=NL 1.262 M=302	750	528	508	315	284	235	233	224	186																
CHLORFENSON	111	175	75	99	113	177	73	63	302	127	12	8			2			3	1			111		
N=NL 1.017 M=302	750	528	508	315	284	235	283	224	186																
2-(3 BUTANONE) 1 CYCLOHEXANONE	43	111	98	55	81	68	58	89	97	70	10	16							2				111		
N=E 5.006 M=148	560	440	320	250	250	240	170	170	160																
1 CHLORO 3 NITRO BENZENE	75	111	157	110	113	99	159	74	76	50	6	4			1			1	2				111		
N=E 1.158 M=157	874	792	438	271	263	229	221	438	125																
1 CHLORO 4 NITRO BENZENE	75	111	157	127	99	113	74	28	50	30	6	4			1			1	2				111		
N=E 1.290 M=157	990	545	364	318	318	259	250	223	177																
4.4'-DICHLORO-BENZOPHENONE	139	111	75	141	250	113	252	50	76	140	13	8			2			1					111		
N=DK 2.185 M=250	480	340	330	190	160	130	100	90	80																
4-CHLOROPHENYL-2.4.5-TRICHLOROPHENYL-																									
... ... -SULFONE	159	111	75	227	229	356	161	187	354	113	14	6			4			2	1			111		
N=DK 2.220 M=354	900	600	520	510	420	390	330	330	300																
2.4.4'.5 TETRACHLORO DIPHENYL SULFONE	159	111	227	229	75	356	161	187	354	113	12	6			4			2	1			111		
N=NL 1.283 M=344	867	988	577	560	443	435	483	323	291																
ACIDE BENZENESULFONIQUE. 4-CHLORO-4-																									
... ... -CHLOROPHENYL ESTER	175	111	177	113	75	302	99	304	127	73	14	8			2			3	1			111		
N=DK 2.203 M=302	930	360	310	360	280	280	240	190	120																
1 BROMO 3 CHLORO BENZENE	192	111	190	75	113	194	50	74	76	38	6	4			1		1						111		
N=EU 1.037 M=190	786	776	410	251	241	230	156	88	69																
1 CHLORO 4 IODO BENZENE	238	111	240	75	113	50	74	76	239	119	6	4			1		1						111		
N=EU 1.042 M=238	581	321	291	187	166	110	70	67	64																
1 CHLORO 2 IODO BENZENE	238	111	240	75	113	50	74	127	239	76	6	4			1		1						111		
N=EU 1.040 M=238	486	349	263	154	131	75	71	67	64																
1 CHLORO 3 IODO BENZENE	238	111	240	75	113	50	74	76	239	127	6	4			1		1						111		
N=EU 1.041 M=238	652	321	286	210	152	93	69	67	59																
DI 1-(2 CYCLOHEXANONE) METHANE	98	189	111	208	67	41	55	190	97	112	13	20							2				111		
N=E 5.012 M=208	310	240	220	180	170	160	140	140	140																
2.4'-DICHLORO-BENZOPHENONE	139	28	111	141	75	31	250	113	252	32	13	8			2			1					111		
N=DK 2.186 M=250	520	390	340	290	200	180	130	110	100																
CHLORO BENZYLATE	139	251	111	141	254	75	252	113	250	44	16	14			2			3					111		
N=E 2.080 M=324	507	394	328	322	156	136	120	93	79																
1.4 DICHLORO BENZENE	146	148	111	75	74	50	113	73	150	147	6	4			2								111		
N=EU 1.020 M=146	660	549	222	141	124	112	165	103	68																
1.3 DICHLORO BENZENE	146	148	111	75	74	113	50	180	73	147	6	4			2								111		
N=EU 1.019 M=146	660	365	218	124	118	116	104	95	68																

- 112 -

Fig. 4 - REPARTITION EN FAMILLES CHIMIQUES DES COMPOSES DE LA COLLECTION

K

1 - COMPOSES ACYCLIQUES

    1.1 Saturés - Alcanes.............................306

    1.2 Insaturés (105)
        Alcènes.....................................90
        Alcadiènes..................................13
        Alcahexaènes.................................2

    1.3. Dérivés d'acides minéraux et organiques (non carboxyliques) (170)
        Aminosulfonique.............................6
        Carbamique.................................24
        Carbonique..................................1
        Dithiocarbamique............................1
        Isothiocyanique.............................4
        Nitrique....................................1
        Phosphonique................................4
        Phosphonochloridique........................1
        Phosphonodithioïque.........................1
        Phosphonothioïque...........................3
        Phosphoramidique............................1
        Phosphoramidothioïque.......................2
        Phosphorique...............................25
        Phosphorodithioïque........................32
        Phosphorothioïque..........................39
        Sulfinique..................................1
        Sulfonique.................................16
        Sulfurique..................................1
        Thiocarbamique..............................4
        Thiophosphinique............................3

    1.4. - Dérivés de l'urée (45)
        Urée.......................................43
        Thiourée....................................2

2 - COMPOSES CYCLIQUES

    2.1 - Non aromatiques et non hétérocyliques (167)
        Cyclanes...................................69
        Cyclènes...................................13
        cycladiènes.................................7
        Bicyclanes.................................14
        Bicyclènes.................................13
        Bicycladiènes...............................3
        Bicyclatriènes..............................1
        Tricyclanes................................16
        Tricyclènes.................................5
        Tricycladiènes..............................3
        Tricyclatriènes.............................1
        Tetracyclanes..............................11
        Tetracyclènes...............................1
        Tetracycladiènes............................4
        Pentacyclanes...............................1
        Pentacyclènes...............................2
        Hexacyclanes................................3

    2.2 - Composés aromatiques (653)
        Non substitués............................199
        Substitués................................454

    2.3. - Composés hétérocycliques (493)
        Azotés....................................288
        Azotés et oxygénés.........................71
        Azotés et soufrés..........................18
        Oxygénés..................................104
        Oxygénés et soufrés.........................5
        Soufrés 7

    2.4. - Stéroïdes.................................33

3 - DIVERS

    3.1 - Composés non carbonés à caractère organique 4
    3.2 - Composés organo-métalliques (25)
    Eléments métalliques :
        Be..1
        Co..1
        Cu..3
        Hg..3
        Mn..1
        Ni..1
        Pb..7
        Sn..7
        Zn..1

IMPROVEMENT IN ESTIMATION OF POLYAROMATIC HYDROCARBONS IN SEAWATER, MARINE SEDIMENTS AND ORGANISMS BY SPECTROFLUOROMETRY BY USING STANDARD ADDITIONS METHOD

M. PICER

Center for Marine Research Zagreb, "Rudjer Bošković" Institute,
Zagreb, Yugoslavia

Summary

For determination of petroleum hydrocarbons (sort of polyaromatic hydrocarbons) in seawater samples, lipophilic material is extracted by using n-pentane, n-hexane or methanol. The extract is afterwards concentrated and cleaned by means of an alumina column. Eluation of the retained fluorescent material is performed by using n-pentane and benzene mixture or methylene chloride. After the fluorescence measurement of the cleaned extracts has been made, known amounts of the standard Kuwait crude oil solution is added in order to evaluate and correct accordingly the observed quenching. By using such a screening method for evaluation of seawater, marine sediments and organisms pollution by polyaromatic hydrocarbons of the Adriatic sea, significantly different correction factors for various types of investigated samples are obtained.

1. INTRODUCTION

In the marine environment monitoring of polyaromatic hydrocarbons or petroleum hydrocarbons taken in larger sense is very often performed by using complicated methods requiring expensive and complex equipment (1). In some cases simpler, though less precise and specific technique using unexpensive equipment to monitor a large number of seawater samples might be preferable. In such cases spectrofluorimetric technique is applied (2).

The aim of this paper is to compare the results of distribution of concentrations of petroleum hydrocarbons polyaromatic fraction in water, sediment and various organisms by using simple and modified spectrofluorimetric methods. Modification of spectrofluorimetric methods consist in using standard addition and correction quenching effect.

2. METHODOLOGY

The scheme of methods for estimation polyaromatic hydrocarbons pollution of seawater samples is presented in Fig. 1a (3). As it can be seen it is possible to obtain three results from one sample. Standard addition is performed by using various dilutions of the standard Kuwait crude oil solution to evaluate corect and accordingly the observed quenching. The obtained values of quenching are persented as the correction factor:

$$\text{Correction factor} = \frac{F_{S(\text{before addition})}}{F_{S(\text{after addition})}} = \frac{F_S}{(F_{a+S}) \text{ measured} - F_a}$$

F_s - is the fluorescence of the standard solution of crude oil,
F_a - is the fluorescence of the sample after alumina cleaning,
$(F_{a+S})_{measured}$ - is the fluorescence of the sample after the addition of the standard solution of crude oil.

For sediment and biota samples two methods of extraction and cleaning are used. Method 1: Lipophilic material is extracted from the sample by using n-hexane, cleaning of extract is performed by using deactivated neutral alumina column. Eluation of petroleum hydrocarbons from the column is obtained by using the mixture of n-pentane and benzene (4). Method 2: Extraction and saponification of the lipophilic material from the sample is performed by using the methanol - water and KOH mixture. The extract is cleaned by using acid alumina column. Eluation of polyaromatic hydrocarbons is obtained by using pentane - methylene chloride mixture (Fraction I) and methylene chloride (Fraction II)(5).

3. RESULTS AND DISCUSSION

From the values of petroleum hydrocarbons concentrations in water samples (Fig. 1b) it can be seen that there are no significant differences between the three methods of analysis when concentraitons of petroleum hydrocarbons are in low ranges (1-5 ppb). At higher ranges the differences are significant, especially when comparing the results of the standard additions method with those of the simple extraction method.

Fig. 2 presents arithmetic means and ranges of fluorescence of sediment and various organisms extracts obtained by using neutral alumina column cleaning procedure (Method 1). As it can be seen, correction for quenching is significant especially for some surface sediment samples collected with correr from heavily polluted area of the Rijeka Bay. It is interesting that correction for quenching effect of samples extracted and cleaned by using Method 2 (Fig. 3) is much lower in comparison with corrections obtained by using Method 1 presented in Fig. 2. It has to be pointed out that the samples extracted and cleaned by using Method 2 were collected from relatively clean open waters of the North Adriatic far away from the local sources of pollution by polyaromatic hydrocarbons and other organics. It might be that such relatively low quenching effect is not only the effect of using Method 2. Another reason could be that such relatively "clean" samples do not contain so much quenching material as the samples heavily polluted not only by petroleum hydrocarbons but also by other organic materials. We shall treat these differences in our future investigations.

4. ACKNOWLEDGEMENT

The authors express their gratitude to the Self-managing Community of Interest for the Scientific Research of S.R. Croatia for the financial support.

REFERENCES

(1)   WHITTLE, K.J., HARDY, R., MACKIE, P.R., and McGILL, A.S. (1982). A Quantitative Assessment of the Sources and Fate of Petroleum Compounds in the Marine Environment. Phil. Trans. R.Soc. Lond. B 297,193.
(2)   PICER, M. (1977). Fate and Abiotic Effects of Polluting Petroleum in the Marine Environment. A review Pomorski zbornik 15, 437 (in croatian).

(3) PICER, M. (1984). Dissolved, Dispersed Petroleum Hydrocarbons in the
     Waters of the Krka River Estuary and the Kornati Archipelago. VII[es]
     Journees Etud. Pollutions, Luzern, C.I.E.S.M. (accepted for publica-
     tion).

(4) PICER, M., and HOCENSKI, V. (1982). Improvement in the Estimation of
     Petroleum Hydrocarbons in Marine Sediments and Organisms by Spectro-
     fluorometry by Using the Standard Additions Method. VI[es] Journees
     Etud. Pollutions, Cannes, C.I.E.S.M., 177.

(5) IOC (Intergovernmental Oceanographic Commission)(1982). The Determi-
     nation of Petroleum Hydrocarbons in Sediments. Manuals and Guides 11,
     UNESCO.

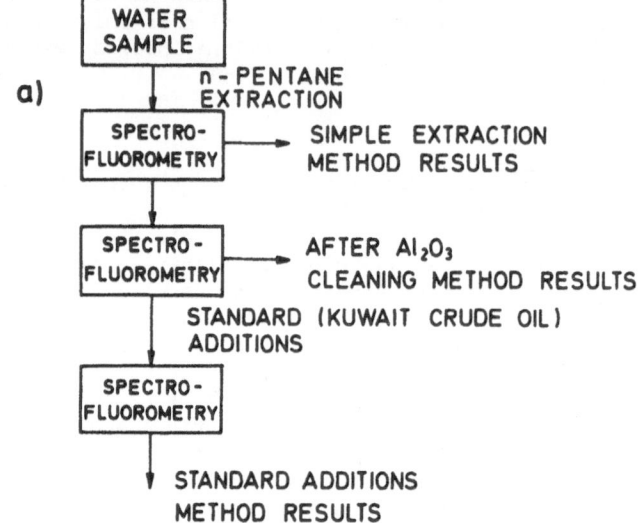

Fig. 1a. The scheme of petroleum hydrocarbons determination
         in seawater samples.

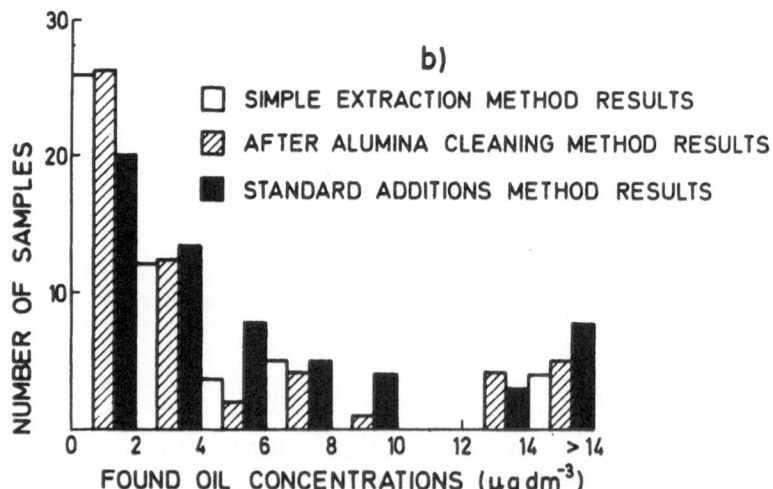

Fig. 1b. The comparison of oil concentrations found in seawater
         samples by using three methods of determination.

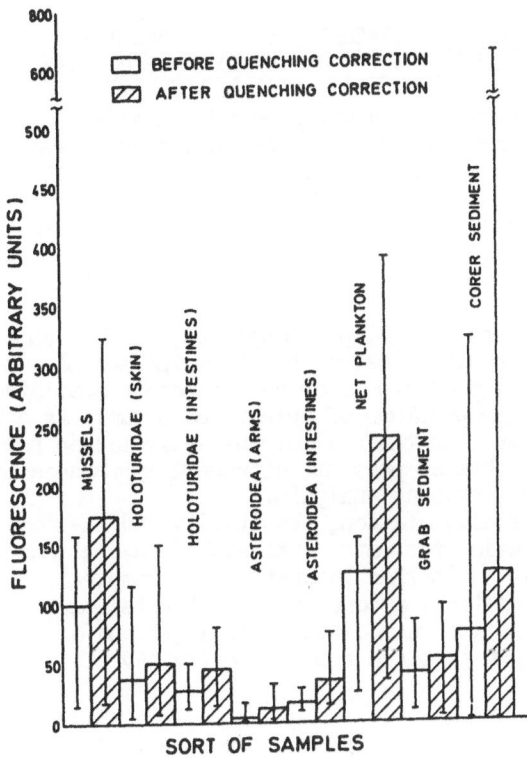

Fig. 2. The comparison of corrected and noncorrected fluorescence of
marine samples extracts. Samples were treated by using Method 1.

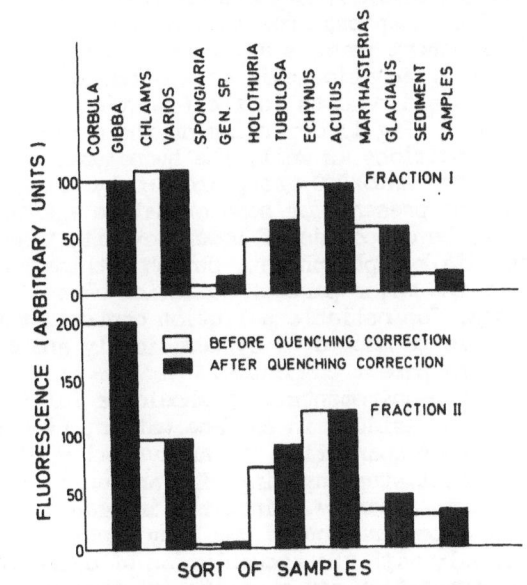

Fig. 3. The comparison of corrected and noncorrected fluorescence of
marine samples extracts. Samples were treated by using Method 2.

A COMPARISON OF PROCEDURES FOR TRACE ENRICHMENT OF ORGANOPHOSPHORUS PESTICIDE RESIDUES FROM WATER

Z. Fröbe, V. Drevenkar and Ž. Vasilić
Institute for Medical Research and Occupational Health,
Zagreb, Yugoslavia

Summary

The trace enrichment of dialkyl phosphates, dialkyl phosphorothioates and dialkyl phosphorodithioates, highly water soluble, polar and acidic degradation products of organophosphorus pesticides, from water, creates some problems of principled nature. As a uniformly efficient procedure for their simultaneous isolation from water samples is not available at present, different approaches have to be combined in a multiresidue analysis.

The efficiency of accumulation, sensitivity and the possibility of practical application for water analysis of several enrichment procedures investigated in this laboratory are compared. The procedures are based on extraction, adsorption on activated carbon, Amberlite XAD-4 resin and octadecyl-modified silica gel and on anion-exchange on an ion-exchanger prepared from Amberlite XAD-4.

1. INTRODUCTION

A considerable amount of dialkylphosphorus residues in surface waters arises from the ability of organophosphorus (OP) pesticides to undergo degradation over relatively short periods after their application. Chemical alteration of OP pesticides in aquatic media most often includes the cleavage of P-O-C or P-O-S bonds and results in the formation of breakdown products of different structures (Fig.1) depending on the structure of parent compounds and hydrolytic conditions as well. The hydrolysis may be accompanied by the oxidation of P=S into P=O group induced by enzyme action, absorption of uv light or the presence of some oxidative agents.

Although considered to be compounds of lower toxicity than the corresponding trialkyl esters, dialkyl phosphates, phosphorothioates and phosphorodithioates represent the major part of pollution caused by the use of OP pesticides. Therefore, for reliable pollution control, their presence in the environment should be monitored systematically and concurrently with the determination of the parent compounds.

There are two important requirements that should be fulfilled when determining dialkylphosphorus residues in surface waters. Firstly, as the pollutants are present in trace quantities the analytical procedure has to include an efficient enrichment step enabling high and uniform recoveries for all compounds of interest. Secondly, since the inorganic phosphates usually interfere with the determination of DMP they should be eliminated from the sample simultaneously with the accumulation of dialkylphosphorus anions or before their GC separation and quantitative analysis.

In this paper we summarize our experience in developing several analytical procedures for the determination of OP pesticide residues in water.

$$
\begin{array}{c}
\text{RO} \diagdown \quad \diagup X \\
\quad P \\
\text{RO} \diagup \quad \diagdown Y\text{--}A
\end{array}
\qquad \xrightarrow{\text{Degradation}} \qquad
\begin{array}{c}
\text{RO} \diagdown \quad \diagup X \\
\quad P \\
\text{RO} \diagup \quad \diagdown Y^{\ominus}
\end{array}
$$

X \ R	-CH$_3$	-C$_2$H$_5$	R \ Y
O	DMP	DEP	O
S	DMTP	DETP	O
S	DMDTP	DEDTP	S

Fig. 1. Structures and abbreviations of the most common hydrolytic products of OP pesticides.

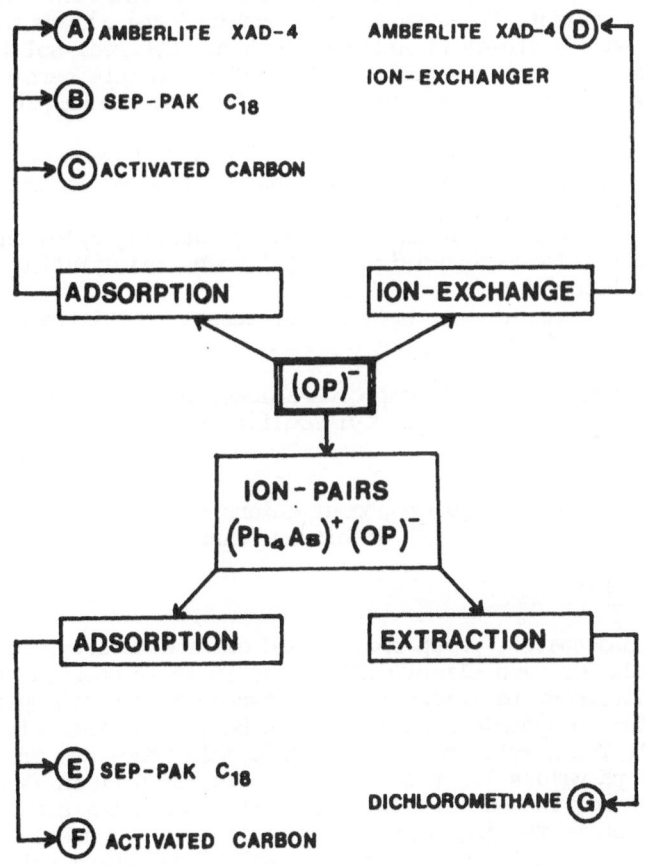

Fig. 2. Analytical procedures for accumulation of organophosphorus pesticide residues, (OP)$^-$, from water.

The procedures discussed (Fig. 2) are based on the accumulation of both free dialkylphosphorus anions and their ion-associates with tetraphenylarsonium cation, $(Ph_4As)^+$, from water by extraction, adsorption on activated carbon, Amberlite XAD-4 resin and octadecyl-modified silica gel (SEP-PAK C_{18} cartridge) and by anion-exchange on an ion-exchanger prepared from Amberlite XAD-4.

2. EXPERIMENTAL

2.1. Accumulation of dialkylphosphorus anions from water by ion-exchange

The ion-exchanger was prepared from Amberlite XAD-4 resin according to the method reported by Richard and Fritz (1). A volume of 100 ml of water solution containing 100 ng/ml of each compound listed in Fig. 1, singly or in a mixture, was passed through a column (4 g of ion-exchanger, 1 x 15 cm) previously converted into OH^-- form by passing 50 ml of 0.1 M NaOH and washed with deionized water to neutral. The residual water was eliminated from the column with 10 ml of methanol and resin was eluted with 15 ml of 0.15 M HCl-methanol (twice the same portion) followed by 10 ml of fresh eluent. The eluates were combined and evaporated in water bath (37°C) under a stream of nitrogen to 5 ml. Ethereal solution of diazomethane was added until the yellow colour of the sample persisted. After 10 min the excess of diazomethane was expelled and the sample evaporated to 5.0 ml.

Standard solutions for quantitative GC analysis were prepared by methylation of known amounts of corresponding standard compounds in methanolic solution.

The HCl-methanol eluent was prepared by bubbling hydrogen chloride gas through distilled methanol for several hours and diluting it to 0.15 M with methanol before use.

All other accumulation procedures discussed in this paper (Fig.2) were described earlier:

2.2. Accumulation of dialkylphosphorus anions from water by adsorption on activated carbon (2), octadecyl-modified silica gel (3) and Amberlite XAD-4 (4).

2.3. Accumulation of dialkylphosphorus anions from water by extraction of $(Ph_4As)^+$ - associates with dichloromethane (5).

3. RESULTS AND DISCUSSION

Dialkyl phosphates, phosphorothio and dithioates are structurally related compounds but even slight differences in their physical properties cause their different behaviour in the water-organic phase system. The acidities of the phosphoric acid esters can be assumed following a few simple rules (6). The alkylation of phosphoric acid increases the acidity,i.e. decreases the pK values for the first dissociation step in 7% alcohol. On the other hand, the acidity decreases with increasing size of the alkyl substituents, while the increasing sulphur content decreases acidity in water but increases acidity in alcoholic solutions. These rules are best illustrated by the following series (6):

```
        Pi > MP > DMP  ;  DMP < DEP < DPrP
  pK 1.97  1.54  1.25 ;  1.25  1.37  1.52
                   DEP     DETP  DEDTP
  pK (7% alc.)     1.37    1.49  1.55
  pK (80% alc.)    3.15    2.84  2.64
```

where DPrP = dipropylphosphate and MP = methylphosphate.

As highly polar and acidic species dialkyl esters of phosphoric, phosphorothioic and phosphorodithioic acids are extremely water-soluble and therefore not easily extractable from water in the organic nonpolar phase, unless they are converted into a suitable form. The recovery of extraction of free anions with pure organic solvents is negligible. Adsorption on Amberlite XAD-4 (Table I, column A) yields high recoveries for phosphate and phosphorothioate derivatives (7), but in our experience considerable difficulties were encountered in desorption of sulphur containing species, especially dialkyl phosphorodithioates. According to reported results (7) adsorption onto Amberlite XAD-4 resin is favoured not only by higher pK but also by increased aliphatic character and sulphur content. On the contrary, the capacity for adsorption on activated carbon (8) becomes lower from the molecular phosphate, such as triethylphosphate, to a more ionic DMP and orthophosphate, whose adsorption behaviour exhibits marked pH-dependence. Therefore, the accumulation efficiencies obtained by adsorption on activated carbon (Table I, column C) were low for all investigated anions, particularly for the most acidic DMP and DEP.

For the isolation of dialkylphosphorus anions from water ion-exchange (1) was assumed to be a promising approach. Our recent results prove that accumulation on ion-exchanger prepared from Amberlite XAD-4 resin yields almost quantitative recoveries for the most acidic DMP and DEP, whereas less polar DMTP and DETP are isolated with lower but still satisfactory recoveries (Table I, column D). At the same time a major part of inorganic phosphates is eliminated allowing determination of DMP in the sample. Dialkyl phosphorodithioates could not be isolated in this way so far probably because they undergo rapid solvolysis in acidic methanol during elution yielding corresponding phosphorothioates. An investigation of the experimental conditions preventing this conversion is still in progress.

Due to their large ionic volume dialkylphosphorus anions are capable of forming ion-pairs in water solutions containing similarly large cations. Such ion-pairing with $(Ph_4As)^+$ results in the formation of more lipophilic

Table I. Accumulation efficiencies of OP pesticide residues from water by procedures listed in Fig. 2.

Compound	Accumulation efficiency (%) by procedure								
	A	A*	B	C	D	D*	E	F	G
DMP	-	50	5	1	97	87	2	13	-
DEP	95	100	47	ND	93	88	15	33	ND
DMTP	-	97	22	8	61	94	29	66	9
DETP	-	85	86	7	52	-	85	69	33
DMDTP	27	-	14	8	-	-	14	91	86
DEDTP	13	-	33	13	-	-	64	81	95
P_i	< 1	3	-	< 1	6	-	-	0	< 1

A* = see ref. 7.   ND = not detected
D* = see ref. 1.   - = not determined

compounds which can be more readily isolated from aqueous media. Extraction efficiencies of ion-pairs with dichloromethane (5) follow the sequence DEDTP>DMDTP>DETP>DMTP, the distribution ratios of more lipophilic dialkyl phosphorodithioates being at least one order of magnitude higher than the values obtained by extraction of dialkyl phosphorothioates. Analogously with the adsorption onto Amberlite XAD-4 resin, the extraction is favoured by increased aliphatic character and sulphur content of the anion. However, this method is not suitable for dialkyl phosphates which are not extracted at all.

The adsorption of ion-pairs of dialkylphosphorus anions with $(Ph_4As)^+$ on activated carbon microcolumn (2) allows the trapping of DMP and DEP but with low recoveries (Table I, column F). The accumulation efficiency of phosphorothioates is considerably improved compared to that obtained by extraction, while for the phosphorodithioates it remains high. Inorganic phosphates are completely eliminated. However, inorganic anions common to polluted waters compete in ion-pairing with $(Ph_4As)^+$ and decrease the elution recoveries of DMTP and dialkyl phosphates.

The reversed-phase adsorption of dialkylphosphorus anions on octadecyl-modified silica gel(3) is affected by the pH value of the water solution and by the volume of the water sample. The best results are achieved when adsorption is performed from 10 ml water samples at pH 4 if the anions are collected as ion-pairs with $(Ph_4As)^+$ (Table I, column E), or pH 1.25 if they are adsorbed in the free form (Table I, column B). In both cases the phosphorothioates are trapped more efficiently than phosphorodithioates, showing an influence of sulphur content opposite to that in extraction or adsorption on activated carbon. The recoveries of the more lipophilic diethyl compounds are still higher than their dimethyl analogues (e.g. 47% for DEP), but the most polar DMP cannot be satisfactorily retained on the adsorbent.

4. CONCLUSION

From the results discussed above it is obvious that so far no single method for accumulation of all dialkyl phosphates, phosphorothio and dithioates is uniformly efficient and applicable for their simultaneous determination in water. Consequently, a combination of procedures is expected to yield the best results in multiresidue analysis. Based on our recent experience dialkyl phosphates are most efficiently trapped from water by ion-exchange. For the accumulation of thio and dithio analogues adsorption of ion-pairs either on activated carbon (detection limit<μg/l) or on octadecyl-modified silica gel (detection limit >1 μg/l) is suggested. If the selective determination of dialkyl phosphorodithioates is desired in samples at concentrations not exceeding 1 μg/ml, the extraction of $(Ph_4As)^+$ - associates from water with dichloromethane is recommended.

REFERENCES

1. RICHARD J.J and FRITZ J.S., J. Chromatog.Sci. 18(1980)35.
2. DREVENKAR V., FRÖBE Z., ŠTENGL B. and TKALČEVIĆ B., Intern.J.Environ.Anal.Chem.(in press).
3. DREVENKAR V., FRÖBE Z., ŠTENGL B. and TKALČEVIĆ B., Mikrochim.Acta (Wien) 1985 I, 143.
4. DREVENKAR V., VASILIĆ Ž., ŠTENGL B., TKALČEVIĆ B. and STILINOVIĆ B., Proceedings of the First European Symposium "Analysis of Organic Micropollutants in Water", Berlin 1979, Vol. 2, p. 341
5. DREVENKAR V., FRÖBE Z. and ŠTENGL B., Anal.Chim.Acta,154(1983)277.
6. FEST C. and SCHMIDT K.J., "The Chemistry of Organophosphorus Pesticides", Springer-Verlag, Berlin, Heidelberg, New York 1982, p. 64.
7. DAUGHTON C.G., CROSBY D.G., GARNAS R.L. and HSIEH D.P.H., J.Agric. Food Chem. 24(1976)236.
8. JAYSON G.G. and LAWLESS T.A., J.Coll.Interface Sci. 86(1982)397.

THE KOVATS INDICES OF SOME ORGANIC MICROPOLLUTANTS ON AN SE54 CAPILLARY COLUMN

B.J. HARLAND, R. I. CUMMING and E. GILLINGS
Brixham Laboratory, ICI PLC, Brixham, Devon

Summary

The Kovats Indices of over 50 organic micro-pollutants, members of either the EPA Priority Pollutants or the EC proposed Black List have been determined on an SE54 fused silica capillary column.

1. INTRODUCTION

The determination of specific organic pollutants, or organic micro-pollutants, in waste and receiving waters is becoming of increasing importance. In the United States, the EPA has published a list of 129 compounds, the so-called Priority Pollutants, of which 114 are organic, whose discharge from industrial sites to receiving waters is likely to be regulated in the future (1). A similar list of 129 compounds, the proposed Black List, has been prepared by the European Community and member states are required to eliminate pollution by these substances in inland surface waters, territorial waters, internal coastal waters and ground water (2).

As discharge limits for these compounds are introduced, then analytical methods will be required to ensure that discharges are in compliance. The technique most applicable to the determination of the majority of organic compounds on these two lists is gas chromatography, or if more specificity is required, gas chromatography-mass spectrometry. The EPA has published such methods for the measurement of its Priority Pollutants (3), and recently some workers have recommended that capillary columns be substituted for packed columns in this analysis, because of their higher resolving power (4,5).

This paper details retention times on 53 organic pollutants which are medium volatiles on either the EPA Priority Pollutants or the EC proposed Black List. The data are given in the form of Kovats Indices and were obtained on an SE54 capillary column under temperature programmed operation.

2. EXPERIMENTAL

The data were obtained using a Varian 6000 gas chromatograph equipped with flame ionisation detection. The column used was a 0.32 mm x 25 m SE54 fused silica open tubular capillary column (Chrompack UK Ltd) operated on a temperature programme of 50°C (held for two minutes) to 250°C (held for 12 minutes) at 8°C/min. The injection temperature was 226°C, and injections were carried out using the Grob splitless technique. Helium was used as carrier gas, and the flow rate was approximately 4.5 ml/min. All chromatographic data was processed by a Trilab 2000 data station.

Kovats Indices were calculated using the formula suggested by Van Den Dool and Kratz (6) for temperature programmed operation, which is

$$\text{Kovats Index} = 100 \cdot \frac{T_{R(substance)} - T_{R(z)}}{T_{R(z+1)} - T_{R(z)}} + 100 \cdot z$$

where $T_{R(z)}$ and $T_{R(z+1)}$ are the retention times of the n-alkanes of carbon chain length z and z+1 respectively, which bracket the compound of interest, which is of retention time $T_{R(substance)}$.

The Kovats Indices for 46 of the compounds were determined using an internal standard procedure, an external standard procedure being used for the remainder. The standard used for each of these procedures contained n-alkanes of carbon chain length from 9-27.

Samples of the EPA Priority Pollutants were obtained in kit form at 200 mg/l concentrations in methylene chloride or methanol (Supelco - Complete water standards kit). Other compounds tested were obtained from various suppliers in analytical or laboratory grade, and were diluted to 200 mg/l with methylene chloride.

3. RESULTS AND DISCUSSION

The Kovats Indices for the organic micropollutants which were considered in this investigation are given in elution order in Table 1, together with information as to whether the compounds are an EPA Priority Pollutant or a member of the EC proposed Black List. All the data presented in the tables were obtained by the internal standard procedure unless otherwise indicated. The majority of the compounds listed in the table are EPA Priority Pollutants, and some of these are also members of the EC proposed Black List. A limited number of other compounds from the Black List have also been included and it is hoped to examine more members in the future. Not all the compounds examined could be assigned retention indices. The late eluting polycyclic aromatic hydrocarbons, eg benzo(b) fluoranthene, benzo(a) pyrene etc, were outside the range of the hydrocarbon standard used (nC9-nC27) and some compounds, eg N-nitrosodimethylamine were not detected. Nevertheless, the tables contain the retention indices of over 50 organic micropollutants.

Other summaries of the retention times of the EPA Priority Pollutants on SE54 capillary columns have been published by Trussel et al (4) and Sauter et al (5). Since this present work began, the EPA has published similar data for a new version of its priority pollutant method (7) using a capillary column of bonded phase (DB5), which is equivalent to the SE54 phase. A comprehensive survey of the Kovats Indices of over 2000 organic compounds (including pollutants and solvents) on capillary columns has also been published, but this is not available in the open literature (8).

The retention time data of Trussel and the EPA has been compared to the Kovats Indices obtained in this present work. Excellent agreement was found as is demonstrated by the graphical comparisons shown in Figures 1 (a) and (b) respectively. (A similar comparison of Sauter's data has not been undertaken as this would have been considerably more complex because of the number of internal standards used by the latter.)

The EPA retention data also includes values for even numbered n-alkanes which should allow calculation of Kovats Indices for the compounds listed. However, the retention times for the even n-alkanes from decane to eicosane were predicted from the Kovats Index data generated in this work, using a polynomial equation which gave a 99.9% fit to the data presented in Figure 1 (b). The values found were significantly different than those quoted by the EPA. In particular, the difference found for eicosane was 91 seconds and that for hexadecane was 42 seconds, the mean difference for all the even n-alkanes considered being 34 seconds. This compares to the mean difference of 8 seconds (range 1-24 seconds) which was found for the other listed compounds of similar retention range (700-1820 seconds) using the same polynomial equation. The differences found between the predicted retention times for the even n-alkanes and those quoted in the EPA method are not unexpected given the irregular retention intervals for the latter.

For an SE54 capillary column run under similar chromatographic conditions, the Kovats Indices obtained should be extremely close to those reported here although it should be noted that changes in injection temperature, temperature program, and variation in SE54 phase could make significant differences to these values. For unknown samples, Kovats Indices on only one column should not be used in isolation as the sole means of identification; confirmation of identity by GC-MS or some other appropriate method is still a prerequisite.

REFERENCES

1   Fed. Reg. 3 December (1979), part 3, vol 44, no 233   p69464;
     KEITH, L.H., LILLARD, W.A., Environ. Sci. Technol. 13 (1979) 416

2   Council of the European Communities, Council Directive of 4 May 1976
     on pollution caused by certain dangerous substances discharged into
     the aquatic environment of the community. Directive 76/464/EEC Off.
     J. Eur. Comm. L129, (1976), 23-29

3   Sampling and analysis procedures for screening of industrial
     effluents for priority pollutants. US Environmental Protection
     Agency (April 1977)

4   TRUSSEL, A.R., MONCUR, J.G., FONG-YI Lieu, LEONG, L.Y.C. J. High
     Resol. Chromatogr. & Chromatogr. Commun., 4, 156-163, (1981)

5   SAUTER, A.D., BETOWSKI, L.D., SMITH, T.R., STRICKLER, V.S.,
     BEIMER, R., COLBY,B.N., WILKINSON, J.E.
     J. High Resol. Chromatogr. & Chromatogr. Commun, 4, 366-384 (1981)

6   VAN DEN DOOL, H., KRATZ, P. J. Chromatogr, Dec 1963, 11, 463-471

7   Fed. Reg., 26 October 1984, vol 49, no 209, pp43416-43429

8   The Sadtler Capillary GC Standard Retention Index Library
     Heyden and Sons Ltd

TABLE I

KOVATS INDICES OF SOME ORGANIC MICROPOLLUTANTS
ON AN SE54 CAPILLARY COLUMN

Kovats index	Compound	EPA priority pollutant	EC black list
955	2-Chlorotoluene*		/
988	Phenol	/	
989	Bis(2-chloroethyl) ether	/	
992	2-Chlorophenol	/	/
1002	1,3-Dichlorobenzene	/	/
1008	1,4-Dichlorobenzene	/	/
1030	1,2-Dichlorobenzene	/	/
1055	Bis(2-chloroisopropyl) ether	/	/
1064	Hexachloroethane	/	/
1071	N-Nitroso-di-n-propylamine	/	
1084	Nitrobenzene	/	
1119	Isophorone	/	
1125	2-Chloroaniline*		/
1126	2-Nitrophenol	/	
1151	2,4-Dimethylphenol	/	
1163	2,4-Dichlorophenol	/	/
1164	Bis(2-chloroethoxy) methane	/	
1170	1,2,4-Trichlorobenzene	/	/
1176	Naphthalene	/	/
1198	3-Chloroaniline*		/
1198	4-Chloroaniline*		/
1207	3-Chlorophenol*		/
1207	4-Chlorophenol*		/
1211	Hexachlorobutadiene	/	/
1294	4-Chloro-3-methyl-phenol	/	/
1324	2,4-Dichloroaniline		/
1328	Hexachlorocyclopentadiene	/	
1348	2,4,6 Trichlorophenol	/	/
1366	2-Chloronaphthalene	/	
1368	1-Chloronaphthalene*		/
1438	Acenaphthylene	/	
1456	Dimethylphthalate	/	
1465	2,6-Dinitrotoluene	/	
1474	Acenaphthene	/	
1500	2,4-Dinitrophenol	/	
1546	4-Nitrophenol	/	
1572	Fluorene	/	
1588	4-Chlorophenyl phenyl ether	/	
1597	Diethylphthalate	/	
1607	2-Methyl-4,6-dinitrophenol	/	
1617	N-Nitrosodiphenylamine	/	
1685	4-Bromophenyl phenyl ether	/	
1699	Hexachlorobenzene	/	/
1748	Pentachlorophenol	/	/
1767	Phenanthrene	/	
1777	Anthracene	/	/
1966	Di-n-butyl phthalate	/	
2045	Fluoranthene	/	
2096	Pyrene	/	
2111	Benzidine	/	/
2350	Butyl benzyl phthalate	/	
2463	3,3'-Dichlorobenzidine	/	/
2552	Bis (2-ethylhexyl) phthalate	/	

* Data obtained by external standard procedure

Figure 1. Plots of Kovats Indices for certain Priority
Pollutants against retention time data for the same compounds
taken from the work of (a) Trussel et al and (b) the EPA.

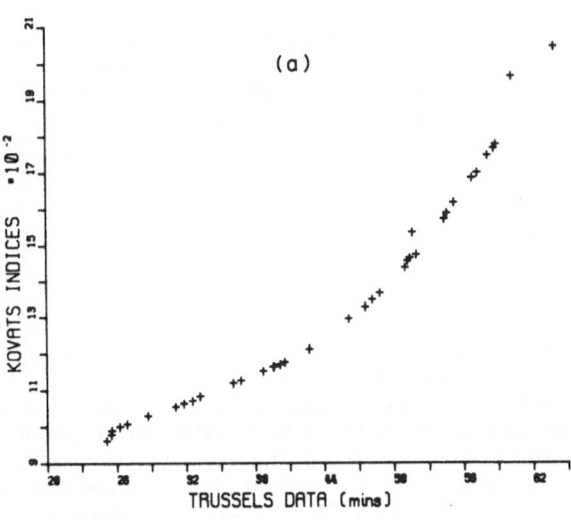

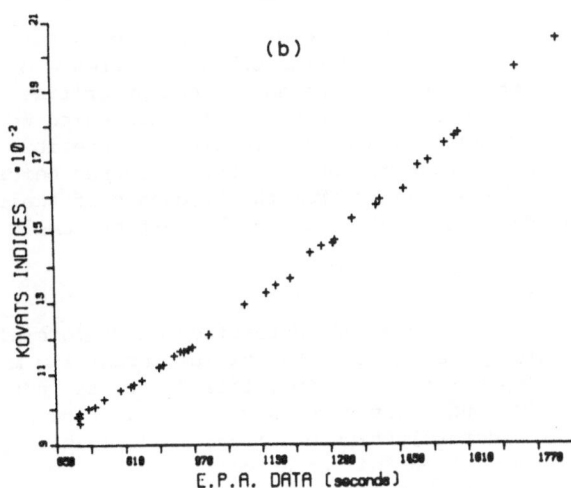

A DUAL DETECTOR GAS CHROMATOGRAPHIC SYSTEM FOR THE ROUTINE DETERMINATION OF RETENTION INDICES

C. O'DONNELL, P. M. LARKIN

Water Resources Division, An Foras Forbartha,
St. Martin's House, Waterloo Road,
Dublin 4, Ireland

Summary

This paper describes a dual detector (ECD/FID) gas chromatographic system for the determination of retention indices of organochlorine compounds in environmental samples. The detector signals are time-selected before feeding into a single-channel integrator and the retention times are transferred to a microcomputer for the index calculations. The indices may be stored for future access so that recurrences of unknown compounds can be detected.

1. INTRODUCTION

Because of the complex nature of environmental extracts, high resolution gas chromatography on capillary columns is required to separate their many components. Certain peaks, from the compounds of immediate interest (the so-called "target" compounds), are then quantified and the remaining information contained in the other peaks may often be lost. This is most likely to occur in routine screening for target compounds. The untargetted peaks may indicate higher concentrations than the target ones, particularly in extracts from waters free of gross pollution. On occasion the major untargetted peaks may demand attention and they can then be followed up, but this is only likely to happen if the analyst regards the matter as a priority.

It is considered that there is a need for some simple system which will facilitate the storage of information on the occurrence of untargetted peaks so that the analyst's decisions on analytical priorities can be based on previous as well as current information. If peaks which would normally be ignored could be compared with a data file based on previous sample runs, in order to detect whether or not they have occurred before, their potential importance could be judged from the frequency of their previous occurrence as well as from their response in the current sample.

2. DISCUSSION

The retention index of a compound, determined on commercially available capillary columns, appears to be the most promising marker of its identity. Unlike a simple retention time, this figure is independent of day-to-day changes in GC conditions such as carrier gas flow rate and injection effects. The determination of linear programmed temperature gas chromatographic (PTGC) retention indices as described by Knoppel et al. (1,2) used polynomial rather than linear interpolation in order to avoid the necessity for the addition of a full series of alkane markers to each sample with inevitable interferences on a GC/MS system. However, in the case of a recently published commercial retention index library (3), the PTGC indices were determined by linear interpolation as the full range of alkane markers were included to bracket the compounds being evaluated. It is noteworthy that the issue of this library reflects the growing interest in the wider application of retention index data.

Both of these systems used single-channel detection systems for the collection of retention data. This is suitable for the creation of an index library using standard compounds at high concentrations as in the latter case. However, in the field of GC/MS work the universal detection capability of the mass spectrometer means that added alkane markers will all be detected and are therefore likely to interfere with the environmental extracts. It is possible when using a dual-channel system to use the selectivity and sensitivity of the electron capture detector (ECD) to analyse many compounds of environmental importance in an extract while the alkane markers and their retention times are detected by flame ionisation detection (FID). Working at standard concentrations of 0.1 ppm chlorinated pesticides on the ECD channel and 2-5 ppm of alkane markers on the FID channel, no cross-channel interference has been found (4), so that a set of ten markers may be added to each sample without any interference with the analysis.

The present contribution describes a low-cost system which can be used to calculate the retention indices of all the peaks in a chromatogram and store them for future reference. A search programme can be used to list previous peaks found within a user-defined retention index interval. The system is based on a dual-detector gas chromatograph with simultaneous flame ionisation and electron capture detection. The signals from the two channels can be blended as required to produce the necessary data for the retention index calculations. The composite chromatogram produced can also be used to provide internal standard responses on the FID channel for ECD analysis using the internal standard calculation routine in the integrator. The retention times are down-loaded into a microcomputer as described by Shaw and Miller (5), and are then compared with standard times from a set-up run. Retention indices are presently calculated to the nearest unit until the reproducibility of the system has been evaluated, the purpose being to produce useful data in the minimum of extra time. The standard GC conditions being adopted initially include temperature programmes of 5 or 8 deg/min; resolution can be improved if required by reducing these ramp rates. The scheme of the analysis is outlined in fig. 1.

3. APPARATUS

Gas Chromatograph   Hewlett-Packard Model 5790 capillary GC fitted with FID and ECD detectors and split, splitless and on-column injection ports.

Column effluent splitter   Vitreous silica outlet splitter (SGE) split ratio 1/1.

Signal processing   Hewlett-Packard Model 3390A integrator with I/O board and sampler/event control module (SECM). The SECM switches a 6-way relay under time control from the 3390A to select the channel to be integrated.

Computer   Apple Model IIc with Super Serial card connected to the report output from the 3390A. Software was based on the transfer programme provided by P. E. Shaw modified as required.

4. PROCEDURE

To maximise reproducibility of retention index measurement, a standard temperature programme is chosen for each column. A sample containing normal alkanes is first run using the FID detector in order to define the standard indices from 1100 to 2600. This information is stored on diskette for reference. One diskette is used for each setup and contains the GC conditions, the standard retention times file and the database for peaks detected under those conditions. The standard retention times are updated every day.

A solution containing one or more alkanes is added to both the standards and sample extracts as internal standards. The standards are then run on ECD detection.

Figure 1. SCHEME OF ANALYSIS

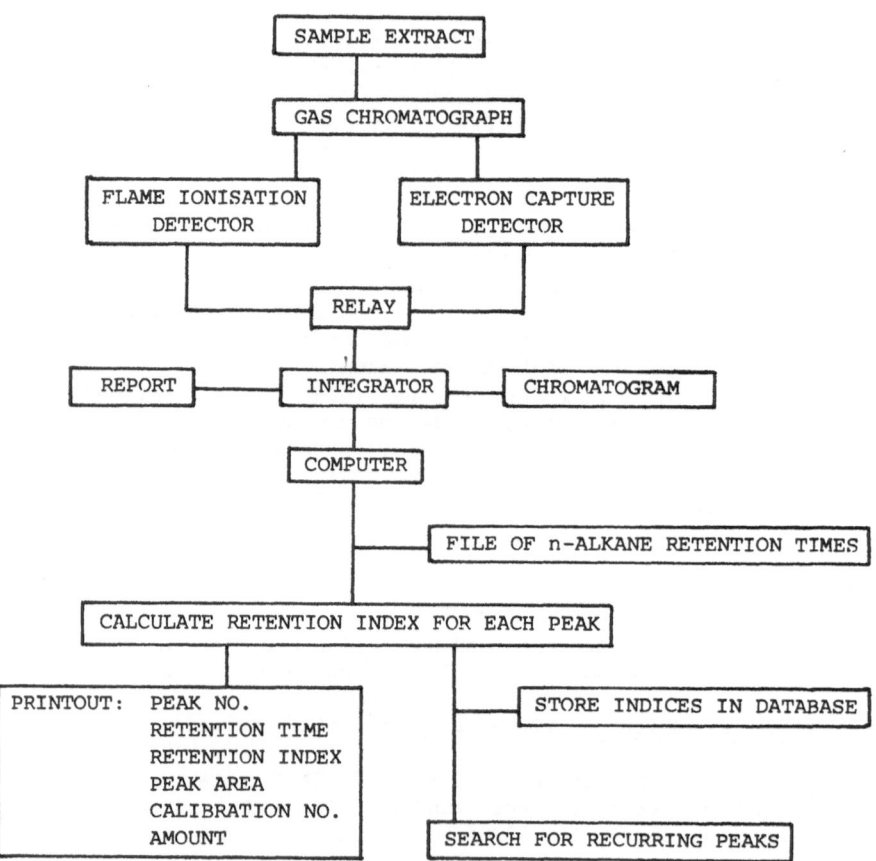

Using the results from these two runs, an alkane internal standard is selected which does not interfere with the target analysis. The channel selection relay is programmed to detect the internal standard while the

rest of the run is scanned on the ECD detector. This involves switching
to the FID detector for about one minute and back again. After the GC
run, the integrator reports all the peak times to the computer. At this
point a correction may be applied to bring the internal standard peak into
line with the standard run on disk. This correction is usually less than
0.02 minutes and is the same for all peaks as it compensates for
differences in retention time due to solvent expansion or injection speed.
The computer corrects all stored peak times and calculates their retention
indices by linear interpolation from the stored standard times.

Using a range of columns of different polarity, a compound can be
characterised so that if it is found at a later date it will be recognised.
With this system it is possible to store indices with further information
on sample type and location in a database. A search programme has been
written to find any peaks in the database within a user-defined interval so
that any pattern of recurrence will be easily detected.

5. CONCLUSION

The system described offers a convenient method for recording the
occurrence of a particular compound in environmental extracts. By
operating it during routine target analysis it will allow the rapid
detection of commonly-occurring compounds which may be of more importance
locally than the target compounds themselves.

REFERENCES

(1) H. Knoppel, M. De Bortoli, A. Peil, H. Schauenburg and H. Vissers:
Proceedings of the Second European Symposium "Physico-chemical Behaviour of
Atmospheric Pollutants", Varese (Italy), 29 September-1 October 1981, H.
Ott and B. Versino (eds.), D. Reidel Publishing Company, Dordrecht,
Holland, 1982, p. 99.
(2) idem: "The Determination of Linear PTGC Retention Indices for use
in Environmental Organics Analysis", Proceedings of the Second European
Symposium on the Analysis of Organic Micropollutants in Water, held in
Killarney (Ireland), November 17-19, 1981, A. Bjorseth and G. Angeletti
(eds.), D. Reidel Publishing Company, Dordrecht, Holland, 1982, pp.133-135.
(3) Sprouse, J. F. and Varano, A. : Development of a GC Retention
Index Library, International Laboratory Vol. 14(9) Nov/Dec. 1984, p. 54.
(4) O'Donnell, C. : Applications in Gas Chromatography, Proceedings of
the Second European Symposium on the Analysis of Organic Micropollutants in
Water held in Killarney (Ireland), November 17-19, 1981, A. Bjorseth and G.
Angeletti (eds.), D. Reidel Publishing Company, Dordrecht, Holland, 1982,
pp. 113-117.
(5) Shaw, P.E., and Miller J.M.: Interfacing an Apple Computer with a
Digital Integrator for Chromatographic Data Storage and Statistical
Calculations, J. Chrom. Sci. Vol. 21 (8), August 1983, p. 372.

Acknowledgements   The authors wish to thank P. E. Shaw for the provision
of a copy of the data transfer programme and P. J. Flanagan for assistance
in preparation of the manuscript.

METHODS FOR THE DETERMINATION OF EEC PRIORITY POLLUTANTS

R.C.C.Wegman
National Institute of Public Health and Environmental Hygiene
3720 BA Bilthoven, the Netherlands

Summary

A questionnaire was sent to all participating countries of the EC
project Cost 641 "Organic micropollutants in the aquatic environment".
Information was asked about the analysis of 129 priority pollutants
in water which are mentioned in the EC publication sheet, number
C 176, July 14, 1982. From six countries, namely Belgium, Finland,
Ireland, Netherlands, Norway and Yugoslavia information was obtained. For 27 of the 129 priority pollutants an analytical method
was not available. In future especially attention has to be given
to the development of analytical methods for these compounds. For
all priority pollutants confiramation criteria had to be stated.

1. INTRODUCTION

The EEC Publication sheet, number C 176, July 1982 includes a list of
129 priority pollutants. Analysis and subsequent monitoring of selected
compounds of the list is a first approach of a more or less extended survey.
In the frame-work of the Cost 641 project "Organic micropollutants in the
aquatic environment" it was planned to have specially attention to the
analysis of these priority pollutants. In order to get an impression of
the state-of-the-art of the analytical methodology a questionnaire was
sent to all participants of the Cost 641 project. Information was asked
about extraction/concentration, clean-up, derivatization, detection and
detectionlimit. From six countries namely Belgium, Finland, Ireland,
Netherlands, Norway and Yugoslavia information was obtained (1-12).
The information obtained by these questionnaires is summarized in this
paper. The methods are very shortly described. A review with respect to
the methods of analysis of organic compounds in water samples is recently
made by Wegman and Melis (13).

2. RESULTS

In table 1 the priority pollutants are summarized for which no analytical method was available (27 compounds). Especially for the organotin
compounds a lack of analytical knowledge is present. For 102 compounds one
or more methods were described. The methods which were used were the following (the numbers behind the compounds correspond with those of the EC-list):

2.1. VOLATILE COMPOUNDS

The more volatile compounds like benzene (7), carbotetrachloride (13),
chlorobenzene (20), chloroform (23), 1-chloro-2-nitrobenzene (28), 1-chloro-
3-nitrobenzene (29), 1-chloro-4-nitrobenzene (30), 4-chloro-2-nitrotoluene
(31), chloronitrotoluenes(32), 1,2-dibromoethane (48), 1,1-dichloroethane
(58), 1,2-dichloroethane (59), 1,1-dichloroethylene (60), 1,2-dichloro-
ethylene (61), dichloromethane (62), dichloronitrobenzene (63), 1,2-di-
chloropropene (67), 2,3-dichloropropene (68), ethylbenzene (79), hexachlo-
roethane (86), isopropylbenzene (87), 1,1,2,2-tetrachlorobenzene (110),

Table 1   List of priority pollutants for which an analytical method
is not available. The numbers correspond with the numbers
of the EC priority pollutant list.

No	Compound	No	Compound
2	2-amino-4-chlorophenol	49	dibutyltindichloride
8	benzidine	50	dibutyltinoxide
14	chloralhydrate	51	dibutyltin salts
16	chloroacetic acid	56	dichlorobenzidines
21	1-chloro-2,4-dinitrobenzene	72	diethylamine
22	2-chloroethanol	74	dimethylamine
24	4-chloro-3-methylphenol	104	propanil
25	1-chloronaphtalene	105	pyrazon
26	chloronaphtalenes	108	tetrabutyltin
27	4-chloro-2-nitroaniline	115	tributyltinoxide
36	chloroprene	125	triphenyltin acetate
41	2-chloro-p-toluidine	126	triphenyltin chloride
42	chlorotoluidines	127	triphenyltin hydroxide
44	cyanuric chloride		

tetrachloroethylene (111), toluene (112), 1,1,1,-trichloroethane (119),
1,1,2-trichloroethane (120), trichloroethylene (121), 1,1,2-trichloro-
trifluoroethane (122), vinylchloride (128) and the xylenes (129) are deter-
mined by the following methods (numbers correspond with reference list):
A closed loop stripping analysis, (CLSA) (1,2,7)
B head-space analysis "static" (2,3,8,10,12)
C purge and trap "dynamic" (1,6,10,12)
D liquid-liquid extraction e.g. with n-pentane (1,2,3,7,8,9,10,11,12)
E steam distillation with cyclohexane/water (3)
F LLE/rotavapor; GC-MS (EI) (12)
    For electron captive compounds like the organochlorine compounds
GC-ECD is used for the others GC-FID or GC-MS.
    An extended review paper about these methods was written by Nuñez
et al. (14).
2.2. ORGANOCHLORINE PESTICIDES AND PCBs
    The organochlorine compounds aldrin (1), chlordane (15), DDT (46),
dieldrin (71), endosulfan (76), endrin (77), heptachlor (82), hexachlorobenzene (83) and hexachlorocyclohexane (85) were analyzed with the
following methods:
A extraction with an apolair extraction solvent e.g. n-hexane, clean-up
   with alumina and silicagel; if necessary a step for removing
   interfering sulphur compounds; quantitation by GC-ECD or GC-MS (1,2,3,
   4,10,12)
B concentration with XAD-2; clean-up with alumina and silicagel; quantitation by GC-ECD or GC-MS (9,12)
C extraction with e.g. pentane; clean-up with sulphuric acid or silica
   H_2SO_4; quantitation by GC-MS or GC-ECD (4,12)
D LLE/rotavapor; clean-up with silica H_2SO_4, GC-ECD (12)
    For determination of the PCBs by GC-ECD a chromatographic separation
of the PCBs and the organochlorine pesticides is necessary (method A and
B). The presence of PCBs can also interfere the organochlorine pesticides.

2.3. AROMATIC AMINES

The aromatic amines and related compounds 2-, 3-, 4-chloroaniline (17,18,19), linuron (88) and monolinuron (95) were isolated by means of a combined hydrolysis/steam distillation/extraction step in a Bleidner-apparatus. The organic extraction phase (n-iso-octane) was extracted with a HBr solution. The aromatic amines are brominated and quantatively determined by GLC-ECD (10).

Another method (8) included hydrolysis with a Lickers-Nickerson apparatus and GC-MS detection.

2.4. CHLOROPHENOLS

2-, 3- and 4-Chlorophenol (33,34,35), 2,4-dichlorophenol (64), penta-chlorophenol (102) and trichlorophenols (122) were mostly determined gas chromatographically after a derivatization reaction with diazomethane, acetic anhydride, heptafluorobutyric anhydride, pentafluorobenzylbromide, 2,6-difluorobenzylbromide, pentafluorobenzoyl chloride or 2,4-dinitro-fluorobenzene. In this study the following methods were used:

A   acidification, extraction with toluene, extraction with K_2CO_3, derivati-zation with acetic acid anhydride; GC-ECD (1,2,10)

B   acidification, extraction with diethylether, ethylation with diazome-thane; GC-MS SIM (7)

C   isolation with Sep-Pak C_{18}; derivatization with acetic acid anhydride or with pentafluoro benzoylchloride; GC-ECD (9,12)

D   extraction with CH_2Cl_2; GC-MS (11)

E   derivatization with acetic acid anhydride in alkaline medium; n-hexane ex-traction; GC-ECD (6)

2.5. ORGANOPHOSPHORUS COMPOUNDS

The organophosphorus compounds azinphos-ethyl (5), azinphos-methyl (6), coumaphos (43), demeton (47), dichlorvos (70), dimethoate (73), disulfoton (75), fenitrothion (80), fenthion (81), methamidophos (93), mevinphos (94), omethoate (97), oxydemeton-methyl (98), parathion (100), phonix (103), triazophos (13) and trichlorfon (16) were determined with the following methods:

A   extraction with $CHCl_3$ or CH_2Cl_2; GC-MS or GC-AFID (P-mode) (2,3,8,10)

B   isolation with XAD-2 column; GC-ECD (9)

C   isolation with Sep-Pak C_{18}, GC-AFID (12)

D   isolation with activated carbon, GC-FID (12)

2.6. CHLOROPHENOXY ACIDS

Chlorophenoxy acids like 2,4-D (45), dichlorprop (69), MCPA (90), mecoprop (91) 2,4,5-T (107) were determined with the following methods:

A   acidification; extraction with diethylether; alkaline hydrolysis (aci-dic alumina), methylation with diazomethane or ethylation with a fu-ming sulphuric acid/ethanol; GC-ECD (5)

B   acidification; extraction with ethylacetate or diethylether, clean-up with desactivated silica and benzene/n-hexane elution (3:1), derivati-zation with PBF-bromide for GC-ECD analysis, HPLC ion-suppression (245 nm) (2)

C   extraction with dichloromethane; GC-MS (11)

D   extraction with chloroform or diethylether; methylation with diazome-thane; GC-ECD (3,8)

E   extraction with $CHCl_3$; derivatization with pentafluorobenzylbromide; GC-ECD (3)

2.7. POLYCYCLIC AROMATIC HYDROCARBONS (PAHs)

The compounds anthracene (3), and 3,4-benzopyrene/3,4-benzofluoran-thene (99) were analyzed with the following methods:

A   extraction with cyclohexane; clean-up with Al_2O_3 column; GC-FID (3)

B   extraction with CH_2Cl_2; GC-MS (1,11)

C extraction with n-hexane; clean-up with Al_2O_3; spectrofluorometric (12)
D extraction with cyclohexane, if needed formation of caffeine complexes;
   GC-MS (8)
E extraction with XAD-2 column, fluorescence detector EX-360, Em > 430
   (9)

2.8. OTHER COMPOUNDS

Arsenic (4), cadmium (12) and mercury (92) and their compounds were
determined by atomic absorption spectrometry (AAS) (4,6,9).

Benzylchloride (9), benzylidene chloride (10), biphenyl (11), 3-
chloropropene (32), 2-chlorotoluene (38), 3-chlorotoluene (39), 4-chloro-
toluene (40), 1,2-dichlorobenzene (53), 1,3-dichlorobenzene (54), 1,4-
dichlorobenzene (55), 1,3-dichloropropanol (66), epochlorohydrin (78),
hexachlorobutadiene (84), 1,2,4,5-tetrachlorobenzene (109), tributylphosphate (114), trichlorobenzene (117), 1,2,4-trichlorobenzene (118), tri-
fluralin (124) were generally determined by a simple liquid-liquid extraction with e.g. CH_2Cl_2 and without clean-up determined by GC-MS (2,11,12).

Dichlorodiisopropylether (57) was extracted with diethylether/n-
pentane (15:85), the extract was cleaned with florisil next GC-FID or
GC-MS/MID was used (2).

Simazine (106) was extracted with CH_2Cl_2 and determined with GC-AFID
(2,13).

3. CONCLUSIONS

For 102 of the 129 priority pollutants an analytical method was
available in one or more of the six contributing countries.

Attention had to be given to development of new analytical methodologies for the 27 compounds.

All the methods had to be compared by means of e.g. Round Robin
tests.

Confirmation criteria had to be stated for all priority pollutants

4. REFERENCES

1) Hoof, F.van, Questionnaire Belgium (1985)
2) Quaghebeur, D,   "   "   "
3) Erkomaa, K.,   "   Finland   "
4) Pontanen, P.,   "   "   "
5) Siltanen, H.,   "   "   "
6) Starck, B.,   "   "   "
7) Toivanen, E.,   "   "   "
8) Wickströur, K.,   "   "   "
9) O'Donnell, C.,   "   Ireland   "
10) Wegman, R.C.C.,   "   Netherlands "
11) Gjøs, Nand Urdal, K., "   Norway   "
12) Drevenkar, V.,   "   Yugoslavia "
13) Wegman, R.C.C. and Melis, P.H.A.M., CRC Press Critical Reviews,
    Anal.Chem., in press
14) Nunez, A.J., Gonzáles, L.F. and Janák, J., J.Chromatogr. 3, 127, 1984

EVALUATION OF SAMPLE PREPARATION PROCEDURES FOR THE DETERMINATION OF ORGANIC MICROPOLLUTANTS IN RIVER SEDIMENTS.

I.TEMMERMAN, W.MATHYS AND D.QUAGHEBEUR.

Institute for Hygiene and Epidemiology, Juliette Wytsmanstraat 14
1050 Brussel, Belgium.

Summary

Different sample preparation methods for the determination of organic
micropollutants in river sediments are compared and evaluated.
Sediment samples were spiked and the wet, chemically dried or freeze-
dried materials were extracted using methods such as Soxhlet extraction, ultrasonic agitation, shaking and stirring with solvent and
steam distillation. The extractions were performed with both
methylene chloride and cyclohexane on different sediments.
This study has shown that for the determination of polyaromatic
hydrocarbons, the best results were obtained from freeze-dried
sediment using ultrasonic agitation (70 % recovery) or stirring
(77 %) with methylene chloride.

Introduction

The fact that organic micropollutants are introduced in surface water
by a number of sources is well known. Due to their low solubility in water
and their specific physical properties, many of these compounds are easily
adsorbed onto suspended particulate matter and settle out into the sediments, establishing a condition of exchange between dissolved and adsorbed
material.

The slow release and in many cases the decreased bioactivity of the
sorbed material make sediments a continuous source of water pollution.
Micropollutant levels in sediments are generally several times higher than
in the water column; which is an advantage to their analysis. It is evident
that, when evaluating the water pollution suspended matter and sediments
should be taken into account.

A wide range of methods is available for the extraction of organics
from sediments, giving rise to a diversity of data so that comparing the
results is rather difficult. Also for sediment, the classical sample
clean-up or pretreatment can't be obviated, in this respect the water
content of sediments is an important factor.

Although the number of relevant papers dealing with sediment pretreatment and extraction methods (1,2) seems limited, the need for a uniform
procedure, especially suitable for routine analysis is evident.

It is a fact that sediments show certain characteristics in a way that
not only interpreting the results but also selecting the strategy to obtain
these results is sometimes difficult.

Surface sediments are frequently non-homogeneous. They are composed of
sands, silts and clays and it has been shown that these fractions accumulate organics in different ways (3,4). Moreover, small changes in extrac-

tion procedure can influence both quantitative and qualitative results. As for the analysis of water there is no method available for the determination of the different organic pollutants at once, also for sediments the extraction procedure is concentrated on classes of organic compounds.

Thus the sediment handling and the choice of extraction method depend on both the compounds to be analysed and on the physical characteristics of the sediment.

The purpose of this study was to compare different extraction methods for the determination of PAH in sediments and to evaluate the influence of the pretreatment, in particular drying, in order to find a procedure applicable to different sediment types and moreover to routine analysis. This study will be continued for the determination of other organic micropollutants such as pesticides, PCB, phenols...

Experimental
      Surface sediments were collected in the heavily polluted Scheldt estuary, and after collection stored at about 4°C.
Two different sediment types, one containing 76-92 % sand (fraction >16μ) and 1-5 % clay (fraction >16μ), the other containing 30-70 % sand and 11-37 % clay, were analysed for polyaromatic hydrocarbons.
Before drying and extraction, the sediments were thoroughly mixed during 1 h in order to make the samples as homogeneous as possible.
The recovery rates from the different extraction procedures were evaluated on samples spiked with six PAH (fluoranthene, benzo(b)fluoranthene, benzo (k)fluoranthene, benzo(a)pyrene, benzo(ghi) perylene, indeno (1,2,3,c,d) pyrene).
A solution of six PAH was prepared : 10 mg of each PAH was dissolved in 100 ml CH$_3$CN. 10 ml of this solution was diluted to 250 ml with distilled water. From this second solution 5 ml, thus containing 20 μg of each PAH was added to 20 g of wet sediment. Further the sample was mixed again during 30 min. Figure 1 gives an illustration of the followed procedure.

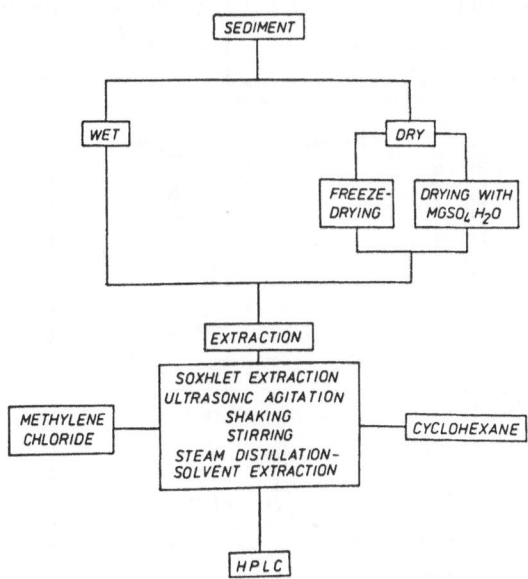

Figure 1.

Some of the HPLC analysis are illustrated in figure 2.

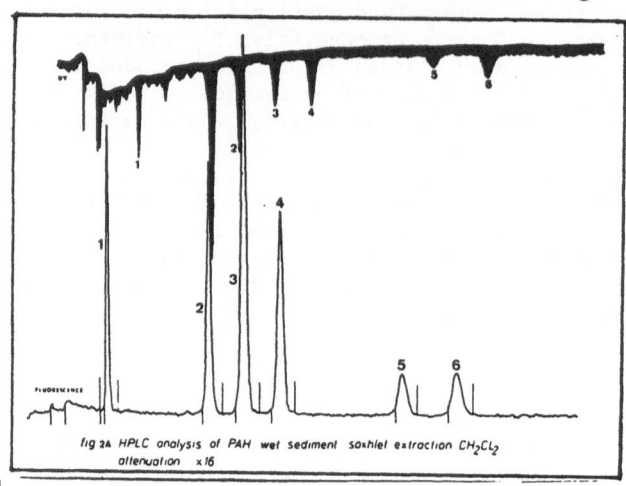

fig 2A HPLC analysis of PAH wet sediment soxhlet extraction CH_2Cl_2
attenuation x16

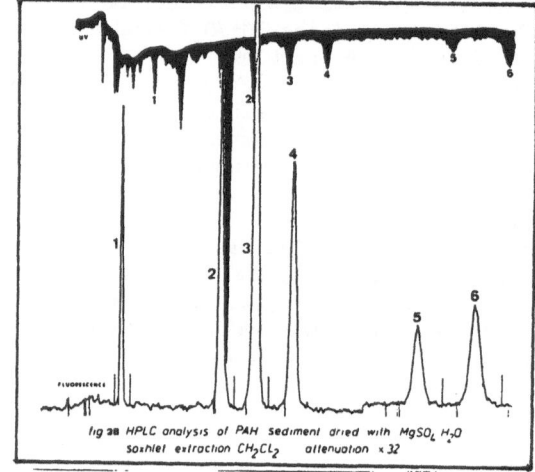

fig 2B HPLC analysis of PAH sediment dried with $MgSO_4$ H_2O
soxhlet extraction CH_2Cl_2 attenuation x32

1. *fluoranthene*

2. *benzo(b)fluoranthene*

3. *benzo(k)fluoranthene*

4. *benzo(a)pyrene*

5. *benzo(ghi)perylene*

6. *indeno(1,2,3,c,d)pyrene*

*for HPLC conditions
see text*

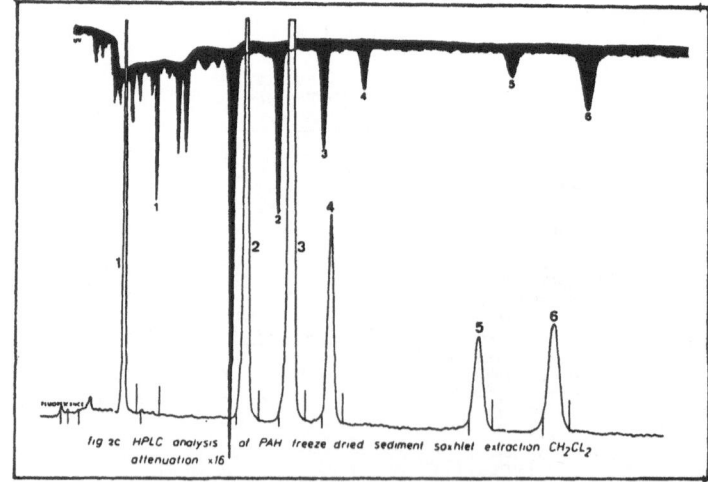

fig 2C HPLC analysis of PAH freeze dried sediment soxhlet extraction CH_2Cl_2
attenuation x16

The extractions were performed on both wet and dried sediments. Dried sediments were obtained in two ways : freeze-drying and drying with Magnesiumsulfate.monohydrate (MgSO$_4$.H$_2$O).
The extraction methods, with both methylene chloride and cyclohexane on wet and dried sediments were the following :
  a. Soxhlet extraction of 20 g sediment using 150 ml of solvent, during 4h.
  b. ultrasonic agitation in glass beakers containing 20 g of sediments and 150 ml of the above mentioned solvents, during 30 min. Afterwards the extract was filtered on filter paper.
  c. stirring of 20 g sediment with 150 ml solvent, during 30 min. and filtration of the extract.
  d. the same procedure as c but in stead of stirring, the sediment was extracted using three-dimensional shaking.
  e. steam distillation-solvent extraction, using the apparatus as described by Godefroot et al (6). To 10 g of sediment 50 ml of distilled water was added and the sediment was extracted with 1 ml of solvent during 1h30.
The resulting extracts were adjusted to 150 ml with the solvent in question. A specific amount of the extract, depending on the concentration of PAH (in most cases 10 ml) was concentrated to near dryness on a rotavapor and to dryness under nitrogen. Prior to HPLC analysis, the residue was dissolved in acetonitrile.
HPLC analysis was performed with a Waters Associates instrument, equipped with a UV detector (254 nm) and a fluorescence detector (excitation 360 nm and emission 430 nm) in parallel. The mobile phase was acetonitrile/water 85 : 15 at a flow rate of 1.5 ml/min.

Results and discussion
     The recovery rates for PAH (total of six) from wet sediment, freeze-dried sediment and sediment dried with MgSO$_4$.H$_2$O using the different extraction methods mentioned above, are shown in Table 1. These results are the mean value of three extractions.

		WET SEDIMENT	FREEZE-DRIED SEDIMENT	SEDIMENT DRIED WITH MgSO$_4$ H$_2$O
SOXHLET EXTRACTION	CH$_2$Cl$_2$	50%	66%	37%
	◯	69	67	31
ULTRASONIC AGITATION	CH$_2$Cl$_2$	17	70	50
	◯	17	38	7
SHAKING	CH$_2$Cl$_2$	25	58	56
	◯	44	22	39
STIRRING	CH$_2$Cl$_2$	23	77	29
	◯	48	40	68

TABLE 1. RECOVERIES OF PAH FROM SEDIMENT

The best results are observed with freeze-dried sediments. Both wet and with $MgSO_4.H_2O$ dried sediments give lower recoveries of PAH.
Regarding the different extraction methods, for freeze-dried sediment, Soxhlet extraction with both methylene chloride and cyclohexane, (66 % and 67 % respectively), ultrasonic agitation with CH_2Cl_2 (70 %) and stirring with CH_2Cl_2 (77 %)give the highest recoveries for PAH. The last two are interesting meaning that they are much less time consuming than Soxhlet extraction.

Also when wet sediment is extracted, although the recoveries are lower as compared with freeze dried samples, Soxhlet extraction is better (CH_2Cl_2 : 50 %, cyclohexane : 69 %).The four extraction methods gave better results, and in one case at least the same result (ultrasonic agitation CH_2Cl_2 : 17 %, cyclohexane : 17 %) with cyclohexane as solvent.

In the case of chemically dried ($MgSO_4.H_2O$) sediment, in contrast to wet and freeze-dried material, ultrasonic agitation (CH_2Cl_2 : 50 %), stirring (cyclohexane : 68 %) and shaking (CH_2Cl_2 : 56 %) give higher recoveries,than Soxhlet extraction (CH_2Cl_2 : 37 %, cyclohexane : 31 %).

As for the individual PAH (the same amount of each PAH was added to the sediment) it was noticed that in the case of freeze-dried and chemically dried sediment the measured amounts of the different PAH are of the same order of magnitude, whereas for the wet sediment the recoveries decrease with increasing molecular weight of the PAH (table 2).

	FLUORANTHENE	BENZO(B) FLUORANTHENE	BENZO(K) FLUORANTHENE	BENZO(A) PYRENE	BENZO(GHI) PERYLENE	INDENO (123CD) PYRENE
SOXHLET EXTRACTION CH_2Cl_2						
WET SEDIMENT	74% 35	59 20	55 18	47 17	38 7	33% 6
SEDIMENT DRIED WITH $MGSO_4 H_2O$	30 51	38 51	37 49	28 49	41 49	47 53
FREEZE-DRIED SEDIMENT	68 65	71 70	88 81	31 69	63 69	72 65
ULTRASONIC AGITATION CH_2Cl_2						

TABLE 2. RECOVERIES OF SIX PAH FROM SEDIMENT

This phenomenon is even stronger in the case of steam distillation-solvent extraction. The recovery of PAH is much lower than for the other extractions and of the heaviest compounds only trace amounts (about 1 %) could be recovered.

All these results were obtained from the sediment containing 76-92 % sand and 1-5 % clay. Only a few experiments were performed on the second type of sediment. The results of these are lower, although the differences are minimal. However, regarding the different extraction methods, the recoveries follow the same tendency as compared with the results of the first type (e.g. Soxhlet extraction CH_2Cl_2 : 55 %, cyclohexane : 55 %, ultrasonic agitation CH_2Cl_2 : 64 %).

Conclusions

This study has shown that extraction methods other than the conventional Soxhlet extraction are suitable techniques for the determination of polyaromatic hydrocarbons in aquatic sediments. It is also shown that pretreatment of the sediment has an important influence on the recovery. In this respect, freeze-drying of the sediment before extraction is preferred.

In the near future, this study will be enlarged for the determination of other organic compounds such as pesticides, PCB in aquatic sediments.

Acknowledgements

The authors wish to thank G. Hiernaux, K. De Greeve, D. De Leersnijder and J. Bruggemans for their assistance.

References

1. J. Grimalt, C. Marfil and J. Albaigés, Intern. J. Environ. Anal. Chem. Vol 18 (1984) p. 183.
2. C.O'Donnell. Analysis of organic micropollutants in water, Proceedings of the Third Eur. Symp., Oslo 1983 (G. Angeletti and A. Bjørseth) p. 36.
3. S. Thompson and G. Eglinton, Geochim. Cosmochim. Acta 42 (1978) 119.
4. S.C. Brasell and G. Eglinton, Analytical Techniques in Environmental Chemistry Vol I (J. Albaigés, Pergamon Press, 1980) 1-22
5. J.C. Monin, R. Pelet and A. Fevrier, Rev. Inst. du Pétrole 33 (1978) 223.
6. Godefroot M., Stechele M., Sandra P. and Verzele M., Analysis of organic micropollutants in water, Proceedings of the Second Eur. Symp., Killarney 1981 (A. Bjørseth and G. Angeletti) p. 16.

TRACING A SOURCE OF POLLUTION BY DETERMINATION OF SPECIFIC POLLUTANTS IN SURFACE- AND GROUNDWATER

D. QUAGHEBEUR, G. HIERNAUX and E. DE WULF
Institute for Hygiene and Epidemiology, Juliette Wytsmanstr. 14
B-1050 Brussels (Belgium)

SUMMARY
On the occasion of the detection of both bis (2-chloroisopropyl) ether (BCIE) and bis (2-chloroethyl) ether (BCEE) in a polluted groundwater it has been possible to trace the source of this specific pollution on a distance of more than 100 km in a heavily polluted area. Furthermore direct infiltration of these pollutants to an aquifer, formerly used for drinking water preparation, could be demonstrated by determining the ratios between these compounds in both the infiltrating surface water and in the receiving groundwater. Multiple Ion Detection Mass Spectrometry has been proven an exellent procedure to trace specific pollution sources in heavily polluted water.

1. INTRODUCTION

Many studies have been carried out to identify and to quantitate organic micropollutants in surface waters and groundwaters. Making use of very powerful and still improving techniques, such as high-resolution gas chromatography, high-performance liquid chromatography and mass-spectrometry, an increasing number of data becomes available and efforts are made to compile them, sometimes combined with information on sampling location and analytical methodology (1). Although this information is very useful in demonstrating the nature and the actual degree of pollution by organic compounds, environmental management obviously is also interested in tracing the source and the fate of a given pollutant. Regarding surface waters in highly industrialized and densely populated areas, this is not so easy. Not only because phenomena like absorption, evaporation, chemical and biological transformation during transportation are an encumbrance for unambiguous indication of the source and the fate of a given pollutant, but also, in many instances, the multiplicity of possible sources and the degree of pollution by a high number of other pollutants make it very difficult to come to a clear conclusion.

The starting-point of the present study was the detection of considerable amounts of both the haloethers bis (2-chloroisopropyl) ether (BCIE) and bis (2-chloroethyl) ether (BCEE) in the groundwater and surface water of a certain region in Belgium during an overall inventory study. As, according to earlier experience, these compounds are not common water pollutants in this country it was decided to determine the extent of this pollution and to trace its sources.

2. EXPERIMENTAL

The extraction procedure was based on the method described by Dressman et al. (2). The water sample (1 litre) was extracted thrice with 50 ml of 15% diethylether in pentane. The combined extracts, together with 125 ng 1-chlorononane as the internal standard, were concentrated on a Kuderna-Danish apparatus to about 0,5 mL. The recoveries were 86% and 53% for BCIE and BCEE respectively. The analysis was performed on a GC-MS (Finnigan 4025) with MID-detection of the fragment ions 77,121,123 for BCIE, 63,65,93,95 for BCEE and 91 for the internal standard; scan time 1.1.

sec and scan rate 0.1 sec. For the gaschromatographic separation a fused
silica capillary column (CP WAX 57, 25 m x 0.22 mm) was programmed from 50
°C to 150 °C at 5 °C min^{-1}.

The detection limits were 0.005 µgL^{-1} and 0.001 µgL^{-1} for both BCIE and
BCEE respectively. However, for instrumental reasons, in the last series of
analyses on groundwater these limits were about 0.040 µgL^{-1} for both
compounds.

3. ORGANISATION

As stated above, this study started in connection with the detection of
important concentrations of haloethers in a certain area. In order to point
out the source(s) of this pollution step-by-step sampling of possibly
contributing rivers was carried out as determined by the preceeding
findings. This means that, due to the lapse of time between the samplings,
the interpretation of the results should be made based on their orders of
magnitude rather than on the absolute values. Detailed data of the present
study have been reported in Dutch (3) and in French (4).

4. RESULTS AND DISCUSSION

In the surroundings of an industrial gypsum waste deposit, situated
between the channel Ghent-Terneuzen and the Scheldt estuary, (fig 1)
concentrations of both BCIE and BCEE ranging from about 300 to 1100 µg.L^{-1}
and 0.010 to 0.035 µg.L^{-1} respectively were found in the groundwater. As
the most apparent reason for this contamination was the infiltration of
polluted water from the precited channel used for wet transportation of the
gypsum waste, and investigation was carried out to determine the real
source(s) or the industry discharging these compounds.

fig.1. SITUATION MAP OF SAMPLED SURFACE WATERS

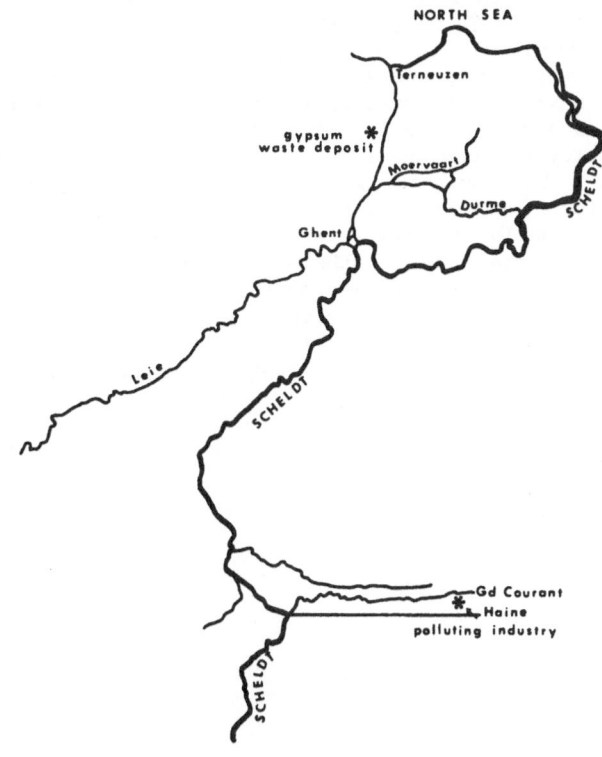

During one particular sampling period in this channel, heavily polluted by adjacent industries, concentrations were found ranging from 0.4 to 1.4 μgL^{-1} for BCIE and from 1.6 to 7.9 for BCEE. However, during an earlier and a later sampling period in general much lower concentrations were observed (0.1 to 1.0 and 0.01 to 0.7 respectively), indicating an irregular or discontinuous supply of these pollutants. Although sampling locations were very close to each other (interjacent distances about 500 m) it was difficult to make conclusions as for a well defined discharge point of these compounds.

The reason for this could be the fact that some adjacent industries use the channel water for cooling and thus concentrate the pollutants in the rejected water, what actually has been observed during control of the effluents. Furthermore the undefined flow directions as normal in many (artifial) channels and in this particular case the complicated and altering system of water supply to the channel by different rivers also disturb a normal distribution pattern.

An investigation in the small side-channel Moervaart (fig. 1) and other smaller creeks directly in connection with it, revealed some traces of BCIE (up to 0.05 μgL^{-1}) demonstrating further dispersion of these pollutants. In the river Durme, situated in the same region, and as a confluent influenced by the tides of the lower course of the river Scheldt, concentrations were found increasing in down-stream direction from < 0.005 to 0.130 μgL^{-1} for BCIE and to trace concentrations for BCEE. In the river Scheldt (Fig. 1) concentrations increased from 0.06 μgL^{-1} BCIE and 0.003 μgL^{-1} BCEE downstream to maximum values of about 2.35 and 0.04 μgL^{-1} respectively at the confluence of the river Haine and upstream this confluent drastically dropped to below the detection limits, meaning that the real source of the haloether pollution should be found in the surroundings of the river Haine.

It should be stressed that the river Scheldt is the major water supplier of the channel Ghent-Terneuzen, thus also the most possible cause of the haloether pollution in this channel. An intensive investigation in the river Haine, its confluents and creeks in the surroundings, clearly indicated an industry having discharges in the river Haine and in a near by creek called Grand Courant as well (Fig. 1). Near these discharges concentrations in the former were about 101 μgL^{-1} BCIE and 7 μgL^{-1} BCEE and in the latter 69 μgL^{-1} BCIE and 58 μgL^{-1} BCEE. The observation that no positive values were found, neither upstream these discharges nor in other near-by channels and confluents nor upstream the Haine in the river Scheldt, clearly indicate the cause of the pollution by haloethers of the downstream part of the river Scheldt and other surface waters directly or indirectly supplied by the water of this river.

Another striking observation also could be made in connection to the permeation of BCIE and BCEE from the surface water into the groundwater.

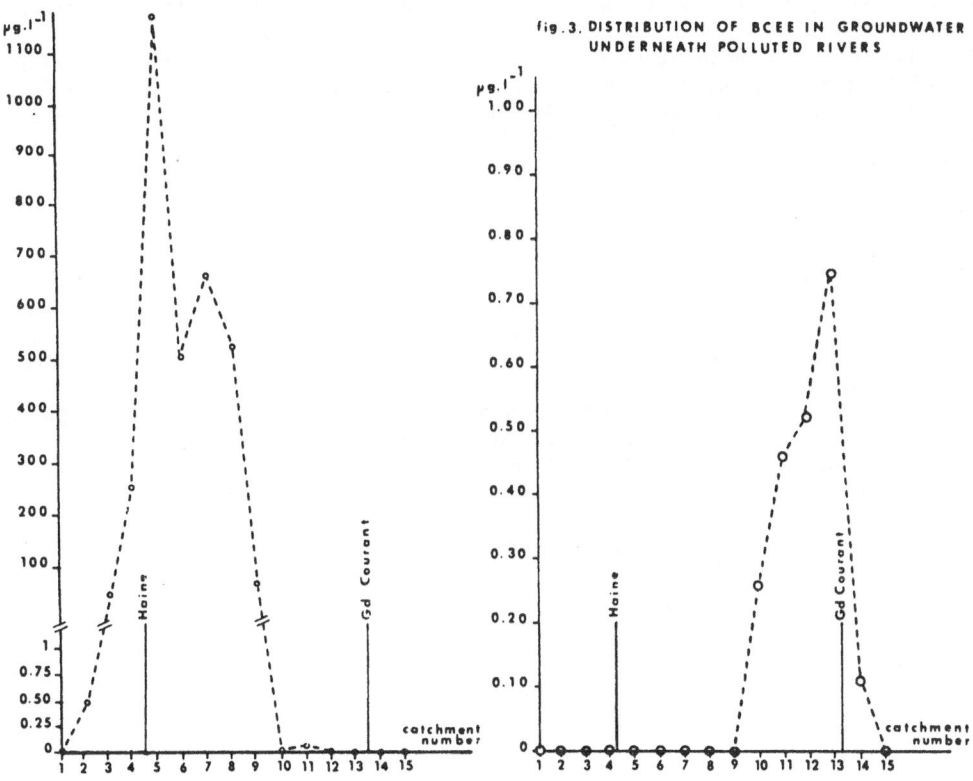

fig.2. DISTRIBUTION OF BCIE IN GROUNDWATER
-UNDERNEATH POLLUTED RIVERS

fig.3. DISTRIBUTION OF BCEE IN GROUNDWATER
UNDERNEATH POLLUTED RIVERS

An aquifer in a geological structure of carboniferous limestone is situated just underneath the rivers haine and Grand Courant, both polluted by BCIE and BCEE as demonstrated above. A series of 15 catchments, on a distance from each other of about 150 m, are located in a straight line perpendicularly on the precited rivers, pumping the water on a depth of 70 to 100 m in the aquifer (Fig. 2 and 3). Concentrations in these catchments ranged from < 0.04 μgL^{-1} to about 1200 μgL^{-1} and from < 0.04 μgL^{-1} to 0.75 μgL^{-1} for BCIE and BCEE respectively. The highest values are observed in the catchments nearest to the polluted rivers, and decrease with increasing distances to them (fig. 2 and 3). The distribution of these pollutants in the groundwater as a function of the distance to the polluting rivers in some degree also reflects the different ratios of BCIE to BCEE in both the rivers Haine and Grand Courant. Only in the groundwater underneath the latter, with relatively important amounts of BCEE, a perceptible maximum for this compound has been found, whereas BCIE-concentrations were predominant in the catchments near the river Haine, also containing the highest amounts of BCIE. This would indicate, at least in this particular geological situation, a direct infiltration of pollutants into the ground-water.

5. CONCLUSION

Using the technique of Multiple Ion Detection (MID) mass spectrometry it has been possible to trace specific organic micropollutants in surface waters on a distance of more than 100 km in a heavily polluted area. Although the compounds studied are not so common water pollutants enabling a relatively easy discrimination among other pollutants and sources of pollution, this technique could also be applied for more common pollutants. The increased sensitivity obtained by the MID-technique also permits to determine trace concentrations present in the groundwater as a consequence of infiltration.

REFERENCES

1. An inventory of polluting substances which have been identified in various fresh waters, effluent discharges, aquatic animals and plants, and bottom sediments. CEC. Concerted Action : Analysis of Organic micropollutants in water (COST 64 b bis) Fourth edition 1984. XII/398/84-EN OMP/51/84.

2. DRESSMAN, R.C., J. FAIR air and E.F. Mc FARREN : J. Environ. Sci. Health, Al. 3(2), 139-147 (1978).

3. QUAGHEBEUR D., G. HIERNAUX and E. DE WULF : Haloethers in enkele oppervlaktewaters en grondwaters van het Scheldebekken. Report IHE-Belgium (1984).

4. QUAGHEBEUR D., G. HIERNAUX and E. DE WULF : Haloethers dans quelques eaux de surface et eaux souterraines du bassin de l'Escaut. Report I.H.E.-Belgium (1984).

DETERMINATION OF EXTRACTABLE ORGANOHALOGENS IN LIVING TISSUE.

C. GRØN and J. FOLKE
VKI, Water Quality Institute
11, Agern Allé, DK-2970 Hørsholm, DENMARK

SUMMARY.

Organohalogens may routinely be determined in water and wastewater samples as group parametres in fractionated analyses such as granular, activated carbon (GAC) adsorbable organic halogens (AOX), extractable organic halogens (EOX), or purgable organic halogens (POX). Generally, organohalogens with propensity for bioaccumulation are found in the EOX-fraction of the water or wastewater and the purpose of this study was to develope and evaluate a method for the determination of EOX in living tissue, *in casu* blue mussels (*Mytilus edulis*). Mussel soft tissue samples were homogenized, grinded with sodium sulphate (sic.) and soxhlet extracted overnight with pentane. A small amount of hexadecane was added, before the volume of extract was gently reduced, until the pentane had evaporated. The content of organohalogens in the reduced extract was determined by combustion of the extract followed by a microcoulometric titration of the evolved hydrogen halide in a Dohrmann DX 20 TOX-analyzer. Using 50 g wet mass of mussel samples the method is applicable in the range of 2-50 Cl µg/g living tissue. Precision is ≈10% (coefficient of variation), and recovery is estimated to 90% from *p*-dichlorobenzene and 2,4,6-trichlorophenol standard additions to the mussel samples. Field samples of mussels from Danish coastal waters gave results of accumulated organohalogens between 5 and 20 Cl µg/g living tissue.

1. INTRODUCTION.

Organic compounds containing halogen substituents (organohalogenes) are mostly xenobiotics, and, consequently, many are considered to be of potential, environmental hazard[1]. Initially, the environmental exposure to organohalogens was assessed by a variety of chromatographic methods, *e.g.* gas chromatography using the electron capture detector (GC/ECD)[2] or combined with a mass spectrometric detector (GC/MS)[3,4]. In order to follow trends, monitoring programs have been conducted in many countries, *e.g.* in US/Canada[5], UK[6,7], Norway[8] and Sweden[9,10] since the sixties to determine the levels in living tissue of selected organohalogenes in the marine environment, *e.g.* DDT, PCB's, HCB *etc.* Today, the concern of the results of these monitoring programs has led to a ban or a restricted use of several of these compounds[11], but other halogen-containing compounds such as toxaphenes, chlorinated alkanes *etc.* have come into use; compounds which can hardly be analysed for specifically[4].

For some years now, chemical screening analyses of environmental samples have included the possibility of determination of group parametres such as total organic carbon (TOC)[12] and total organic halogenes (TOX)[12]. General steps of an OX analysis are *(i)* isolation of organohalogens from the aqueous matrix, *(ii)* removal of inorganic chloride from the sample, and *(iii)* detection of the organohalogen content in the sample. Commonly used isolation/concentration steps are adsorption on granular, activated carbon (GAC)[13] or crosslinked polystyrene-divinylbenzene resin (XAD)[14], solvent-extraction of hydrophobic organohalogenes[15] or purging of volatile organohalogenes[16]. The determination of halogens is mainly achieved by combustion or chemical destruction, followed by detection of the formed hydrogenhalides by coulometry, potentiometric titration or by using ion-selective electrodes [12]. Neutron-activation analysis has been used in some instances, too [17].

Determination of total organic halogenes (TOX) is merely a theoretical concept, *i.e.* the result depends very much on operational conditions. Operational parametres have been defined as mentioned (*vide supra*), such as adsorbable organohalogens (AOX) [13,14,18], extractable organohalogens (EOX) [15], and purgable organohalogens (POX) [16], and they have all been used in a variety of investigations, *e.g.* drinking water quality assurance [19], control of wastewater treatment processes [20], characterization of wastewaters from bleach plants of pulp manufacturing [21,22] and ground water pollution monitoring [23,24]. Recently, the EOX-methodology has been extended to include determinations in tissue from fish and mussels. Detected levels reported range from 500 ppm (Cl μg/g lipid) in perchs [25] to 500-2.000 ppm in fish oil [26].

Over the past few years several impact assessment studies have been conducted for the marine environment of several different emissions of organochlorines [27]. In several cases immissions [28] in the marine environment have proved to be very difficult to establish [29,30], partly owing to a lack of knowledge of background concentrations of organohalogens both in terms of specific compounds and in terms of overall levels. The aim of this study has been to establish a level of reference of extractable organochlorines (EOX) in blue mussels (*Mytilus edulis*) in Danish coastal waters.

Mussel soft tissue samples were homogenized, grinded with sodium sulphate (sic.) and soxhlet extracted overnight with pentane. A small amount of hexadecane was added, before the volume of extract was gently reduced until the pentane had evaporated. The content of organohalogenes in the reduced extract was determined by combustion of the extract followed by a microcoulometric titration of the evolved hydrogen halide in a Dohrmann DX 20 TOX-analyzer. Using 50 g wet mass of mussel samples the method is applicable in the range of 2-50 Cl μg/g living tissue, *fig.1*. Precision is ≈10% (coefficient of variation), and recovery is estimated to 90% from *p*-dichlorobenzene and 2,4,6-trichlorophenol standard additions to the mussel samples. Field samples of mussels from Danish coastal waters gave results of accumulated organohalogens between 5 and 20 Cl μg/g living tissue.

2. EXPERIMENTAL.

2.1 SAMPLING.

A mussel sample from the commercial fish market in Copenhagen was used initially for set-up and test-performance of the EOX-method. Then, soft tissue materials, collected from the geographical sites shown on *fig.2*, were used in the further study. All mussels from each site were subject to shell length determinations, and their soft tissues were pooled from each site and frozen (–20°C) for later analyses of dry matter, lipid, and EOX contents (*vide infra*).

2.2 ANALYSIS.

The total shell length of mussels were determined with an accuracy better than 0.1 mm. Dry matter content of mussels soft tissue was determined gravimetrically after drying homogenized sub-samples at 105°C for two days.

Lipid content of mussels was determined by analysing 5-10g of each homogenized sample [29]. The homogenate was mixed with anhydrous sodium sulphate (100 g Na_2SO_4, *sic.*, soxhlet extracted with *n*-pentane overnight) and allowed to equilibrate overnight, when the sample was grinded with additional dry sodiumsulphate in a mortar to give a dry consistency of the sample. The sample was then soxhlet extracted overnight with 150-200 ml *n*-pentane; the extract was filtered and the pentane was evaporated to sample dryness on a vacuum rotor-vapour apparatus, dried at 60°C for 30 min., cooled in an excicator, and the lipid content determined gravimetrically.

Extractable organohalogenes (EOX) were isolated from approximately 50 g of soft tissue homogenate. The lipid phase containing the EOX-fraction was isolated from the dried homogenate by a procedure similar to the lipid extraction (*vide supra*). Exactly 200 ml *n*-

pentane (p.a. distilled) were used. The volume of extract was then reduced *in vacuo* at 10°C to approximately 10 ml before 500µl of hexadecane (p.a.) was added and the residual pentane was evaporated. The extract volume was adjusted to amount to 1000 µl, and 25 µl of the extract were transferred to a 40 mg granular, activated carbon bed (100-200 mesh granular activated carbon P/N 511-877 from Dohrmann) at the boat inlet of the TOX-analyzer (Dohrmann DX-20). The content of organohalogenes was determined as chlorine by combustion of extracted organics, solvents, and carbon bed, followed by microcoulometric titration with silver ions of the evolved hydrogen halides. The analyzer had been calibrated prior to the determinations using 2,4,6-trichlorophenol standards.

3. RESULTS AND DISCUSSION.

A method has been developed and evaluated for the determination of extractable organohalogenes (EOX) in living tissue, *in casu* blue mussels (*Mytilus edulis*), for the purpose of being able to establish a level of reference of EOX in biota in Danish coastal waters.

3.1 METHOD EVALUATION.

The performance of the method is summarized in **table I**.

Table I. Performance characteristics for the determination of EOX in mussels.

Standard addition[†] (Cl µg/g wett mass)	Number of samples	Mean concentration found (Cl µg/g wett mass)	Recovery of standard[*] (%)	Standard deviation (Cl µg/g wett mass)	Coefficient of variation (%)
0	5	9.89	-	1.22	12.3
22	4	29.62	90.3	2.92	9.9

[†]1.069 mg organic chlorine is added to each portion of approximately 50 g of mussels.
[*]Recovery of standard is calculated separately for each spiked sample, and the mean is given.

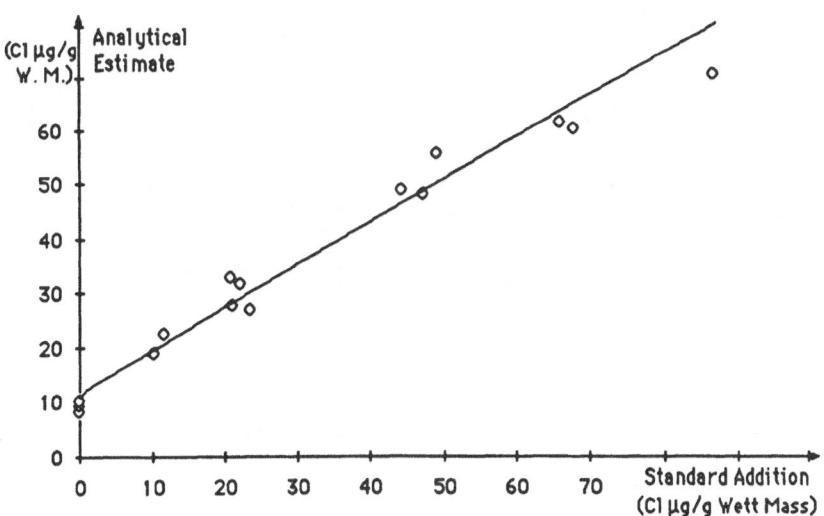

Fig. 1. The analytical results plotted against standard additions to the sample proves the linear dynamic range of the method. Concentrations are in Cl µg/g wett mass.

Recovery and linearity estimates of the method were based on standard additions of an ethanolic solution of p-dichlorobenzene (1.125 g/l) and 2,4,6-trichlorophenol (0.979 g/l), making up 1.069 Cl g/l. Chlorine of the standard solution was recovered 98.5% after combustion and detection steps of the TOX-analyzer.

The limit of detection may be estimated to 1.75 Cl µg/g wett mass based on 50 g wett sample mass, by calculating the standard deviation of blank sample analyses of appropriate amounts of sodium sulphate. The dynamic range of the method was determined from standard additions to the homogenized mussel sample; results are depicted on *fig.1*. The best linear relationship between concentrations and responses was obtained in the range below 50 Cl µg/g wett mass.

From **table 1** can be seen, that the method within its dynamic range has a precision of ≈10% (coefficient of variation); the recovery was found to be ≈90% as estimated from p-dichlorobenzene and 2,4,6-trichlorophenol; and from *fig. 1* can be seen that the dynamics of the method range from 2 to 50 Cl µg/g wett mass. The EOX content of this commercial fish market sample was estimately 10 ppm (wett mass).

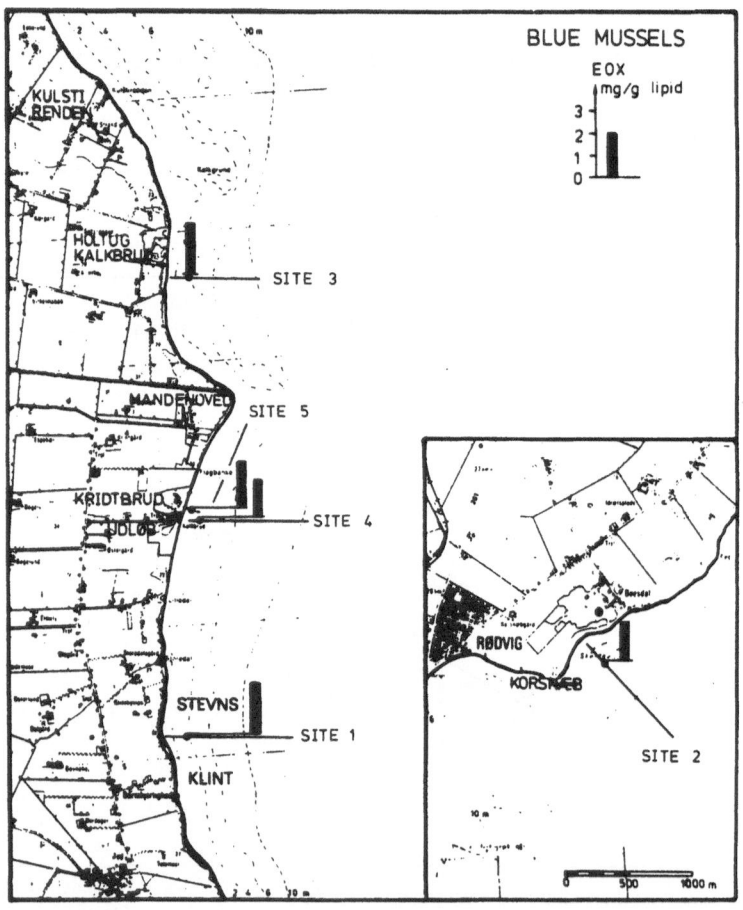

Fig. 2.   *Geographical sampling sites and plot of EOX-content (Cl mg/g lipid) of mussel populations, the shell length distributions of which are plotted on fig. 3.*

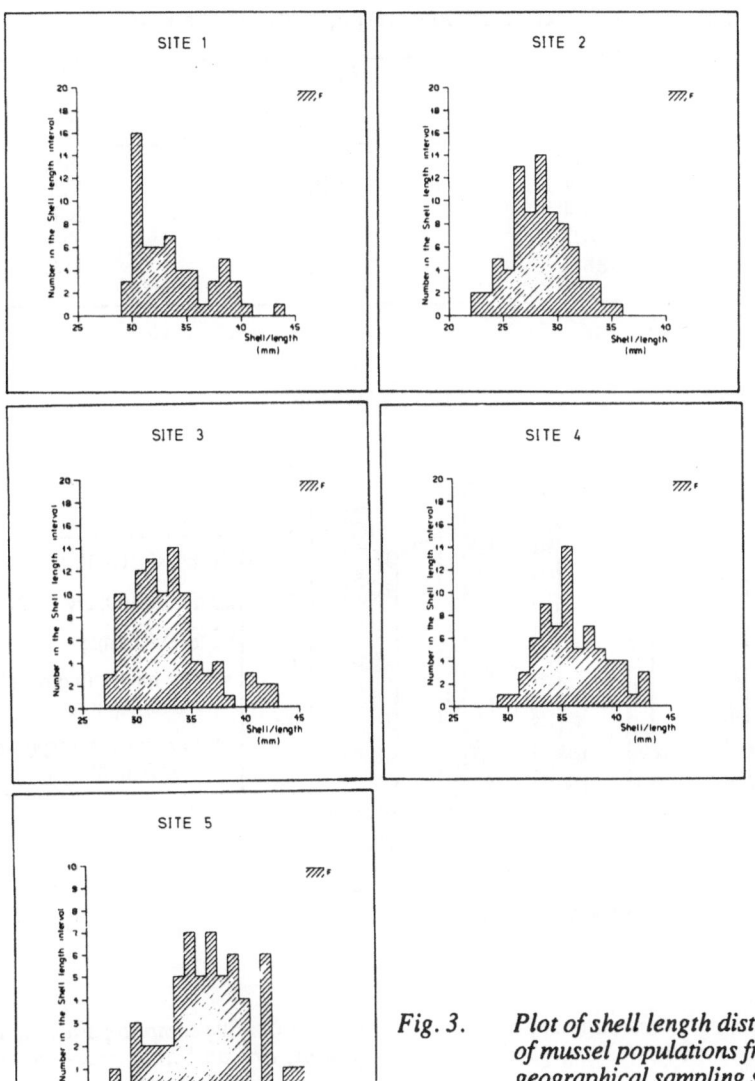

Fig. 3. *Plot of shell length distributions of mussel populations from each geographical sampling site.*

3.2 MONITORING RESULTS.

The analytical monitoring results are summarized in **table II**, and the EOX-results are then related to the geographical sites as illustrated on *fig. 2*. The distributions of shell length variations of the mussel samples from each geographical site are shown, *fig.3*. The basic results of **table II** are further graphically evaluated on *fig. 4*.

Table II. Characteristics of collected mussel populations from the "Køge Bugt" including EOX-determinations.

Site #	Dry matter mass mg/g wett mass	Lipid mass mg/g wett mass	EOX µg/g wett mass
1	41.2	2.65	5.9
2	36.1	3.06	5.3
3	35.4	3.85	8.7
4	102	10.9	20
5	88.3	7.99	16

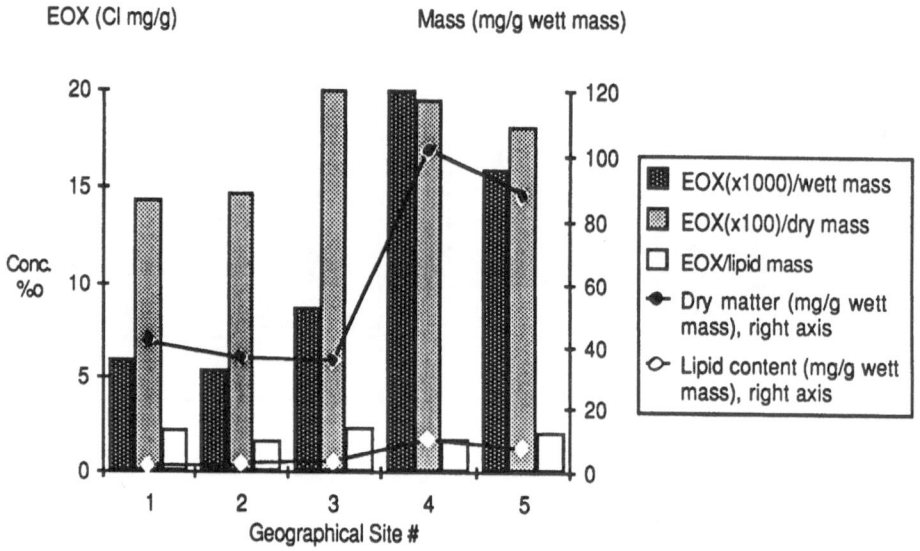

Fig. 4.   *Graphical presentation of data given in table II from five sampling sites including EOX-contents (Cl mg/g) on the basis of wett, dry and lipid mass of mussel populations.*

From **table II** can be seen that the collected wild mussel populations showed a three-fold variation in dry matter and lipid mass contents, and EOX-content variations were from 5 to 20 ppm (wett mass). To some extent these variations could be explained by the difficulty of differentiating the amount of tissue liquids associated with the frossen mussel samples from pure sea water. Calculated on the basis of the lipid mass content EOX variations of the five sites are only from 1,700 to 2,300 ppm (lipid mass), and no great variation is thus observed between the geographical sites, *fig.2*.

The geographical site chosen has no recognized, substantial sources of emissions (industrial or municipal) of organohalogenes. The area is dominated by several white cliffs (calcium carbonates) and, therefore, proved to be very poor on nutrients for mussels according to an evaluation performed by a VKI-biologist (cf. acknowledgement), who thinks that the age of the mussel populations is quite possibly higher than average for this size of mussels, *fig.3*. However, no accurate estimate of the age of the mussel populations has been made, only that they are expected to be a few years old, whereas mussels this size normally are no more than one or two years in Danish coastal waters.

It is quite evident from *fig.4* that the mussel populations show great variations in dry matter and lipid contents. Whether or not that is due to analytical uncertainties as discussed (*vide supra*), it has the consequence, that EOX-contents vary a great deal depending on the mass of reference being wet, dry matter or lipid. However, the fact that EOX-contents quantified with reference to the lipid mass of the mussel populations are quite constant, roughly 2,000 ppm, proves that the organohalogenes are associated with the lipids of the mussels, *i.e.* if they are not bioaccumulated intra-cellulary they are certainly very closely related to the cell membranes of the animal. It may be speculated that a great part of the organohalogenes is more or less persistent, high molecular mass compounds (*e.g.* from the pulp and paper industry discharging into the Baltic Sea[31]), that have not really penetrated the cell membranes of the animals, but are only closely associated or adsorbed to this lipid compartment. But it may also be possible that these indeed very high extractable organohalogen contents of wild mussels are only summing up the environmental exposure to a lot of different low molecular mass compounds, which are in fact bioaccumulated and, therefore, are of concern, *e.g.* the sum of PCB's, toxaphenes, chloroalkanes *etc*. Further experimental evidence is needed to conclude on the above hypotheses, especially concerning molecular mass distributions of the EOX-fraction; also further monitoring results are needed and in progress at our laboratory of samples of mussels and marine sediments.

4. ACKNOWLEDGEMENTS.

Jørgen Birklund has been in charge of mussel samplings and of biological surveillance of the geographical, marine area where the sampling sites were situated. Lone B. Christensen has skilfully performed all the analytical work in the laboratory. This publication was financed by part of a basic grant to the VKI from the *Danish Council of Technology*.

REFERENCES.

1. Keith, L.H., Advances in the Identification & Analysis of Organic Pollutants in Water, Vol. 2. (L.H. Keith (ed.). Ann Arbor Science, Ann Arbor, 1981), pp 1165-1170.
2. Zlatkis, A. & C.F. Poole (eds.), Electron Capture - theory and practice in chromatography (Elsevier Scientific Publ. Com., Amsterdam, 1981).
3. Jansson, B. & U. Wideqvist, Intern J. Environ. Anal. Chem., **13**, 309-321 (1983).
4. Bergman, Å. , A. Hagman, S. Jacobsen, B. Jansson & M. Åhlman, Chemosphere, **13**, 237-250 (1984).
5. Morrison, P.F., J.F. Leatherland & R.A. Sonstegard, Aquatic Toxicology, **6**, 73-86 (1985).
6. Rickard,D.G. & M.E.R. Dully, Environ. Pol. (Series B), **5**, 101-119 (1983).
7. Murray, A.J. & J.E. Portmann, Metals and Organochlorine Pesticide and PCB Residues in Fish and Shellfish in Ergland and Wales in 1976 and Trends since 1970 (Ministry of Agriculture Fisheries and Food Directorate of Fisheries Research, Publ. No. 10, United Kingdom, 1984).
8. Knutzen, J., K. Martinsen & K. Næs, Preliminary Note on Organochlorines in Organisms and sediments from the Kristiansfjord (S. Norway) in 1982-83 (in Norwegian), Vann, **3**, 392-400 (1984).
9. Jansson, B., R. Vaz, G. Blomqvist, S. Jensen & M. Olsson, Chemosphere, **8**, 181-190 (1979).

10. National Swedish Environment Protection Board, Heavy Metals and Organic, Xenobiotic Toxicants in the Swedish Environment (in Swedish), Monitor, 86-95 (1982).
11. EEC, Council Directive of the 4th of May 1976 Concerning Pollution, (EEC-directive, The Aquatic Environment, 76/464/EEC, 1976).
12. Mineau, R.A. & L.H. Keith (eds.), Water Analysis, Vol III, Organic Species, (Academic Press, New York, 1984).
13. Oake, R.J & I.M. Pinder, The Determination of Carbon Adsorbable Organo-halide in waters, (Water Research Centre, Technical Report TR 217, October, 1984).
14. Glaze, W.H., G.R. Peyton & R. Rawley, Environ. Sci. Technol.,11, 685-690 (1977).
15. Deutsches Institut für Normung, Bestimmung der extrahierbaren organisch gebundenen Halogene (EOX) (DIN 38409, November, 1982).
16. Cooper, W.J. (ed.), Chemistry in Water Reuse, (Ann Arbor Science Publ. Inc., Vol. 2, 1981), Chap. 7.
17. Ahnhoff, M., B. Josefsson, G. Lunde & G. Anderson, Water Research, 13, 1233-1237 (1979).
18. EPA, Methods for Evaluating Waste: Test Method 9020 for Total Organic Halides (TOX), (U.S. Environmental Protection Agency).
19. Kool, A.J. & C.F. van Kreijl, Water Research, 18, 1011-1016 (1984).
20. Jäger, V.W. & H. Hagenmaier, Z. Wasser Abwasser Forsch., 13, 66-69 (1980).
21. Sjöström, L., Determination of inorganic chlorine compounds, total organically bound chlorine and total chlorine in pulp mill spent bleach liquors. (Swedish Forest Products Research Laboratory, Stockholm, 1981).
22. Folke, J., Environmental Impact Assessment for Marine Ecosystems of Wheat and Rye Straw Pulp Bleaching and Combined Mill Effluents. (Chalmers University of Technology & University of Gothenburg, Göteborg, 1985).
23. Harper, F., History and Implementation of Total Organic Halogen Methodology as an Indicator of Ground Water Quality, Ground Water Monitoring Review, 46-48 (1984).
24. Williams, D.T., J.A. Coburn & J.J. Bancsi, Environment International, 10, 39-43 (1984).
25. Dahlquist, Fish with High Contents of Organic Chlorine in Fatt (in Swedish), Miljöaktuelt, 13(3), 3 (1985).
26. Carlberg, G., (Centre of Industrial Research, Oslo, 1984). Personal communication.
27. Wegman, R.C.C. & P.A. Greve, Science of the Total Environment, 7, 235-245 (1977).
28. Folke, J., Toxicol. Environ. Chem., 10, 201-224 (1985)
29. Folke, J. & J. Birklund, Toxicol. Environ. Chem., 10, 41-65 (1985).
30. Folke, J. ,J. Birklund, A. K. Sørensen & U. Lund, Chemosphere, 12, 1169-1181 (1983).
31. Kringstad, K.P. & K. Lindström, Environ. Sci. Tech., 18, 236A-248A (1984).

DETERMINATION OF ACIDIC AND NEUTRAL RESIDUES OF ALKYLPHENOL POLYETHOXYLATE
SURFACTANTS USING GC/MS ANALYSIS OF THEIR TMS DERIVATIVES

EURIPIDES STEPHANOU
Section of Chemical Technology and Environmental Chemistry,
Dep. of Chemistry, University of Crete, Iraklion, Greece

Summary

Alkylphenols, alkylphenol ethoxyates and alkylphenol carboxylates with
one and two oxyethylene groups can be simultaneously determined in the
water. The analytical procedure used in this study was the following:
Liquid-liquid extraction (with CH_2Cl_2), concentration, silylation
(with BSTFA) and GC/MS analysis.

1. INTRODUCTION
    One distiusuishes three main categories of surfactants: (I) cationic,
(II) anionic and (III) non-ionic. In the 1950s the release of surfactants
(specially of branched alkylbenzene sulfonates) led to sharp environmental
problems manifested by foaming. As the cause of this problem was the
nonbiodegradability of the tetrapropylenebenzenesulfonates, restrictions on
its use have been established. Since then, non-ionic surfactants have
enjoyed an ever-increasing market in Europe as well as in the USA (1). Two
particular classes of these compounds, the alkylphenol ethoxylates (APE)
and the linear alcohol ethoxylates (LAE) (Figure 1) have been well studied
in order to determine their biodegradability (2,3,4).

$$C_nH_{2n+1}-O-(CH_2CH_2O)_m-H \qquad : LAE$$

$$R-C_6H_4-O-(CH_2CH_2O)_m-H \qquad : APE$$

R = C_8H_{17} : Octylphenol Ethoxylates   : OPE

R = C_9H_{19} : Nonylphenol Ethoxylates   : NPE

n = 12-16, m = 4-20, C_6H_4 :

Fig. 1 Major categories of non-ionic surfactants

The LAE are considered to be biodegradable compounds. Although the recalcitrance of APE was disputed for some time, this class of surfactants is also accepted as biodegradable. Nevertheless residues of APE have been detected in wastewaters, sewage sludges, surface waters and in a leachate plume of a municipal landfill (4,5,6,7). The major findings concerning the fate of APE in the environment are summarized on Fig. 2. Considering the environmental relavance of these compounds sensitive analytical methods, for their determination at trace levels, are needed. Various methods have been developped (5,8,9) aiming the determination of the neutral residues of APE. Although the acidic metabolites (Fig.2) appear also to be recalcitrant (10) less attention has been paid for their qualitative and quantitative determination.

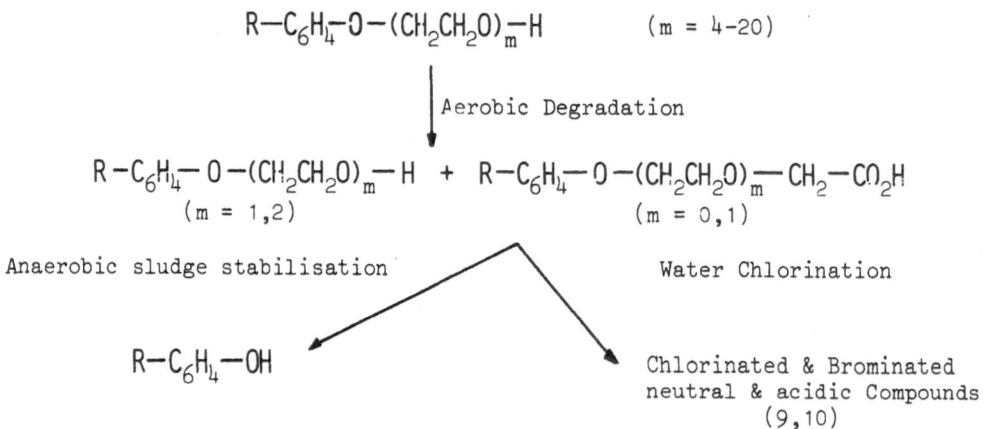

Fig. 2   Fate of APE in the environment (3,4,10)

In this paper is presented a first attempt of determination of APE carboxylates (APES) simultaneously with the corresponding neutral metabolites without prior separation. For this scope GC/MS analysis has been applied to the trimethylsilyl (TMS) derivatives of both acidic & neutral residues.

3. EXPERIMENTAL
Materials   The solvents used were freshly distilled from the "technical" grade material (Fluka, Buchs, Switzerland. Hexamethybenzene (HMB) and BSTFA were used as received from the supplier (Fluka). Tert.-octylphenol polyethoxylate (SU 145) was provided by Chem. Service, West Chester, PA, USA. Anhydrous Na_2SO_4 was prewashed with methylene chlordide and baked at 500°C overnight prior to use.

Extraction and sample preparation. Removal of APE and APES was accomplished by adjustment of pH to 2 with H_2SO_4 and subsequent extraction with methylene chloride in a liquid-liquid extractor or in a separatory funnel (5). The extracts were dried with Na_2SO_4, filtered through a pre-extracted paper filter and concentrated to 1-2 ml on a rotary evaporator under reduced pressure. Using a nitrogen stream the sample was evoporated to dryness in a 5 ml vial. 30 µl of BSTFA were added, the vial was sealed and heated at 100°C for 1-2 min. Prior to GC/MS analysis the internal standard (HMB) and 0,5 ml of methylene chloride were added to the sample. The internal standard was prepared also in methylene chloride at a known concentration at a range 5-10 mg/ml. The samples from laboratory biodegradation studies were obtained as in (3).

GC/MS analysis. A Finnigan mass spectrometer (Model 4000) with an INCOS 2000 data system was used for mass spectrometric identifications. A Carlo Erba gas chromatograph (Model 4160) was connected to the mass spectrometer. The gas-chromatograph contained a fused silica capillary column (ORION, SE-54, 25 m x 0.25 mm) connected to the ion source by another fused silica capillary. The temperature programm used was 50°C (1 min), 50-280 (3°C/min) and 280°C (20 min). Helium was used as a carrier gas with a back pressure of 0.8 bar. The injector temperature was 250°C. The ion source was operated at 70 eV with a nominal temperature of 250°C.

3. RESULTS AND DISCUSSION

On Fig. 3 and 4 are presented the most representive ions of mass spectra of the trimethylsilyl derivatives of octylphenol (OPTMS), octylphenol monoethoxylate (OP1TMS), octylphenol diethoxylate (OP2TMS), octylphenol acetic acide (OPS1TMS) and octylphenoxyethoxy acetic acid (OPS2TMS), as well as the main fragmentation pathways generating these ions. Fig. 5 shows the GC/MS analysis of an extract of a laboratory biodegradation study applied to OPE surfactants (11). On the same figure the inserted traces present the mass spectra of OPS1TMS (left) and OPS2TMS (right).

The main fragmentation processes occuring in the mass spectra of non silylated compounds namely benzylic clivage and degradation of the ethoxy chain, producing the typical alkylphenolic ions, are also observed when OPE & OPES are silylated (Fig. 3 and 4). Nevertheless some ions produced through an electron impact induced intramolecular silylation (12) are present in our study. The ion at m/z 207 present in the spectra of OP1TMS, OP2TMS, OPS1TMS and OPS2TMS is a prove for the above reaction. It seems also that the TMS group adds some stability to the ionic fragmemts (at m/z 133, 177 (Fig.3, 147, 191 and 175 (Fig. 4)), because these fragmentations cannot be observed by the non-silylated compounds. Degradation of the ethoxy chain generates the ions at m/z 117, 161 (Fig. 3) and 117 (Fig.4).

The mass spectra of the silylated OPE metabolites are conclusive for their mass and structure and may be used for their identification. They also include ions which could be picked out to be used for a selected ion monitoring (SIM) GC/MS analysis. For this purpose the following ions are proposed:

- Those generated through benzylic clivage :ions A (Fig. 3) and A´(Fig. 4)
- The ions at m/z 117, which have a different structure and elemental formula for the neutrals ($C_5H_{13}OSi$ and for the acids ($C_4H_9O_2Si$).

For a relialable GC/SIM-MS analysis one should not exceed 12 selected ions. Here are proposed 7 ions, which offer molecular weight (A+71 and A´+71) and also stuctural information. The ion at m/z 207 is present all spectra and is also a good diagnostic species.

The fact that the acidic compounds are easily simultaneously determined with the neutral ones is shown in Fig. 5. Extracting the samples at a low pH (2-3) we avoid the formation of emulsion occuring during the separation of neutrals from acids (5) at a high pH (11-12) and obtain at once both classes of residues. The silylation is, easily and briefly, yielding a 100% derivatization. The injection into GC is realized without any purification and also without causing any damage to the column.

Some quantitative work has been also realized. The internal standard was HMB. The response factor for OPTMS is 1.0 and for OPS1TMS is 0.2. More quantitative studies are needed in order to achieve the simultaneous qualitative and quantitative GC/MS analysis of these environmentally relevant compounds (13).

Concluding, for this study at this point, it can be said that the developped precedure is relialable and fast, allowing doubtless determination of APE residues in the water.

ACKNOWLEDGEMENTS

I am gratefull to Dr. W. Giger (EAWAG for allowing me to work in his laboratory for a part of this study, as well as to Dr. T. Conrad (EAWAG) from whom I could receive some samples. I would also like to thank Mr. C. Schaffner (EAWAG) for his kind collaboration and Mrs. Evangelia Badjakaki for typing this manuscript.

REFERENCES

1. CHEM. and ENG. NEWS, Jan. 23 (1984).
2. HAUPT, D.E. Tenside Detergents, 20, 332 (1983).
3. GEISER, R. Ph.D. Thesis No. 6678 (1980), ETH, Zurich, Switzerland.
4. GIGER, W., BRUNNER, P.H., SCHAFFNER, C. Science, 225, 623 (1984).
5. STEPHANOU, E., GIGER, W. Environ. Sci. Tech., 16, 800 (1982).
6. REINHARD, M., BARKER, J.F., GOODMAN, N.L. Environ. Sci. Tech. 1985, in press.
7. SHELDON, L.S., HITES, R.A. Environ. Sci. Tech., 13, 574 (1982).
8. LIEONARDO, R.A., NEUBECKER, T.A. Anal. Chem., 55:93 R (1983).
9. STEPHANOU, E. Intern. J. Environ. Anal. Chem., 1985, in press.
10. REINHARD, M., GOODMAN, N.L., NORTELMANS. K.E. Environ. Sci. Tech., 16, 351 (1982).
11. CONRAD, T. unpublished results.
12. VOUROS, P. J. Org. Chem., 38, 3555 (1973).
13. STEPHANOU, E. in preparation.

m/z	(ION)	−(NEUTRAL FRAGMENT)
m=0 : 278 m=1 : 322 m=2 : 366	$(CH_3)_3CCH_2(CH_3)_2C-C_6H_4-O(CH_2CH_2O)_m$ TMS	
m=1 : 117(x=1) m=2 : 161(x=2)	$(CH_2CH_2O)_x$ TMS　(B)	$C_8H_{17}-C_6H_4-O(CH_2CH_2O)_{m-1}$
m=1 : 133 m=2 : 177	$O-(CH_2CH_2O)_m$ TMS	$C_8H_{17}-C_6H_4-O$
m=0 : 207 m=1 : 251 m=2 : 295	$(CH_3)_2C-C_6H_4-O(CH_2CH_2O)_m-TMS$　(A)	C_5H_{11}
m=1,2 : 207	$(CH_3)_2C-C_6H_4-OTMS$	$(CH_2CH_2O)_{1\ or\ 2}$

Fig. 3　Fragmentation mode of OPE − TMS compounds

-(NEUTRAL FRAGMENT)$^{\bullet}$

(ION)$^{+}$

m/z

m/z	(ION)$^+$	-(NEUTRAL FRAGMENT)$^\bullet$
m=1 : 336 m=2 : 380	$C_8H_{17}-C_6H_4-O(CH_2CH_2O)_{m-1}CH_2CO_2$TMS	$C_8H_{17}-C_6H_4-O(CH_2CH_2O)_{m-1}$
m=1,2 :117	CO_2TMS	$C_8H_{17}-C_6H_4$
m=1 : 147 m=2 : 191	$O(CH_2CH_2O)_{m-1}-CH_2-CO_2$TMS	
m=2 : 175	$CH_2CH_2O-CH_2CO_2$TMS	$C_8H_{17}-C_6H_4-O$
m=1 : 265 m=2 : 309	$(CH_3)_2C-C_6H_4-O(CH_2CH_2O)_{m-1}-CH_2CO_2$TMS (A´)	C_5H_{11}
m=1,2 :207	$(CH_3)_2-C_6H_4-$OTMS	CH_2CO_2 and/or CH_2CH_2O

Fig. 4 Fragmentation mode of OPES-TMS compounds

- 160 -

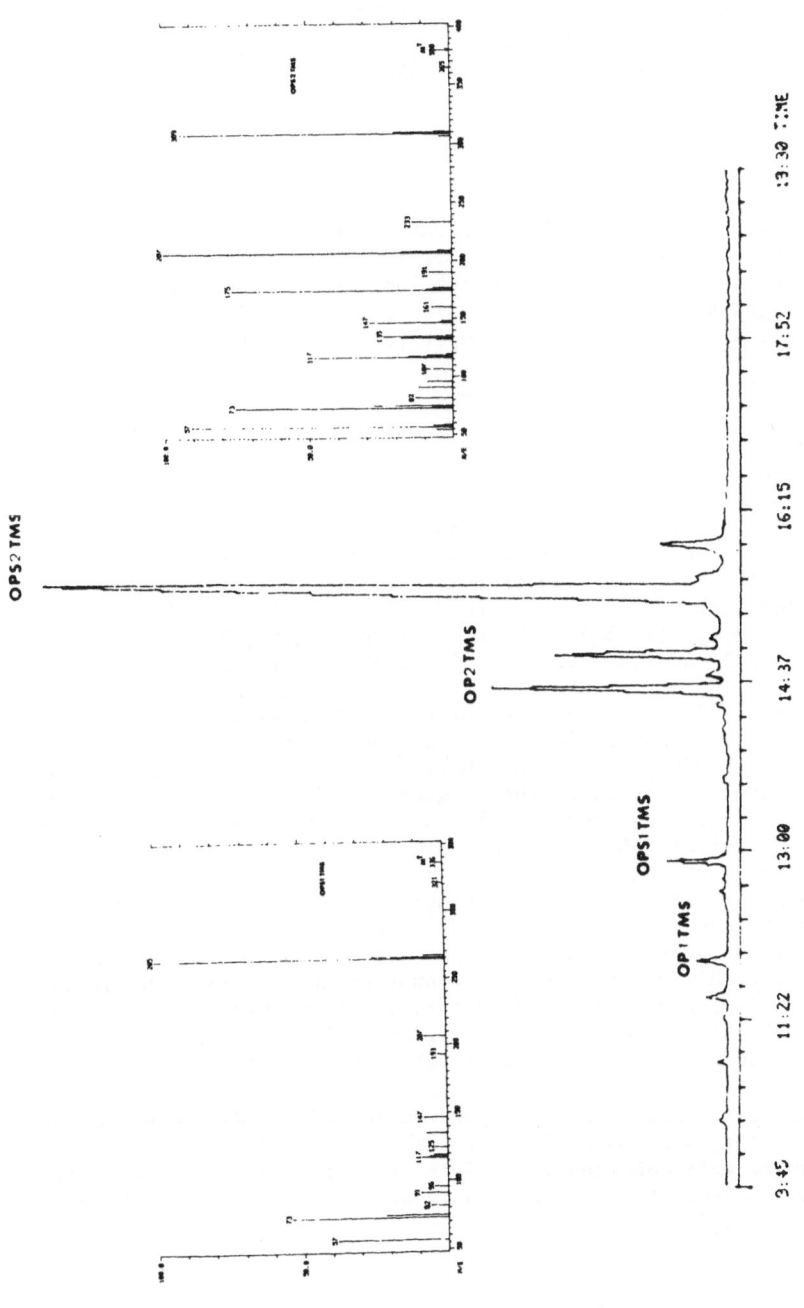

Fig. 5 GC/MC determination of OPE and OPES through their TMS derivatives. The inserted traces show the mass spectra of OPS1TMS (left) and OPS2TMS (right) TMS derivatives.

DETERMINATION OF PCB-CONGENERS IN SEDIMENT SAMPLES USING A RETENTION INDEX SYSTEM

R.C.C. WEGMAN AND A.W.M. HOFSTEE
National Institute of Public Health and Environmental Hygiene,
P.O.Box 1, 3720 BA Bilthoven, The Netherlands

Summary

A method for the determination of PCB congeners in sediment and particulate matter in described. For identification of the peaks in the gas chromatogram a retention index system was used based on n-alkyltrichloroacetates (ACTAs). The retention index system could also applied if the GC column temperature was programmed.

1. INTRODUCTION

For the determination of PCBs in sediment and particulate matter samples a method had to be developed that includes the determination of the individual congeners by capillary GC-ECD. Especially attention had to be given to the determination of the so-called indicator congeners: 2,4,4'-trichlorobiphenyl, 2,2',5,5'-tetrachlorobiphenyl, 2,2',4,5,5'-pentachlorobiphenyl, 2,2',3,4,4',5'-hexachlorobiphenyl, 2,2',4,4',5,5'-hexachlorobiphenyl and 2,2',3,3,4',5,5'-heptachlorobiphenyl with the corresponding IUPAC numbers 28, 52, 101, 138, 153 and 180. Lateron in the Netherlands also 2,3',4,4',5-pentachlorobiphenyl (118) was selected as indicator biphenyl. These seven compounds give a good impression about the contamination with PCBs, their contribution to the total amount of PCBs is about 10-25%. Identification of the congeners is based on correct retention time. For a correct interpretation of the capillary gas chromatogram with a large amount of PCBs and other peaks use of a retention index system will give more reliable results.

In 1958 Kovats (1) introduced the use of n-alkanes for calculation of GC retention indices. The precision and reproducability of these indices led to general use of these indices especially in the identification of complex mixtures. However this system cannot be applied in combination with detectors e.g. ECD, which are insensitive for alkanes. Neu (2) and Ballschmiter (3) used a series of homologous n-alkyl trichloroacetates (ATCAs). Pacholec and Poole (4) used a series of homologous n-monobromo-alkanes.

Neu compared a couple of homologous series of α-w-dichloro-n-alkanes, α-chloro-n-alkanes, α-bromo-n-alkanes, chlorophenylalkyl ether and ATCAs. The best results were obtained with ATCAs. In this paper the use of ATCAs for the determination of PCB congeners in sediment and particulate matter samples is described.

2. PROCEDURE
2.1. Reagents

For extraction purposes acetone ("Zur Analyse", Merck Darmstadt, FRG) and petroleum ether, distilled (boiling range 40-60°C, Ph.Ned.Brocacef, Maarssen, The Netherlands) were used. Na_2SO_4 (Analar, BDH Chemicals Ltd,

Poole, U.K.) was heated at 500°C for 3 h before use. NaCl ("Zur Analyse", Merck), saturated solution in water. For removal of elemental sulfur Na$_2$SO$_3$ ("Zur Analyse", Merck), saturated solution in water, was used.

For clean-up basic alumina (W-200, activity Super I, Woelm, Eschwege, FRG) activated before use for 16 h at 150°C next deactivated with 11% w/w water, silica gel (Merck, Kieselgel No 7754, 70-230 mesh) activated 15 h at 200°C before use and quartz wool, extracted 8 h with a mixture of petroleum ether and acetone (4 + 1, v/v).

The individual PCB congeners were obtained from Foxboros Analabs (North Haven, CT, USA) and Promochem G.mb.H. (Wesel, FRG), the technical PCB mixtures Arclor 1232 and Aroclor 1254 from Chrompack Nederland B.V. (Middelburg, The Netherlands).

The ATCAs were not for sale so they had to be synthezised. The synthesis and purification were as follows. 0.04 Mol of an alcohol were dissolved in 50 ml of toluene, 0.04 mol of pyridine and 0.045 mol trichloroacetyl chloride were added and the mixture was heated on a waterbath for about 2 h. After cooling the mixture was transferred into a separation funnel containing 25 ml of a 4% (w/w) solution of NaHCO$_3$. The mixture was gently mixed, the water phase decanted and the toluene phase was washed several times with 25 ml of a 4% (w/w) solution of NaHCO$_3$ till pH 7 was reached. The toluene was dried over anh. Na$_2$SO$_4$ and removed by distillation. After cooling 5 ml of petroleum ether were added to the residue and this was brought on a column (40 x 2.0 cm) filled with 75 g of neutral alumina. The column was eluted with 150 ml of petroleum ether. The petroleum ether was removed by distillation.

The ATCAs used for identification were n-octyltrichloroacetate, n-decyltrichloroacetate, n-dodecyltrichloroacetate, n-tetradecyltrichloroacetate, n-hexadecyltrichloroacetate, n-octadecyltrichloroacetate and n-eicocyltrichloroacetate. Octanol, undecanol and pyridine were obtained from Baker, Deventer, The Netherlands, nonanol from BDH, octadecanol and tetradecanol from Aldrich, Brussel, Belgium, pentadecanol, nonadecanol, NaHCO$_3$ and toluene from Merck, decanol, dodecanol, tridecanol, hexadecanol, heptadecanol, eicosanol and trichloroacetylchloride from Fluka, Buchs, Switzerland.

2.2. Extraction

In a centrifuge tube with screw cap 5 ml of water were added to 25 g of sediment or particulate matter. The mixture was shaken with 40 ml of acetone for 20 min. by means of a mechanical shaker. The tube was centrifuged for 10 min. and the acetone phase was decanted into a 1 l separation funnel. The sample in the tube was shaken again with 40 ml of acetone for 20 min. After centrifugation for 10 min. the acetone phase was decanted. To the combined acetone extracts 100 ml of water and 5 ml of a saturated solution of Na$_2$SO$_3$ were added next the solution was shaken for 1 min. Next 700 ml of water and 5 ml of a saturated solution of NaCl were added. The solution was shaken 2 times with 50 ml of petroleum ether for 20 min. by a mechanical shaker. The petroleum ether phases were decanted and the combined petroleum ether extracts were dried over anh. Na$_2$SO$_4$ and concentrated in a Kuderna-Danish apparatus to about 5 ml. The extract was brought in a volumetric flask of 25 ml and filled up with petroleum ether.

2.3. Clean-up

One ml of the extract was added to a column (23 x 0.6 cm I.D.) containing a plug of quartz wool and 2.0 g of basic alumina. For rinsing 3 portions of 0.5 ml of petroleum ether were used. The column was eluted with 10 ml of petroleum ether to produce Eluate A. The eluate was concentrated to 1 ml by a gentle stream of nitrogen at room temperature. The concentrated eluate was added to a column (40 x 0.9 cm I.D.) containing a plug of

quartz wool, 0.5 g of anh. Na_2SO_4, 5.0 g silicagel and 0.5 g of anh.
Na_2SO_4. For rinsing 3 portions of 0.5 ml of petroleum ether were used.
The column was eluted with 46 ml of petroleum ether to produce Eluate C,
containing HCB, heptachlor, aldrin, isodrin, p.p'-DDE, o.p'-DDT, p.p'-DDT
(60%) and PCBs. The eluate was concentrated to 1 ml by a gentle stream of
nitrogen at room temperature.

2.4. GC

One µl of the concentrated eluate was injected into a Hewlett Packard
Model 5880 GC-^{63}Ni ECD. Instrument parameters and operational conditions
were as follows: separation column: fused silica WCOT (50 m x 0.33 mm I.D.,
Chrompack; stationary phase: CP Sil 5, film thickness 0.15 µm; temperatures: oven isothermal at 80°C for 3 min, next programmed 80-270°C at
2.5°C min^{-1}; injection block 200°C; detector 300°C; carrier gas helium
111.5 kPa, constant pressure; purge gas argon +10% methane, flowrate
40 ml min^{-1}; splitless injection; automatic liquid sampler; electronic
integrator level four with data storage on a cartridge tape unit and BASIC
programming for data processing.

2.5. Recovery and detection limit

Sediment samples were spiked with Aroclor 1232 and Aroclor 1254.
Seven days after spiking the samples were extracted. The mean recovery
± rel. S.D. was 84 ± 9%. The detection limit was 0.002-0.005 mg kg^{-1} on a
dry weight base (sediment sample was dried for 4 h at 105°C).

2.6. Calculation retention indices

In literature use of n-alkane is described for calculation of retention indices also if a temperature program was used during the GC analysis
(5). In this paper the use of ATCAs in combination with a GC temperature
program is described. In Fig.1 the calculation formula is given. After
data storage of chromatographic runs from injections of ATCA reference
standards, PCB congeners standards and samples collected for a period of
24 h the following calculations were made with the integrator of the GC
by means of a computer program in basic:
- PCB reference standard: the mean and standard deviation of the retention
  index for each PCB congener and a retention index window equal to mean ±
  3 times the standard deviation of the retention index
- samples: the retention index and the amount of the PCB congener and a
  remark if the retention index of the PCB congener is outside the window.

3. RESULTS

In Fig.2 retention time (RT) versus alkyl chain length are given for
n-alkanes and ATCAs with a same column temperature program. In Fig.3 GC
separation of ATCAs and PCBs are given. In Table I the reproducability of
RT of ATCA at different time intervals are summarized. The results are
obtained with the same chromatographic column and with the same GC.

For PCB congeners the mean standard deviation is 0.29 retention index
units (0.62 s retention time) in the area 1400-1600 retention index units.
The GC peak width at half peak height is 6 s (2.8 retention index units)
which means that the resolving power of the ATCA index system is enough.

The ATCA index system is used during the analysis of PCB congeners in
sediment and particulate matter samples for a monitoring program of surface water of the Netherlands. In Table 2 the retention indices of 6 PCB
congeners determined in sediment samples are summarized. From these
results it follows that the retention index of peak 52 is out of the 3
S.D. limit. In Table 3 as an example the sum of 5 PCB congeners for a
couple of sampling sites are given (28, 101, 138, 153 and 180).

4. CONCLUSION

Use of ATCA as external retention standard for determination of retention indices at temperature programmed gas chromatography gives an adequate resolution for identification of peaks in complex PCB mixtures.

5. REFERENCES

(1) KOVATS, E., Helv.Chim.Acta 41, 1915–1920 (1958)
(2) NEU, H.J., Fres.Z.Anal.Chem. 293, 193–200 (1978)
(3) BALLSCHMITER, K., Chemosphere 1, 51–56 (1977)
(4) PACHOLER, F. and POOLE, C.F., Anal.Chem. 54, 1019–1021 (1982)
(5) BÁRTŮ, V., J.Chromatogr. 260, 225–264 (1983)

Figure 1   *CALCULATION OF RETENTION INDEX (I)*
AND TEMPERATURE PROGRAMMED GLC

FORMULA:
$$I(X)=I(1)+[I(2)-I(1)] [t(X)-t(1)]/[t(2)-t(1)]$$

I(X)=RET.INDEX OF PCB COMPONENT
I(1),I(2)=RET.INDEX OF ATCA(1) AND (2)
t(X)=RET.TIME OF PCB COMPONENT
t(1),t(2)=RET.TIME OF ATCA (1) AND (2)

PREDICTIONS:
–INITIAL TEMP. LOW ENOUGH TO IMMOBILIZE THE SOLUTE
–I(ATCA)=100X (C NUMBER ALKYL GROUP)
–A LINEAR RELATIONSHIP BETWEEN I AND t.
(IN FACT THE RELATIONSHIP IS NOT PURELY LINEAR.
WHEN USING ATCAs WITH A DIFFERENCE IN C NUMBER OF 1
THE ERROR WILL BE SMALL).

Figure 2

C NUMBER VERSUS RETENTION TIME.
temp prog 80-280°C, 25°C min⁻¹

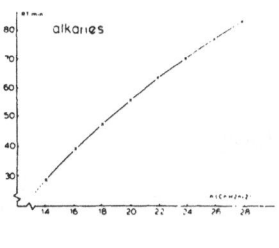

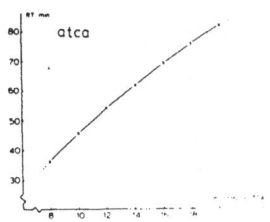

Figure 3

GLC-SEPARATION OF ATCAs AND PCBs
25m WCOT fused silica, CP Sil-8

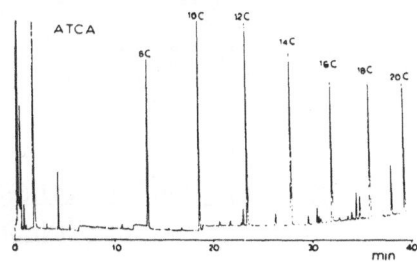

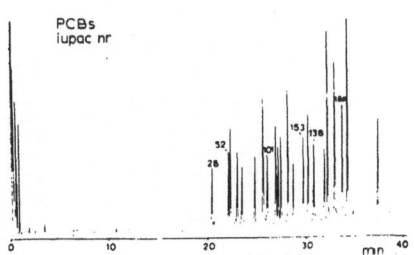

Table 1

REPRODUCIBILITY OF RT OF ATCA
$\bar{X}$ AND RELATIVE S.D., N=2
AT DIFFERENT TIME INTERVALS
(ΔI= S.D. OF RET.INDEX I)

ATCA	TIME INTERVAL	
	3h	24h
C8	30.889±0.0125	30.919±0.0472
I=800	ΔI=0.32	ΔI=1.22
C10	40.161±0.0142	40.182±0.0507
I=1000	ΔI=0.35	ΔI=1.26
C12	48.689±0.0160	48.709±0.0507
I=1200	ΔI=0.39	ΔI=1.25
C14	56.499±0.0178	56.518±0.0516
I=1400	ΔI=0.44	ΔI=1.28
C16	63.683±0.0214	63.701±0.0534
I=1600	ΔI=0.54	ΔI=1.34
C18	70.310±0.0276	70.330±0.0632
I=1800	ΔI=0.71	ΔI=1.62
C20	76.457±0.0258	76.477±0.0623
I=2000	ΔI=0.67	ΔI=1.63

RET.INDICES OF PCBs IN SEDIMENT

Table 2

IUPAC NUMBER	STANDARD(n=4) $\bar{X}$ S.D.	3. S.D.LIMIT	SAMPLING SITE		
			1.RHINE RIVER AT LOBITH	4.WAAL RIVER AT TIEL	12.NIEUWE MAAS R.
28	1077.77±0.420	1076.51–1079.03	1077.16	1077.72	1078.17
52	1143.70±0.442	1142.38–1145.02	1145.04	1145.51	1146.05
101	1313.82±0.459	1312.44–1315.20	1313.37	1313.88	1314.49
138	1536.12±0.419	1534.86–1537.38	1535.77	1536.24	1536.96
153	1485.24±0.465	1483.85–1486.64	1484.68	1485.21	1485.96
180	1683.06±0.475	1681.84–1684.49	1682.65	1683.10	1684.09

SUM OF CONCENTRATIONS OF SIX INDIVIDUAL PCBs IN SEDIMENT

Table 3

SAMPLING SITE:	CONCENTRATION (mg/kg ON DRY WEIGHT BASIS)
1.RHINE RIVER AT LOBITH	●●●●●●●●●●●●●●●●●●●●●●●●●●●●●●●●●●●(0.32)
12.NIEUWE MAAS R. AT ROTTERDAM	●●●●●●●●●●●●●●●(0.14)
14.MAAS AT EYSDEN	●●●●●●●●●●●●●●●●●●(0.17)
16.MAAS AT VENLO	●●●●●●●●●●(0.10)
8.BIESBOSCH AT SPIJKERBOOR	●●●●●●●●●●●●●●●(0.15)
28.IJSSEL RIVER AT KAMPEN	●●●●●●(0.06)
26.IJSSEL LAKE AT MUNNIKPLAAT	●(<0.01)
20.WESTERSCHELDE AT VLISSINGEN	●(<0.01)
31.WADDEN SEA	●(<0.01)

TRANSPORT OF ORGANIC MICROPOLLUTANTS IN THE

AQUATIC ENVIRONMENT

Sorption behaviour of neutral and ionizable hydrophobic organic compounds

Properties of humic material in relation to transport phenomena

PCB congeners in the marine environment - A review

Dispersion et évolution des hydrocarbures pétroliers dans l'environnement marin. Cas de référence : la pollution de l'Amoco Cadiz

Bioaccumulation of chlorinated paraffins - A review

Investigation of the influence of aquatic humus on the bioavailability of chlorinated micropollutants towards fish

Mixtures of hydrophobic chemicals in aqueous environments : Aqueous solubility and bioconcentration by fish of Aroclor 1254

Comparative analysis at the molecular level of the interactions between organic and inorganic mercury compounds and membrane lipidic bilayers

Experimental systemic study of the accumulation and transfer processes of organic and inorganic mercury in freshwater environment - Methodology and results

Bioaccumulation by fish in relationship to the oxygen concentration in water

SORPTION BEHAVIOR OF NEUTRAL AND IONIZABLE HYDROPHOBIC ORGANIC COMPOUNDS

R.P. Schwarzenbach
Swiss Federal Institute for Water Resources
and Water Pollution Control (EAWAG),
8600 Dübendorf

Summary

Factors influencing the equilibrium distribution of neutral and ionizable hydrophobic organic pollutants between water and natural solid
phases are discussed. For neutral hydrophobic compounds, the sorption
process can be viewed as a simple partitioning of the organic solute
between the water and the natural organic matter present in a given
solid phase, if the organic matter exceeds $\sim$ 0.1%. The degree of
sorption can be expressed by a partition coefficient, which may be
estimated from the octanol/water partition constant of the compound
and the organic carbon content of the sorbent. For ionizable (anionic)
hydrophobic compounds (represented by polychlorinated phenols), the
distribution ratio depends on both the pH and on the major ion composition of the aqueous phase, in contrast to the partitioning model for
neutral compounds in which the chemical composition of the aqueous
phase is relatively unimportant.

1. INTRODUCTION

The interaction of an organic pollutant with natural solids is one of
the key factors influencing its transport and fate as well as its bioavailability in the environment. Binding of a compound to solid materials as
occurs in the aquatic environment is generally referred to as sorption.
Distinction can be made between adsorption, or binding at a two-dimensional
surface, and absorption, or partitioning into the three- dimensional bulk
of the sorbing phase. Depending on the chemical structure of the compound,
the nature of the solid surfaces present, and the chemical composition of
the water (i.e., pH, ionic strength, presence of other dissolved organic
compounds), one or several binding mechanisms may contribute significantly
to the overall sorption of the compound. Therefore, in order to estimate
the sorption behavior of a given compound in a variety of different environments, mechanistically based models for decribing the various binding
processes are needed.

In this paper, we discuss concepts for describing the sorption behavior of neutral and anionic organic compounds for which hydrophobic
partitioning is the predominant sorption mechanism. The term "hydrophobic
partitioning" is adopted to indicate that sorption of such compounds is
determined primarily by their incompatibility with the solvent water rather
than by their specific affinity to the sorbent. We will confine ourselves
to sorption equilibrium considerations, although some remarks on kinetic
aspects will be made.

2. SORPTION OF NEUTRAL HYDROPHOBIC COMPOUNDS

In recent years, many studies have been conducted to investigate the sorption behavior of various classes of neutral hydrophobic organic compounds including polycyclic aromatic hydrocarbons (1,2), halogenated hydrocarbons (3-5), and various pesticides (6-8). From these and other studies it can be concluded that sorption of such compounds can be described in terms of a simple partitioning model, that is, the phase transfer reaction is viewed as transfer of the neutral compound from the aqueous phase to a bulk non-aqueous phase (4,9):

$$A \rightleftharpoons \overline{A} \tag{1}$$

where A represents a neutral compound, and the overbar indicates species in the non-aqueous phase. In these cases, sorption equilibrium can be described by an equilibrium partition coefficient for a particular species:

$$K_p = \frac{[\overline{A}]}{[A]} \qquad [\text{e.g., } cm^3 g_s^{-1}] \tag{2}$$

As demonstrated for 1,4-dichlorobenzene in Figure 1, it is the organic material present in natural sorbents that is primarily responsible for sorption of hydrophobic organic compounds. In particular for nonpolar solutes and for sorbents with organic carbon (oc) contents greater than $\sim 0.1\%$, the partition coefficient K_p can thus be expressed simply in terms of the organic carbon content of the sorbent, and in terms of a partition constant, K_{oc}, between water and a hypothetical sorbent of 100% organic carbon representing an average natural organic material:

$$K_p = f_{oc} \cdot K_{oc} \tag{3}$$

where f_{oc} is the fraction of organic carbon $[g_{oc} g_s^{-1}]$. The dimensions of K_{oc} are, e.g., $[cm^3 g_{oc}^{-1}]$. Figure 1 indicates that for nonpolar compounds, Equation 3 holds over a wide range of organic carbon contents, and that K_{oc} values obtained from sorbents of very different origin do not differ substantially. This is not too surprising since the sorption of such compounds is primarily controlled by their incompatibility with water and not by specific interactions with the organic sorbent. The later point is illustrated in Figure 2, which shows that nonpolar organic compounds have very similar hexane/water (K_{hw}) and n-octanol/water (K_{ow}) partition constants, although the two solvents differ significantly in their polarity. However, for polar compounds (e.g., compounds with functional groups such as -OH, -NH$_2$), the compatibility of the organic solute molecules with the organic solvent plays a more important role, and substantial differences between partition constants for different solvent/water systems can be expected. At this point we note that n-octanol has become the most popular organic reference phase mostly on account of the similarity of its solvent properties to many kinds of environmental and physiological organic matter.

As described by Leo et al. (10) and recently discussed by Westall (11), linear free-energy relationships (LFER's) can be used to relate

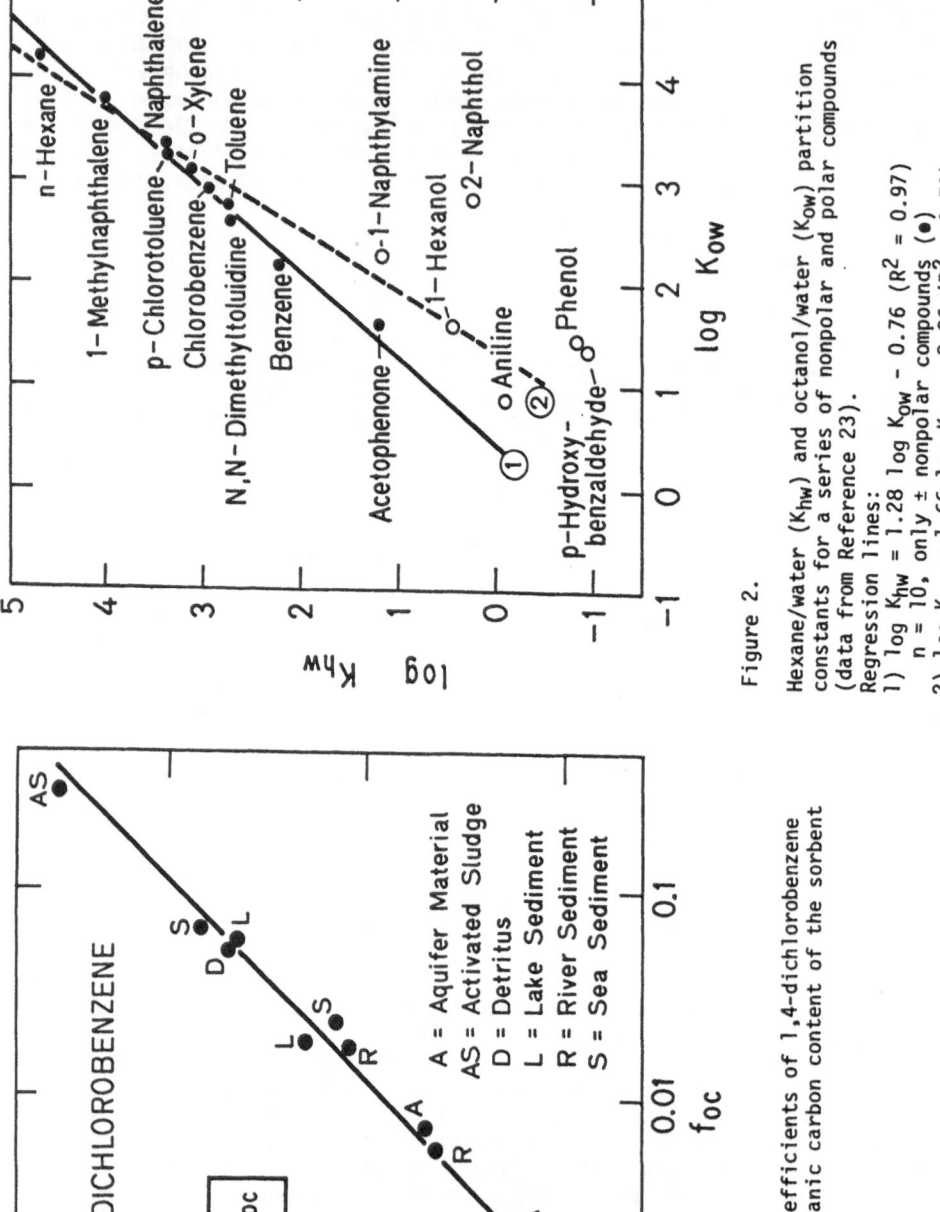

Figure 1.

Solid/water partition coefficients of 1,4-dichlorobenzene as a function of the organic carbon content of the sorbent

Figure 2.

Hexane/water (K_{hw}) and octanol/water (K_{ow}) partition constants for a series of nonpolar and polar compounds (data from Reference 23).
Regression lines:
1) log K_{hw} = 1.28 log K_{ow} - 0.76 (R^2 = 0.97)
   n = 10, only ± nonpolar compounds (●)
2) log K_{hw} = 1.66 log K_{ow} - 2.26 (R^2 = 0.79)
   n = 16, all compounds (● + o)

partition constants in different solvent/water and sorbent/water systems (see example given in Figure 2). Thus, we can, for example, relate natural organic carbon/water partition constants with octanol/water partition constants, and use such relationships to predict K_{oc} from K_{ow}:

$$\log K_{oc} = a \log K_{ow} + b \qquad (4)$$

For nonpolar compounds, the parameters a and b in Equation 4 are determined primarily by the type of compounds (i.e., compound class(es), range of hydrophobicity) based on which a linear free-energy relationship is established, and only to a much smaller degree by the type of natural sorbents used. Such relationships with empirically derived parameters a and b are, therefore, very useful for predicting partition coefficients between water and natural sorbents of very different origins:

$$\log K_p = \log f_{oc} + \log K_{oc} = \log f_{oc} + a \log K_{ow} + b \qquad (5)$$

Values reported for a and b are summarized in Table I. Best correlations are usually obtained for structurally related compounds (homologues series, analogues), in particular, for compounds that do not specifically interact with the organic matrix, that is, nonpolar compounds. For more polar compounds (e.g., some pesticides), much poorer correlations may be found. It should be noted that Equation 5 is applicable only to those cases in which the organic material predominates sorption. When dealing with sorbents of low organic carbon content (i.e., $f_{oc} < \sim 0.001$), mineral surfaces (especially of clay minerals) may become more important with respect to sorption. However, for such sorbents, relatively small K_p values are usually found even for highly hydrophobic compounds. In addition, the dependence of K_p on hydrophobicity of the compound is much less pronounced for organic poor sorbents compared to sorbents with higher organic carbon content. For example, with γ-Al$_2$O$_3$ (specific surface area = 120 m^2 g$_s^{-1}$) the K_p of 1,2,4,5-tetrachlorobenzene was found to be only on the order of 2 cm^3 g$_s^{-1}$, or 4 times greater than the K_p of chlorobenzene; with an aquifer material of 0.7% organic carbon and a specific surface area much smaller than that of γ-Al$_2$O$_3$ (4 m^2 g$_s^{-1}$), the K_p of the tetrachlorobenzene was 40 cm^3 g$_s^{-1}$ or 40 times greater than that of chlorobenzene. For more details of this study, see Reference 5. Further work is necessary to establish quantitative relationships for estimating partition coefficients of non-polar organic compounds with organic-poor sorbents, in particular with clay minerals (9).

Another area which needs more research, is the influence of colloidal and/or dissolved organic material (e.g., fulvic and humic materials, solvents, detergents, etc.) on the speciation and thus on the transport and distribution of organic pollutants in aquatic systems. The few studies conducted so far (12-14) strongly indicate that, in particular for highly hydrophobic compounds, interaction of the pollutant with "dissolved" organic matter has to be evaluated if the concentration of the dissolved organic matter is high (i.e., > $\sim$ 10 mg L^{-1}). Such situations may be found, for example, in pore waters or highly contaminated groundwaters. It should also be noted that dissolved or colloidal material may interfere with experimental determinations of partition coefficients of hydrophobic organic pollutants (15).

Table I. Estimation of K_{oc} based on K_{ow}: $\log K_{oc} = a \log K_{ow} + b$

Coefficients Correlation a	b	Correlation Coefficient r^2	Number of Compounds	Range of $\log K_{ow}$	Type of Chemicals	Reference
0.544	1.377	0.74	45	-3.0-6.6	primarily agricultural chemicals	6
1.00	-0.21	1.00	10	2.1-6.3	polycyclic aromatic hydrocarbons	1
0.937	-0.006	0.95	19	2.1-6.3	compounds of Karickhoff et al. (1), triazines nitroanilines	8
1.029	-0.19	0.91	13	0.4-6.3	herbicides, insecticides	7
1.00	-0.317	0.98	13	1.6-6.5	compounds of Karickhoff et al. (1), heterocyclic aromatic compounds	2
0.72	0.49	0.95	13	2.6-4.7	chlorinated hydrocarbons, alkylbenzenes	5
0.52	0.64	0.84	30	0.5-3.3	substituted phenyl ureas and alkyl-N-phenyl carbamates	22

Before we now turn to a more complicated case, we summarize that soption equilibrium of neutral hydrophobic pollutants can, in many cases, be described by a partition coefficient, K_p, which can be estimated from the organic carbon content of the sorbent and from the octanol/water partition constant of the compound. Linear sorption isotherms (i.e., constant K_p as a function of concentration) are observed up to aqueous phase pollutant concentrations of 10^{-5} M or less than one half of the solute water solubility (whichever is lower, (9)), and competitive interactions between dilute co-contaminants seem to be of minor importance (5,9). Finally, we note that slow sorption kinetics, in particular with respect to desorption reactions, may play a role for larger molecules (9). To date, various attempts including mostly empirical approaches have been made to describe sorption kinetics of hydrophobic compounds (e.g., 7,9). The most promising recent conceptual model is the radial diffusion model discussed by Wu and Gschwend (16).

3. SORPTION OF IONIZABLE HYDROPHOBIC COMPOUNDS: CHLORINATED PHENOLS
The simple partitioning model used to describe the sorption of neutral hydrophobic organic chemicals (Equation 2) is applicable only to a limited degree to compounds which are fully or partially ionized at natural pH values. Such compounds include amines, carboxylic acids and phenols. For example, the sorption of benzidine was found to be controlled largely by the pH of the aqueous phase, and non-linear sorption isotherms were obtained which were interpreted to be the result of the superposition of several different sorption processes (17). Also, a significant enhancement of the sorption above that expected based on simple partitioning, i.e., predicted from Equation 5 (derived for polycyclic aromatic hydrocarbons), was observed for two polycyclic aromatic amines (18). For anthracene-9-carboxylic acid, however, which is present predominantly as anion at the pH of natural waters, the same authors found no significant differences between predicted and experimentally determined distribution ratios. In any case, when dealing with the sorption of hydrophobic compounds containing functional groups which ionize or which may strongly interact with the various organic and inorganic constituents of natural sorbents, processes such as ion exchange, ligand exchange, formation of ion pairs or ion complexes (that may be transferred into the organic phase), etc., have to be considered in addition to simple partitioning.
In the following section, we discuss the sorption behavior of chlorinated phenols, a class of compounds which is of great environmmental concern. The chlorinated phenols (HA) are hydrophobic weak acids exhibiting octanol/water partition constants (for the non-ionized form) between 10^2 and 10^5 and pK_a values in the range of 4.75 to 9 (K_a = acidity constant). Thus, at typical ambient pH values, in particular the highly chlorinated phenols, e.g., pentachlorophenol, pK_a = 4.75; 2,3,4,6-tetrachlorophenol, pK_a = 5.40; 2,3,4,5-tetrachlorophenol, pK_a = 6.35 are present in the water predominantly as phenolate anions (A^-). From the results of a recent study by Schellenberg et al. (19), the following conclusions concerning the sorption behavior of chlorinated phenols can be drawn:

(1) For sorbents with organic carbon contents of greater then about 0.1%, the sorption of both the neutral (HA) and anionic (A^-) species can be viewed as a partitioning process between the aqueous phase and the organic phase present in a natural sorbent. At fixed pH and ionic strength, a linear isotherm can be used to describe the overall distribution between water and a given sorbent up to aqueous phase

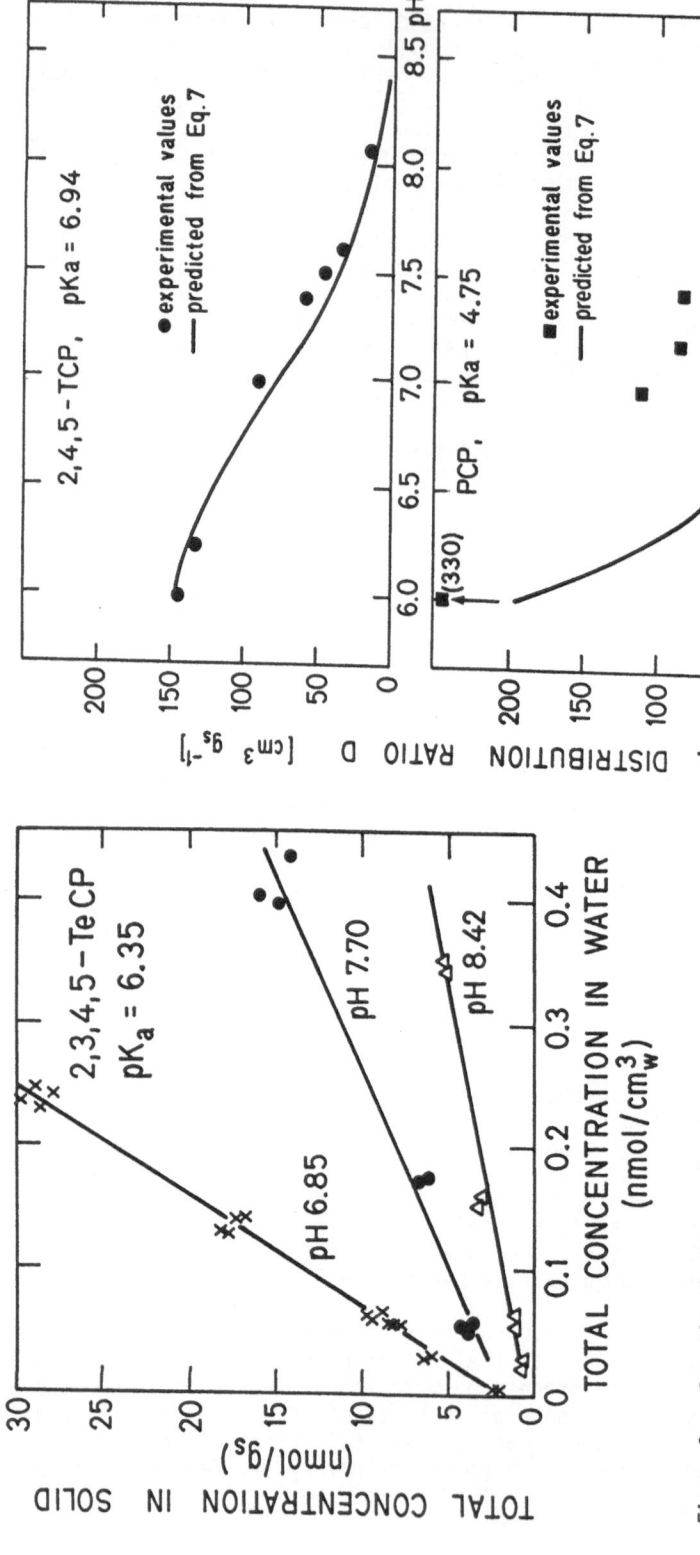

Figure 4: Predicted (Eq. 7) vs. experimentally determined overall distribution ratios for 2,4,5-trichlorophenol (top), and pentachlorophenol (bottom), as a function of pH. Sorbent: lake sediment (f_{oc} = 0.094). The pH is controlled by a CaCO$_3$/CO$_2$ buffer. Note that the Ca^{2+} concentration decreases with increasing pH. (Adapted from Reference 19).

Figure 3: Sorption isotherms for 2,3,4,5-tetrachlorophenol at different pH values. Sorbent: river sediment (f_{oc} = 0.026). (adapted from Reference 19).

total solute concentrations of at least 10^{-6} M (see example given in Figure 3). The distribution ratio, D, is defined for the total analytical concentrations:

$$D = \frac{\overline{[HA]} + \overline{[A^-]}}{[HA] + [A^-]} \qquad (6)$$

where the oberbar indicates species in the non-aqueous phase. As one can see in Figure 3, the value of D is strongly pH dependent.

(2) For chlorinated phenols containing three or less chlorine atoms, the contribution of phenolate (A^-) sorption may be neglected and the distribution ratio may be calculated from the partition coefficient K_p of the non-dissociated phenol, its acid dissociation constant, K_a, and the hydrogen ion activity, $[H^+]_w$, in the aqueous phase (see example given in Figure 4, top):

$$D = K_p \cdot \frac{1}{1 + K_a/[H^+]_w} \qquad (7)$$

Allthough K_{oc} values obtained for neutral chlorinated phenol species vary more significantly than for nonpolar compounds between different sorbents, as a first approximation, K_p values may be estimated using the linear free-energy relationship derived by Schellenberg et al. (19):

$$K_p = f_{oc} \cdot 1.05 \ (K_{ow})^{0.82} \qquad (8)$$

(3) For highly chlorinated phenols, i.e., for tetra- and pentachlorophenol, the sorption of the anion can usually not be neglected, since, at ambient pH values, these compounds are present predominantly in the ionized form. Furthermore, the results of recent (unpublished) work in our laboratory suggest that polychlorinated phenolate ions are sorbed quite significantly by natural organic matter, in particular, in the presence of Ca^{2+} or Mg^{2+} ions. Some of the mechanisms involved in the distribution of hydrophobic ionizable organic compounds between aqueous phases and non-aqueous phases are discussed in a recent paper by Westall et al. (20). Thus, for the highly chlorinated phenols, predictions of overall distribution ratios based on simple partitioning of the non-dissociated species are generally in error. These findings are illustrated by the example given in Figure 4 (bottom) and are corroborated by the results of a very recent field study in which substantial retention of tetra- and pentachlorophenol has been observed at pH 10 (at which practically all polychlorinated phenolic species are present as anions) in an aquifer (20). The results of our present study of the sorption behavior of hydrophobic anions will be published in the near future.

In summary, to date, only very few investigations of the sorption behavior of ionizable organic compounds have been reported and equilibrium based equations have satisfactorily described partitioning when pH and major ion composition of solutions are incorporated. It is clear, that future studies need to be expanded to a variety of organic pollutants exhibiting polar, in particular, ionizable functional groups.

REFERENCES

(1) KARICKHOFF, S., BROWN, D.S. and SCOTT, T. (1979). Sorption of hydrophobic pollutants on natural sediments. Water Res., 13, 241.
(2) MEANS, J.C., WOOD, S.G., HASSETT, J.J. and BANWART, W.L. (1980). Sorption of polynuclear aromatic hydrocarbons by sediments and soils. Environ. Sci. Technol., 14, 1524.
(3) CHIOU, C.T., PETERS, L.J. and FREED, V.J. (1979). A physical concept of soil-water equilibria for non-ionic organic compounds. Science, 206, 831.
(4) CHIOU, C.T., PORTER, P.E. and SCHMEDDING, D.W. (1983). Partition equilibria of nonionic organic compounds between soil organic matter and water. Environ. Sci. Technol., 17, 227.
(5) SCHWARZENBACH, R.P. and WESTALL, J. (1981). Transport of nonpolar organic compounds from surface water to groundwater. Laboratory sorption studies. Environ. Sci. Technol., 15, 1360.
(6) KENAGA, E.E. and GORING, C.A.I. (1980). In: Aquatic Toxicology, J.G. Eaton, P.R. Parrish and A.C. Hendricks (Eds.), ASTM Special Publication 707, Philadelphia.
(7) RAO, P.S.C. and DAVIDSON, J.M. (1980). Estimation of pesticide retention and transformation parameters required in nonpoint source pollution models. In: Environmental Impact of Nonpoint Source Pollution, M.R. Overcash and J.M. Davidson (Eds.), Ann Arbor Science Publishers, Ann Arbor, p.23.
(8) BROWN, D.S. and FLAG, E.W. (1981). Empirial prediction of organic pollutant sorption in natural sediments. J. Environ. Qual., 10, 382.
(9) KARICKHOFF, S.W. (1984). Organic pollutant sorption in aquatic systems. J. Hydraulic Eng., 110, 707.
(10) LEO, A., HANSCH, E. and ELKINS, D. (1971). Partition coefficients and their uses. Chem. Rev., 71, 525.
(11) WESTALL, J. (1984). Properties of organic compounds in relation to binding by natural materials. Proceedings of the Conference on Biofilm Processes in Ground Water Research, Stockholm, The Ecological Research Committee of NFR, Box 6711, 113 85 Stockholm, Sweden.
(12) CARTER, C.W. and SUFFET, I.H. (1982). Binding of DDT to dissolved humic materials. Environ. Sci. Technol., 16, 735.
(13) WIJAYARATNE, R.D. and MEANS, J.C. (1984). Affinity of hydrophobic pollutants for natural estuarine colloids in aquatic environments. Environ. Sci. Technol., 18, 121.
(14) HASSETT, J.P. and MILICIC, E. (1985). Determination of equilibrium and rate constants for binding of a polychlorinated biphenyl congener by dissolved humic substances. Environ. Sci. Technol., 19, 638.
(15) GSCHWEND, P.M. and WU, S.-C. (1985). On the constancy of sediment-water partition coefficients of hydrophobic organic pollutants. Environ. Sci. Technol., 19, 90.
(16) WU, S.-C. and GSCHWEND, P.M. (1986). Sorption kinetics of hydrophobic organic compounds to natural sediments and soils. Environ. Sci. Technol., in press.

(17) ZIERATH, D., HASSETT, J.J., BANWART, W.L., WOOD, S.A. and MEANS, J.C. (1982). Sorption of benzidine by sediments and soils. Soil Sci., 129, 277.
(18) MEANS, J.C., WOOD, S.G., HASSETT, J.J. and BANWART, W.L. (1982). Sorption of amino- and carboxy-substituted polynuclear armatic hydrocarbons by sediments and soils. Environ. Sci. Technol., 16, 93.
(19) SCHELLENBERG, K., LEUENBERGER, C. and SCHWARZENBACH, R.P. (1984) Sorption of chlorinated phenols by sediments and aquifer materials. Environ. Sci. Technol., 18, 652.
(20) WESTALL, J., LEUENBERGER, C. and SCHWARZENBACH, R.P. (1985). Influence of pH and ionic strength on the aqueous-nonaqueous distribution of chlorinated phenols. Environ. Sci. Technol., 19, 193.
(21) JOHNSON, R.L., BRILLANTE, S.M., ISABELLE, L.M., HOUCK, J.E. and PANKOW, J.F. (1985). Migration of chlorophenolic compounds stored at the Chemical Waste Disposal Site at Alkali Lake, Oregon. 2. Contaminant distributions, transport, and retardation. Groundwater, in press.
(22) BRIGGS, A.A. (1981). Theoretical and experimental relationships between soil adsorption, octanol/water partition coefficients, water solubilities, bioconcentration factors, and the parachor. J. Agric. Food Chem., 29, 1050.
(23) HANSCH, C. and LEO, A. (1979). Substituent Constants for Correlation Analysis in Chemistry and Biology, Elsevier, Amsterdam.

PROPERTIES OF HUMIC MATERIAL IN RELATION TO TRANSPORT PHENOMENA

E. GJESSING
Norwegian Institute for Water Research, Oslo, Norway

Summary

Humic material, originating from soil, is present in all surface water;
the world mean in the range of 5-6 mg DOC/l. Aquatic humus as such
will "contain" minute amount of most all elements. Characteristic
properties are
- Brownish-yellow coloured
- Weak acid behaviour
- Macromolecules with a broad MW-spectrum
- Change of molecular size-distribution with pH
- Negatively charged under natural pH-conditions
- Change of charge at low pH
- Vehicle for heavy metals and organic micropollutants
- Assorted kitchen micronutrients
- Formation of haloforms by chlorination
- Ion exchange and complexing properties.
Having the ability to complex with metal ions and organic chemicals
it appears relevant to introduce the term "polluted humus". The con-
sequences of the properties of humus to increase the water solubility
of inorganic and organic chemicals, for external and internal trans-
port of humus are not yet quantified.

1. INTRODUCTION

Perhaps chemists lack persistance or maybe the research findings have
not been sufficiently encouraging, but whatever the reason, relatively few
scientists in the last century have devoted their energies to research on
the nature and properties of aquatic and terrestrial humic materials. Those
who have struggled valiantly in the last decade have found their patience
rewarded by renewed interest in this very ancient subject.

In 1974, it was suggested that aquatic humic materials were ubiquitous reaction precursors of halogenated methanes via reaction with chlorine
disinfectant (1). The number of humic chemistry research reports in the
literature has risen dramatically since the recognition of this problem.

The benefits from increased understanding that will accrue from this
renewed interest are only partially evident at present. New experimental
directions in isolation and purification of natural humic materials, increased identification of structural moieties resulting from controlled degradation (including chlorine degradation) and revised concepts of the mechanisms by which metal ions interact with these complex natural products are
among the initial scientific findings.

2. AQUATIC HUMIC MATERIALS IN THE CARBON CYCLE

Major research questions pertaining to the extent of occurrence,
qualitative nature and role of humic materials in natural aquatic systems

have been explored in the last decade and are still largely unanswered. During this period factual evidence of the levels of natural organic carbon carried by terrestrial streams has been slowly gathering, but increases in our understanding of the chemical characteristics of these complex natural products have been more modest.

The recent discovery of halogenated organic reaction products in drinking water has called attention to the ubiquitous occurrence of complex forms of organic carbon in terrestrial water systems. The quantities of carbon transported annually to the earth's oceans are large (approximately 1.0 billion tons/yr) compared to annual chemical industry production, but small compared to other carbon transfer routes to the oceans (Table I).

Dissolved organic carbon (DOC) values vary considerably with climatic zone, ranging from 3 mg/l in temperate and arid zones to 6 mg/l in tropical zones and 10 mg/l in subarctic zones. Meybeck (2) has reported a stream volume weighted world average of 5.75 mg DOC/l. Thus, our understanding of natural carbon flux in terrestrial streams has broadened significantly within the last decade. In the 1960s and 1970s research reports noted the presence and significance of complex organic carbon in terrestrial surface waters and groundwaters and today the presence of these organic forms is seen as an important part of a natural global process.

Table I. Selected Annual Carbon Fluxes to Oceans (x 10^9 tons (3))

Route	Annual Flux
Terrestrial streams	8
Ocean photosynthesis	450
Net dissolution from atmosphere	19

The data will undoubtedly continue to improve, as we still have limited knowledge of the seasonal variations in DOC values among climatic zones. More important to this discussion is the fraction of DOC that is comprised of humic materials. It is attractive to assume that most, if not all, of the DOC in terrestrial streams is natural product organic material. Multiple opportunities exist in the earth's water and exogenic cycles for the transfer of virtually every form of terrestrial carbon to various environmental reservoirs (oceans, atmosphere, sediments, etc.) through dissolution or suspension in water. Qualitatively, the organic carbon pool of terrestrial streams may contain a myriad of structural moieties arising not only from the extraction and/or sorption of components of terrestrial vegetation in all stages of growth and decay, the organic portion of soil and any transformation products formed in subsequent stream transit, but also from the organic structures of soil and aquatic microorganisms and their numerous metabolic and natural degradation products (4),(5). Considering the facts that 90% of the terrestrial biospheric carbon (standing biomass) is tied up in woody/grassy tissue (6) and that microbial communities have had hundreds of millions of years to adapt to the transfer process, it is reasonable to assume that these sources are ubiquitous in nature.

3. STRUCTURES OF AQUATIC HUMIC MATERIALS

Most of our knowledge of the structural characteristics of aquatic humic materials has come from degradative methods (5), (7) which have been patterned after similar work on closely related, natural products such as lignin and soil organic matter (8), (9).

Liao et al. (7) recently reported an extensive investigation of the NaOH hydrolysis and $KMnO_4$ oxidation degradation products of the aquatic and fulvic acid fractions of two geographically separated surface waters. The maximum yield of products identified by GC/MS was approximately 12 to 17 wt % of starting material for base hydrolysis and 20 to 25 wt % for alkaline $KMnO_4$ oxidation.

The fact that the majority of starting material was not identified makes it difficult to generalize about the original macromolecular structures of aquatic humic material. It can be said, however, that the presence of both aromatic and aliphatic structures is indicated in the data of Liao et al., as is the presence of oxygen in ring-bound aliphatic side-chains (carboxyphenylglyoxylic acids) and in heterocyclic rings (furancarboxylic acids). It is attractive to assume that the predominance of the aromatic polycarboxylic acidity after $KMnO_4$ reaction is due to cleavage of inter-aromatic carbon to carbon linkages since the base hydrolysis yield of these polyacids is much lower. It this is true, the data of Liao et al. suggest that the aliphatic chains connecting aromatic rings in their aquatic humic and fulvic material are relatively short (C_2-C_4) to account for the predominance of these dibasic acids in product mixtures. Some (less than 3%) of the aromatic acids must be bound originally in ester linkages to account for the base hydrolysis yields and some of the aliphatic and/or aromatic carboxyl groups must be unbound to account for the well documented acidity of aquatic humic material (5), (10), (11). Unfortunately, the present literature does not contain reports of degradative studies of sufficient scope to permit comment on temporal and spatial variations in structural composition.

4. REACTIONS OF HUMIC MATERIALS

Two specific reaction properties of aquatic humic materials have been of intensive interest to research workers. The first property, metal ion interactions, has been recognized for many years as being of ecologic significance as well as a means of elucidating reactive functional group properties. The other, being of relatively new date, is the association between organic micropollutants and aquatic humic materials.

Introducing the potensial of inorganic and organic micropollutants to react or associate with the aquatic humus, one also introduces an uncertainty about what is natural, genuine humus and polluted humus.

a) Metal-Ion Interactions - Consequences for Transport

Advances in the last few decades in soil and water chemistry, as well as the development of analytical techniques such as anodic stripping voltammetry and atomic absorption spectrophotometry, have extended the sensitivity and accuracy of trace metal data of interest to biologists and chemists. The role of humic materials as a complicating factor in the uptake and utilization of metal ions has been long recognized. With increasing knowledge of the structural features of humic materials and more available trace metal data, recent research has focused on the relation of chemical environmental measurements to the measurements of responses of biological systems.

The knowledge of the reaction mechanism between metals and humus is limited, most probably there is a series of different reactions taking place depending on the physio-chemical conditions and the elements presently involved. In the present paper the different views will not be considered. It is, however, important to emphasize that most all elements are capable to associate with humus, as indicated in Table II.

It appears from Table II that humus will play an important role for the speciation of metals in water and may act as a vehicle for heavily soluble metals in the aquatic environment.

Table II. Composition of water from Lake Trehørningen (Baerum, Norway) and the percentage of the different elements retained by dialyzing membranes (mw > 10,000) - according to Benes et al. (12).

Element	Concentration $(\mu g\ 1^{-1})$	Retained by dialysis membrane (%)
Na	1800	0
Rb	1.03	12
Ca	2085	6
Ba	13.0	16
Sc	0.025	82
La	0.145	72
Ce	0.331	92
Al	113	78
Cr	1.4	88
Fe	109	64
Th	0.023	>75
Mn	27.0	25
Co	0.126	66
Zn	18.5	26
Cl	1290	0
Br	7.3	11
I	2.5	73

To illustrate the potential of aquatic humus to increase the solubility of metal, two examples are refered to below:

Metallic iron, as finely powdered metallic iron, was added to samples of concentrated humic water. The particle size of iron ranged from 1 to 10 μm. The natural humic water was concentrated ten times by evaporation under reduced pressure at 35 °C and then filtered through Whatman GF/C glass fiber filters to remove particles. In some experiments samples of unconcentrated humic water were also used. The tests were carried out at different pH-values and hydrochloric acid or sodium hydroxide was used to adjust the pH in the range 2.8-9.0. The colour (420 nm) was measured after the pH adjustment. To each sample of 50 ml humic water, in 100 ml polyethylene bottles, 0.5 g of powdered metallic iron was added and mechanically shaken for three days. After the treatment excess of metallic iron powder was removed by filtering through GF/C filters. In most of the humic samples a distinct increase of visible colour was observed. The experiments included natural and concentrated humic water, city tap water and distilled water. The final pH, absorbance and iron were measured. For determination of total iron a Perkin-Elmer model 303 atomic absorption spectrophotometer was used.

As Figure 1 indicates, the relationship between iron and absorbances is almost linear.

It seems clear that humic water can dissolve metallic iron, at least when iron is present as small particles and has a large surface. The humic substances also seem to keep iron in solution for a long time. it is

possible that iron in solution exists partly as colloidal particles of iron
hydroxides which may be stabilized by negatively charged organic matter,
and partly as humus iron complexes.

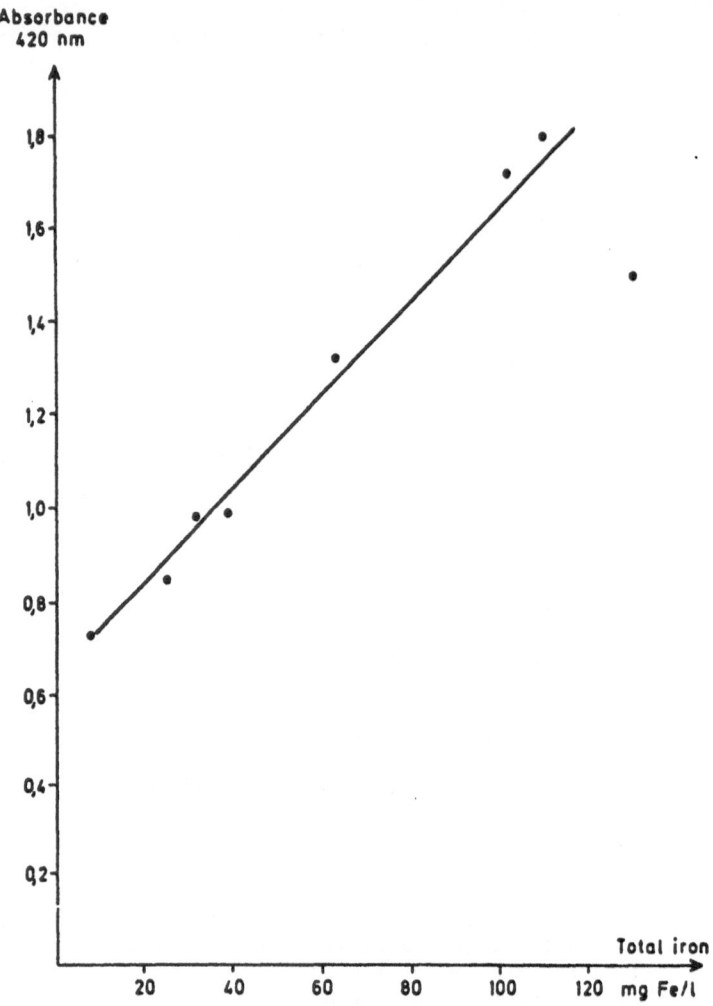

Figure 1. Correlation between iron and absorbance at 420 mm
after shaking powdered iron with concentrated humic
water (13).

The second experiment was performed by adding an aliquot amount of
pure mercury to natural humus water samples from several different locations. Some of the samples were diluted with artificial "surface water"
without organic matter; the experiment thus covered a wide range of humus
concentrations. All samples were filtered through glassfiber filters
(Whatman GF/C) before adding the mercury. The samples were left on a shaking table for 19 h and the supernatant analysed for mercury by atomic
absorption spectrophotometer (Perkin-Elmer 460).

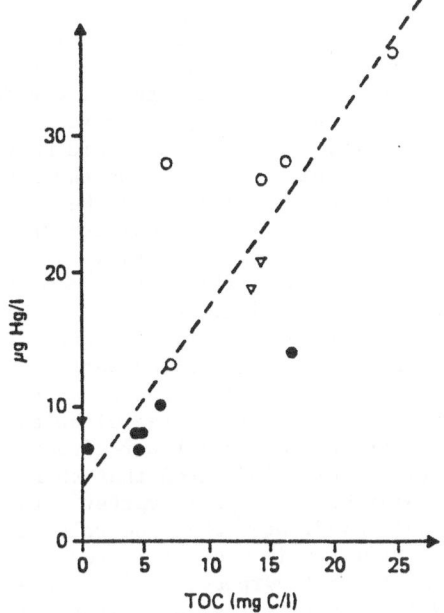

Figure 2. Relationship between "dissolvable" mercury and
the concentration of humus in surface water expressed
as colour and TOC (14).

The results given on Figure 2 suggest a relationship between the content of mercury in the supernatant and the humus concentration. This indicates then that humus has a "dissolving action" on metallic mercury, which
may be considerable compared to distilled water (7-9 µg Hg/1) and compared
to water without humus (13 µg Hg/1). It appears from the figure that there
is som discrepancy from a complete correlation. This might be due partly
to small differences in the amount of metallic mercury added (because two
different mercury droppers were used during the experiment), partly to differences in the surface area of the mercury during shaking. It was observed that the number of droplets differed during the experiment. Finally
the pH of the samples can have some side-effects on the mercury both directly and indirectly due to changes of the character of the humus.

b) Humus as a Vehicle for Organic Micropollutants
     The association between humic material and micro-organic matter has
received limited attention during the last decade. Ogner and Schnitzer (15)
indicated 15 years ago that there was a relationship between dialkylphtalate
and humus and pointed out the pollution potentials. Since then there have
been relatively few papers on these important aspects. The most serious
consequences of complexation between humus and micro-organic matter, particularly the hydrophobic organic micropollutants, are the increased transport and spreading potentials, both in the water itself and into the aquatic organisms.
     In 1981 Gjessing and Berglind (16) stated that there appeared to be
a strong association between several of the PAH's and the humic material
in the water. They showed that up to 50% of the PAH's present in a natural
humic water were not available with the most used analytical method. By
adding 1 µg/1 of benzo(a)pyrene (B(a)P) they showed that less than 10% was
recovered.
     Similarly it was shown that with hexachlorobenzene (HCB) only 50-80%
of the added 10 ng HCB/1 was recovered from water containing humus, whereas

from pure distilled water all the added HCB was recovered (17). Carter and
Suffet (18) used equilibrium dialysis technique to study the extent of
binding between DDT and dissolved humic material and found that a significant fraction of the DDT in water may be bound to the humus. The extent
of the binding depends on the source of the humic material, the pH, the
calcium concentration, the ionic strength, and the concentration of the
humic materials. Wershaw et al. (19) found that 500 mg/l of humic material extracted from soil increased the solubility of DDT 20 to 40 times.
        Hassett and Anderson (20) found that radiolabeled cholesterol added
to water samples was recovered less efficiently from river water than from
distilled water by solvent extraction. However, if organic matter in river
water was destroyed with UV radiation prior to cholesterol addition, recovery was equivalent to that from distilled water. This demonstrated that
cholesterol could be bound by dissolved organic matter and stabilized in the
aqueous phase. Water samples containing radiolabeled cholesterol or the
PCB isomer 2.2´, 5.5´-tetrachlorobiphenyl were fractionated by gel permeation chromatography. Results indicated that these compounds were associated with high molecular weight organic matter present in the samples.
        The work cited above indicates that the role of aquatic humic material
in analytical organic chemistry is underestimated. As all analytical
struggle does have a biological purpose, it is pertinent also to relate the
chemical laboratory work to the biological availability of the chemical in
question. To achieve this the chemist and the biologist have to work more
tightly together in the future.

5. CONCLUDING REMARKS

        Research in this important area is of course heavily influenced by
developments in our fundamental understanding of aquatic and terrestrial
humic chemistry. New information on chemical structures and functionali-
ties, especially those of strong binding potential, remains the principal
determinant of research directions in this field. Environmental application of these findings also depends on information on how these properties
vary spatially and temporally in nature. Humic molecules are polydisperse
and molecular size and shape may bear upon the degree of binding of metals
and organic chemicals, the number of sites available or the ease of uptake
biologically. In many environments concentrations of micropollutants are
quite low compared to humic concentrations, and research on relatively high
humic-to-pollutant concentration ratios needs to be done. Chemists must
consider the applicability of their data to environmental systems and not
assume that measurements made at high pollutant loading are significant in
the actual environment. In addition, bonding at extremely low saturation
may take place at sites in humic materials that include relatively minor
electron donor sites (N, S) rather than at oxygenated sites. Analytical
developments in the area of fluorescence, ionselective electrodes and separation chromatography may yield new methods to detect bonding at extremely
low saturation and would be extremely important to this area of inquiry.
        Our understanding of the nature of environmental effects associated
with metal ion-organic interactions is still primitive. Examples of both
positive and negative effects on the uptake rate of metal ions by living
organisms can be found in the literature. The so-called "Heidermoore"
illness to plants and organisms is thought to result from copper deficiency
due to humic interactions. There is also ample evidence that organic ligands including humic substances lower metal ion activity, which inhibits
phytoplanktonic growth (21-24). Metal complexation thus may act in a positive way by decreasing acute toxic effects of heavy metals on aquatic organisms. The same relatively beneficial effect may extend to mammals exposed

to heavy metals in the presence of humic materials. although this effect as well as the effects of chronic exposures to metal-bound humics are still uncertain. On the other hand, George and Coombs (25) suggest that accumulation of cadmium by Mytilus edulis is facilitated by the presence of humic material. No general conclusions can be made about the effect of metal ion-organic interactions because of the different organisms involved in these various research reports.

All the reaction-potentials between humic material and aquatic micro-pollutants justify the introduction of a new term, "polluted humus".

REFERENCES

(1) ROOK, J.J. (1974). Formation of haloforms during chlorination of natural waters. Water Treatment Exam. 23(2): 234-243.
(2) MEYBECK, M. River transport of organic carbon to the ocean, in: Carbon dioxide effects research and assessment program. Flux of organic carbon by rivers to the ocean. Prepared by the committee on flux of organic carbon to the ocean (G.E. Likens, Chairman).
(3) CARBON DIOXIDE EFFECTS RESEARCH AND ASSESSMENT PROGRAM (1981). Flux of organic carbon by rivers to the ocean. Prepared by the committee on flux of organic carbon to the ocean (G.E. Likens, Chairman), division of biol. sciences, national research council, published by U.S. Dept. of energy, office of energy research, Washington, DC 20545, Reference: CONF-8009140 UC 11.
(4) STEELINK, C. (1977). Humates and other natural organic substances in the aquatic environment. J. Chem. Ed. 54: 599-603.
(5) CHRISTMAN, R.F. and GHASSEMI, M. (1966). Chemical nature of organic color in water. J. Am. Water Works Assoc. 58(6): 723-741.
(6) WHITTAKER, R.H. and LIKENS, G.H. (1973). Carbon and the biosphere. U.S. Atomic Energy Commission, AEC Symposium Series, No. 30, p.281.
(7) LIAO, W., CHRISTMAN, R.F., JOHNSON, J.D., MILLINGTON, D.S. and HASS, J.R. (1982). Structural characterization of aquatic humic material. Environ. Sci. Tech. 16: 403-410.
(8) SCHNITZER, M. and KHAN, S.U. (1972). Humic substances in the environment. (New York: Marcel Dekker, Inc.).
(9) GREEN, G. and STEELINK, C. (1962). Structure of soil humic acid. J. Org. Chem. 27: 170-174.
(10) SHAPIRO, J. (1957). Chemical and biological studies on the yellow organic acids of lake water. Limnol. Oceanog. 2: 161.
(11) GJESSING, E.(1976). Physical and chemical characteristics of aquatic humus. (Ann Arbor, MI: Ann Arbor Science Publishers, Inc.).
(12) BENES, P., GJESSING, E.T. and STEINNES, E. (1976). Interaction between humus and trace elements in freshwater. Water Research 10: 711-716.
(13) TRYLAND, Ø. (1976).Humic water dissolves metallic iron. Vatten 3: 271-273.
(14) GJESSING, E.T. and ROGNE, Å.K. (1982). Solubility of metallic mercury in water containing humus. Vatten 38: 406-408.
(15) OGNER, G. and SCHNITZER, M. (1970). Humic substances: Fulvic acid-dialkylphtalate complexes, and their role in pollution. Science 170: 317-318.
(16) GJESSING, E.T. and BERGLIND, L. (1981). Adsorption of PAH to aquatic humus. Arch. Hydrobiol. 92: 24-30.
(17) GJESSING, E.T. and BERGLIND, L. (1982). Analytical availability of hexachlorobenzene (HCB) in water containing humus. Vatten 38: 402-405.

(18) CARTER, C.W. and SUFFET, I.H. (1982). Binding of DDT to dissolved humic materials. Environ. Sci. Technol. 16: 735-740.

(19) WERSHAW, R.L., BURCAR, P.J. and GOLDBERG, M.C. (1969). Interaction of pesticides with natural organic material. Environ. Sci. Technol. 3: 271-273.

(20) HASSETT, J.P. and ANDERSON, M.A. (1979). Association of hydrophobic organic compounds with dissolved organic matter in aquatic systems. Environ. Sci. Technol. 13: 1526-1529.

(21) MANTOURA, R.F.C., DICKSON, A. and RILEY, J.P. (1978). The complexation of metals with humic materials in natural waters. Estuar. Coast. Mar. Sci. 6: 387-408.

(22) BENES, P., GJESSING, E.T. and STEINNES, E. (1976). Interaction between humus and trace elements in fresh water. Water Res. 10: 711-716. (See (12).)

(23) HART, B.T. and DAVIES, S.H.R. (1979). Speciation of Fe, Cd, Cu, Pb and Zn in the Yarra River and estuary, Australia, In: International conference management and control of heavy metals in the environment (London: CEP Consultants) 466-471.

(24) GJESSING, E. (1981). The effect of aquatic humus on the biological availability of cadmium. Arch. Hydrobiol. 91: 144-149.

(25) GEORGE, S.G. and COOMBS, T.H. (1977). The effect of chelating agents on the uptake and accumulation of cadmium by Mytilus edulis. Marine Biol. 39: 261-268.

PCB CONGENERS IN THE MARINE ENVIRONMENT - A REVIEW

J.C. DUINKER[1] and J.P. BOON[2]

[1] Institut für Meereskunde an der Universität Kiel, Düsternbrookerweg 20, 2300 Kiel 1, Federal Republic of Germany

[2] DGW-Rijkswaterstaat, P.O. Box 8039, 4330 MA Middelburg, the Netherlands

Summary

Different molecular structures cause differences in physical-chemical characteristics between the 209 theoretically possible PCB congeners. The general processes and differences between congeners in environmental behaviour are discussed; analysis, atmospheric input into the marine environment, estuarine and marine transport processes in the water column, sedimentation, desorption from sediments, bioaccumulation, biotransformation and congener specific induction of mixed function oxygenase enzyme systems are considered.

1. INTRODUCTION

The industrial chlorination of biphenyl leads to complex mixtures containing many out of the 209 theoretically possible chlorinated biphenyl congeners in the various technical mixtures produced with different overall mass percentages of chlorine in the 20-60% range (e.g. Aroclors, Clophens, Phenochlors, Kanechlors) (1,2). Fig. 1 represents the molecular skeleton. The numbers 2-6 and 2'-6' in this figure indicate binding positions at each of the aromatic rings to which either a hydrogen or a chlorine atom can be attached. The phenylrings are joined by a single carbon-carbon bond al-

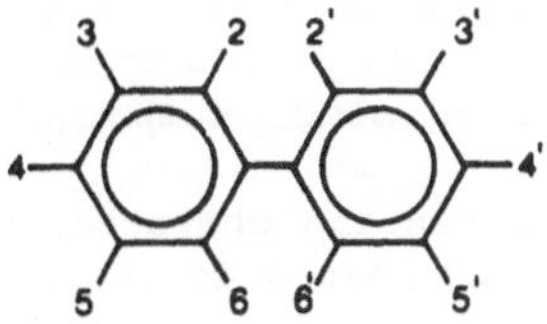

Fig. 1. The general molecular frame work of PCB congeners.

lowing, in principal, a free rotation of both rings. The presence of one or more ortho-chlorine atoms (i.e. in the 2,2', 6 or 6' position) results in an inter-ring angle of 90° (3).

The presence of PCB's in the environment was discovered by Jensen in 1966 (4). The observation that PCB's are acutely toxic to humans (5, 6), their presence in practically all samples of air, water, and organisms even from remote areas (1), as well as the improved possibilities for studying details of their environmental distribution, transport and effects, have resulted in a tremendous research effort to understand the environmental behaviour of this class of contaminants.

This review deals with general processes and differences between congeners in environmental behaviour in different compartments of the marine environment (sources, transport/tranformation processes and sinks) and with differences in toxicological effects in marine animals. Observations made in freshwater environments are included whenever interesting for the marine environment as well.

2. ANALYTICAL ASPECTS

The presence of 1-10 chlorine atoms in the biphenyl molecular frame work allows the detection and quantification of trace amounts of individual congeners (10-12g) with the aid of an electron capture detector (ECD), after separation on a gaschromatigraphic column. The detailed analysis of PCB mixtures as they occur in technical formulations and in environmental samples requires the elution of the constituents from the column as well-separated peaks. Most work in the literature on PCB's in the environment, however, has been reported on the basis of packed column chromatographic separations with concentrations expressed, both qualitatively and quantitatively, in terms of technical formulation equivalents. Although the resulting low-resolution chromatograms have supplied information on hot spots and on distribution trends, their usefulness for evaluating important aspects such as sources, sinks, transport mechanisms, biological uptake and toxicity is limited (7).

Serious attempts have been made by a number of groups to improve the accuracy of PCB analyses by using capillary columns, characterized by greatly increased separation capabilities. The need to distinguish individual congeners was realized about a decade ago by some groups (8-10), who analyzed the composition of various technical formulations with the use of up to 60 chlorinated biphenyls as references. Additional information on their compositions has been obtained with the aid of columns with different stationary phases (11) and additional techniques such as infrared (IR) spectrometry, nuclear magnetic resonance (NMR) spectrometry and in particular mass spectrometry (12 and references therein), calculation of retention times to assist in identifying congeners which were not available as reference compounds (13, 14) and the use of more individual congeners (i.e. 102) as references (12). A rigoruous individual PCB congener approach in the environment offers the advantage that this class of lipophilic compounds with a wide range of substitution patterns of chlorine atoms in the biphenyl molecular frame work, and thus with a wide range of physico-chemical characteristics, can be used as model compounds to study fundamental processes in the environment.

The lack of information on retention characteristics of all theoretically possible 209 PCB congeners has been a limiting factor in all qualitative and quantitative analyses mentioned above. Recently all components were synthesized and characterized (15). It appears from the latter data that not all components can yet be separated by the use of one column only (i.e. SE-54 fused silica column), but many components can now be well separated from neighbouring ones on this column, and can thus be identified and quantified unambiguously. In this paper, we adopt and recommend the use of the systematic numbering system proposed by Ballschmiter and Zell (14). In this numbering system, congeners 1-3 represent monochlorobiphenyls (mono-CBs); 4-15 di-CBs; 16-39 tri-CBs; 40-81 tetra-CBs; 82-127 penta-CBs; 128-169 hexa-CBs; 170-193 hepta-CBs; 194-205 octa-CBs; 206-208 nona-CBs and 209 deca-CB.

3. ENVIRONMENTAL PATHWAYS OF PCBs

Input into lakes and the marine environment originates from freshwater and from land run-off, dumping activities and municipal and industrial waste water. It has become evident in recent years, that the atmosphere is also a significant source of PCB and other organic contaminants in both lakes (16) as well as in ocean waters (17-21). For example, 60-90% of the total PCB input to lakes Superior and Michigan have been accounted for by atmospheric input (16).

3.1. The role of the atmosphere

In the atmosphere, PCBs occur in the vapour phase and in particulate forms (aerosols). There is no generally accepted sampling method to distinguish these forms. The usual methods, involving high- or low-volume samplers which collect the vapours on an adsorbent trap after removal of particulates on a filter, do not necessarily distinguish the aerosol and vapor phases accurately. Due to losses from the filter material, the importance of the vapour phase may be overestimated (22); also, PCBs may be associated with particles which are smaller than the particle diameter cutoff (0.3 μm) (23), and adsorbent materials most commonly used do not collect individual components with uniform efficiency (16).

It appears that the vapour phase operationally defined by the filter characteristics, contains by far the largest fraction of PCBs in total atmospheric samples (16, 18, 19, 23). This is supported by theoretical expectations for the vapour-aerosol distribution of organics based on saturation vapour pressure (P_O) and available surface area of atmospheric particulate matter (24). Compounds with $P_O > 10^{-4}$ mm Hg, should exist almost entirely in the vapour phase and those with $P_O < 10^{-8}$ mm Hg almost entirely in the particulate phase. PCBs roughly cover the intermediate range (16).
The vapour-aerosol partition may be different for the various individual PCB congeners. Thus, the filter may contain relatively more of the high-molecular and low-vapour pressure components, as has been demonstrated for PAHs (25, 26). A further complicating factor for PCB may result from the differences in fractionation characteristics between congeners in relation to particle size distribution, which is usually poorly known. Recent studies indicate that large-size fractions are deposited near land. Thus, smaller size fractions dominate off-shore (27). It has been suggested that PCBs associate with the submicrometer particles (23), which have also been reported to contain the dominant fraction of organic carbon (28). Aerosol PAHs in various environments were reported to have mass median diameters of 0.7 - 1.4 μm. Between 30 and 70% of the mass was found in the $< 1\mu$m size fraction. The larger particle size fractions ($>1\mu$m) were found to contain a significant fraction of the higher molecular weight components. Because of differences in deposition velocities and washout ratios the depositional flux of the higher molecular weight fractions of PCBs in particulates may be dominated by the larger particles (29).

PCBs present in the atmosphere can enter the aquatic environment by vapour partitioning across the air-water interface, dry deposition of aerosol-associated froms and by wet deposition involving the scavenging of particles and rain-atmosphere partitioning of vapours. Theoretical models for dry and wet deposition form the atmosphere are available (references in 16, 23).

For wet removal processes, washout factors of aerosols ($10^5 < W < 10^6$) exceed those for gases considerably ($10^0 < W < 10^4$), but other factors are also important for removal rates. Comparison of experimentally determined values of W (corresponding to washout of vapours plus particulates) with theoretical values corresponding to gases only, may assist in estimating the role of particle scavenging.

Dry deposition of aerosol-associated PCBs can be calculated for different conditions of receptor surface, particle size, meteorology etc.

The wet flux of airborne PCBs can be calculated from data on the amount of precipitation and PCB concentration levels in rain or snow. Dry deposition fluxes are less easily accessible both theoretically and experimentally; the deposition velocity V_d being the critical variable. The calculated flux of airborne PCBs in the vapour phase and associated with submicron particles (depending on the V_d value selected) has been estimated to exceed the wet deposition flux to the Great Lakes (16).

3.2. PCB in the aquatic environment

3.2.1. Distribution and Processes

The resistance of PCBs against hydrolysis, oxidation, reduction, photodegradation and biodegradation by microorganisms (30) has been investigated under laboratory conditions for many individual PCB congeners (1, 31-33). Of these mechanisms, only microbial activities may result in significant modifications of PCB mixtures under environmental conditions, affecting mainly components with 1-4 chlorine atoms, penta- and higher-chlorinated congeners being hardly affected or not at all.

PCBs present in abiotic compartments in the fresh- and seawater environment occur in solution and associated with particles. Since transport mechanisms in the environment are different for water and particulates, solution-particulate partitioning characteristics of organics are important in relation to their transport and chemical/microbiological transformation processes as well as to bioavailability to higher organisms.

Various attempts have been made to relate water solubility, octanol-water distribution coefficients and partition coefficients between water and particulates for PCBs (34-36).

PCBs are present in natural waters in extremely low concentrations. Sorption to particles may be predominantly governed by equilibrium partitioning of the constituents between suspended particles and ambient water (36-39). The distribution coefficient K_d (concentration ratio) is defined as

$$K_d = \frac{C^i{}_s}{C^i{}_w} \qquad \text{(eq. 1)}$$

where $C^i{}_s$ = concentration of congener i in suspended matter (ng g^{-1}) and
$C^i{}_w$ = concentration of congener i in water (ng g^{-1} water)

K_d values can be obtained from concentration measurements; they reflect the net effect of transport and transformation steps and can be used as universal indices to predict the distribution of chemicals in the aquatic environment. Distribution coefficients appear to depend on the fraction of organic carbon in the particulate forms (f_{oc}) as well on the composition of the organic fraction (quantified by the hydrophobic partitioning constant K_{oc}). Since K_{oc} values have been found to vary between different sediments by not more then a factor 2, K_d values depend largely on f_{oc}, according to equation 2 (40):

$$K_d = f_{oc} \cdot K_{oc} \qquad \text{(eq. 2)}$$

However, evidence is also presented that the mechanism described above is an oversimplification. Other experiments showed that $K_{desorption} > K_{sorption}$, indicating that sediment sorbed PCBs were comprised of both reversibly and strongly bound or resistant fractions (41, 42). The equilibrium constants K_d and K_{oc} were also shown to decrease as the ratio of solids to water increased (43, 44). However, the results might have been affected by the presence of nonsettling microparticles, colloids and dissolved organic matter (39).

Regional K_d values were calculated for seawater and suspended matter in Puget Sound, for PCB congeners grouped according to their number of chlorine atoms (N_{Cl})(36). Average values were 4.36×10^4($N_{Cl}=3$), 4.43×10^4 ($N_{Cl}=4$), 6.27×10^4 ($N_{Cl}=5$), 9.18×10^4 ($N_{Cl}=6$), and 9.67×10^4 ($N_{Cl}=7$). Thus, the average distribution coefficients increased for each group of congeners with increasing chlorine numbers.

The concept of equilibrium partitioning is useful for estimating the relative contributions of the dissolved and suspended particulate phases to the total concentration per unit volume of the compounds under consideration. For instance, compounds with $K < 10^5$ will reside mainly in the water phase at SPM concentrations <10 mg dm^{-3} (Fig. 2).

We have tested the concept of equilibrium partitioning between dissolved and SPM phases for some individual PCB congeners in some rivers of NW Europe and the adjacent North Sea (Table I). The individual K_d values show the same trend as the data on groups of components with the same chlorine number. Values of individual congeners generally increase with increasing chlorine number and are between 6×10^3 and 3×10^5 for the rivers. Despite the significant differences in the composition of both water and SPM in the various rivers (both in terms of natural and anthropogenic variables), the K_d values for each PCB congener in the various rivers appear to be very similar. The values for K_d are higher for offshore seawater (between 4×10^4 and 5×10^6). The reason is discussed in § 3.2.2.

Partitioning of each PCB congener over the dissolved and SPM phases in lakes, rivers, estuaries and the marine environment is determined by its distribution coefficient (K_d) and the concentration of SPM. Concentrations of suspended matter in natural waters vary with place and time. Large variations may occur in shallow waters where particulates are eroded from the bottom into the overlying water by tidal action, wind, and during periods of high river discharge (49).

SPM concentrations in estuaries where river- and seawater are mixed, usually exceed those in the source waters significantly as a consequence of hydrodynamical and physicochemical processes (49,50). Under these conditions, the equilibrium partitioning model would require an increased residence of each PCB component on suspended particles. For example, in the Rhine estuary, total concentrations of PCB in suspension (in terms of technical formulations) exceeded those dissolved in the same volume of water, while in the river, PCBs were roughly equally distributed over dissolved and particulate suspended forms (51). However, the transitions from solution to particulate suspended forms, which are required to satisfy the equilibrium partitioning model, were not obvious from the relation between dissolved PCB concentrations and salinity, where a negative deviation from the ideal mixing line would have supported removal processes of PCBs from solution into particulate forms. However, the relation appeared to be practically linear. The same observation on conservative behaviour of PCBs during estuarine mixing was also made in some other estuaries (36, 52). The complexity of estuarine processes and the extremely small amounts of individual components involved may be the main factors why such phase transitions have not yet been observed, in particular with the use of packed columns for separations.

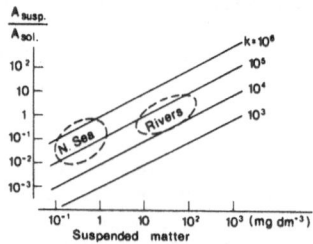

Fig. 2. Relation between suspended matter concentration (mg dm⁻3) and
the ratio between the amounts in suspension (Asusp) and in
solution (Asol) for various values of the distribution coefficient (K) of any chemical constituent ($10^3 < K < 10^6$).
The values for individual PCB congeners in the North Sea and
in some Nortwest European rivers are contained in the encir-
cled regions (see text).

Table I. Distribution coefficients for water-suspended matter partition of individual PCB congeners
in fresh water of the Scheldt, Rhine, Weser and Elbe rivers and in the offshore region of
the North Sea.

The latter samples were obtained at positions 52°01'N, 2°47'E, 53°00'N, 1°31'.5E, 55°20'N,
5°30'E (October 1982) at salinities $> 35 \times 10^{-3}$.

PCB-congener			d)	d)	e)	f)	g)	h)
a)	b)	c)	Scheldt	Rhine	Weser	Elbe	Ems	North Sea
18	2,2',5	3	$8 \times 10^3 - 5 \times 10^4$	2×10^4	2×10^4	2×10^4	2×10^4	$4 \times 10^4 - 6 \times 10^5$
15	4,4'	2	$6 \times 10^3 - 6 \times 10^4$	4×10^4	2×10^4	2×10^4	3×10^4	5×10^5
26	2,3',5	3	$2 \times 10^4 - 2 \times 10^5$	7×10^4	2×10^5	1×10^5	1×10^5	$8 \times 10^4 - 2 \times 10^5$
31	2,4',5	3	$1 \times 10^4 - 2 \times 10^5$	7×10^4	-i)	-i)	-i)	$1 \times 10^5 - 3 \times 10^5$
28	2,4,4'	3	$2 \times 10^4 - 1 \times 10^5$	7×10^4	-	-	-	$7 \times 10^4 - 2 \times 10^5$
52	2,2',5,5'	4	$3 \times 10^4 - 7 \times 10^4$	8×10^4	4×10^4	5×10^4	5×10^4	$3 \times 10^5 - 2 \times 10^6$
84	2,2',3,3',6	5	5×10^4	8×10^4	8×10^4	8×10^4	5×10^4	$7 \times 10^4 - 4 \times 10^5$
101	2,2',4,5,5'	5	$8 \times 10^4 - 1 \times 10^5$	1×10^5	1×10^5	1×10^5	6×10^4	$8 \times 10^4 - 1 \times 10^6$
99	2,2',4,4',5	5	$1 \times 10^4 - 4 \times 10^5$	1×10^5	1×10^5	1×10^5	6×10^4	$6 \times 10^4 - 2 \times 10^6$
153	2,2',4,4',5,5'	6	$1 \times 10^5 - 3 \times 10^5$	1×10^5	2×10^5	2×10^5	1×10^5	$5 \times 10^5 - 5 \times 10^6$
138	2,2',3,4,4',5'	6	$7 \times 10^4 - 1 \times 10^5$	6×10^4	2×10^5	2×10^5	1×10^5	$2 \times 10^5 - 5 \times 10^6$
SPM concentration (mg dm⁻3)			45	19	30	35	40	0.5-1.0

a) systematic numbering (14), b) chlorine substitution pattern, c) number of chlorine atoms (N_{Cl})
d) ref. 45, e) ref. 46, f) ref. 47, g) ref. 48, h) unpublished results, i) -: individual peaks of
31 and 28 not separated.

3.2.2. Sedimentation in estuaries

Large amounts of fine riverborne particulates can be deposited in estuaries. Concentrations of PCBs in fine sediments have been reported to be similar to those in fine SPM and orders of magnitude larger than in coarse sediments (53, 54). Deposition in the freshwater section of the estuary may involve larger size fractions than in the mixing zone of the estuary (51, 55, 56). A significant fraction of the riverborn SPM, in particular the particles with smaller size/density characteristics, may escape the more or less effective estuarine trapping mechanism and escape to the open sea, along with congeners in solution (57). These smaller particles have higher contrations of PCBs (per unit SPM weight) than the larger/denser particles, remaining in the estuary.

This accounts for the higher distribution coefficients found for several individual PCB congeners in the North Sea with respect to values in adjacent rivers (Table I). With decreasing concentrations of suspended matter offshore, the dissolved phase may become the dominant phase in the water-particulate partition, despite higher partition coefficients than the rivers. For example, this will be the case for SPM concentrations < 1 mg dm^{-3} and $K_d < 10^6$ (Fig. 2).

In addition to the dominance of the dissolved phase over the SPM phase for the partition of PCBs in seawater at SPM concentrations as high as 10 mg dm^{-3}, the other surprising fact was the high contribution of early eluting PCB congeners in seawater (51, 58). This can be explained by their lower K_{ow} values (34). They may be derived from rivers and, especially in remote areas, from the atmosphere (17, 59).

3.2.3. Sediments

Sediments are part of the potential sinks for PCBs. Contents of PCBs in sediments may differ from those in SPM of the overlying water because of dilution with existing sediments with different PCB contents (with different organic carbon contents and/or grain size distribution), or because of (bio)degradation of particular components within the sediments. The significance of biological degradation of PCB components with lower degrees of chlorination was suggested to occur in sediments of the North Sea on the basis of packed column analyses (60). Equilibrium partitioning with the overlying water has been suggested as the controlling mechanism for individual PCB (including di- and trichloro-components) in sediments of the Wadden Sea and for average Arochlor 1242 and 1254 compositions in sediments of the tidal Hudson River (53, 54). Total PCB contents in fine coastal sediments are in the μg g^{-1} range, and in the ng g^{-1} or sub ng g^{-1} range in total sediments offshore due to dilution with coarse sediments. In sediments by far the largest fraction of PCBs is in the particulate form rather then in the interstitial water (Fig. 2).

Sediments may also act as source of PCBs for the overlying water. Transport of PCBs from sediments to the overlying water by molecular diffusion has been considered in field and laboratory studies (61, 62). Its role in dynamic sediments in estuaries and in shallow marine regions is probably overshadowed by a episodic wind- and current induced erosion of the upper sediment layer into the overlying water and by bioturbation effects (61, 63-68). As a result of such processes, (net-) sedimentation type and erosion type bottoms can be distinguished in different areas of the North Sea (66-68).

Penta- and hexachlorobenzene, show a similar environmental behaviour as PCBs. The increase of their desorption rates was the result of two counteracting processes: Sediment circulation by the worms strongly increased the transport of chemicals to the sediment surface, facilitating mobilization, while sorption to intact faecal pellets retarded contaminant release to the water phase. The net difference between these two processes and molecular diffusion only, was a 4-6 fold increase in the presence of organisms (63).

Gas production under anaerobic circumstances was also shown to increase the transfer of PCBs from sediments to water and air when compared to aerobic sediments without macroinvertebrates (64).

Dredging of contaminated sediments from a Swedish lake did not result in lower PCB concentrations in sediments or biota even ten years after the completion of the operation (69). This might be attributed to internal circulation of PCBs from areas not dredged and/or import from outside sources.

Changes in PCB patterns in vertical sediment profiles were observed between different depths of aerobic sediments from the Oslo-fjord, but not in anaerobic sediments (70). Especially the contribution of penta-CB's increased at the cost of hexa-CBs in the aerobic cores.

Aroclor 1254 desorbed from clay particles inhibited fotosynthesis and chlorophyl-a content in a natural phytplankton community more than a similar concentration added directly to water, indicating differences in bioavailability of congeners (71).

3.2.4. Open ocean waters

PCBs are present in open ocean waters mainly in dissolved form. Total PCB concentrations are in the (sub)ng dm^{-3} range. Some di- tri- and tetra-CBs (18, 15, 31, 28, 52, 49, 44, 41) are present in relatively large concentrations (IOC, 1984). Atmospheric deposition is a significant source in remote areas. Their presence has been detected in the water column at greater depths (72-75 and references in 76), probably transported downward from the surface layer with rapidly sinking particles, from which they can dissolve at greater depths (77, 78).

It has been suggested that the residence time of chlorinated hydrocarbons in the water columns will differ significantly among oceans that differ in primary productivity (78). The residence time of PCB in the eupotic zone was estimated to be less than 1 year. The deep ocean would thus be a significant sink of PCB components; the North Atlantic may represent the dominant sink (1).

4. BIOACCUMULATION

The possible pathways of PCB congeners in an organism (e.g. a fish) are illustrated in Fig. 3. Uptake can occur from solution via surfaces in

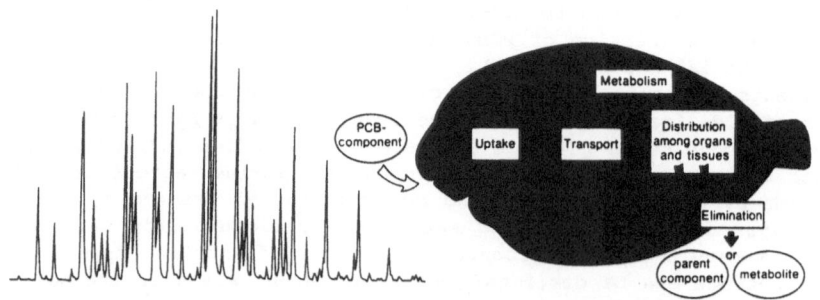

Fig. 3. Capillary gaschromatogram of a technical PCB mixture, and a scheme showing the processes determining the kinetics of PCB congeners in organisms.

contact with the ambient (interstitial) seawater, such as gills and skin (79-87). Marine teleost fishes drink large amounts of seawater and excrete the excess of electrolytes to prevent dehydration in their hyperosmotic environment (88). Therefore uptake of PCBs via the intestine is also a likely mechanism in these animals. Uptake from sediment (82,84-86) and uptake from food (85,89-94) is also reported.

Once present in the interior of an organism, PCBs are transported via the circulating internal fluids (blood or haemolymph). The distribution of PCB's among different organs and tissues in fish was shown to be closely linked to lipid contents (93-97), although relatively small differences in lipid based concentrations were found between organs; brain generally possessed the lowest concentrations (93,94,97). However the PCB patterns (i.e. the contribution of individual congeners to total ($\leq$)-PCB) were virtually identical in different organs and tissues of the same fishes.

In general, organic contaminants may undergo enzyme mediated reactions, resulting in the formation of metabolites which differ in physical and chemical properties (e.g. lipophilicity) from the parent compound (phase I reactions). These metabolites can be conjugated to compounds naturally present in organisms, such as glucuronic-acid (phase II reactions). Metabolism takes place mainly in liver or hepatopancreas, but also to a lesser degree in other organs. Phase I reactions can result in metabolites with lower elimination rates than the parent compounds, but result more often in a faster elimination; the latter is always the case with phase II reactions. Phase I reactions can either increase or decrease the toxicity of a compound, while phase II reactions generally decrease the toxicity.

The metabolic capacity of enzyme systems for PCB congeners appears to be low in gill-breathing aquatic animals compared to birds and mammals (98,99). However, phenolic metabolites and their conjugates of congeners 4, 31 and 99 were actually found in the polychaete Nereis virens (100). A conjugated 4-hydroxy metabolite of congener 52 was found in Salmo gairdneri (rainbow trout (101)). Phenolic metabolites of several diCB's were found in Poecilia reticulata (guppy). However, even for these congeners with a low degree of chlorine substitution, facilitating metabolism, the net bioconcentration was found to be mainly regulated by the excretion rates of the parent congeners (102). In bivalves, no metabolism of PCBs has been reported. Therefore, the kinetics of PCB congeners in bivalve species is likely to represent the behaviour of the parent congeners only. Lipid based concentrations did not increase with age in Macoma balthica, while seasonal variations were also shown to occur, showing the occurence of elimination in environmental samples (54,103).

Of the theoretically possible 209 congeners, many are hardly synthetized. Since each congener will show its own bio-accumulation and toxicity, pattern analyses of PCB (congeners) in organisms is of great interest. The patterns were highly similar over a range of $\leq$ -PCB concentrations of about an order of magnitude in the bivalve Macoma balthica, the polychaetes Arenicola marina (lugworm) and Nepthys spp. and the flatfish species Pleuronectes platessa (plaice) and Solea solea (sole); all originating from the Dutch Wadden Sea or the southern North Sea (54,97,104 and fig. 4). Different patterns were found in the shrimp Crangon crangon and in the harbour porpoise Phocoena phocoena from the same areas (54,105 and fig. 4) Patterns in liver oil of Brevoortia tyrannus (menhaden) and Gadus morhua (cod) from areas in the North Atlantic influenced by the Gulf Stream were also highly similar (106). Different patterns were found in cod from the upwelling area west of Portugal and in Cephylopholis sp. (Coral brass) from Bermuda. In case of shrimp, porpoise and coral brass, especially congeners that posses vicinal protons and lack a 4,4'-chlorine substitution showed relatively low

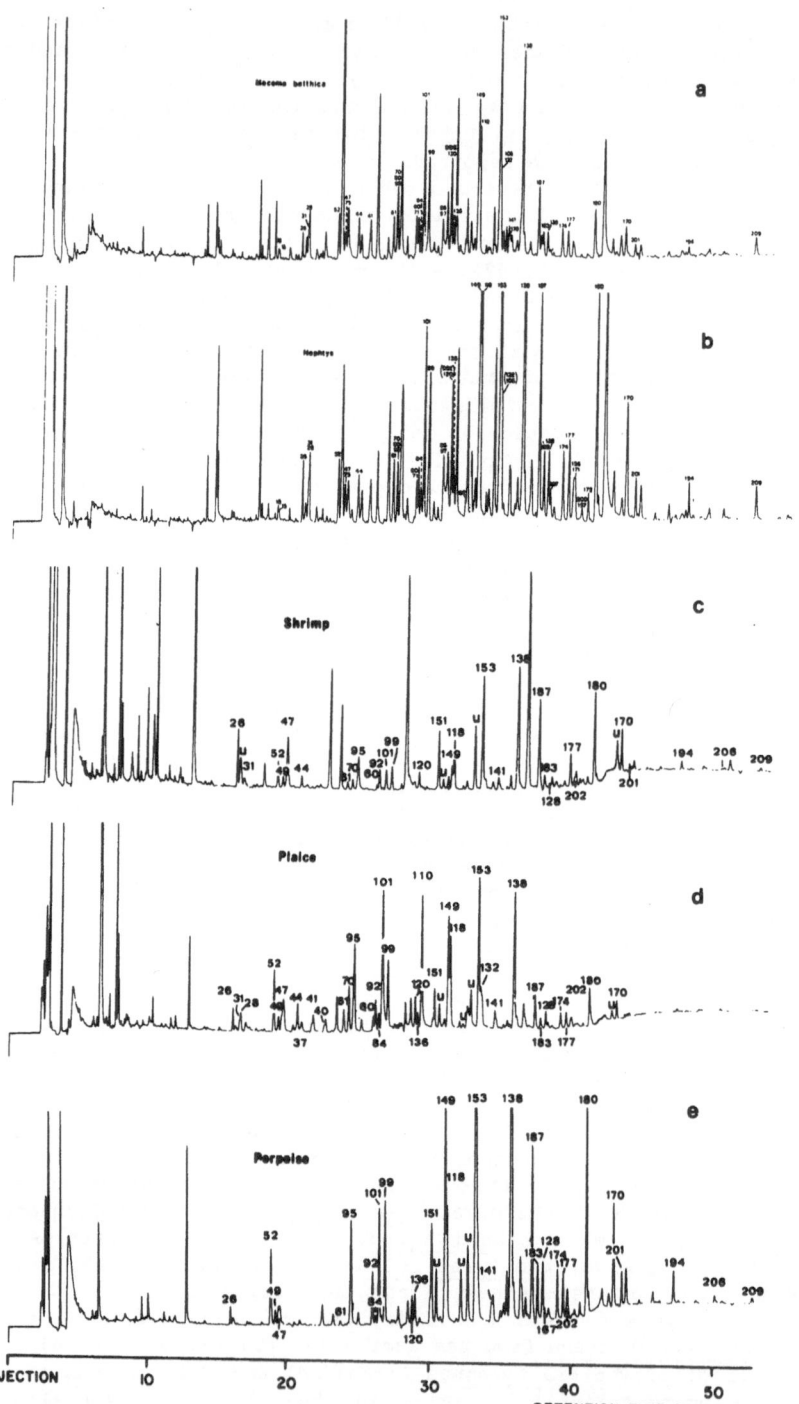

Fig. 4. Gaschromatograms of PCB congeners in a bivalve(a), a polychaete(b), a crustacean(c),a fish(d) and a marine mammal(e) from the southern North Sea or the Wadden Sea.u=unknown.Since chromatograms 'a'and'b' were obtained under different operational conditions, retention times differ from c-e.

contributions to $\lesssim$ -PCB compared to the other species. These molecular features are reported to facilitate metabolism (99,107,108). Extensive metabolism of PCB congeners up to hexa-CB's to hydroxy-compounds occurred in guillemot and to hydroxy- and methylsulfone metabolites in seal (109,110).

Differences in the degree of bioaccumulation between congeners were caused entirely by differences in elimination rates in the fish species Carassius auratus (goldfish), Solea solea and Salmo gairdneri because uptake rates did not differ (85,94,111).

The biological half-lifes for the same congeners were generally much higher in rainbow trout than in goldfish and sole. In all species, the elimination of single congeners could be described by first order exponential processes. Biological half-lifes of congeners were well correlated with their hydrophobicity/lipophilicity, as quantified by their octonol-water partitioning coëfficiënts (K_{OW}'s). The uptake rates of hexa-CB's and higher chlorinated congeners have been reported to be slower than those of mono- to penta-CB's (93,112-114). However, once absorbed, their biological half-lifes and bioconcentration factors also showed high correlations with their respective K_{OW}'s (97,114). A high correlation between K_{OW} values of individual congeners and bioconcentration factors was also found in the more theoretical study (115).

Perca fluviatilis (perch) accumulated twice as much of congener 153 at 15° C than at 5° C in a 48 hr uptake experiment, while elimination rates did not differ between these temperatures (116). This effect was also demonstrated in the field (117).

Even at apparently constant contamination levels of food and water and at a constant water temperature, internal physiological processes may also cause differences in PCB kinetics: Seasonal changes in total lipid contents and in the contribution of different lipid classes to total lipids in some organs influenced the plateau-levels of equilibrium concentrations of PCBs in sole (93,97). Spawning was also shown to alter the distribution of PCBs in fish considerably (96,118).

In the benthic invertebrates Arenicola marina, Macoma balthica and Crangon crangon from the same area (Dutch Wadden Sea), PCB concentrations expressed on the basis of extractable lipids were highly similar despite large differences between these species in morphology and feeding-type. For instance, the contribution of uptake of organochlorines from sediment is verly likely to be higher in Arenicola than in Crangon, but this did not show up in the lipid based concentrations (54). Concentrations on a lipid basis in flat-fishes from this area were also similar and their PCB patterns did not show a selective increase in the contribution of highly chlorinated PCBs, indicating the absence of a food-chain magnification even for highly chlorinated congeners. It is therefore concluded, that equilibrium partitioning between body-lipids, blood and the ambient water is likely to be the dominant proces regulating the concentrations of PCBs in gill-breathing animals. The actual source of uptake (water, seston, sediment or food) might thus be of less importance. The absence of some congeners in a PCB pattern, may indicate their metabolism.

5. INDUCTION OF MIXED FUNCTION OXYGENASES

Although the literature on mixed function oxygenase (MFO) enzyme systems in aquatic animals was recently reviewed (119), this paper would be incomplete without a discussion of the main aspects of what is probably the most important effect of PCBs.

MFO systems are located in the microsomal fraction (endoplasmatic reticulum) of mainly the liver in vertebrates and in the hepatopancreas, the digestive gland and other tissues in marine invertebrates (120). The enzyme systems catalyse oxidation reactions of the general type given in equation 5 (phase I metabolism):

AH (substrate) + NADPH + $O_2 \longrightarrow$ A-OH (oxidized metabolite) + NADP$^+$ + H$_2$O. (eq. 5).

The oxidation of the substrate is caused by an electron transport chain and requires phospholipid, NADPH cytochromes P-450 and reductase for activity. Endogenous steroids as well as foreign compounds can serve as substrates.

Low levels of MFO enzyme systems are constitutively present, but they can be increased by the presence of certain xenobiotics. PCB mixtures are strong inducers of MFO systems (119-122); they may also act as a substrate to be hydroxylated, but more often they only induce the systems resulting in an increased and/or altered metabolism of other compounds (e.g. 123). An increased activity of MFO systems by PCBs and/or PAHs was also demonstrated under field conditions in marine fishes and invertebrates (124,125).

Compounds inducing MFO systems are usually divided into a group indu-cing the same enzymes as the model compounds 3-methylcholonthrene and ß-naphtoflavone (3-MC type), a group acting similar to the model compound phenobarbital (PB-type) and a group causing an MFO induction pattern simi-lar to a mixture of 3-MC + PB. However, MFO systems induced by phenobarbi-tal have not been reported in gill-breathing animals (119,121). These dif-ferent patterns of MFO induction can be correlated with multiple forms of cytochrome P-450. The formes induced by 3-MC are usually referred to as cy-tochromes P-448.

Among PCB congeners, 3-MC type, PB-type and mixed type inducers were shown in rats (126). Structural requirements for a pure 3-MC type inducer were a 4, 4'-chlorine substitution and a Cl$^-$ substitution of at least two meta-positions (3(1) and 5(1)-postions), but no ortho Cl$^-$ substitution at all. These congeners (77, 81, 126 and 169) are hardly found in technical PCB mixtures (8,11-14). However, the attachment of one or two ortho chlo-rines to these pure 3-MC type inducers changed several of them into mixed type and others into PB-type inducers. Several of the mixed type inducers were found in environmental samples in significant concentrations (105, 118, 156, 128, 138 and 170; see fig. 4). All congeners with other struc-tural characteristics than mentioned above were PB-type inducers. 2, 2', 4, 4'-tetraCB did not cause an increase in any parameter of MFO-activity in rainbow-trout (122). Since this is a PB-inducer in rats, this is consistent with the absence of a PB-inducable MFO system in fish.

The 3-MC type induction caused by the mixed type inducers in rats, can thus influence the metabolism of steroid hormones and xenobiotics in gill-breathing aquatic animals.

REFERENCES

(1) N.A.S. (1979). Polychlorinated Biphenyls. Report prepared by the committee on the assessment of polychlorinated biphenyls in the environment. National Academy of Sciences, Washington D.C.: 182 pp.
(2) GORTER, J. and SCHAAKE, P (1980). PCBs in the Netherlands. Rap. Cen-traal Bureau voor de Statistiek 803-021-00. Staatsuitgeverij, the Hague: 108 pp. (in Dutch).
(3) MCKINNEY, J.D., GOTTSCHALK, K.E. and PEDERSEN, L. (1983). The polari-zability of planar aromatic systems. An application to polychlorinated biphenyls (PCB's). J. Mol. Struct. 105: 427-438.
(4) JENSEN, S. (1966). Report of a new chemical hazard. New Scient. 32: 612.
(5) UMEDA, G. (1972). PCB poisoning in Japan. AMBIO 1: 132-134.

(6) CHEN, P.H., CHANG, K.T. and LU, Y.D. (1981). Polychlorinated biphenyls and polychlorinated dibenzofurans in the toxic rice-bran oil that caused PCB poisoning in Taichung. Bull. Environ. Contam. Toxicol. 26: 489-495.

(7) DUINKER, J.C., HILLEBRAND, M.T.J., PALMORK, K.H. and WILHELMSEN, S. (1980). An evaluation of existing methods for quantitation of poly-chlorinated biphenyls in environmental samples and suggestions for an improved method based on measurement of individual components. Bull. Environ. Contam. Toxicol. 25: 956-964.

(8) SISSONS, D. and WELTI, D.J. (1971). Structural identification of PCBs in commercial mixtures by GLC, NMR and MS. J. Chromat. 61: 15-32.

(9) JENSEN, S. and SUNDSTRÖM, G. (1974). Structures and levels of most chlorobiphenyls in two technical PCB products and in human adipose tissue. AMBIO 3: 70-76.

(10) SCHULTE, E. and ACKER, L. (1974). Identifizierung und Metabolisier barkeit von Polychlorierten Biphenylen. Naturwiss. 61: 79-80.

(11) ALBRO, P.W., CORBETT, J.T. and SCHROEDER, J.L. (1981). Quantitative characterization of polychlorinated biphenyl mixtures (Aroclors (B) 1248, 1254 and 1260) by gas chromatography using capillary columns. J. Chromat. 205: 103-111.

(12) DUINKER, J.C. and HILLEBRAND, M.T.J. (1983). Characterization of PCB components in Clophen formulations by capillary GC-MS and GC-ECD tech-niques. Environ. Sci. Technol. 17: 449-456.

(13) ZELL, M., NEU, H.J. and BALLSCHMITER, K. (1977). Identifizierung der PCB-Komponenten durch Retentions-indexvergleich nach Kapillar-Gaschro-matographie. Chemosphere 6: 69-76.

(14) BALLSCHMITER, K. and ZELL, M. (1980). Analysis of polychlorinated bi-phenyls (PCB) by glass capillary gas chromatography. Composition of technical Aroclor and Clophen-PCB mixtures. Z. Anal. Chem. Vol. 302: 20-31.

(15) MULLIN, M.D., POCHINI, C.M., MCCRINDLE, S., ROMKES, M., SAFE, S.H. and SAFE, L.M. (1984). High-Resolution PCB Analysis: Synthesis and chroma-tographic properties of all 209 PCB congeners. Environ. Sci. Technol. 18: 468-476.

(16) EISENREICH, S.J., LOONEY, B.B. and THORNTON, J.D. (1981). Airborne or-ganic contaminants in the Great Lakes ecosystem. Environ. Sci. Tech-nol. 15: 30-38.

(17) BIDLEMAN, T.F. and OLNEY, C.E. (1974). High-volume collection of at-mospheric polychlorinated biphenyls. Bull. Environm. Contam. Toxicol. 11: 442-450.

(18) HARVEY, G.R. and STEINHAUER, W.G. (1974). Atmospheric transport of polychlorobiphenyls to the North Atlantic. Atmos. Environ. 9: 777-782.

(19) HARVEY, G.R. and STEINHAUER, W.G. (1976). Transport pathways of polychlorinated biphenyls in Atlantic water. J. Mar. Res. 34: 561-575

(20) MC CLURE, V.E. (1976). Transport of heavy chlorinated hydrocarbons in the atmosphere. Environ. Sci. Technol. 10: 1223-1229.

(21) TANABE, S., TATSUKAWA, R., KAWANO, M. and HIDAKA, H. (1982). Global distribution and atmospheric transport of chlorinated hydrocarbons HCH (BHC) isomers and DDT compounds in the Western Pacific, Eastern Indian and Antarctic Oceans. J Oceanogr. Soc. Japan 38: 137-148.

(22) VAN VAECK, L., VAN CAUWENBERGHE, K. and JANSSENS, J. (1984). The gas-particle distribution of organic aerosol constituents: Measurement of the volatization artifact in high-vol cascade impactor sampling. Atmos. Environ. 18: 417-430.

(23) DOSKEY, P.V. and ANDREN, A.W. (1981). Modeling the flux of atmospheric polychlorinated biphenyls across the air/water interface. Environ.

Sci. Technol. 15: 705-711.

(24) JUNGE, C.E. (1977). In: Fate of Pollutants in the Air and Water Environment. I.H. Suffet (Ed) Wiley-Interscience New York: 7-25.

(25) CAUTREELS, W. and VAN CAUWENBERGHE, K. (1978). Experiments on the distribution of organic pollutants between airborne particulate matter and the corresponding gas phase. Atmos. environ. 12: 1133-1141.

(26) VAN VAECK, L. and VAN CAUWENBERGHE, K. (1978). Cascade impactor measurements of the size distribution of the major classes of organic pollutants in the atmospheric particulate matter. Atmos. Environ. 12: 2229-2239.

(27) HOPPEL, W.A., FITZGERALD, J.W. and LARSON, R.E. (1985). Aerosol size distributions in air masses advecting off the coast of the United States. J. Geophys. Res. 90: 2365-2379.

(28) HOFFMAN, E.J. and DUCE, R.A. (1977). Organic carbon in marine atmospheric particulate matter: Concentration and particle size distribution. Geophys. Res. Lett. 4: 449-452.

(29) SLINN, W.G.N., HASSE, L., HICKS, B.B., HOGAN, A.W., LAL, D., LISS, P.S., MUNNICH, K.O., SEHMEL, G.A. and VITTORI, O. (1978). Some aspects of the transfer of atmospheric trace constituents past the air-sea interface. Atmos. Environ. 12: 2055-2087.

(30) NISBET, I.C.T. and SAROFIM, A.F. (1972). Rates and routes of transport of PCBs in the environment. Environ. Health perspect. 1: 21-

(31) BAXTER, R.M. and SUTHERLAND, D.A. (1984). Biochemical and photochemical processes in the degradation of chlorinated biphenyls. Environ. Sci. Technol. 18: 608-610.

(32) BUNCE, N.V., KUMAR, Y. and BROWNLEE, B.G. (1978). An assessment of the impact of solar degendation of polychlorinated biphenyls in the aquatic environment. Chemosphere 7: 155-164.

(33) HUTZINGER, O., SAFE, S. and ZITKO, V. (1971). The chemistry of PCB's. CRC Press, Cleveland, Ohio: 269 pp.

(34) RAPAPORT, R.A., EISENREICH, S.J. (1984). Chromatographic determination of octanol-water partition coefficients (K_{ow}'s) for 58 polychlorinated biphenyl congeners. Environ. Sci. Technol. 18: 163-170 + additions and corrections vol. 19: 376 (1985).

(35) WOODBURN, K.B., DOUCETTE, W.J. and ANDREN, A.W. (1984). Generator column determination of octanol/water partition coefficients for selected polychlorinated biphenyl congeners. Environ. Sci. Technol. 18: 457-459.

(36) PAVLOU, S.P. and DEXTER, R.N. (1979). Distribution of polychlorinated biphenyls (PCB) in estuarine ecosystems. Testing the concept of equilibrium partitioning in the marine environment. Amer. Chem. Soc. 13: 65-71.

(37) STEEN, W.C., PARIS, D.F. and BAUGHMAN, G.L. (1978). Partitioning of selected polychlorinated biphenyls to natural sediments. Water Res. 12: 655-657.

(38) CHIOU, C.T., PORTER, P.E. and SCHMEDDING, D.W. (1983). Partition equilibria of nonionic organic compounds between soil organic matter and water. Environ. Sci. Technol. 17: 227-231.

(39) GSCHWEND, P.M. and WU, S-C. (1985). On the constancy of sediment-water partition coefficients of hydrophobic organic pollutants. Environ. Sci. Technol. 19: 90-96.

(40) MEANS, J.C., WOOD, S.G., HASSETT, J.J. and BANWART, W.L. (1980). Sorption of polynuclear aromatic hydrocarbons by sediments and soils. Environ. Sci. Technol. 14: 1524-1528.

(41) WILDISH, D.J., METCALFE, C.D., AGAKI, H.M. and MCLEESE, D.W. (1980). Flux of Aroclor 1254 between estuarine sediment and water. Bull. Envi-

ron. Contam. Toxicol. 24: 20-26.

(42) DI TORO, D.M. and HORZEMPA, L.M. (1982). Reversible and resistant components of PCB absorption-desorption isotherms. Environ. Sci. Technol. 16: 594-602.

(43) O'CONNOR, D.J. and CONNOLY, J.P. (1980). The effect of concentration of adsorbing solids on the partitation coefficient. Water Res. 14: 1517-1523.

(44) VOICE, T.C., RICE, C.P. and WEBER Jr., W.J. (1983). Effect of solids concentration on the sorptive partitioning of hydrophobic pollutants in aquatic systems. Environ. Sci. Technol. 17: 513-518.

(45) DUINKER, J.C. and HILLEBRAND, M.T.J. (1983). Analyses van geselecteerde chloorkoolwaterstoffen in zeewateroplossing en suspensie. Rep. to RIZA, Neth. Inst. for Sea Res., mimeo: 18 pp.

(46) DUINKER, J.C., HILLEBRAND, M.T.J., NOLTING, R.F. and WELLERSHAUS, S. (1982). The river Weser: processes affecting the behaviour of metal-sand organochlorines during estuarine mixing. Neth. J. Sea Res. 15: 170-195.

(47) DUINKER, J.C., HILLEBRAND, M.T.J., NOLTING, R.F. and WELLERSHAUS, S. (1982). The river Elbe: processes affecting the behaviour of metals and organochlorines during estuarine mixing. Neth. J. Sea Res. 15: 141-169.

(48) DUINKER, J.C., HILLEBRAND, M.T.J. NOLTING, R.F., and WELLERSHAUS, S. (1985). The river Ems: processes affecting the behaviour of metals and organochlorines during estuarine mixing. Neth. J. Sea Res. 19(1) (in press).

(49) WOLLAST, R. and DUINKER, J.C. (1982). General methodology and sampling strategy for studies on the behaviour of chemicals in estuaries. Thalass. Yugosl. 18: 471-491.

(50) POSTMA, H. (1980). Sediment transport and sedimentation. In: Chemistry and biochemistry of estuaries. Olausson, E. and Cato, I. (eds) Wiley: 153-186.

(51) DUINKER, J.C. and HILLEBRAND, M.T.J. (1979). Behaviour of PCB, penta-chlorobenzene, hexachlorobenzene, $\propto$ -HCH, γ -HCH, β -HCH, dieldrin, endrin and p,p'-DDD in the Rhine-Meuse estuary and the adjacent coastal area. Neth. J. Sea Res. 13: 255-281.

(52) HERRMANN, R. and THOMAS, W. (1984). Behaviour of some PAH, PCB and organochlorine pesticides in an estuary, a comparison. Exe, Devon. Z. Anal. Chem. 319: 152-159.

(53) BOPP, R.F., SIMPSON, H.J., OLSEN, C.R. and KOSTYK, N. (1981). Polychlorinated biphenyls in sediments of the tidal Hudson river, New York. Environ. Sci. Technol. 15: 210-216.

(54) DUINKER, J.C., HILLEBRAND, M.T.J. and BOON, J.P. (1983). Organochlorines in benthic invertebrates and sediments from the Dutch Wadden Sea; identification of individual PCB components. Neth. J. Sea Res. 17, 19-38.

(55) EISENREICH, S.J., HOLLOD, G.J. and JOHNSON, T.C. (1979). Accumulation of polychlorinated biphenyls (PCBs) in surfacial Lake Superior sediments. Atmospheric deposition. Environ. Sci. Technol. 13: 569-573.

(56) MALISCH, R., SCHULTE, E. and ACKER, L. (1981). Chlororganische Pestizide, polychlorierte Biphenyle und Phtalate in Sedimenten aus Rhein und Neckar. Chemiker-Zeitung, 105: 187-194.

(57) DUINKER, J.C. (1980). Particulate matter in estuaries: Adsorption and desorption processes. In: Chemistry and biochemistry of estuaries, Olausson, E. and Cato, I (eds) Wiley: 121-151.

(58) IOC (1984). The determination of polychlorinated biphenyls in open ocean waters. Unesco Techn. Rept. Series no. 26: 48 pp.

(59) ATLAS, E. and GIAM, C.S. (1981). Global transport of organic pollutants: Ambient concentrations in the remote marine atmosphere. Science 211: 163-165.

(60) EDER, G. (1976). Polychlorinated biphenyls and compounds of the DDT-group in sediments of the central North Sea and the Norwegian De-pression. Chemosphere 5: 101-106.

(61) LARSSON, P. (1983). Transport of ^{14}C-labelled PCB compounds from sediment to water and from water to air in laboratory model systems. Water Res. 17: 1317-1326.

(62) FISHER, J.B., PETTY, R.L. and LICK, W. (1983). Release of polychlorinated biphenyls from contaminated lake sediments: Flux and apparent diffusivities of four individual PCB's. Environ. Pollut. (Ser.B) 5: 121-132.

(63) KARICKHOFF, S.W. and MORRIS, K.R. (1985). Impact of tubificid oligochaetes on pollutant transport in bottom sediments. Environ. Sci. Technol. 19: 51-56.

(64) SODERGREN, A. and LARSSON, P. (1982). Transport of PCB's in aquatic laboratory model ecosystems from sediment to the atmosphere via the surface microlayer. AMBIO 11: 41-45.

(65) LARSSON, P. (1985). PCB's (Clophen A 50) change composition when transported from sediment to air in aquatic model systems. Environ. Pollut. (Ser. B) 9: 81-94.

(66) CREUTZBERG, F. and POSTMA, H. (1979). An experimental approach to the distribution of mud in the southern North Sea. Neth. J. Sea Res. 13: 99-116.

(67) MC CAVE, I.N. (1981). Location of coastal accumulations of fine sediments around the southern North Sea. Rapp. P.-v. Réun. Cons. int. Explor. Mèr. 181: 15-27.

(68) EISMA, D. (1981). The mass-balance of suspended matter and associated pollutants in the North Sea. Rapp. P.-v. Réun. Cons. int. Explor. Mèr 181: 7-14.

(69) SODERGREN, A. (1984). The effect of sediment dredging on the distribution of organochlorine residues in a lake ecosystem. AMBIO 13: 206-210.

(70) ABDULLAH, M.I., RINGSTAD, O. and KVESETH, N.J. (1982). Polychlorinated biphenyls in the sediments of the inner Oslo fjord. Wat. Air Soil Pollut. 18: 485-497.

(71) NAU-RITTER, G.M., WURSTER, C.F. and ROWLAND, R.G. (1982). Polychlorinated biphenyls (PCB) desorbed from clay particles inhibit photosynthesis by natural phytoplankton communities. Environ. Pollut. (Ser. A) 28: 177-182.

(72) HARVEY, G.R.. STEINHAUER, W.G. and TEAL, J.M. (1973). Polychlorobiphenyls in North Atlantic oceanwater. Science 180: 643.

(73) SCURA, E.D. and MCCLURE, V.E. (1975). Chlorinated hydrocarbons in seawater: Analytical method and levels in the northeastern Pacific (p,p' DDE; p,p'-DDT. Mar. Chem. 3: 337-346.

(74) HARVEY, G.R. and STEINHAUER, W.G. (1976). Biochemistry of PCB and DDT in the North Atlantic. In: Nriagu, J.O. (ed). Environ. Biochem. 1, Anne Arbor Science Publishers, Michigan: 203-221.

(75) JONAS, R.B. and PFAENDER, F.K. (1976). Chlorinated hydrocarbon pesticides in western North Atlantic Ocean. Environ. Sci. Technol. 10: 770-773.

(76) BRÜGMANN, L. and LUCKAS, B. (1979). Polychlorierte Biphenyle und DDT-Metaboliten im Oberflächenwasser des Atlantik. Acta Hydrochim. Hydrobiol. 7: 435-447.

(77) TANABE, S., KAWANO, M. and TATSUKAWA, R. (1982). Chlorinated hydrocar-

bons in the Antarctic, western Pacific and eastern Indian Oceans. Trans. Tokyo Univ. Fisher 5: 97-109.

(78) TANABE, S. and TATSUKAWA, R. (1983). Vertical transport and residence time of chlorinated hydrocarbons in the open ocean water column. J. Oceanogr. Soc. Japan 39: 53-62.

(79) LOWE, J.I., PARRISH, P.R., PATRICK, J.M. and FORESTER, J. (1972). Effects of the polychlorinated biphenyls Aroclor 1254 on the American oyster Crassostrea virginica. Mar. Biol. 17: 209-214.

(80) GUINEY, P.D., PETERSON, R.E., MELANCON, M.J. and LECH, J.J. (1977). The distribution and elimination of 2,5,2',5'-(^{14}C) tetrachlorobiphenyl in rainbow trout (Salmo gairdneri). Toxicol. Appl. Pharmacol. 39: 329-338.

(81) HANSEN, D.J., PARRISH, P.R. and FORESTER, J. (1974). Aroclor 1016: Toxicity to and uptake by estuarine animals. Environ. Res. 7: 363-373.

(82) COURTNEY, W.A.M. and LANGSTON, W.J. (1978). Uptake of polychlorinated biphenyl (A 1254) from sediment and from seawater in two intertidal polychaetes. Environ. Pollut. 15: 303-309.

(83) FOWLER, S.W., POLIKARPOV, G.G., ELDER, D.L., PARSI, P. and VILLENEUVE, J.P. (1978). Polychlorinated biphenyls: Accumulation from contaminated sediments and water by the polychaete Nereis virens. Mar. Biol. 48: 303-309.

(84) ELDER, D.L., FOWLER, S.W., POLIKARPOV, G.G. (1979). Remobilization of sediment associated PCB by the worm Nereis diversicolor. Bull. Env. Contam. Toxicol. 21: 448-452.

(85) BRUGGEMAN, W.A., MARTRON, L.B.J.M., KOOIMAN, D. and HUTZINGER, O. (1981). Accumulation and elimination of di-, tri-, and tetrachlorobiphenyls by goldfish after dietary and aquous exposure. Chemosphere 10: 811-832.

(86) MARINUCCI, A.C. and BARTHA, R. (1982). Accumulation of the polychlorinated biphenyl Aroclor 1242 from contaminated detritus and water by the saltmarsh detrivore, Uca Pugnax. Bull. Environ, contam. Toxicol. 29: 326-333.

(87) LARSSON, P. (1984). Uptake of sediment released PCB's by the eel Anguilla anguilla in static model systems. Ecol. Bull. 36: 62-67.

(88) HOAR, W.S. (1975). General and comparative physiology, 2nd ed., Prentice-Hall inc.; New Yersey: 848 pp.

(89) GOERKE, H. and ERNST, W. (1977). Fate of ^{14}C-labelled di-, tri- and pentachlorebiphenyl in the marine annelid Nereis virens. I. Accumulation and elimination after oral administration. Chemosphere 9: 55-558.

(90) MCLEESE, D.W., METCALFE, C.D. and PEZZACK, D.S. (1980). Bioaccumulation of chlorobiphenyls and endrin from food by lobsters (Homerus americanus). Bull. Environ. Contam. Toxicol. 25: 161-168.

(91) GUINEY, P.D. and PETERSON, R.E. (1980). Distribution and elimination of a polychlorinated biphenyl after acute dietary exposure in yellow perch and rainbow trout. Arch. Environ. Contam. Toxicol. 9: 667-674.

(92) PIZZA, J.S. and O'CONNOR, J.M. (1983). PCB dynamics in Hudson river striped bass. II. Accumulation from dietary sources. Aquat. Toxicol. 3: 313-327.

(93) BOON, J.P., OUDEJANS, R.C.H.M. and DUINKER, J.C. (1984). Kinetics of individual polychlorinated biphenyl (PCB) components in juvenile sole (Solea solea) in relation to concentrations in food and to lipid metabolism. Comp. Biochem. Physiol. 79C: 131-142.

(94) BOON, J.P. (1985). Uptake, distribution and elimination of selected polychlorinated biphenyl components of Clophen A40 in juvenile sole (Solea solea) and effects on growth. In: Eighteenth European Marine Biology Symposium, "Marine Biology of Polar Regions" and "Effects of

Stress on Marine Organisms", eds. J.S. Gray and M.E. Christiansen. John Wiley & Sons Ltd, Chichester: 493-512.

(95) YOSHIDA, T., TAKASHIMA, F. and WATANABE, T. (1973). Distribution of ^{14}C PCB's in carp. AMBIO 2: 111-113.

(96) SOLBAKKEN, J.E., PALMORK, K.H. and INGEBRIGTSEN, K. (1984). An autoradiographic study of the chlorinated biphenyl 2,4,5,2',4',5' hexachlorobiphenyl and the polycyclic aromatic hydrocarbon phenanthrene in flounder (Platichthys flesus) accompanied by liquid scintillation counting. Mar. Environ. Res. 14: 446-447.

(97) BOON, J.P. and DUINKER, J.C. (1985). Kinetics of individual polychlorinated biphenyl (PCB) components in juvenile sole (Solea solea) in relation to concentrations in water and to lipid metabolism under conditions of starvation. Aquat. Toxicol. 7 (in press).

(98) HUTZINGER, O., NASH, D.M., SAFE, S., DEFREITAS, A.S.W., NORSTROM, R.J., WILDISH, D.J. and ZITKO, V. (1972). Polychlorinated biphenyls: Behaviour of pure isomers in pigeons, rats and brook trout. Science 178: 312-314.

(99) SUNDSTROM, G., HUTZINGER, O. and SAFE, S. (1976). The metabolism of chlorobiphenyls - A review. Chemosphere 5: 267-298.

(100) ERNST, W., GOERKE, H. and WEBER, K. (1977). Fate of ^{14}C-labelled di-, tri- and pentachlorobiphenyl in the marine annelid Nereis virens. II: Degradation and faecal elimination. Chemosphere 9: 559-568.

(101) MELANCON, M.J. and LECH, J.J. (1976). Isolation and identification of a polar metabolite of tetrachlorobiphenyl from bile of rainbow trout exposed to ^{14}C-tetrachlorobiphenyl. Bull. Environ. Contam. Toxicol. 15: 181-188.

(102) BRUGGEMAN, W.A., DIJKHUIZEN, W.A. and HUTZINGER, O. (1983). Bioaccumulation and transformation of dichlorobiphenyls in fish. In: Bruggeman, W.A. (1983). Bioaccumulation of polychlorobiphenyls and related hydrophobic chemicals in fish. Thesis, Univ. Amsterdam: 151-158.

(103) BOON, J.P. and DUINKER, J.C. (in press). Monitoring of cyclic organochlorines in the marine environment. Environ. Monit. Assess.

(104) BOON, J.P., VAN ZANTVOORT, M.W., GOVAERT, M.J.M.A. and DUINKER, J.C. (1985). Organochlorines in benthic polychaetes (Nephtys spp.) and sediments from the southern North Sea. Identification of individual PCB components. Neth. J. Sea Res 19(2) (in press).

(105) DUINKER, J.C. and HILLEBRAND, M.T.J. (1983). Composition of PCB mixtures in biotic and abiotic marine compartments (Dutch Wadden Sea). Bull. Environ. Contam. Toxicol. 31: 25-32.

(106) BALLSCHMITER, K., BUCHERT, H., BIHLER, S. and ZELL, M. (1981). Baseline studies of the global pollution. IV: The pattern of pollution by organochlorine compounds in the North Atlantic as accumulated by fish. Z. Anal. Chem. 306: 323-339.

(107) BALLSCHMITER, K., ZELL, M. and NEU, H.J. (1978). Persistence of PCB's in the ecosphere: Will some PCB isomers never degrade? Chemosphere 2: 173-176.

(108) ZELL, M. and BALLSCHMITER, K. (1980). Baseline studies of the global pollution. III: Trace analysis of polychlorinated biphenyls (PCB) by ECD Glass Capillary Gaschromatography in environmental samples of different trophic levels. Z. Anal. Chem. 304: 337-349.

(109) JANSSON, B., JENSEN, S., OLSSON, L., RENBERG, L., SUNDSTROM, G. and VAZ, R. (1975). Identification by GC/MS of phenolic metabolites of PCB and p,p'-DDE isolated from baltic guillemot and seal. AMBIO 4: 93-97.

(110) JENSEN, S. and JANSSON, B. (1976). Anthropogenic substances in seal from the Baltic: Methyl sulfone metabolites of PCB and DDE. AMBIO 5:

257-260.

(111) NIIMI, A.J. and OLIVER, B.G. (1983). Biological half-lives of polychlorinated biphenyl (PCB) congeners in whole fish and muscle of rainbow trout (Salmo Gairdneri). Can. J. Fish. Aquat. Sci. 40: 1388-1394.

(112) LANGSTON, W.J. (1978). Accumulation of polychlorinated biphenyls in the cockle Cerastoderma edule and the tellin Macoma balthica. Mar. Biol. 45: 265-272.

(113) SHAW, G.R., CONNELL, D.W. (1984). Physicochemical properties controlling polychlorinated biphenyl (PCB) concentrations in aquatic organisms. Environ. Sci. Technol. 18: 18-23.

(114) BRUGGEMAN, W.A., OPPERHUIZEN, A., WIJBENGA, A. and HUTZINGER, O. (1984). Bioaccumulation of super-lipophilic chemicals in fish. Toxicol. Environ. Chem. 7: 173-190.

(115) CHIOU, C.T. (1985). Partition coefficients of organic compounds in lipid-water systems and correlations with fish bioconcentration factors. Environ. Sci. Technol. 19: 57-62.

(116) EDGREN, M., OLSSON, M. and RENBERG, L. (1979). Preliminary results on uptake and elimination at different temperatures of p,p'-DDT and two chlorobiphenyls in perch from brackish waters. AMBIO 8: 270-272.

(117) EDGREN, M., OLSSON, M. and REUTERGARDH, L. (1981). A one year study of the seasonal variations of DDT and PCB levels in fish from heated and unheated areas near a nuclear power plant. Chemosphere 10: 447-452.

(118) GUINEY, P.D., MELANCON, M.J., LECH, J.J. and PETERSON, R.E. (1979). Effects of egg and sperm maturation and spawning on the distribution and elimination of polychlorinated biphenyl in rainbow trout (Salmo gairdneri). Toxicol. Appl. Pharmacol. 47: 261-272.

(119) PAYNE, J.F. (1984). Mixed-function oxygenases in biological monitoring programs: Review of potential usage in different phyla of aquatic animals. In: Ecotoxicological testing of the marine environment. Eds. G. Persoone, E. Jaspers and C. Claus. State Univ. Ghent and Inst. Mar. Scient. Res., Bredene, Belgium; Vol. 1: 625-655.

(120) LEE, R.F. (1981). Mixed function oxygenases (MFO) in marine invertebrates. Mar. Biol. Lett. 2: 87-105.

(121) ELCOMBE, C.R. and LECH, J.J. (1979). Induction and characterization of hemoprotein(s) P-450 and monooxygenation in rainbow trout (Salmo gairdneri). Toxicol. Appl. Pharmacol. 49: 437-450.

(122) MELANCON, M.J., ELCOMBE, C.R., VODICNIK, M.J. and LECH, J.J. (1981). Induction of cytochromes P-450 and mixed-function oxydase activity by polychlorinated biphenyls and ß-Napthoflavone in carp (Cyprinus carpio). Comp. Biochem. Physiol. 69C: 219-226.

(123) FÖRLIN, L. and ANDERSSON, T. (1981). Effects of Clophen A50 on the metabolism of paranitro anisole in an in vitro perfused rainbow trout liver. Comp. Biochem. Physiol. 68C: 239-242.

(124) SPIES, R.B., FELTON, J.S. and DILLARD, L. (1982). Hepatic mixed-function oxidases in California flatfishes are increased in contaminated environments and by oil and PCB ingestion. Mar. Biol. 70: 117-127.

(125) FRIES, C.R. and LEE, R.F. (1984). Pollutant effects on the mixed function oxygenase (MFO) and reproductive systems of the marine polychaete Nereis virens. Mar. Biol. 79: 187-193.

(126) SAFE, S.A., PARKINSON A, ROBERTSON, L., SAFE, L. and BANDIERA, S. (1982). Polychlorinated biphenyls: Effects of structure on biologic and toxic activity. In: Workshop on the combined effects of xenobiotics, Ottawa, Canada; June 22-23. Publ. NRCC/CNRC no. 18978 of the environmental secretariat, Natl. Res. Council Canada: 151-176.

DISPERSION ET EVOLUTION DES HYDROCARBURES PETROLIERS DANS L'ENVIRONNEMENT MARIN CAS DE REFERENCE : LA POLLUTION DE L'AMOCO CADIZ

M. H. MARCHAND
IFREMER - Centre de Brest
BP 337 29273 - BREST Cédex (France)

Summary

The shipwreck of the AMOCO CADIZ supertanker on the rocks of the Brittany Coast of France (March, 1978) was the largest and best studied tanker spill in history. The crude oils discharged in the marine environment (223,000 metric tons) were light petroleums. The distribution and the evolution of the oil pollution in the marine environment was examined. This oil spill has affected a very large section of the western English Channel. The diffusion of the hydrocarbons into the water column was observed. The evolution of the sea water pollution was followed, and the half-life of hydrocarbons in subsurface water was found to be between 11 and 28 days in different areas. One month after the disaster, the marine sediments were contaminated in the areas reached by the drifting slicks. The highest accumulation of petroleum in the sediments was located in the sheltered coastal environments. The natural decontamination process was found to be related to the nature of the sediment and the energy level of the geographic zone. After 3 years, the most obvious effects of the spill have passed, although hydrocarbon concentrations remain elevated in those estuaries and marshes that were initially most heavily oiled. A quantitative estimate of AMOCO CADIZ oil dispersal components (atmosphere, sea water, onshore, sediments) for the first month of the spill is presented.

1. INTRODUCTION

Les pollutions accidentelles par hydrocarbures constituent des événements qui peuvent bouleverser les équilibres écologiques et par extension les activités économiques d'une région sinistrée. De tels événements, ayant lieu le plus souvent à proximité des côtes, nécessitent de la part des pouvoirs publics une intervention rapide, non seulement pour mettre en place une stratégie de lutte anti-pétrole, mais également pour évaluer les effets immédiats et à long terme sur l'environnement marin. L'année 1967 restera longtemps gravée dans les mémoires comme l'année de la "marée noire" du TORREY CANYON. Avant cet accident, on n'avait pas encore imaginé que ce type de pollution pouvait atteindre l'ampleur d'une catastrophe. Cet accident attira l'attention de l'opinion publique et des Administrations sur les dangers présentés par les accidents des navires pétroliers et des plates-formes d'exploitation pétrolière, et sur la nécessité de prévoir à l'avance la conduite à tenir au cas où de tels sinistres se renouvelleraient.

Les années suivantes montrèrent à quel point les craintes de voir se reproduire de nouveaux accidents étaient fondées. En France, le cas du littoral breton est explicite. La dérive des nappes de pétrole s'échappant

Tableau I

Etudes des principales pollutions accidentelles par hydrocarbures dans le milieu marin

Accident	Date	Pays	Navire ou Plate-forme	Nature pétrole	Quantité (tonnes)	Impact côtier	Climat	Références
TORREY CANYON	1967	G.B.	Navire	Brut Koweit	120 000	Oui	Tempéré froid	(29, 30)
SANTA BARBARA	1969	USA	Plate-forme	Brut	25 000	Oui	Tempéré chaud	(33)
FLORIDA	1969	USA	Navire	Fuel # 2	600	Oui	Tempéré froid	(28)
ARROW	1970	Canada	Navire	Bunker C	9 000	Oui	Froid	(17)
GENERAL M.C. MEIGGS	1972	USA	Navire	Fuel	Faible	Oui	Tempéré froid	(8, 9)
METULA	1974	Chili	Navire	Brut	53 000		Tempéré froid	(34)
					46 000	Oui	Tempéré froid	(11)
ARGO MERCHANT	1976	USA	Navire	Fuel # 6	28 000	Oui	Tempéré froid	(22)
URQUIOLA	1976	Espagne	Navire	Brut Koweit	101 000	Oui	Tempéré chaud	(15)
EKOFISK	1977	Norvège	Plate-forme	Brut	21 000	Non	Froid	(14, 20)
TSESIS	1977	Suède	Navire	Fuel # 5	1 100	Oui	Froid	(23)
AMOCO CADIZ	1978	France	Navire	Brut (Arabie/Iran)	223 000	Oui	Tempéré froid	(10, 16, 26)
IXTOC-1	1979	Mexique	Plate-forme	Brut	550 000	Non	Tropical	(24)

- 207 -

de l'épave du TORREY CANYON, échoué sur les côtes anglaises de la Cornouaille, et atteignant les côtes bretonnes préfigurait une longue série de catastrophes touchant ce même littoral quelques années plus tard : OLYMPIC BRAVERY (1976), BOHLEN (1976), AMOCO CADIZ (1978), GINO (1979), TANIO (1980) (fig. 1). L'exemple de ces marées noires démontre un premier fait significatif : le caractère distinct de chaque pollution, fonction de la nature des produits pétroliers déversés bruts ou raffinés. La pollution de l'AMOCO CADIZ, exceptionnelle par son ampleur (223 000 tonnes de pétrole brut), constitue un cas exemplaire d'étude compte tenu de l'effort de recherche consenti, non seulement pour évaluer les effets sur la faune et la flore marines, mais également pour estimer le coût social et économique d'une telle catastrophe (25). Les craintes fondées à la suite de l'accident du TORREY CANYON amenèrent la communauté scientifique internationale à s'interroger et à étudier le comportement des hydrocarbures pétroliers dans l'environnement marin et les effets à court et long terme sur les écosystèmes pélagiques et benthiques.

Trois questions essentielles motivent de telles études :
- Quels sont les processus biogéochimiques essentiels et leur importance respective qui déterminent la dispersion, l'évolution et l'altération des hydrocarbures pétroliers déversés dans le milieu marin ?
- Quels sont les effets biologiques sur les organismes marins exposés à une pollution par hydrocarbures, effets mesurés en termes de bioaccumulation, de mortalités, d'altérations physiologiques ?
- Quels sont les effets à long terme sur les communautés marines atteintes, définis en terme de retour à l'équilibre des écosystèmes perturbés ?

Le tableau I résume les accidents pétroliers qui ont été principalement étudiés les quinze dernières années. Il existe un point essentiel sur lequel converge l'ensemble des études réalisées : l'importance de la relation existant entre d'une part l'énergie du site atteint, d'autre part le degré de dégradation et de disparition du pétrole et la cinétique de retour à l'équilibre des écosystèmes perturbés (15, 27). Les plus grandes catastrophes survenues en zones côtières (METULA, TORREY CANYON, AMOCO CADIZ, URQUIOLA) eurent lieu dans des secteurs fortement brassés, de haute énergie. Les observations s'accordent pour montrer la surprenante capacité de récupération du milieu dans les secteurs non abrités. Cette relation montre que l'altération naturelle et la dispersion du pétrole échoué en zone littorale dépend non seulement du type et du volume de pétrole déversé, mais pour beaucoup de la quantité d'énergie disponible. Les apports énergétiques peuvent être biologiques, chimiques, mécaniques ou thermiques. Cependant le facteur le plus important semble bien être l'énergie mécanique (vents, vagues, marées, niveaux d'eau, glace). Les principaux apports d'énergie mécanique proviennent de l'action des vagues et les niveaux d'énergie varient selon le régime des vents et donc de l'exposition locale d'un secteur de la côte. Dans les climats froids et polaires, la présence de glace sur la mer abaisse ces niveaux d'énergie dans la zone côtière. Ce concept de niveau d'énergie détermine l'importance des processus de dispersion, dissolution, photo-oxydation, dégradation microbiologique. Il fournit une base essentielle d'estimation de la persistance du pétrole sur les rivages et dans les sédiments subtidaux.

2. CARACTERISTIQUES D'UNE POLLUTION PAR HYDROCARBURES

A la suite d'une pollution accidentelle par hydrocarbures, le déplacement du pétrole à la surface de la mer, sa dispersion dans le milieu marin, au sein de la masse d'eau et vers les fonds marins, dépendent d'un certain nombre de facteurs qui sont étroitement liés à la nature du produit déversé et aux conditions locales de l'environnement.

Les pétroles bruts, sans parler des produits raffinés, peuvent être très différents selon leurs caractéristiques physiques et chimiques. Les hydrocarbures aromatiques représentent la fraction la plus toxique d'un pétrole. Les composés aromatiques légers s'évaporent facilement mais sont également les plus solubles dans l'eau et peuvent se trouver à des concentrations létales pour de nombreux organismes. Une nappe d'huile plus épaisse sera beaucoup moins toxique mais causera des dommages "mécaniques", par engluement. D'autre part, les conditions locales de l'environnement, géographiques, climatiques, météorologiques (vents, tempêtes) et océanographiques (vagues, courants) sont des données essentielles pour évaluer l'extension d'une pollution. Le vent apparaît être un facteur très important de la dérive des nappes en mer. De fortes conditions locales favorisent le mélange du pétrole dans la masse d'eau.

A partir de ces caractéristiques, plusieurs processus déterminent le comportement du pétrole en mer : son déplacement en surface, sa dispersion dans le milieu et son "vieillissement".

- L'étalement du pétrole à la surface de l'eau aboutit à un film d'une épaisseur variable (10^{-3} mm à 1 mm). Ce premier processus est rapide et dépend étroitement des caractéristiques physiques du pétrole et des conditions locales (vents, vagues).

- L'évaporation agit sur les fractions légères et est d'autant plus intense que l'étalement est important. Les fractions légères inférieures au $n-C_{15}$ (température d'ébullition < 250°C) peuvent être totalement volatilisées de la surface de l'eau en 10 jours. Le processus est plus limité pour les hydrocarbures situés entre le $n-C_{15}$ et le $n-C_{25}$. La fraction lourde au-dessus de $n-C_{25}$ ($T > 400°C$) ne subit pas ce processus. Des conditions locales fortes (mer agitée, vents forts) et des températures élevées favorisent ce phénomène.

- La dissolution dans la masse d'eau intervient comme processus compétitif à celui de l'évaporation. Il concerne surtout les fractions légères et certains composés polaires. Il est à noter qu'à température d'ébullition égale, les hydrocarbures aromatiques sont beaucoup plus solubles que les hydrocarbures aliphatiques (tableau II). L'importance du phénomène dépend de la composition du pétrole, de sa viscosité, des conditions locales et de l'état d'oxydation des hydrocarbures. Ce processus dure durant toute la phase d'évolution du pétrole parce que les phénomènes d'auto-oxydation et de dégradation biologique produisent des composés polaires qui sont généralement solubles dans l'eau.

- L'émulsification concerne le mélange des deux fractions insolubles pétrole/eau. On distingue deux types d'émulsions : "huile-dans-l'eau" et "eau-dans-huile". Les émulsions "huile-dans-l'eau" peuvent être dispersées facilement par des courants et les mouvements de turbulence. Les émulsions du type "eau-dans-huile" (ou "mousse au chocolat") tendent vers une consistance semi-solide et sont à l'origine des boules de goudron que l'on retrouve échouées à la côte.

- L'oxydation résulte de réactions chimiques et photo-chimiques qui aboutissent à la formation de dérivés d'oxydation, généralement polaires et par conséquent solubles dans l'eau.

- La sédimentation. Le "vieillissement" du pétrole, son évolution chimique et physique, l'adsorption sur les matières en suspension aboutit à la sédimentation, phénomène majeur de l'évolution à moyen et long terme de la majeure partie des hydrocarbures rémanents.

- La dégradation microbiologique est un phénomène complexe, étape finale de l'évolution et de l'altération du pétrole déversé dans le milieu marin. Deux types de processus interviennent : la dégradation en milieu aérobie et en milieu anaérobie. Le premier processus nécessite de l'oxygène et intervient à l'interface air/eau, dans la masse d'eau et dans les sédiments oxygénés. Le second processus s'opère par des organismes qui utilisent, à la place de l'oxygène, les sulfates et les nitrates, comme sources oxydantes ; il intervient dans les sédiments côtiers où l'échange surface/fond est très faible.

Tableau II

Solubilité de quelques hydrocarbures aliphatiques et aromatiques (ppm) (réf. 7)

Température d'ébullition (°C)	Hydrocarbures aliphatiques	Hydrocarbures aromatiques
·36	n-pentane : 39	–
69	n-hexane 9,5	–
80	–	benzène : 1780
98	n-heptane : 2,9	–
111	–	toluène : 515
126	n-octane : 0,66	–
136	–	éthyl benzène : 152
174	n-décane : 0,052	–
216	n-dodécane : 0,0037	–
218	–	naphtalène : 31
245	–	méthylnaphtalène : 25
263–270	–	diméthylnaphtalènes : 1,3–2,8
287	n-hexadécane : 0,0009	–
340	–	phénanthrène : 1,07
343	n-eicosane : 0,0019	–
448	–	chrysène : 0,002

A l'échelle spatiale et temporelle, comment peut-on caractériser une pollution par hydrocarbures du type de celle de l'AMOCO CADIZ ? On distingue généralement deux phases :

- Une phase de déploiement et d'extension de la pollution en mer et vers la côte. Durant cette période, la pollution est massive, la progression des nappes d'hydrocarbures, la fixation du pétrole à la côte sont des phénomènes rapides mais également fluctuants, soumis au régime des vents et des courants. Conjointement, interviennent les grands processus d'évolution du pétrole, à savoir : l'évaporation des composés légers vers l'atmosphère, mais également leur dissolution dans la masse d'eau, la formation des émulsions, la sédimentation du pétrole vers les fonds marins et les premiers processus d'altération (photo-oxydation, biodégradation). C'est durant cette phase que les effets immédiats de la pollution sur les organismes vivants sont les plus brutaux et les plus spectaculaires, par toxicité directe et engluement. Cette phase est généralement courte de quelques jours à quelques semaines.

- La seconde phase intervient lorsque l'extension maximale de la pollution est atteinte. Après stabilisation du phénomène, le processus de décontamination est amorcé, évoluant favorablement sur les sites fortement battus ou au contraire évoluant très lentement dans les zones abritées. Les mortalités enregistrées durant la première phase engendrent des déséquilibres écologiques. L'étude de la restauration du milieu nécessite un suivi écologique à long terme de plusieurs années.

3. POLLUTION DE L'AMOCO CADIZ
3.1. L'événement

Au matin du 16 mars 1978, le pétrolier AMOCO CADIZ se trouve au nord de l'île d'Ouessant. Ce navire de 334 m de long fait route vers Rotterdam. Il transporte, pour le compte de la Compagnie SHELL, 223 000 tonnes de pétrole brut, de qualité légère, en provenance d'Arabie et d'Iran. Le navire bat pavillon libérien mais appartient à la Compagnie AMOCO dont le siège est à Chicago aux Etats-Unis. A 10h45, heure française, l'homme de barre prévient le commandant que le navire a perdu sa direction. Le gouvernail est bloqué. A partir de cet instant et durant 11 heures, le pétrolier va dériver de façon quasi-linéaire jusqu'à la côte, et ceci malgré deux tentatives d'intervention d'un remorqueur spécialisé de 10 000 CV. A 23h00, le navire s'échoue sur les hauts fonds rocheux, face au petit port breton de Portsall. La marée noire de l'AMOCO CADIZ commence. La totalité de la cargaison va s'écouler sans interruption dans la mer durant 15 jours et occasionner une catastrophe sans précédent dans le domaine des pollutions pétrolières. Plus de 300 km de côtes seront pollués.

3.2. La lutte

Le plan POLMAR est engagé dès le 17 mars. L'éventualité d'une mise à feu de l'épave n'est pas retenue devant les risques d'une pollution à terre par projection de gouttelettes de goudron. L'allégement du pétrolier par pompage de la cargaison vers un navire allégeur ou vers la terre est envisagé. Mais les moyens nécessaires à la mise en oeuvre de ce projet ne seront disponibles que 10 à 15 jours après le naufrage. Toutes les actions destinées à maîtriser la pollution au niveau du navire se soldent par un échec.

Les autorités tentent alors d'agir sur les nappes d'hydrocarbures désormais répandues en mer. L'état des techniques disponibles et les conditions météorologiques conduisent principalement à l'emploi des dispersants. Leur utilisation est interdite sur des fonds inférieurs à 50 m, de façon à éviter que vents et courants ne rabattent sur la côte du pétrole trop récemment dispersé. Au total, environ 500 tonnes de produits concentrés (dits de 2e génération) et 1 000 tonnes de produits concentrés (3e génération) seront utilisés en mer. Les actions sur les nappes d'hydrocarbures en pleine mer ne peuvent en réalité empêcher leur extension.

Les efforts se portent alors sur la protection et le nettoiement des côtes. A la date du 25 mars, 1 400 m de barrages flottants sont posés en divers secteurs sensibles. Mais ces barrages ne peuvent assurer la protection escomptée. Les conditions météorologiques défavorables constituent l'élément déterminant de leur inefficacité. Le nettoiement des plages devient l'objectif majeur. Ces opérations menées en deux temps, d'abord par pompage du pétrole encore liquide, ensuite par déblaiement des déchets souillés d'hydrocarbures. Les tonnes à lisier des agriculteurs, nombreuses dans cette région d'élevage porcin, jouent un rôle déterminant durant les premières semaines. A la fin du mois de mai, un volume global de 65 000 tonnes de produits sera pompé ; cette masse contient environ 30 000 tonnes de pétrole. Quant aux produits solides collectés sur les grèves, ils représentent 185 000 tonnes, soit environ 15 000 tonnes d'hydrocarbures (5).

Cette rapide chronologie de l'événement et de la lutte démontre que si, lors d'un naufrage, la masse de pétrole est trop importante, il devient impossible de lutter. L'essentiel est donc la prévention.

Il faut à présent souligner le caractère sans précédent de la pollution provoquée par l'AMOCO CADIZ, du fait :
- de la nature du pétrole brut transporté qui contenait un tiers de substances aromatiques, admises comme toxiques,
- de la faible distance entre l'emplacement de l'épave et la côte (moins de 2 milles nautiques),
- de la quantité massive d'hydrocarbures déversée en peu de jours : 223 000 tonnes en moins de deux semaines, soit un écoulement de 18 000 à 20 000 tonnes par jour.

3.3. Extension de la pollution

La phase d'extension de la pollution dure un mois et demi, essentiellement contrôlée par la direction et l'intensité des vents. On admet généralement une vitesse de déplacement des nappes d'hydrocarbures égale à 3 % de la vitesse des vents. En mars, le régime des vents de secteur ouest provoque une large dérive côtière vers l'est. Les nappes atteignent l'Aber Wrac'h (19 mars), Roscoff (20 mars), la baie de Lannion (21 mars), les Sept Iles (22 mars), le sillon de Talbert (23 mars). A partir du 2 avril, le passage des vents à l'est renverse le sens de dérive des nappes qui atteignent Le Conquet et Ouessant (11 avril), le Raz de Sein (13 avril), Douarnenez (22 avril). Quelques traces d'hydrocarbures arrivent à la côte dans le secteur de la baie d'Audierne en mai. A cette date, l'extension maximale des nappes d'hydrocarbures en mer est atteinte (fig. 2). La phase de décontamination et d'évolution à long terme débute schématiquement à partir de mai 1978.

3.4. Atmosphère

Les émanations de pétrole dans l'atmosphère, liées à l'évaporation des hydrocarbures légers, sont ressenties dans les jours qui suivent le naufrage, non seulement par la population bordant le littoral touché, mais également par celle vivant à l'intérieur des terres, et même dans le Finistère Sud, jusqu'à Audierne. Ceci atteste de l'importance du phénomène. La quantité de pétrole évaporée dans l'atmosphère se situe environ à 67 000 tonnes, dont 40 000 tonnes d'aromatiques légers, composés admis comme toxiques, tels que benzène, toluène, xylènes, ...

3.5. Littoral

La dérive côtière des nappes d'hydrocarbures entraîne une pollution du littoral sur près de 360 km de côtes. Par ailleurs cette dérive intervient au moment des grandes marées d'équinoxe de la fin mars qui, associées à des vents puissants, provoquent la pollution de zones rarement recouvertes par la mer, comme le marais de l'Ile Grande, en baie de Lannion. A la fin mars, 62 000 tonnes de pétrole sont fixées à la côte.

La phase de décontamination débute en mai, indépendamment des efforts entrepris pour le nettoiement. Cette décontamination est liée étroitement à la géomorphologie côtière. Dans les secteurs abrités des vents et des courants, tels les abers et les marais maritimes, il n'y a pratiquement aucune évolution. Par contre, l'auto-nettoyage des côtes rocheuses par les éléments naturels est très actif dès les premières semaines et se poursuivra pendant tout l'hiver suivant.

Sur les plages, le phénomène est plus complexe (fig. 3). La pollution de l'AMOCO CADIZ intervient à la fin de la période hivernale, qui se caractérise par un fort amaigrissement des plages. A partir du mois d'avril, la dynamique sédimentaire s'inverse et permet l'accumulation de sable sur les plages (phénomène d'engraissement). De ce fait, les hydrocarbures déposés à marée descendante se retrouvent interstratifiés dans le sable à la marée montante suivante. La profondeur d'enfouissement

Tableau III
Indice de vulnérabilité morpho-sédimentaire (12)

	Index	Types morphosédimentaires	Accumulation des hydrocarbures	Durée de la pollution	Persistance des hydrocarbures
Zones à haute énergie	1	Côtes rocheuses et plates-formes d'abrasion	- Partie supérieure de l'estran	Quelques mois	
	2	Plages de sable fin à moyen	- Interstratification dans le sédiment - Migration lente en profondeur	1 à 2 ans	
	3	Plages de sable grossier à graviers	- Interstratification dans le sédiment - Migration rapide en profondeur	1 à 3 ans	
	4	Plages de galets, cailloux et champs de blocs	- Migration rapide des hydrocarbures en profondeur; Peu ou pas de dépôts en surface	3 à 5 ans	
	5	Côtes rocheuses	- Accumulation des hydrocarbures dans les anfractuosités des rochers - Rochers recouverts d'une fine pellicule	3 à 5 ans	
	6	Plages de sable fin à moyen	- Percolation en profondeur - Pollution de la zone subtidale par les marées (mélange des hydrocarbures et des sédiments fins) - Formation en surface d'une couche durcie après un an	> 5 ans	
Zones à faible énergie	7	Plages de sable grossier à gravier	- Percolation rapide en profondeur après un an - Formation de couche durcie après un an	> 5 ans	
	8	Plages de galets	- Percolation rapide en profondeur jusqu'au substratum - Formation d'une croûte de galets et d'hydrocarbures après un an	> 5 ans	
	9	Estuaires et "Tidal flat" à sédiments vaseux	- Percolation en profondeur due aux organismes fouisseurs et mouvements d'eau interstitielle	> 10 ans	
	10	Marais maritimes	- Encroûtements en surface - Migration dans le sédiment	> 10 ans	

est en moyenne de 15 à 20 cm, mais peut atteindre 70 cm localement. Un
second processus est également observé sur les plages de sable grossier et
de galets, celui d'une migration verticale des hydrocarbures par simple
percolation, phénomène favorisé par la faible viscosité du pétrole de
l'AMOCO CADIZ. Actuellement, si la plupart des plages sont propres en
surface, quatre ans après la catastrophe, l'imprégnation des sédiments
subsiste, tout particulièrement dans les terrasses de basse mer.

De cette étude, il a été possible d'établir, selon la méthodologie
développée par (15), un index de vulnérabilité du littoral breton à la
pollution par hydrocarbures. Deux critères ont été retenus : d'une part,
la notion d'énergie à laquelle est exposée une section de côte, d'autre
part, la granulométrie du sédiment qui permet ou non la migration des
hydrocarbures en profondeur (tableau III).

3.6. Eau de mer

La pollution de l'eau de mer par les hydrocarbures couvre un large
secteur de la Manche Occidentale, de la mer d'Iroise à la baie de
St Brieuc (fig. 4, tableau IV).

Tableau IV

Concentrations moyennes des hydrocarbures dans l'eau de mer (μg/l)
(mesures effectuées par spectrofluorescence UV)

Secteur géographique		Avril 1978	Période de décontamination T (jours)	Mars 1979
Au large	L	1,6 + 0,5		
Mer d'Iroise	A	1,8 + 0,8		
A l'ouest de Portsall	B	2,2 + 0,9		
A l'est de Portsall	C	9,1 + 8,6	11	
Face aux Abers	D	38,9 + 6,7	14	1,2 + 1,0
Aber Wrac'h	–	120	40	
Plouguerneau-Roscoff	E	15,6 + 21,2		
Face à Roscoff	F	4,8 + 8,4		
Baie de Morlaix	G	11,5 + 5,1	28	1,0 + 0,05
Baie de Lannion	H	10,7 + 3,0	28	1,3 + 0,6
Sept Iles-Le Trieux	I	5,2 + 0,8		0,8 + 0,2
Zone du Trieux	J	3,9 + 0,8		0,7 + 0,2
Baie de St Brieuc	K	1,0 + 0,5		

Nous avons adopté comme seuil de pollution une teneur de 2 μg/l
d'hydrocarbures dans l'eau, mesurée par spectrofluorescence UV. Les
concentrations observées un mois après l'accident sont en moyenne de
120 μg/l dans l'Aber Wrac'h, 40 μg/l en face des Abers, 10 μg/l dans les
baies de Morlaix et Lannion. Ces valeurs sont comparables à celles
trouvées pour d'autres cas de pollution par hydrocarbures : EKOFISK,
30 μg/l (14) ; ARGO MERCHANT, 450 μg/l (22) ; ARROW, 45 μg/l (19), mais
sont plus faibles que les 7 000 μg/l observées à proximité de la
plate-forme IXTOC-1 (24). Les mesures effectuées montrent par ailleurs une
diffusion du pétrole sur l'ensemble de la colonne d'eau, de la surface
vers le fond. Cette diffusion verticale des hydrocarbures permet déjà de
présumer que les fonds marins seront pollués dans les secteurs atteints
par la dérive des nappes.

Dans les mois qui suivent, les teneurs en hydrocarbures dans l'eau diminuent, de façon générale. La période de décontamination, durée durant laquelle la concentration en hydrocarbures diminue de moitié, est de 11 à 14 jours en zones côtières non abritées, 28 jours dans les baies de Morlaix et Lannion et de 40 jours dans l'Aber Wrac'h (fig. 5). Un an après la catastrophe, les niveaux sont revenus à la "normale". De légers taux résiduels sont toutefois encore observés à proximité des Abers et au fond de la Baie de Lannion.

3.7. Sédiments

Les sédiments en Manche occidentale sont constitués généralement de sables grossiers à moyennement fins. Les prélèvements effectués en avril 1978 montrent une large pollution des fonds marins, entre Portsall et les Sept Iles. La zone touchée est sensiblement la même que celle correspondant à la contamination des eaux (fig. 6). Près de la côte, les plus fortes teneurs, supérieures à 100 ppm, sont observées dans les baies de Morlaix et Lannion (200 à 600 ppm) et dans les deux Abers (Aber Benoît et Aber Wrac'h) (700 à > 10 000 ppm) qui sont des petits estuaires de 10 à 15 km de long, comprenant des zones de sable et de vase. Ces zones constituent des sites privilégiés d'exploitation ostréicole (fig. 7).

Le processus naturel de décontamination est lié à deux facteurs essentiels : la nature du sédiment, le caractère abrité ou battu du secteur considéré (tableau V).

Tableau V
Évolution des hydrocarbures dans les sédiments
(concentrations moyennes exprimées en ppm)

Zones côtières		Avril 1978	Mars 1979	Mars 1981
Les Abers				
Aber-Benoît	(vasières)	> 10 000	> 10 000	
Aber-Benoît	(sables fins)	700	27	
Aber-Wrac'h	(sables légèrement vaseux)	3 300	1 700	600
Aber-Wrac'h	(sortie) (sables fins)	1 000	100	25
Baies de Morlaix et de Lannion				
Baie de Morlaix	(sables vaseux)	311	172	
Baie de Lannion	(sables fins)	281	126	
Pte de Primel	(sédiments grossiers)	600	19	

Dans les baies de Morlaix et Lannion, le processus évolue dans l'ensemble favorablement. Les quantités de pétrole piégées dans les sédiments passent de 7 600 tonnes en avril 1978 à 1 800 tonnes en août 1978 et à 800 tonnes en août 1979 (16). En août 1980, la plupart des sédiments prélevés ne présentent plus de traces de pollution. Dans les sédiments sableux de la partie aval de l'Aber Benoît, le processus d'auto-épuration est très sensible, la contamination des sédiments décroît en moyenne de 700 ppm à 27 ppm en un an. Le cas extrême est observé dans les vasières de l'Aber Benoît. Ces vases constituent de véritables pièges à hydrocarbures, avec des teneurs observées supérieures à 10 000 ppm et aucun processus de décontamination n'est enregistré un an après la catastrophe.

Le cas intermédiaire et le plus intéressant est constitué par les sédiments de l'Aber Wrac'h. Ce sont pour l'essentiel, autant dans la

partie aval que dans la partie amont de l'estuaire, des sables plus ou
moins vaseux. A la sortie de l'Aber, on trouve, en zone de forte énergie,
des sables fins non vaseux. L'étude du processus de décontamination de cet
estuaire s'est poursuivie jusqu'en 1984 (21) et est illustrée par la
figure 8. Les sables fins à l'embouchure de l'Aber se sont rapidement
décontaminés, de 1 000 ppm en 1978 à 25 ppm en 1981. Dans les sables
vaseux de la partie aval de l'estuaire, la contamination initiale est plus
importante, en moyenne de 5 000 ppm. La décontamination dans cette zone,
où le caractère marin est prononcé, est également observée mais reste plus
faible. En 1981, les teneurs résiduelles restent en moyenne de l'ordre de
800 ppm. Dans la partie amont, les sédiments sablo-vaseux sont
initialement moins pollués que ceux de la partie aval, en moyenne
1 500 ppm, et le processus de décontamination n'est pas observé jusqu'en
1981. Il faut attendre 1984, soit six ans après la catastrophe, pour
observer une décontamination générale des sédiments de l'Aber Wrac'h.
Selon la nature du sédiment et le degré d'énergie du site estuarien, on
peut estimer les périodes de décontamination, à 9 mois pour les sables
fins à l'entrée de l'Aber, à 11 mois pour les sables vaseux de la partie
aval et à 24 mois pour ceux de la partie amont.

3.8. Evolution chimique des hydrocarbures

Le pétrole de l'AMOCO CADIZ est un mélange de deux pétroles bruts
légers d'Arabie et d'Iran. Il contient 39 % d'hydrocarbures saturés, 34 %
d'aromatiques, 24 % de composés polaires et 3 % de composés résiduels (6).
Les hydrocarbures saturés sont identifiés du n-C$_8$ au n-C$_{37}$. La fraction
aromatique contient des composés allant des dérivés du benzène aux dérivés
du phénanthrène. Cette fraction est dominée par les dérivés du naphtalène.
Les composés du dibenzothiophène (dérivés aromatiques soufrés) sont
également identifiés.

Dans les jours, les semaines et les mois qui suivent l'accident, la
composition chimique de ce pétrole va considérablement évoluer, sous
l'influence des processus d'évaporation, de dissolution, de
photo-oxydation et de dégradation biologique par les micro-organismes dans
l'eau et dans les sédiments. La vitesse d'évolution est variable selon les
sites et dépend de nombreux facteurs liés à l'environnement : le degré
d'énergie du site, l'état d'oxydation du sédiment, la disponibilité en
éléments nutritifs. Les études qui sont réalisées permettent toutefois de
décrire un scénario type (16).

L'évaporation provoque une perte des hydrocarbures légers saturés et
aromatiques (dérivés du benzène). La diminution des alcanes linéaires par
rapport aux alcanes ramifiés permet d'enregistrer l'existence d'une
dégradation microbiologique des hydrocarbures ; le rapport n-C$_{17}$/pristane
et n-C$_{18}$/phytane passe de 4 à 0,5.

Cette première phase de l'évolution commence dans l'eau et se
poursuit lorsque le pétrole se dépose sur les sédiments. Elle se déroule
dans les premiers jours et semaines qui suivent l'accident. Elle démontre
un fait important, à savoir que le processus de biodégradation des
hydrocarbures se manifeste aussi rapidement que le processus d'évaporation
des fractions légères.

L'évolution chimique se poursuit ensuite plus lentement par
disparition progressive des alcanes linéaires et ramifiés. L'enveloppe
chromatographique non résolue, tant pour les hydrocarbures saturés que
pour les hydrocarbures aromatiques, augmente en proportion relative. Cette
enveloppe correspond à des composés qui résistent à la dégradation
biologique. Au-delà de un à deux ans, les triterpanes pentacycliques
(alcanes cycliques de C$_{27}$ à C$_{33}$) deviennent les marqueurs chimiques pour
la fraction des hydrocarbures saturés du pétrole de l'AMOCO CADIZ
(fig. 9).

Pour les hydrocarbures aromatiques (fig. 10), la première phase d'évolution se caractérise par la perte des composés légers, dérivés du benzène, par évaporation et dissolution. Les dérivés du naphtalène, initialement dominants, disparaissent ensuite par processus physiques et d'oxydation. Les substances persistantes sont constituées en dernière phase d'évolution par les dérivés du phénanthrène et du dibenzothiophène et par les composés naphténo-aromatiques qui correspondent à l'enveloppe chromatographique non résolue.

3.9. Biodégradation

S'il existe de nombreux travaux expérimentaux qui montrent la capacité des micro-organismes de dégrader les hydrocarbures, c'est la première fois que, dans un milieu naturel, cette biodégradation des hydrocarbures dans l'eau de mer est mise en évidence et quantifiée. Le phénomène est observé fin mars dans une zone située au large de Plouguerneau, où l'on enregistre sur toute la colonne d'eau des déficits en oxygène et en éléments nutritifs d'azote et de phosphore. Les quantités d'hydrocarbures dégradés sont estimées à 0,4 mg/l en surface et 0,15 mg/l au fond. Ceci représente une quantité totale de 10 000 tonnes d'hydrocarbures biologiquement dégradés en mer pendant les deux semaines qui suivent l'accident (1).

Sur le littoral, la pollution des sédiments entraîne un enrichissement des bactéries utilisant les hydrocarbures. La quantité d'hydrocarbures biodégradés dans les sédiments côtiers est estimée à 0,5 µg/g de sédiment sec/jour, soit pour une zone intertidale touchée de 320 km de long et 500 m de large, une dégradation microbiologique de l'ordre de 8 tonnes/jour (3).

La biodégradation apparaît donc avoir joué un rôle, à la fois immédiat et très important dans le processus d'altération du pétrole de l'AMOCO CADIZ. La modification du rapport n-alcanes/alcanes isoprénoïdes, caractéristique d'une action microbiologique, a été mesurée dans les jours et les semaines qui ont immédiatement suivi la catastrophe (3). De tels changements constrastent avec ceux observés à l'échelle de mois et d'années après la pollution du ARROW, dans les eaux également froides de la baie de Chedabucto au Canada (17). Dans les eaux chaudes tropicales du golfe du Mexique et malgré l'enrichissemnet observé de la flore bactérienne utilisant les hydrocarbures, l'altération du pétrole de IXTOC-1 par processus de biodégradation n'est pas plus rapide et se mesure à l'échelle de mois. L'importance du phénomène a été évaluée d'un ordre de grandeur inférieur à celui mesuré pour le pétrole de l'AMOCO CADIZ (2, 24). Le processus de biodégradation des hydrocarbures dans l'environnement marin est un phénomène complexe. Si certains facteurs limitants sont à présent connus (oxygène, éléments nutritifs), la nature des interactions entre la flore microbienne, les conditions du milieu environnant et les produits pétroliers est encore mal maîtrisée pour prédire et modéliser la cinétique et l'importance d'un tel phénomène.

3.10. Contamination des organismes marins

La plupart des organismes marins accumulent les hydrocarbures, soit par leur mode alimentaire, soit directement à partir de l'eau et des sédiments pollués. Ainsi, la totalité des invertébrés marins, mobiles ou sédentaires, prélevés dans les zones atteintes par la pollution, ont été contaminés. Le comportement de certaines espèces marines leur a permis d'échapper à cette contamination : réaction de fuite des poissons aussi longtemps que la mer a été fortement polluée, éloignement initial des araignées des zones polluées du fait qu'elles n'avaient pas encore entamé leur migration saisonnière vers le littoral au moment de la catastrophe. Les études réalisées sur les invertébrés marins montrent que le processus d'accumulation des hydrocarbures n'est pas sélectif. Il en résulte, d'une part, un taux d'accumulation lié au degré de pollution du milieu environnant, d'autre part, une étroite relation entre la nature des hydrocarbures accumulés dans l'organisme et l'état de vieillissement du pétrole. Les hydrocarbures polyaromatiques, plus résistants à la biodégradation, en particulier les dérivés du phénanthrène et du dibenzothiophène, se retrouvent donc dans les invertébrés, notamment dans les huîtres, plusieurs mois après la catastrophe.

Les sites d'accumulation préférentiels des hydrocarbures dans les organismes existent et varient selon les espèces : tissus riches en lipides endogènes pour certains invertébrés comme les mollusques, branchies et ovaires pour les poissons, hépatopancréas pour les crustacés, tissus musculaires pour les oiseaux (l'absence dans le foie reflète plutôt la conséquence d'une métabolisation importante dans cet organe) (18).

Les travaux les plus importants ont été consacrés aux **huîtres**, dont les activités sont concentrées dans les deux Abers et la baie de Morlaix. Face à la pollution massive du pétrole de l'AMOCO CADIZ, les mortalités ont été faibles, principalement observées dans l'Aber Benoît, site le plus proche du lieu du naufrage. Les mortalités ont affecté uniquement les huîtres recouvertes d'une épaisse couche de pétrole. Les conséquences de la pollution sur l'ostréiculture n'ont donc pas porté sur la survie des huîtres vivant en sites pollués mais sur leur insalubrité vis-à-vis de la consommation en raison des fortes teneurs d'hydrocarbures bio-accumulés. Ainsi, dès les premières semaines qui suivirent l'accident, la Direction des Affaires Maritimes décidait la destruction d'importants stocks en place : 5 000 tonnes d'huîtres en baie de Morlaix et 1 000 tonnes dans les Abers, et le transfert d'environ 1 600 tonnes d'huîtres contaminées vers des centres ostréicoles afin de permettre l'auto-épuration des hydrocarbures accumulés.

Il est à présent admis que la bioaccumulation des hydrocarbures dans l'huître est un phénomène purement passif dans la mesure où ce phénomène est fonction du coefficient de partage entre l'eau et les lipides biotiques. Le facteur de bio-concentration, défini comme le rapport de la concentration des hydrocarbures dans l'huître (par rapport au poids humide) sur la concentration dans l'eau, a été évalué environ à 4 000 dans les Abers (4). Cette valeur est en accord avec celles évaluées en expérimentation in "vitro", entre 1 000 et 3 000 (13, 32). Les teneurs les plus fortes d'hydrocarbures totaux déterminées par spectrofluorescence UV (SFUV) sont de l'ordre de 1 500 mg/kg (poids sec). Dans les huîtres provenant de sites non pollués, considérés comme référence, les niveaux sont en moyenne de 22 mg/kg.

Les expériences de transfert d'huîtres contaminées vers des zones salubres mettent en relief plusieurs faits significatifs. On observe en premier lieu que le processus d'auto-épuration des hydrocarbures totaux (mesurés par SFUV) est d'autant plus rapide que le séjour dans le site contaminé a été bref. Le maintien en milieu fortement pollué, pendant une période de 8 ou 18 mois, se traduit par un blocage du processus de décontamination, même si l'huître est immergée dans des eaux salubres (fig. 11). Un tel phénomène peut s'expliquer par une rétention des hydrocarbures dans certains compartiments du mollusque. L'analyse fine des hydrocarbures accumulés montre que la cinétique d'épuration des hydrocarbures aliphatiques est beaucoup plus rapide que celle des hydrocarbures polyaromatiques (HPA). La composition de la fraction aromatique évolue elle-même durant la phase d'épuration. Parmi les HPA persistants dans l'huître, figurent les composés soufrés, les dibenzothiophènes substitués, particulièrement les homologues polyalkylés. Le processus d'épuration des hydrocarbures chez l'huître apparaît être de nature biphasique : une phase rapide concernant surtout les hydrocarbures aliphatiques et une phase lente concernant les hydrocarbures polyaromatiques, cette cinétique étant elle-même influencée par la durée du séjour de l'huître en zone polluée. A la différence des poissons (31), l'huître (comme la moule) ne possède pas de systèmes enzymatiques de détoxification par hydroxylation d'une fraction des hydrocarbures bioaccumulés (35). L'huître ne peut donc s'épurer que grâce au phénomène chimique du coefficient de partage entre les tissus lipidiques et l'eau. Cette observation justifie l'hypothèse que le taux d'hydrocarbures accumulés dans les mollusques reflète le niveau de pollution en milieu environnant. Six ans après la pollution de l'AMOCO CADIZ, les teneurs résiduelles en hydrocarbures dans les huîtres de l'Aber Benoît restent encore importantes, d'un niveau environ cinq fois supérieur à la valeur moyenne observée dans les huîtres non contaminées (22 ppm par SFUV). Le niveau de contamination résiduelle dans les huîtres de la baie de Morlaix, à Carantec, est plus faible, environ deux fois la valeur de référence.

3.11. Bilan massique

Si l'on s'interroge sur la finalité de telles études, une des questions intéressante est sans doute celle de savoir, en terme quantitatif, comment se sont dispersés les 223 000 tonnes de pétrole de l'AMOCO CADIZ dans l'environnement.

Si l'on intègre l'ensemble des observations réalisées dans le cadre de ce programme, on peut dresser un bilan dans le mois qui a suivi le naufrage du pétrolier. Nous arrivons aux chiffres suivants :

- 67 000 tonnes se sont évaporées dans l'atmosphère,
- 30 000 tonnes étaient présentes dans la masse d'eau, dont 10 000 tonnes ont été dégradées par les micro-organismes,
- 18 000 tonnes se sont déposées sur les fonds marins,
- 62 000 tonnes se sont fixées à la côte.

Ce bilan fait apparaître un déficit de 45 000 tonnes, soit 20 % de la cargaison, correspondant probablement à une dispersion au large par les courants de nappes de pétrole et de résidus goudronneux. En toute rigueur, ces valeurs résultent d'estimations qui doivent être prises avec prudence. Un tel bilan doit être interprété en terme d'ordre de grandeur.

REFERENCES

(1) AMINOT, A. (1981). Anomalies du système hydrobiologique côtier après l'échouage de l'AMOCO CADIZ. Considérations qualitatives et quantitatives sur la biodégradation "in situ" des hydrocarbures. In : "AMOCO CADIZ. Conséquences d'une pollution accidentelle par hydrocarbures". Publ. CNEXO, Paris : 223-42.

(2) ATLAS, R.M. (1981). Microbial degradation of petroleum hydrocarbons : an environmental perspective. Microbiological Reviews 45 (1) : 180-209.

(3) ATLAS, R.M. et BRONNER, A. (1981). Microbial hydrocrbon degradation with the intertidal zones impacted by the AMOCO CADIZ oil spillage. In : "AMOCO CADIZ. Conséquences d'une pollution accidentelle par hydrocarbures". Publ. CNEXO, Paris : 251-56.

(4) BALOUET, G., BERTHOU, F., BODENNEC, G. et MARCHAND, M. (1985). Effets de la pollution par les hydrocarbures de l'AMOCO CADIZ sur l'ostréiculture (1978-1984). Rapp. IFREMER et UBO, Brest : 95 pp.

(5) BELLIER, P. (1979). Lutte contre les pollutions accidentelles par les hydrocarbures. L'expérience de l'AMOCO CADIZ. Rapp. CEDRE, Direction des Ports et de la Navigation Maritime : 193 pp.

(6) CALDER, J.A. et BOEHM, P. (1981). The chemistry of AMOCO CADIZ oil in the Aber-Wrac'h. In : "AMOCO CADIZ. Conséquences d'une pollution accidentelle par hydrocarbures". Publ. CNEXO, Paris : 149-158.

(7) CLARK, R.C.Jr. et McLEOD, D.Jr. (1977). Inputs, transport mechanisms and observed concentrations of petroleum in the marine environment. In : "Effects of petroleum on Artic and Subartic marine environments and organisms". Vol. I. Nature and fate of petroleum. Malins D.C. ed., Acad Press, New-York : 91-223.

(8) CLARK, R.C.Jr., FINLEY, J.S., PATTEN, B.G. et NIKE, E.E. (1975). Long-term chemical and biological effects of a persistent oil spill following the grounding of the GENERAL M.C. MEIGGS. In : "Proc. Conf. Prev. Cont. Oil Poll.", San Francisco, Am. Petrol. Inst., Washington D.C. : 479-87.

(9) CLARK, R.C.Jr., PATTEN, B.G. et NIKE, E.E. (1978). Observations of a cold-water intertidal community after 5 years of a lox-level, persistent oil spill from the GENERAL M.C. MEIGGS. J. Fish. Res. Bd. Can. 35 (5) : 754-65.

(10) CNEXO (1981). AMOCO CADIZ. Conséquences d'une pollution accidentelle par hydrocarbures. Publ. CNEXO, Paris : 881 pp.

(11) COLWELL, R.R., MILLS, A.L., WALKER, J.D., GARCIA-TELLO, P. et CAMPOS-P, V. (1978). Microbial ecology studies of the METULA spill in the Straits of Magellan. J. Fish. Res. Bd. Can., 35 (5) : 573-80.

(12) D'OZOUVILLE, L., BERNE, S., GUNDLACH, E.R. et HAYES, M.O. (1981). Evolution de la pollution du littoral breton par les hydrocarbures de l'AMOCO CADIZ entre mars 1978 et novembre 1979. In : "AMOCO CADIZ. Conséquences d'une pollution accidentelle par hydrocarbures". Publ. CNEXO, Paris : 55-78.

(13) FOSSATO, V.U. et CANZONIER, W.J. (1976). Hydrocarbon uptake and loss by the mussel Mytilus edulis. Mar. Biol., 36 : 243-50.

(14) GRAHL-NIELSEN, O. (1978). The EKOFISK BRAVO blow-out. Petroleum hydrocarbons in the sea. In : "Proceedings of the conference on assessment of ecological impacts of oil spills". AIBS, 14-17 june 1978, Keystone, Co. (USA).

(15) GUNLACH, E.R., RUBY, C.H., HAYES, M.O. et BLOUNT, A.E. (1978). The URQUIOLA oil spill, La Coruna, Spain. Initial impact and reaction on beaches and rocky coasts. Envir. Geol. 2 (3) : 131-43.

(16) GUNDLACH, E.R., BOEHM, P.D., MARCHAND, M., ATLAS, R.M., WARD, D.M. et WOLFE, D.A. (1983). The fate of the AMOCO CADIZ oil. Science 221 : 122-29.

(17) KEIZER, P.D., AHERN, T.P., DALE, J. et VANDERMEULEN, J.H. (1978). Residue of Bunker C oil in Chedabucto bay, Nova Scotia, 6 years after the ARROW spill. J. Fish. Res. Bd. Can. 35 (5) : 528-35.

(18) LAWLER, G.C., HOLMES, J.P., ADAMKIEWICZ, D.M., SHIELDS, M.I., MONNAT, J.Y. et LASETER, J.L. (1981). Characterization of petroleum hydrocarbons in tissues of birds killed by the AMOCO CADIZ oil spill. In : "AMOCO CADIZ. Conséquences d'une pollution accidentelle par hydrocarbures". Publ. CNEXO, Paris : 573-584.

(19) LEVY, E.M. (1971). The presence of petroleum hydrocarbon residues off the east coast of Nova Scotia in the Gulf of St Lawrence and the St Lawrence river. Wat. Res. J. : 723-33.

(20) MACKIE, P.R., HARDY, R. et WHITTLE, K.J. (1978). Preliminary assessment of the presence of oil in the ecosystem at EKOFISK after the blow-out, april 22-30, 1977. J. Fish. Res. Bd. Can. 35 (5) : 541-51.

(21) MARCHAND, M. et BODENNEC, G. (1985). Evolution de la pollution des hydrocarbures de l'AMOCO CADIZ dans les sédiments de l'Aber-Wrac'h. Processus de décontamination en site estuarien. Rapp. IFREMER, Brest : 10 pp.

(22) NOAA (1977). The ARGO MERCHANT oil spill. A preliminary scientific report. US Dept. Commerce, NOAA, Boulder, Co. (USA) : 133 pp.

(23) NOAA (1980). The TSESIS oil spill. Us Dept. Commerce, NOAA, Boulder, Co. (USA) : 296 pp.

(24) NOAA (1980). Preliminary results from the September 1979 Researcher/ Pierce IXTOC-1 cruise. US Dept. Commerce, NOAA, Boulder, Co. (USDA) : 591 pp.

(25) NOAA (1983). Assessing the social costs of oil spills : the AMOCO CADIZ case study. NOAA, US Dept. of Commerce, Boulder, Co. : 144 pp.

(26) NOAA/CNEXO (1982). Ecological study of the AMOCO CADIZ oil spill. US Dept. Commerce, NOAA, Boulder, Co. (USA) : 479 pp.

(27) OWENS, E.H. (1978). Mechanical dispersal of oil stranded in the littoral zone. J. Fish. Res. Bd. Can., 35 (5) : 563-72.

(28) SANDERS, H.L., GRASSLE, J.F., HAMPSON, G.R., MORSE, L.S., GARNER PRICE, S. et JONES, C.C. (1980). Anatomy of an oil spill : long term effects from the grounding of the barge FLORIDA off West Falmouth, Massachusetts. J. Mar. Res., 38 (2) : 265-380.

(29) SMITH, J.E. (1968). TORREY CANYON. Pollution and marine life. University Printing House, Cambridge, U.K. : 196 pp.

(30) SOUTHWARD, A.J. et SOUTHWARD, E.C. (1978). Recolonization of rocky shores in Cornwall after use of toxic dispersants to clean up the TORREY CANYON spill. J. Fish. Res. Bd. Can., 35 (5) : 682-706.

(31) STEGEMAN, J.J. (1978). Influence of environmental contamination on cytochrome P-450 mixed-function oxygenases in fish : implication for recovery in the wild harbor marsh. J. Fish. Res. Bd. Can., 35 (5) : 668-74.

(32) STEGEMAN, J.J. et TEAL, J.M. (1973). Accumulation, release and retention of petroleum hydrocarbons by the oyster Crassostrea virginica. Mar. Biol., 22 : 37-44.

(33) STRAUGHAN, D. (ed.) (1971). Biological and oceanographical survey of the SANTA BARBARA oil spill, 1969-1970. Vol. 1, Biology and Bacteriology. Sea Grant n° 2. Allen Hancock Foundation, University of Southern California, Los Angeles.

(34) STRAUGHAN, D. (1977). Biological survey of intertidal areas in the Straits of Magellan in january 1975, five months after the METULA oil spill. In : "Fate and effects of petroleum hydrocarbons in marine ecosystems and organisms". Ed. D.A. Wolfe, Pergamon Press, New-York : 247-60.

(35) VANDERMEULEN, J.H. et PENROSE, W.R. (1978). Absence of arylhydrocarbon hydroxylase (AHH) in three marine bivalves. J. Fish. Res. Bd. Can., 35 (5) : 643-47.

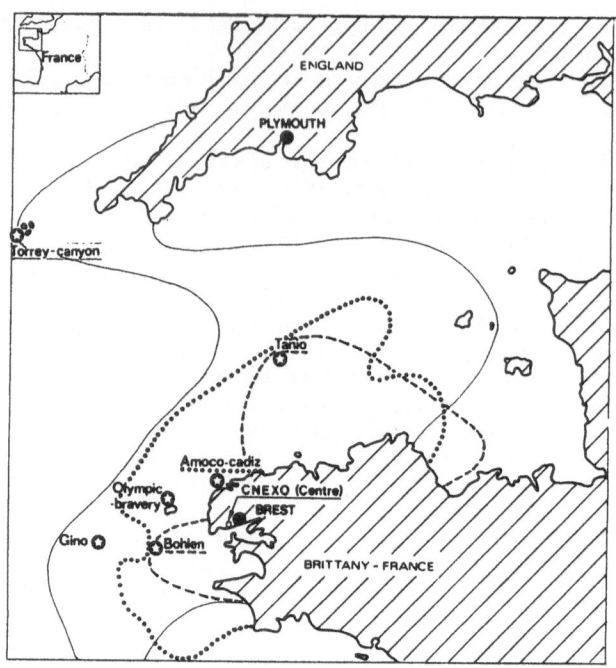

Figure 1 - Localisation des principaux accidents pétroliers survenus
à proximité des côtes bretonnes.

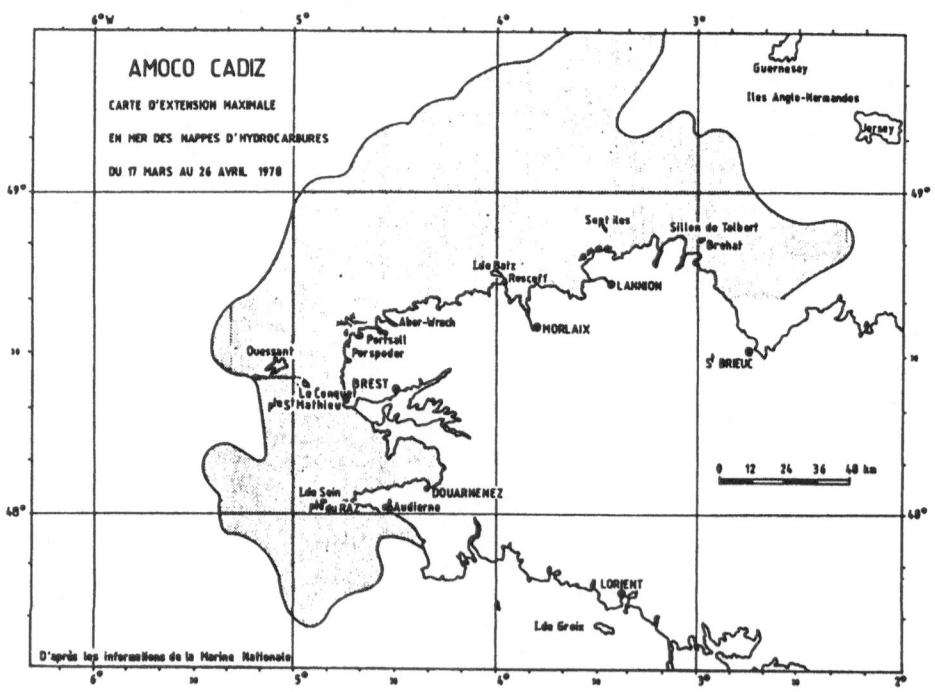

Figure 2 - Carte d'extension maximale en mer des nappes d'hydrocarbures
de l'AMOCO CADIZ, du 17 mars au 26 avril 1978.

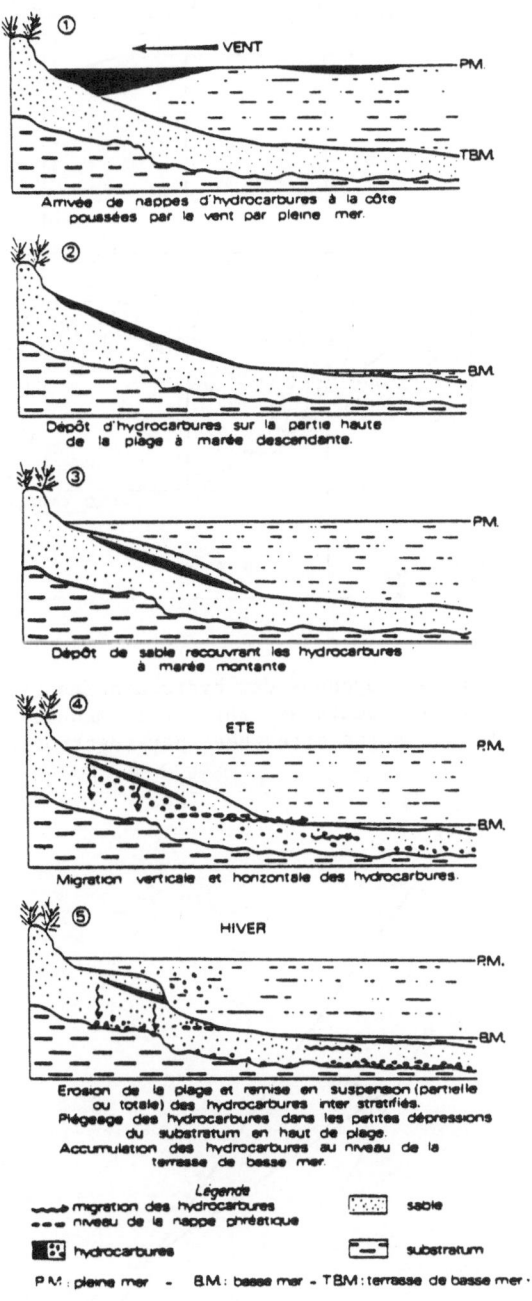

Figure 3 – Dépot et migration des hydrocarbures sur une plage.

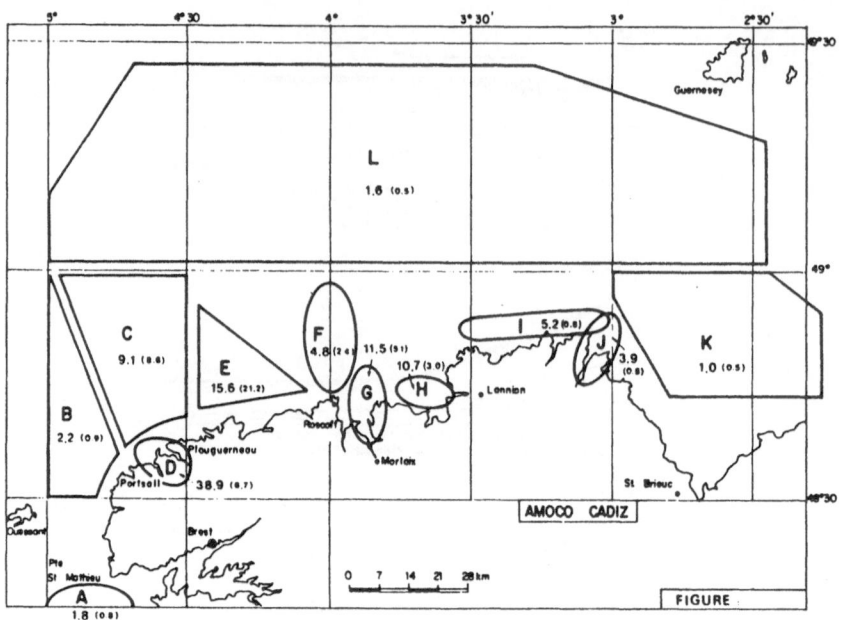

Figure 4 - Concentrations moyennes des hydrocarbures dans l'eau de mer
(ug/l) en différents secteurs de la Manche occidentale, en
avril 1978 (mesures effectuées par spectrofluorimétrie UV).

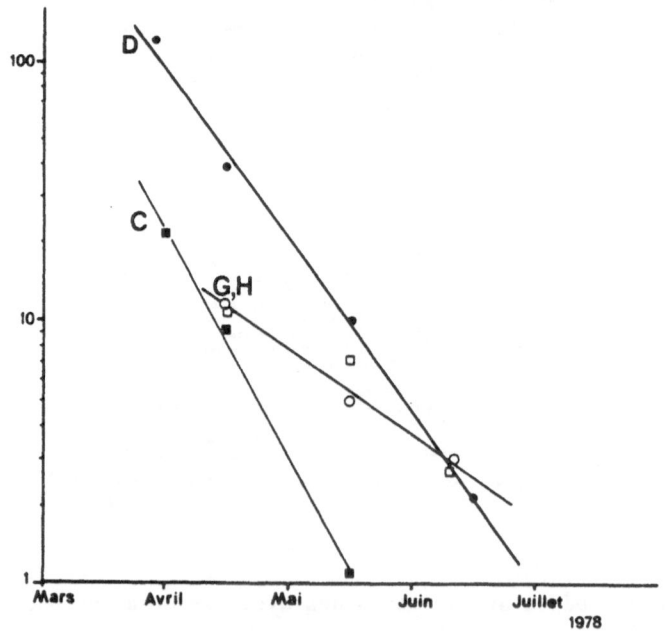

Figure 5 - Evolution de la concentration en hydrocarbures dans l'eau de
mer, de mars à juin 1978, en différents secteurs de la Manche
occidentale (cf. fig. 4).

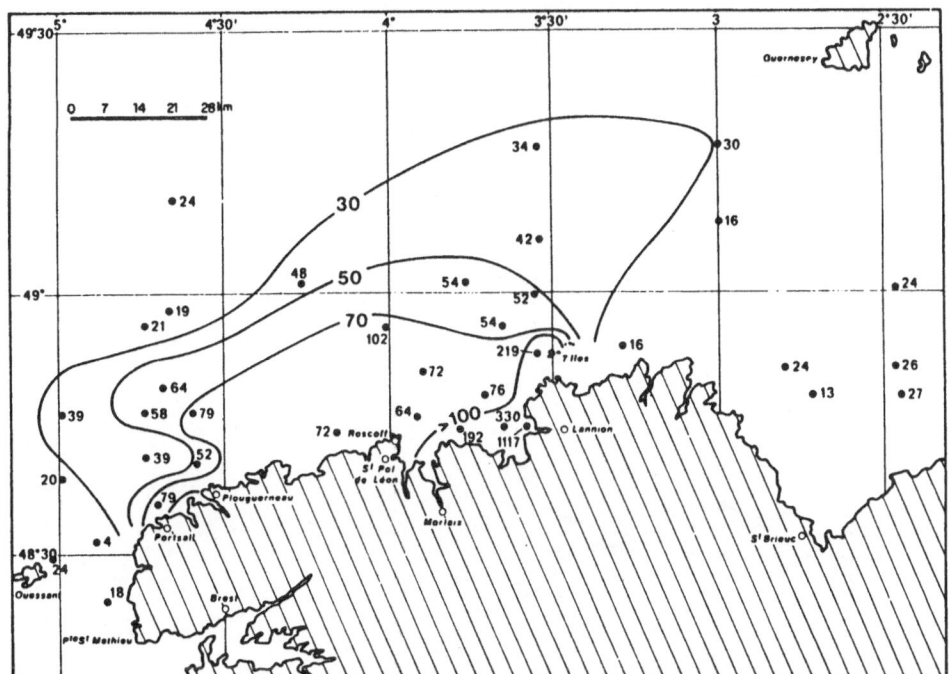

Figure 6 – Contamination des fonds marins par les hydrocarbures, en
avril 1978 (concentrations exprimées en ,ug/g).

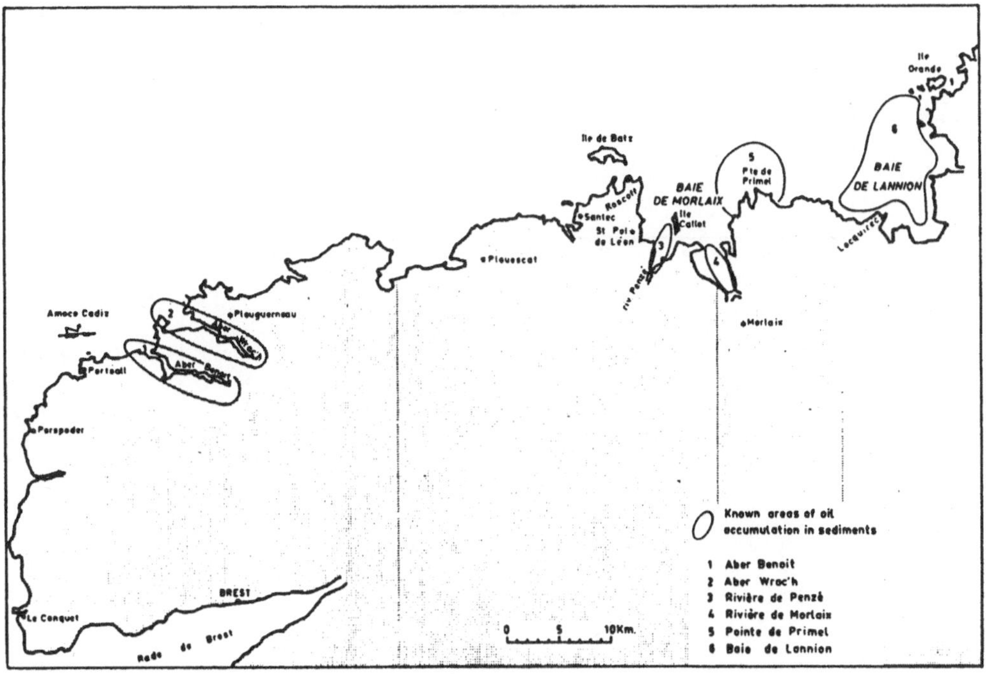

Figure 7 – Zones d'accumulation des hydrocarbures dans les sédiments.

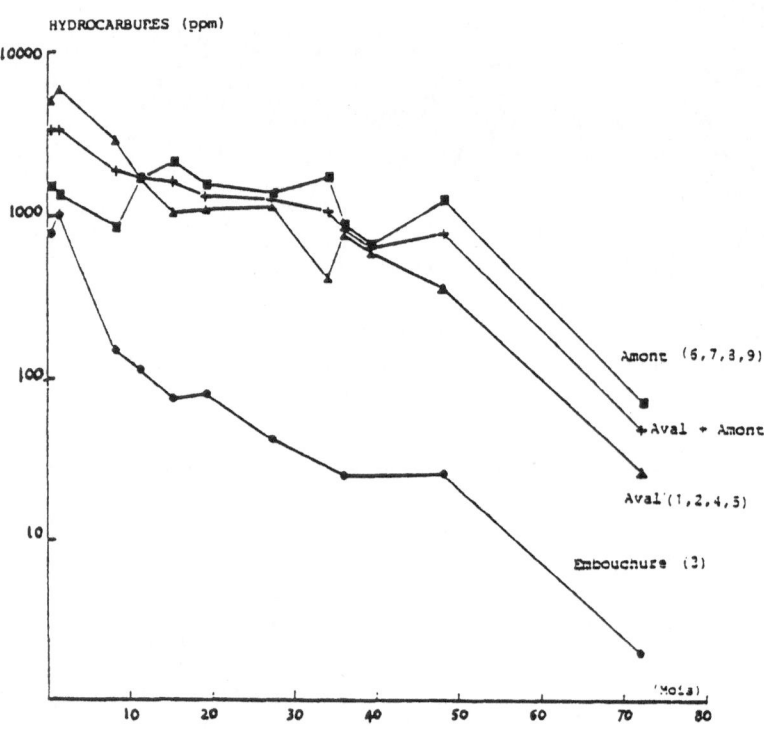

Figure 8 – Evolution des teneurs en produits pétroliers dans les sédiments
de l'ABER Wrac'h, de 1978 à 1984.

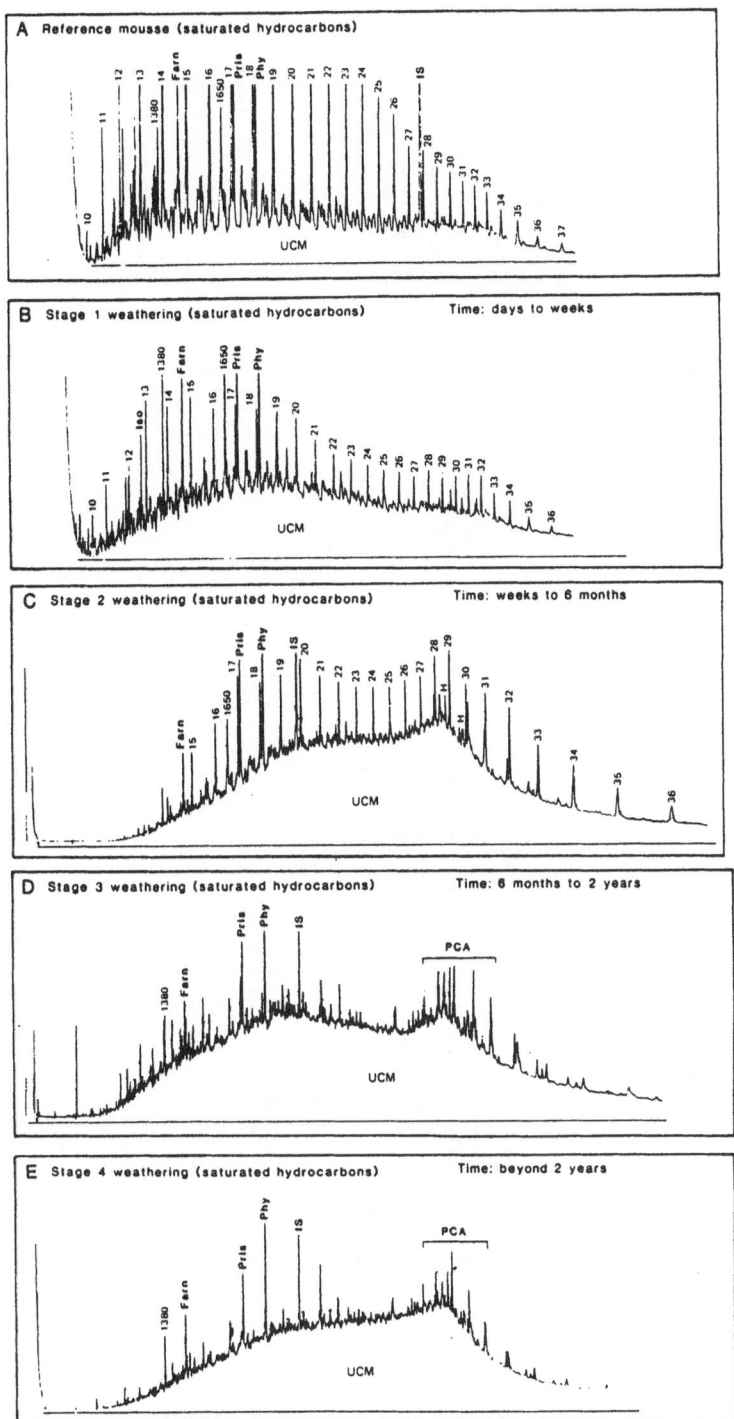

Figure 9 – Evolution chimique des hydrocarbures saturés du pétrole de l'AMOCO CADIZ.

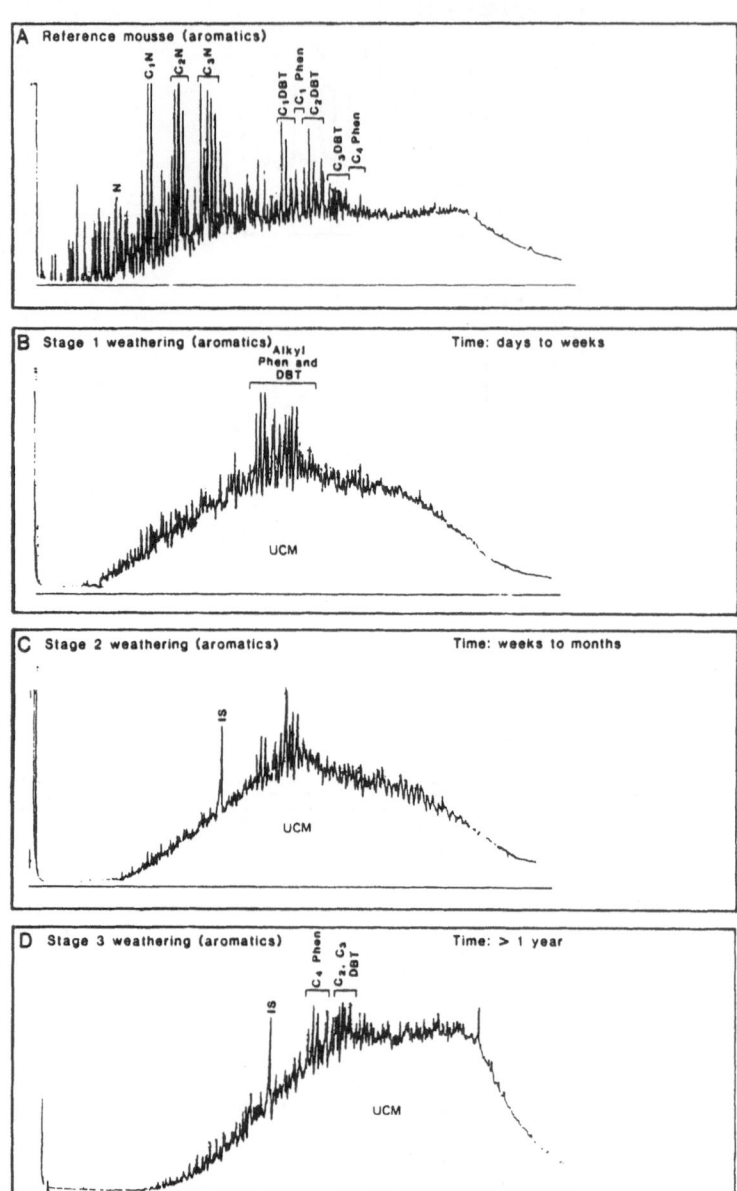

Figure 10- Evolution chimique des hydrocarbures aromatiques du pétrole de l'AMOCO CADIZ.

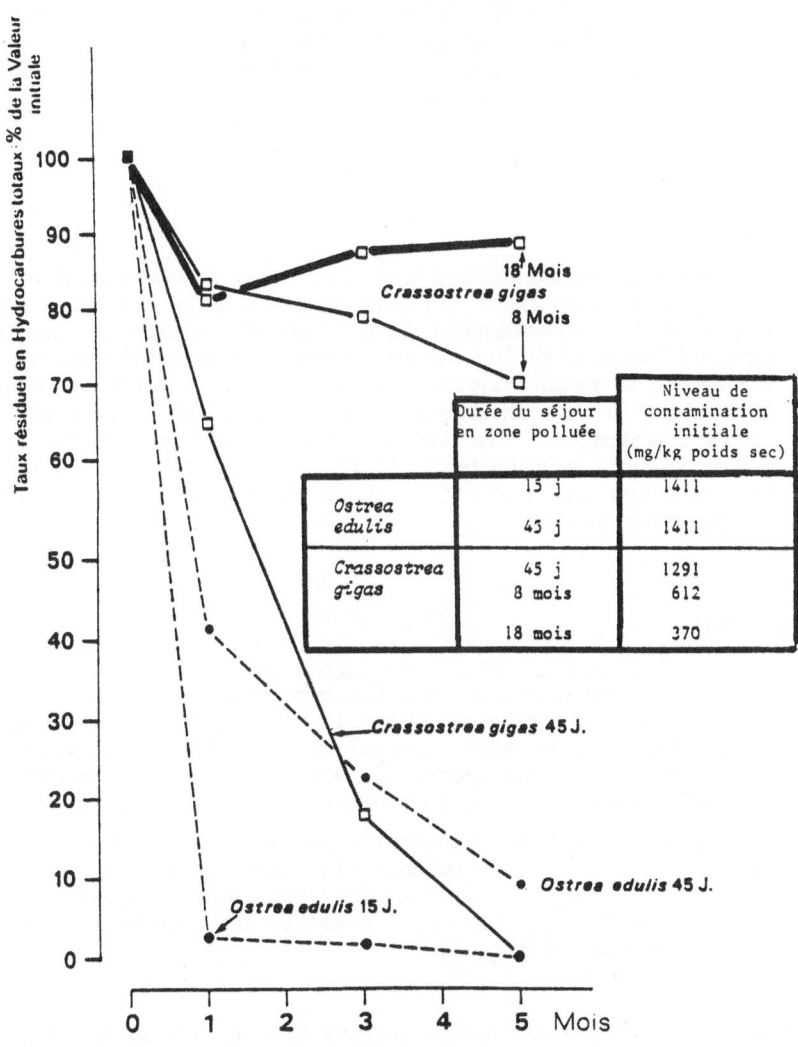

Figure 11- Taux d'épuration des hydrocarbures accumulés dans les huitres, en fonction du temps de séjour sur les sites pollués,

BIOACCUMULATION OF CHLORINATED PARAFFINS - A REVIEW

G. SUNDSTRÖM and L. RENBERG
Special Analytical Laboratory
National Environmental Protection Board
Box 1302, S-171 25 SOLNA
Sweden

Summary

Laboratory bioconcentration experiments with chlorinated paraffins (CPs) in aquatic biota are reviewed. The collected data indicates that most CPs investigated have a bioaccumulation potential, but actual bioconcentration factors have been determined experimentally for only few products. Most CP types are also excreted rapidly when CP exposure is terminated, except high chlorinated/short carbon chain products which are retained for long periods in fish and mussels. The knowledge about CP levels in the environment is still too limited to allow any conclusions about the environmental significance of the experimental data.

1. INTRODUCTION

The chlorinated paraffins (CPs) have during the last decade attracted a great interest as to their effects and behaviour in the environment. This is due to their relative biological stability (although some technical uses actually are based on their relative *chemical instability*), their wide use on a large scale and their physical/chemical similarity to other organic chlorinated compounds, known to cause environmental and health effects.

CPs are manufactured by the chlorination of straight-chain paraffin oils and waxes in the C_{10-30} range to chlorination degrees of 40 to 70 %. Their uses span over a wide range and include fire retardants and plasticizers for PVC, rubber, other polymers, varnishes, sealants and adhesives, and additives in lubricants and cutting oils.

TABLE 1

Western World consumption and use pattern of CPs in 1977 (after [6]).

n-paraffin stock	chlorination %	consumption kton/year	main use
C_{10-13}	50 - 70	60	extreme pressure lubricants fire retardant additives sealants etc.
C_{14-17}	45 - 60	110	secondary PVC-plasticizers
C_{20-30}	40 - 70	60	paints extreme pressure lubricants fire retardant additives

The consumption of CPs in the Western World was estimated to 230 ktons per year in 1977, with a yearly growth rate of 5 % [6]. The main uses of CPs distributed over different types of CP products are given in Table 1. A number of reviews covering the environmental aspects, as well as other relevant background material have been published during recent years [5,6, 19,29,33].

When it comes to studies of the possible presence and distribution of CPs in the environment, including many types of laboratory experiments (such as bioaccumulation experiments, the main subject of this paper) they have so far been hampered by the lack of sensitive and specific analytical methods. Although a number of methods have been recorded in the literature many of them are indirect and/or laborious, and only few have been used on a large scale.

The present review is intended to comment on the present knowledge about the bioaccumulating properties of CPs, a key parameter when trying to predict the environmental behaviour of organic chemicals. However, in order to correctly interpret bioaccumulation data knowledge about CP levels and distribution in the environment as well as the analytical methods must also be aquired. Such background data will therefore also be shortly treated.

2. ANALYTICAL METHODS

In Table 2 are summarized the methods so far used for the analysis of CPs, from pure technical CP products, to CPs in biological material from laboratory experiments and the environment. Omitted are analyses of CPs labelled with radioactive isotopes.

The earliest described analytical methods for CPs were based on relatively insensitive and non-specific methods. They involved e.g. dechlorination of technical CP products or CPs extracted from experimental animals and characterization of the resulting n-paraffins, or determination of organic chlorine in extracted material.

The first more general method used on a large scale is the one described by Hollies and coworkers [14]. Their analytical method, based on densitometric or visual evaluations of thin layer chromatograms, have been used for determination of CPs of C_{13-30} type with a chlorine content of $42-45$ percent. Thus, also this analytical method seems to be restricted in use, i.e. to specific types of CPs. Still, by using this technique the environmental presence of CPs have been proven in a large number of samples in the United Kingdom [6].

Gas chromatography (GC), the most common analytical method today for organic environmental analysis, have not been used for CP analysis on a broad scale until very recently. This is due to the fact that many types of CPs decompose in the gas chromatograph and only the modern, strongly deactivated and inert gas chromatographic columns seem to be suitable for this purpose. Schmid and Müller [20,27] recently described CP analyses using GC in combination with mass spectrometry (MS) with negative ion chemical ionization (NCI) detection. The MS-NCI detection method has been described earlier by Gjøs and Gustavsen [12], but not in combination with GC. In our laboratory we have used both the direct MS-NCI method, the combined GC-MS-NCI method and also GC with electron capture detection to determine CPs in many types of samples [15], Figure 1.

Finally, in cases where CPs labelled with radioactive isotopes have been used, which is mainly in biological experiments, the analytical method is

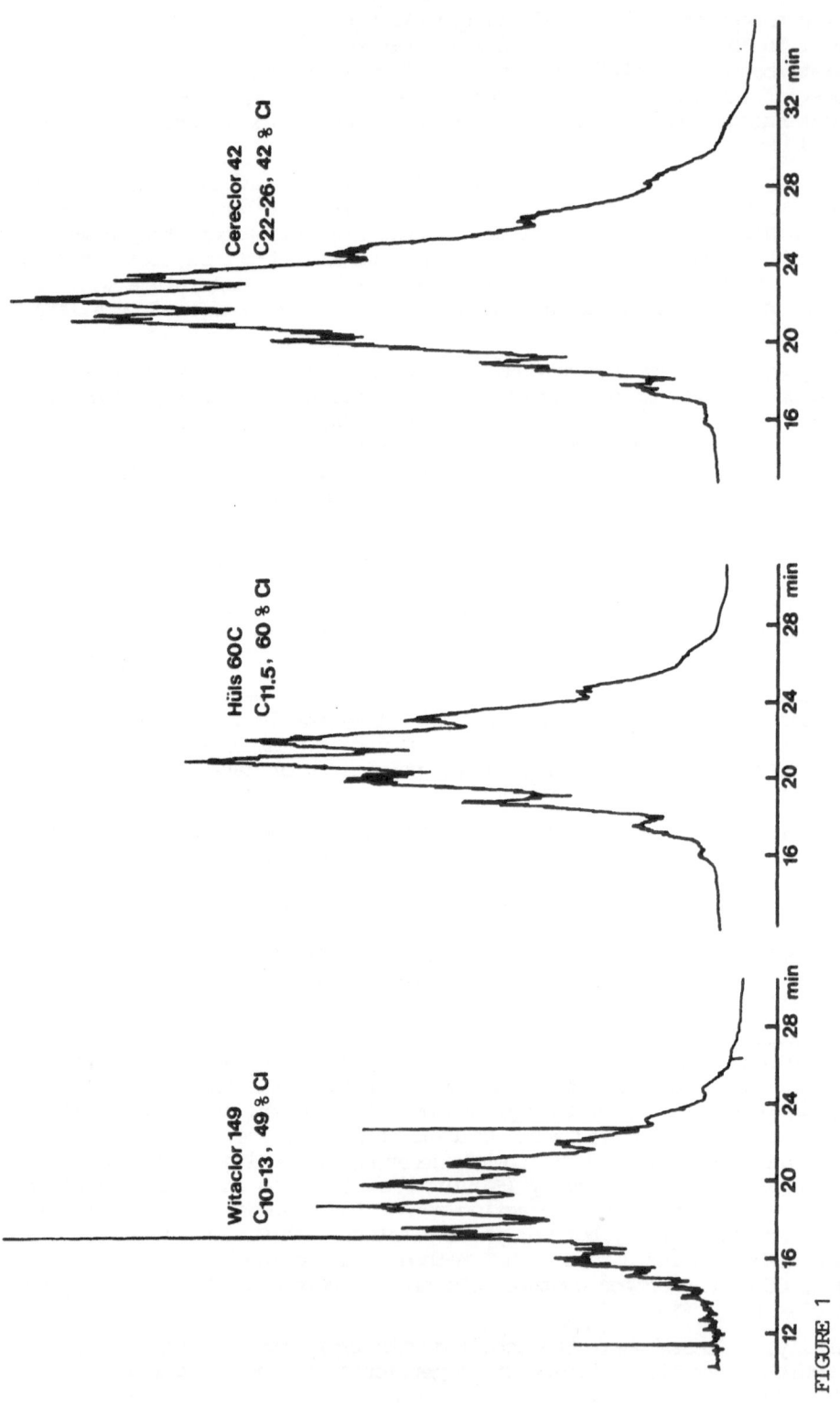

FIGURE 1

Gas chromatograms of three CP products. The column used was a 24 m BP-1 (SGE, Australia) programmed from 80 °C (1 min) to 280 °C at 10 °C/min in a Varian 3700 instrument equipped with a 63Ni electron capture detector (300 °C)

TABLE 2

Summary of analytical methods for chlorinated paraffins

Analytical method	Aim of study	Refs
Neutron activation analysis of extractable organic chlorine	Uptake studies of CPs in Bleaks (*Alburnus alburnus*) via water and feed	2,3,18,28
GC-MS characterization	Identification and characterization of pure CPs and dechlorination products	16
GC-microcoulometry	Uptake studies of CPs in Rainbow Trout (*Salmo gairdneri*) via feed	10,17
Direct microcoulometry	Analysis of spiked samples of seal, eggyolk and fish food	30
	Analysis of antifouling paint, eggyolk and pure CPs	31
	Uptake studies of CPs in Atlantic Salmon (*Salmo salar*) from suspended solids	32
	Biodegradation studies of CPs in sediments. Analyses combined with TLC and IR	34
MS, chemical ionization negative ion detection	Analysis of pure CPs, spiked oils and fish extracts	12,13
GC-MS, chemical ionization negative ion detection	Analysis of pure CPs, human tissue, sewage sludge and sediments	20,27
GC of dehydrochlorinated CPs	Analysis of pure CPs, spiked sediments and polymers	26
Miscellaneous methods	Characterization of CPs and dechlorinated CPs by IR, NMR, TLC and by GC	21
	Characterization of CPs in presence of chloroaromatics after UV-treatment of extracts	10
Thin layer chromatography with densitometric or visual detection	Analysis of environmental material (water, sediments, biota) and experimental material (bioaccumulation experiments)	14

obvious - liquid scintillation counting. However, unless separation methods are used, radioactive measurements must be considered rather non-specific, as metabolites and degradation products will not be distinguished separately (see below).

3. BIOCONCENTRATION STUDIES

In the first report on the uptake of CPs by fish, Atlantic Salmon (*Salmo salar*), the CP levels in the tissues were determined by direct microcoulometry on tissue extracts [32]. The CPs, Cereclor 42 (C_{22-26}, 42 % chlorine) and Chlorez 700 ($C_{\sim 20}$, 70 % chlorine), were dosed adsorbed on suspended solids (silica) and in the feed. Parallel experiments were performed with a PCB mixture, Aroclor 1254 (54 % chlorine). It was concluded that CPs are much less, if at all, accumulated by juvenile Atlantic Salmon as compared with PCB, via the administration routes used.

Lombardo and coworkers [17] made bioaccumulation experiments using Rainbow Trout (*Salmo gairdneri*) which were fed Chlorowax 500C ($C_{\sim 12}$, 60 % chlorine) in the diet (10 ppm) for up to 82 days. Their analytical technique was based on common extraction and partition procedures followed by GC-microcoulometric quantification. CP levels of up to 1.1 ppm whole body wet weight were determined in the fish, corresponding to 18 ppm on a fat weight basis.

Due to some failiure in the experimental setup the experiments had to be interrupted and it could not be established whether a steady state had been reached, nor could the excretion of CPs be studied. A slightly changed chromatographic pattern of the extracted CPs, as compared to the standard Chlorowax 500C, was observed. This indicated either a lesser uptake of early eluting components, or differential metabolism/elimination. No adverse physiological or behavioural effects could be positively proven among the fish.

Since the middle of the 1970's a number of investigations on the environmental behaviour and biological effects of CPs have been undertaken in Sweden (see [3, 9, 29] and references therein). The first publication on fish experiments in 1978 [28], indicated that short chain/high chlorinated CPs (Chlorparaffin Hüls 70C, $C_{\sim 11.5}$, 70 % chlorine) was taken up from water by Bleaks (*Alburnus alburnus*) to a considerable degree and also caused neurotoxic effects during the exposure. The uptake of CP in the fish was recorded by measurement of extractable organic chlorine by neutron activation analysis. Uptake from 0.1 and 1.0 mg/L water solutions did not differ significantly but compared to another notorious bioaccumulating compound mixture, PCB, the uptake was judged to be slow. It could not be established whether a steady state had been reached in the uptake of CPs during the 29 days of the experiment. At that time levels of chlorine were estimated to about 35 ug/g whole body wet weight.

These experiments were followed by comparative experiments using different types of CP products, dosed via the water and the feed [2, 3]. When given via the water (static experiments, 14 days accumulation period) [2], it was shown that there is a relation between composition of CP product and uptake. It was concluded that short carbon chains and a low degree of chlorination give the most effective uptake. The highest levels of chlorine, determined by neutron activation analysis, in the fish were recorded with Witaclor 149 (C_{10-13}, 49 % chlorine), Witaclor 159 (C_{10-13}, 59 % chlorine) and Witaclor 171P (C_{10-13}, 71 % chlorine). The latter CP is equivalent to the product originally studied, Hüls 70C. The uptake rates were for all CPs slower than for the PCB included in the test, but again it could not be established whether a steady state had been reached.

On the basis of the above experiments long term studies were performed on the uptake and elimination of CPs in the Bleak [3]. The fish were exposed to CPs in the feed as given in Figure 2, which also schematically gives the result of the investigation. Levels of CPs in the tissues were estimated as before by neutron activation analysis. Being an indirect method of analysis one cannot immediately conclude that measured chlorine content is present as parent CPs or incorporated into the extracted fatty material or constitute metabolites. However, it should be pointed out that later analyses of the extracts by the MS-NCI method of Gjøs and Gustavsen [12] have shown that at least part of the residual chlorine emanates from the parent CPs given the fish, even after the extremely long period of 316 days of depuration [13].

Madeley and Birtley [19] studied the uptake by Rainbow Trout (*Salmo gairdneri*) of a ^{14}C-labelled CP, Cereclor 42 (C_{20-30}, 42 % chlorine) from the feed. Two dose levels were used, 47 ppm and 385 ppm CP (dry weight) fed during 35 days, followed by a depuration period of 49 days. Quantities of CP found in the tissues after 35 days were clearly related to the dose levels, and total concentrations of ^{14}C in whole body were estimated to 10.3 and 100.6 ppm dery weight at the respective dose levels. The uptake and depuration of ^{14}C in the tissues of the fish fed 47 ppm CP is illustrated in Figure 3.

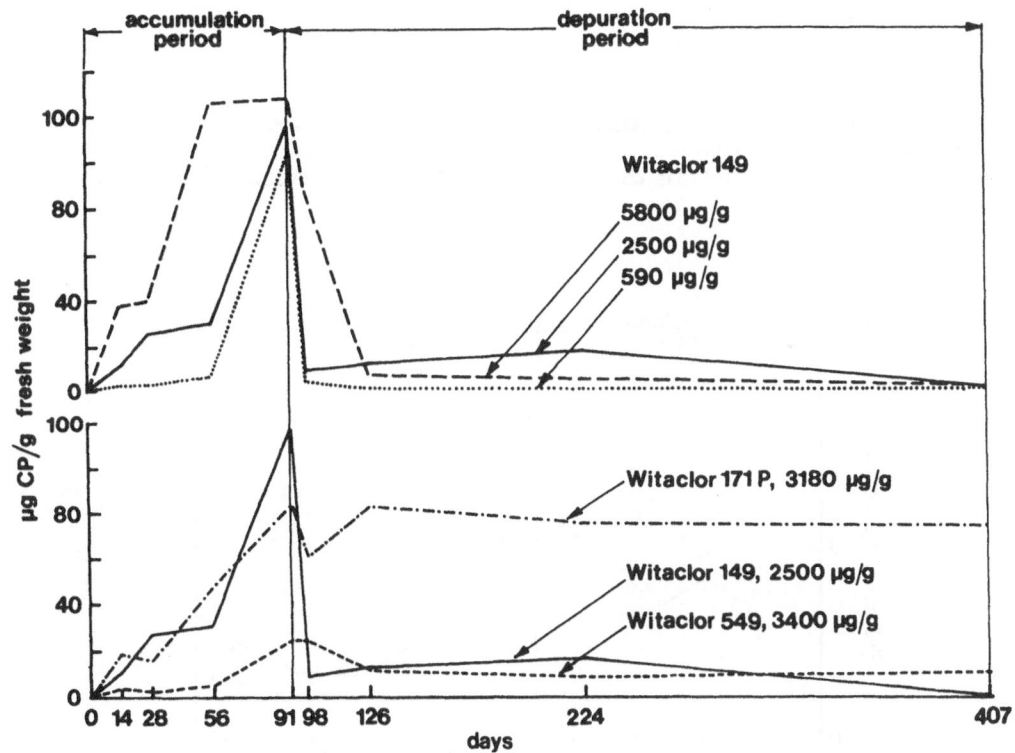

FIGURE 2

Whole body levels of CPs in Bleaks (*Alburnus alburnus*) after exposure to technical products at different levels in the feed. CP levels calculated from content of extractable organic chlorine determined by neutron activation analysis (from [3]).

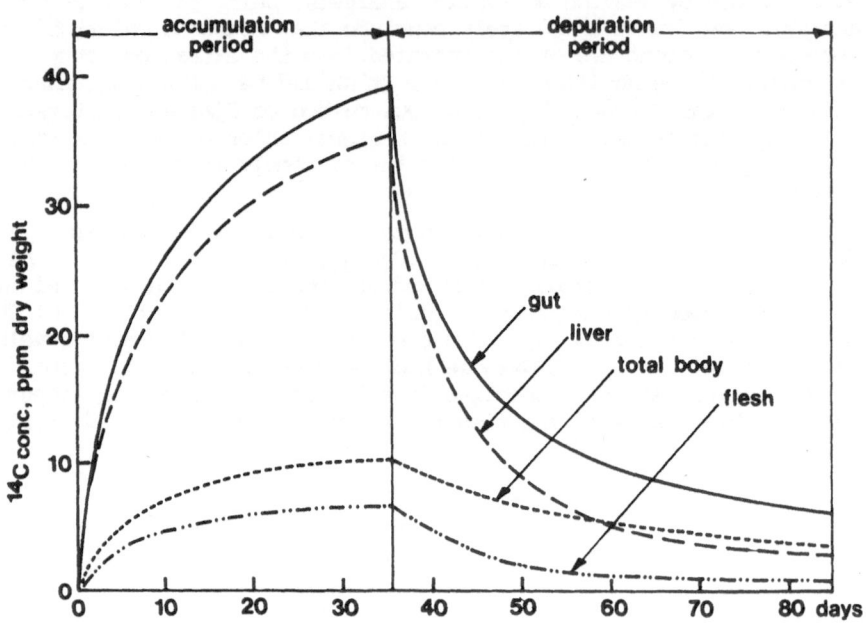

FIGURE 3

Uptake/excretion of ^{14}C by Rainbow Trout (*Salmo gairdneri*) in different tissues and total body given a ^{14}C-labelled CP corresponding to Cereclor 42 (C_{20-30}, 42 % chlorine) in the diet (47 mg/kg). Simplified after [19].

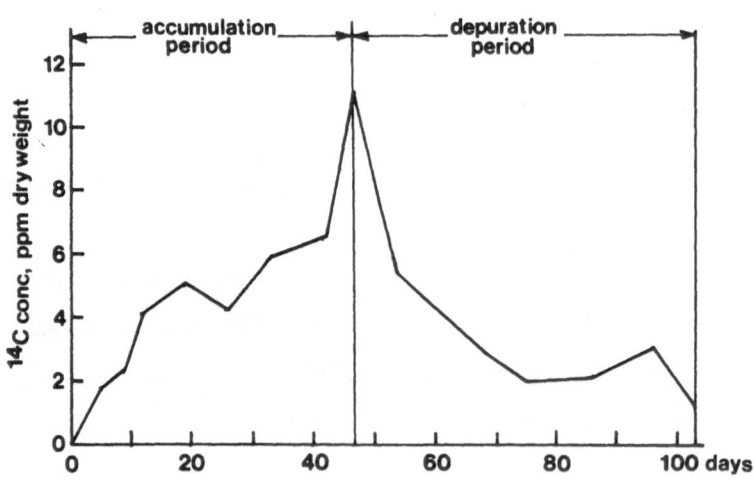

FIGURE 4

Whole body uptake/excretion of ^{14}C by Mussles (*Mytilus edulis*) given baker's yeast containing a ^{14}C-labelled CP (524 ppm dry weight) corresponding to Cereclor 42 (C_{20-30}, 42 % chlorine). Figure drawn after tabular data in [19].

A breakdown/metabolism of CPs in the fish experiments were indicated by Madeley and Birtley by parallel analyses of extracts by radioactivity measurements and thin layer chromatographic analyses combined with specific extraction and separation schemes. Dehydrochlorination, metabolism to smaller molecules (below C_{10}) and incorporation of fragments into degradative pathways within the animal was suggested as degradation routes for the CP in fish.

The uptake by Common Mussels (*Mytilus edulis*) of ^{14}C-Cereclor 42 dosed as contaminated baker's yeast (524 ppm CP dry weight) was also studied by Madeley and Birtley [19]. The uptake and excretion data obtained for whole mussel body are schematically given in Figure 4. By far the highest ^{14}C levels in individual tissues were recorded in the digestive glands. In this case analyses showed that metabolism was probably not a major route for CP removal from the mussel tissues.

We have in our laboratories had access to several radiolabelled CP-preparations [4], the uptake/excretion of which have been studied in a continuous-flow exposure system using Common Mussels (*Mytilus edulis*) [22]. Three experiments were conducted with dose levels as given in Figure 5 [23]. The uptake phases were of 28 days duration and obviously this was not enough to reach steady state in one low dose experiment. On the basis of radioactivity measurements the levels of CPs in the tissues were calculated and bioaccumulation factors (BCF) determined. The following BCFs were obtained (wet weight basis with confidence limits):

Chloroparaffin	*Bioconcentration factor*
C_{16}, 34 % chlorine	7 000 (4 600 - 9 500)
C_{12}, 69 % chlorine	140 000 (123 000 - 155 000)

The depuration period was studied only for one CP preparation, representing the same type of CP previously found to be retained in the experiments with the Bleak as discussed above. The excretion was found to be very slow also from the mussel.

In order to study the possible incorporation of radiolabel in the tissues, which might be the reason for the slow excretion, the activity was also recorded in the extraction residues, Figure 6. The level of ^{14}C in those residues were only some percent of the total activity recovered and showed an uptake/excretion curve similar to that of the activity in the lipid extract.

The octanol/water partition coefficients (P, given as log P below) of the CP preparations used in our experiments have been estimated by a thin layer chromatographic method [24] to 6.4 - 10.3 (C_{16}, 34 % chlorine) and 6.3 - 9.3 (C_{12}, 69 % chlorine) [1]. These values are in close agreement with the estimated values for similar technical products [25] and suggest that both CPs should have reasonably similar bioaccumulation tendencies in aquatic fauna. According to the relation between bioconcentration factors in *Mytilus edulis* and log P, as determined by Geyer and coworkers [11], the bioconcentration factors (log BCF) of both CPs should be in the range 4.6 - 7.7. The values experimentally found by us, 3.85 (3.66 - 3.98) and 5.15 (5.09 - 5.19), show that only the short chain/high chlorinated CP falls within the calculated range.

A few years ago an international group of chlorinated paraffin manufacturers formed a consortium in order to thoroughly investgate the health and environmental effects of CPs. The test programme involved two tests of the bioconcentration of CPs in aquatic animals - Rainbow Trout (*Salmo gairdneri*)

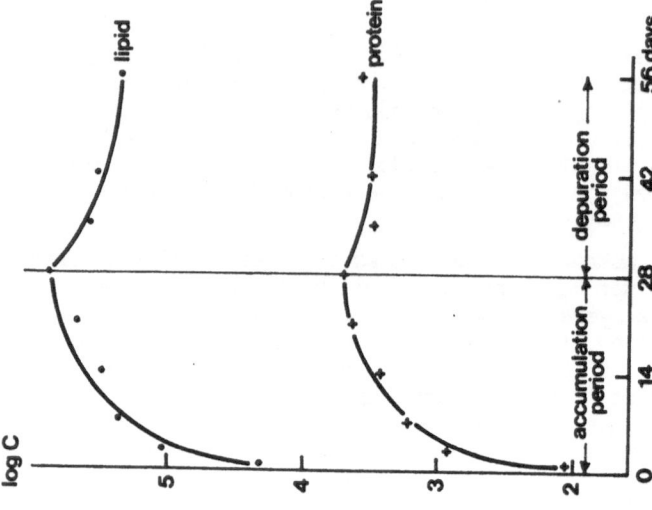

FIGURE 6

Concentration (C ug/kg) of a radiolabelled CP (C_{12}, 69 % chlorine) in the extractable fat and non-extractable protein residues during an uptake/excretion experiment with Mussels (*Mytilus edulis*). Exposure level was 0.13 ug/ /L water, continuous flow. From [23].

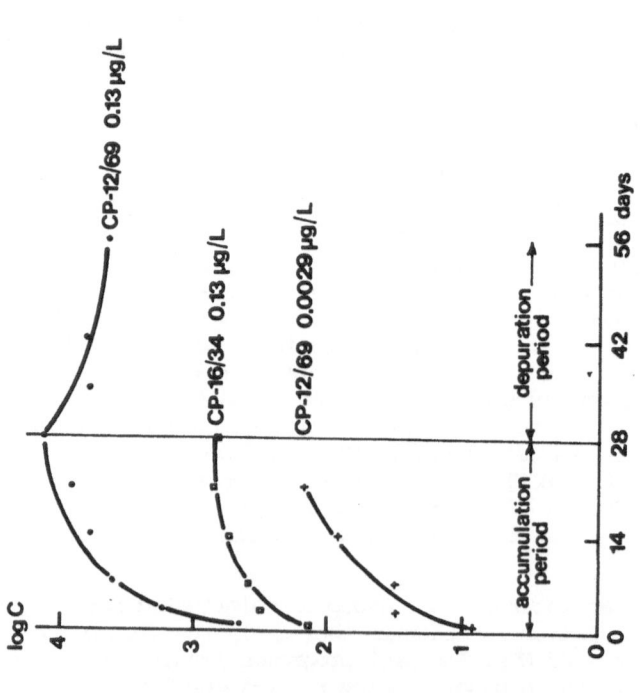

FIGURE 5

Concentration (C ug/kg wet weight) of radiolabelled CPs in Mussels (*Mytilus edulis*) during uptake/ /excretion experiments. Exposure via the water at continuous-flow conditions and levels as indicated. CP-12/69 = C_{12}, 69 % chlorine, CP-16/34 = C_{16}, 34 % chlorine. From [23].

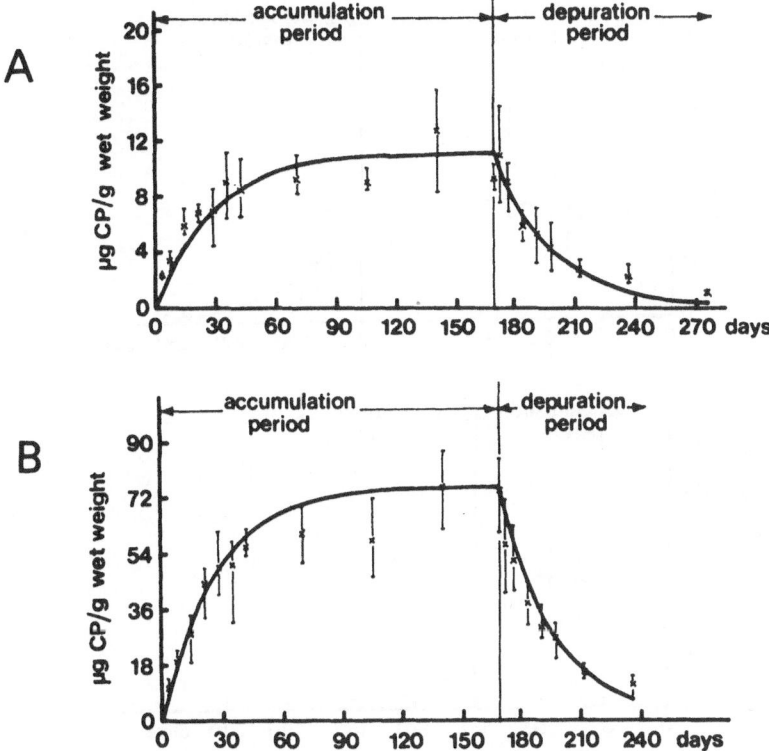

FIGURE 7

Concentration of a chlorinated paraffin (C_{12}, 58 % chlorine) in whole fish
(calculated from ^{14}C-levels) during exposure and depuration experiments with
Rainbow Trout (*Salmo gairdneri*). Nominal exposure levels of 3 ug/L (A) and
15 ug/L (B) chloroparaffin in the water. From [7].

and Common Mussel (*Mytilus edulis*) - the results of which were recently released [7,8].

The test substance was in both cases a centrally ^{14}C labelled undecane,
with a chlorination degree of 58 %. This test chemical resembles technical
CP mixtures of short chain length and high, but not highest possible (70 %),
degree of chlorination. The results of the experiments, performed at two
exposure levels for each species, are given in Figures 7 and 8.

In the experiments with the Rainbow Trout [7] the bioconcentration factor of the ^{14}C-CP was estimated to 3 600 - 5 300 (range in whole fish) in
the two experiments. Highest levels in specific tissues were recorded in the
viscera with a BCF of 11 700 - 15 500. The elimination half-life was estimated to 18 - 20 days (range in whole fish). By analytical separation schemes as discussed above [19] some evidence was obtained for metabolic breakdown of the CP in the liver and viscera. During the depuration period a significant mortality was observed among the fish after both exposure levels.

In the bioconcentration experiments with Mussels [8] a BCF (whole body)
of 40 900 was recorded at an exposure level of 3 ug CP/L and 24 800 at 15
ug CP/L. As shown before with another CP (see above [19]) the highest

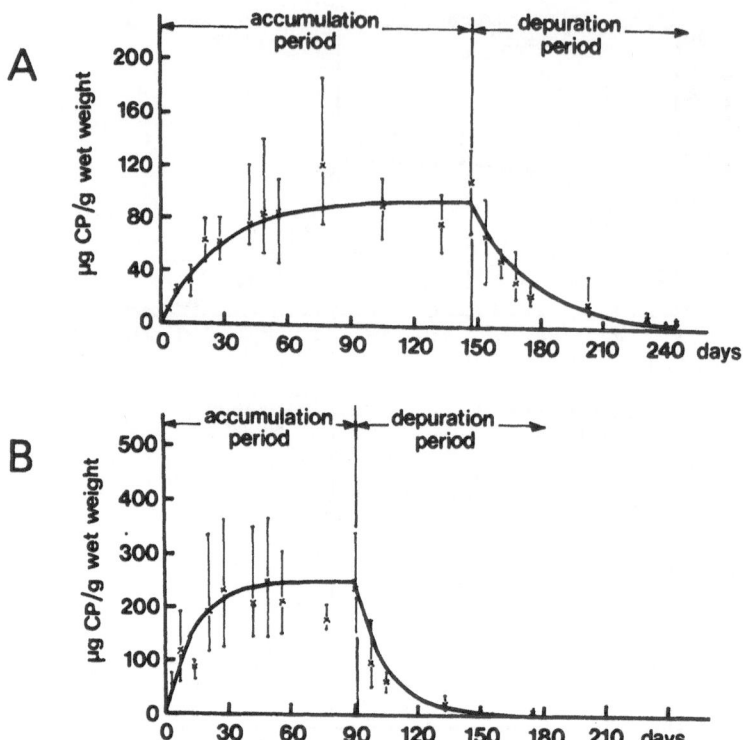

FIGURE 8

Concentration of a chlorinated paraffin (C_{12}, 58 % chlorine) in whole mussel (calculated from ^{14}C-levels) during exposure and depuration experiments with Common Mussel (*Mytilus edulis*). Nominal exposure levels of 3 ug/L (A) and 15 ug/L (B) chlorparaffin in the water. From [8].

levels in individual tissues were found in the digestive glands. The elimination half-life in Mussels was estimated to about 10 and 13 - 20 days at the high and low exposure levels, respectively. A significant mortality among the test animals was recorded both during the exposure and elimination period in the high CP level experiment.

4. LEVELS IN THE ENVIRONMENT

There are still too few figures on environmental CP levels, especially in biota, to allow any relevant comparisons with the laboratory data on bioconcentration factors. The data published by Campbell and McConnell [6] indicate low levels of CPs in areas remote from industry, but in industrial areas levels of up to 6 ug/L of CPs were determined in fresh waters, and up to 15 mg/kg in bottom sediments, analyses made by the TLC method [14]. Generally, they indicated a concentration factor of about 1 000 between CP levels in the water phase and in the sediments.

Seabird eggs contained C_{10-20} CPs at levels up to 2 mg/kg and C_{20-30} CPs up to 0.1 mg/kg, but in most samples CPs of the latter type could not be detected at all (detection limit 0.05 mg/kg). Among aquatic animals CPs were most frequently detected in Mussels (*Mytilus edulis*). It was concluded in this investigation that there is a general predominance of C_{10-20} CPs over C_{20-30} CPs in all samples analyzed, and that this might reflect the production rates.

For human related material the same authors reported the following data on CP levels: Dairy products 0.3 mg/kg, vegetable oils 0.15 mg/kg, fruit and vegaetables 0.025 mg/kg. In human post-mortem tissues long chain CPs were found in only a few samples, while C_{10-20} CPs were frequent in liver, fat and kidneys. The highest individual figure was reported for liver, 1.5 mg/kg but mostly samples contained less than 0.09 mg/kg. Schmid and Müller [27] recently reported a CP level of 200 ug/kg in one sample of human adipose tissue from Switzerland.

Interestingly, Campbell and McConnell [6] found CP levels of 6 - 12 mg//kg wet weight in Mussels outside a CP plant. This level is of the same magnitude as those we have recorded in our bioaccumulating experiments in the laboratory [23] with exposure levels below 1 ug/L. At exposure levels of 15 ug/L the CP levels in whole Mussels (wet weight) were as high as 200 - 300 mg/kg when steady state was reached, as recorded in the investigations by the Chloroparaffin Consortium [8].

5. CONCLUSIONS

Although not all types of CP products have been properly tested it is obvious so far that some CPs do have a bioconcentration potential in aquatic animals. This is especially true for the short carbon chain products of all degrees of chlorination. Increasing carbon chain length seem to decrease this potential irrespective of chlorination. Taking into account the increasing molecular weight this behaviour should, however, be expected at least for the long chain/highly chlorinated CPs whose molecular weight (on the average) is around 1 000.

The analytical data also indicate that at least the fish are able to metabolize the CPs taken up. This has been investigated independently by measurement of exhaled $^{14}CO_2$ from Carp given labelled CPs (see discussion and references in [9]).

The experimental data show, in addition, that the CPs taken up are readily excreted during the depuration period, with the exception of the short chain/high chlorinated products. Such CPs are retained for extensive periods of time both in the mussels and the fish tested. It is remarkable that this behaviour is drastically changed by decreasing the chlorine content as little as 10 %. I.e. the C_{12}, 58 % chlorinated CP is rapidly excreted, while the C_{12}, 69 % chlorinated CP is retained in the Mussel. A parallel behaviour has been observed in the Rainbow Trout and the Bleak with the exception that the routes of administration were different (via water and feed respectively). At present no explanation can be found for this behaviour.

In conclusion it must be stated, as was already done by Bengtsson and Baumann Ofstad [3], that certain chlorinated paraffins seem to be more persistent and bioaccumulating than others. Therefore CPs should not be regarded as a homogenous group of substances when discussing their potential environmental and health effects. As shown in this review this is especially true when considering the bioconcentration potential of CPs.

In Tables 3 and 4 the most reliable data on bioconcentration and bioaccumulation of CPs in aquatic animals are summarized.

TABLE 3

Bioaccumulation of CPs in fish

CP type and level in diet	Species	BAF* (total body)	Ref
C23-30, 42 % chlorine			
47 ppm	Rainbow Trout	0.2	[19]
385 ppm	"	0.3	[19]
C~12, 60 % chlorine			
10 ppm	"	0.1	[17]
C10-13, 49 % chlorine			
590 ppm	Bleak	0.18	[3]
2 500 ppm	"	0.05	[3]
5 800 ppm	"	0.02	[3]
C10-13, 71 % chlorine			
3 180 ppm	"	0.03	[3]
C18-26, 49 % chlorine			
3 400 ppm	"	0.007	[3]

* Bioaccumulation factor = concentration of CP in organism/concentration of CP in diet at steady state

BAF of DDT in fish have been determined to 0.3 - 1.2; BAF of PCB in rat to 1.3 - 2

TABLE 4

Bioconcentration of CPs in aquatic animals

CP type and level in water	Species	BCF* (fresh weight)	Ref
C12, 58 % chlorine			
3 ug/L	Rainbow Trout	3 600	[7]
15 ug/L	"	5 300	[7]
3 ug/L	Mussel	40 900	[8]
15 ug/L	"	24 800	[8]
C16, 34 % chlorine			
0.13 ug/L	"	7 000	[23]
C12, 69 % chlorine			
0.13 ug/L	"	139 000	[23]

* Bioconcentration factor = concentration of CP in organism (whole body fresh weight)/concentration of CP in water at steady state

BCF of PCB determined to ~43 000 - 200 000; BCF of DDT in fish to ~60 000 - 85 000

6. ACKNOWLEDGEMENT

We are indebted to the Chlorinated Paraffin Consortium, through J R Madeley, Imperial Chemical Industries, UK, for being able to cite their bioaccumulation studies before publication.

7. REFERENCES

1. ÅHLMAN, M and BERGMAN, Å. Wallenberg Laboratory, University of Stockholm, S-106 91 Stockholm, Sweden. Personal communication.
2. BENGTSSON, B-E, SVANBERG, O, LINDÉN, E, LUNDE, G, BAUMANN OFSTAD, E. Ambio 8 (1979) 121.
3. BENGTSSON, B-E and BAUMANN OFSTAD, E. Ambio 11 (1982) 38.
4. BERGMAN, Å, LEONARDSSON, I, WACHTMEISTER, C A. Chemosphere 10 (1981) 857.
5. BIRTLEY, R D N, CONNING, D M, DANIEL, J W, FERGUSON, D M, LONGSTAFF, E, SWAN, A A B. Toxicol. Appl. Pharmacol 54 (1980) 514.
6. CAMPBELL, I and McCONNELL, G. Environ. Sci. Technol. 14 (1980) 1209.
7. CHLORINATED PARAFFIN CONSORTIUM. The bioconcentration of a chlorinated paraffin in the tissues and organs of Rainbow Trout (Salmo gairdneri). Report BL/B/2310, October 6, 1983.
8. CHLORINATED PARAFFIN CONSORTIUM. The bioconcentration of a chlorinated paraffin by the Common Mussel (Mytilus edulis). Report BL/B/2351, September 30, 1983.
9. DARNERUD, P O. Comparative disposition and metabolism of some ^{14}C-labelled chlorinated paraffins and polychlorinated biphenyls. Department of Pharmacology and Toxicology, Swedish University of Agricultural Sciences, Box 573, S-751 23 Uppsala, Sweden. Thesis.
10. FRIEDMAN, D and LOMBARDO, P. J. Assoc. Offic. Anal. Chemists 58 (1975) 703.
11. GEYER, H, SHEEHAN, P, KOTZIAS, D, FREITAG, D, KORTE, F. Chemosphere 11 (1982) 1121.
12. GJØS, N and GUSTAVSEN, K O. Anal. Chem. 54 (1982) 1316.
13. GJØS, N. Central Institute for Industrial Research, P.B. 350 Blindern, Oslo 3, Norway. Report nr. 81 06 12 - 2 (Sept 30, 1983).
14. HOLLIES, J I, PINNINGTON, D F, HANDLEY, A J, BALDWIN, M K, BENNETT, D. Anal. Chim. Acta 111 (1979) 201.
15. JANSSON, B. Special Analytical Laboratory, National Environmental Protection Board, Box 1302, S-171 25 Solna, Sweden. Personal communication.
16. LAHANIATIS, E S, PARLAR, H, KLEIN, W, KORTE, F. Chemosphere 4 (1975) 83.
17. LOMBARDO, P, DENNISON, J L, JOHNSON, W W. J. Assoc. Offic. Anal. Chemists 58 (1975) 707.
18. LUNDE, G and STEINNES, E. Environ. Sci. Technol. 9 (1975) 155.
19. MADELEY, J R and BIRTLEY, R D N. Environ. Sci. Technol. 14 (1980) 1215.
20. MÜLLER, M D and SCHMID, P P. J. High Resol. Chrom. & Chrom. Commun. 7 (1984) 33.
21. PANZEL, H and BALLSCHMITER, K. Z. Anal. Chem. 271 (1974) 182.
22. RENBERG, L, TARKPEA, M, LINDÉN, E. Ecotoxicol. Environ. Safety 9 (1985) 171.
23. RENBERG, L, TARKPEA, M, SUNDSTRÖM, G. Manuscript.
24. RENBERG, L and SUNDSTRÖM, G. Chemosphere 8 (1979) 449.
25. RENBERG, L, SUNDSTRÖM, G, SUNDH-NYGÅRD, K. Chemosphere 9 (1980) 683.
26. ROBERTS, D J, COOKE, M, NICKLESS, G. J. Chromatog. 213 (1981) 73.
27. SCHMID, P P and MÜLLER, M D. J. Assoc. Offic. Anal. Chemists 68 (1985) 427.

28. SVANBERG, O, BENGTSSON, B-E, LINDÉN, E, LUNDE, G, BAUMANN OFSTAD, E. Ambio 7 (1978) 65.
29. SVANBERG, O (editor). Chlorinated paraffins. A review of environmental behaviour and effects. National Environmental Protection Board, Report SNV PM 1614 (1983).
30. ZITKO, V. J. Chromatog. 81 (1973) 152.
31. ZITKO, V. J. Assoc. Offic. Anal. Chemists 57 (1974) 1253.
32. ZITKO, V. Bull. Environ. Contam. Toxicol. 12 (1974) 406.
33. ZITKO, V. Chlorinated paraffins in Hutzinger, O (editor), The Handbook of Environmental Chemistry, Vol 3 Part A: Anthropogenic Compounds. Springer-Verlag (1980) 149.
34. ZITKO, V and ARSENAULT, E. Chlorinated paraffins: Properties, uses and pollution potential. Environment Canada Fisheries and Marine Services Technical Report no 491 (1974).

INVESTIGATION OF THE INFLUENCE OF AQUATIC HUMUS ON THE BIOAVAILABILITY OF CHLORINATED MICROPOLLUTANTS TOWARDS FISH

G.E. Carlberg[*], K. Martinsen[*], A. Kringstad[*], E. Gjessing[**], M. Grande[**], T. Källqvist[**], J.U. Skåre[***]

[*] Center for Industrial Research
Box 350, 0314 Blindern, OSLO 3, NORWAY

[**] Norwegian Institute for Water Research
Box 333 Blindern, 0314 OSLO 3, NORWAY

[***] Norwegian College of Veterinary medicine
Box 8146, Dep. 0033 OSLO 1, NORWAY

Summary

The possible influence of natural humic water on the uptake of 2,4,6-
-trichlorophenol, lindane and 2,4',5-trichlorobiphenyl in Atlantic
salmon underyearlings has been studied using a semistatic test procedure. 2,4,6-trichlorophenol and lindane were added to diluted natural
humic water (springtime) twentyfour hours before the fishes were
introduced. In a second experiment lindane and 2,4',5-trichlorobi-
phenyl were added to diluted natural humic water (autumn sampling) and
the mixture equilibrated 3 months before fish exposure. Control
experiments were carried out with lakewater. The exposure lasted
fourteen days and fish and water were taken for analysis every third
day and everyday respectively. Bioconcentration factors were calculated for each experiment. The springtime humus water reduced the
bioconcentration factors for both the 2,4,6-trichlorophenol and
lindane with about 18 percent compared to the lake water. The autumn
humus water reduced the bioconcentration factor of 2,4',5trichloro-
biphenyl with 30 percent compared to the lake water, while no reduction was observed for lindane in this case.

1. INTRODUCTION

Humic substances represent a major fraction of the organic matter in
natural waters. Aquatic humus consists of very complex organic compounds
of terrestrial origin, probably with no toxic effects.

Recent recognition, however, signify that the humus macromolecules are
able to chelate inorganic and organic compounds, thereby affecting the
state of these chemicals (1,2,3,4).

It has been well established that aquatic humus is associated with
heavy metal ions and organic micropollutants in natural waters thereby
changing the state of the pollutants and affecting the analytical methods
used for their determination (5,6,7).

Recently there have been several reports on the influence of humic
substances on the uptake, accumulation and toxicity of heavy metals to
living organisms. Humus has been found to reduce the toxicity of heavy
metals to algae, salmon and Daphnia magna (8,9,10).

So far little is known about the influence of natural organic substances on the bioavailability of organic micropollutants. The aim of this
study was to investigate the possible influence of natural humic water on
the bioavailability of selected micropollutants in fish. The contact time
between the pollutants and the humic water before the exposure has been
varied. An extended version of this investigation has been published elsewhere (11).

2. EXPERIMENTAL
2.1 Chemicals
Natural humic water collected in April 1982 and October 1983 from a
bog (Hellerudmyra) just outside Oslo and surface water from Lake Maridals-
vannet in Oslo were used. The lake and humic water mixture used in the
experiments had a colour of 45 mg Pt/L.
A stock solution was made in 0.4 N NaOH of 2,4,6-trichlorophenol
(3CPH, 21.2 µg/ml). Three stock solutions were made in acetone, γ-hexa-
chlorocyclohexane (lindane, 2.33 µg/µl and 1.83 µg/µl) and 2,4',5-tri-
chlorobiphenyl (3CB, 1.034 µg/µl).

2.2 Bioconcentration test
The bioconcentration studies were carried out as semistatic tests and
each chemical was investigated in separate experiments.
Fifteen underyearlings (0+) of Atlantic salmon (Salmo salar L.) were
exposed to the test compound for fourteen days in two exposure glass
aquariums (10 l), one containing diluted natural humic water and the other
lake water. The fishes weighed 1-1.5 g and were acclimated to the test
water for several months. They were not fed during the experiments. Dupli-
cate aquariums without fish were used as control (Fig. 1).

Semistatic system

The fishes were transferred every third day
to a new aquarium.

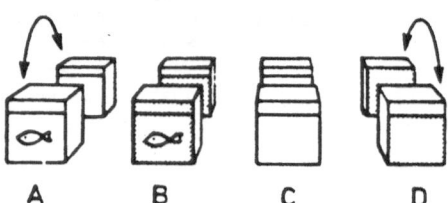

A   B   C   D

A Lake water + fish

B Humic water + fish

C Lake water   ⎫
                 ⎬ controls
D Humic water   ⎭

Fig. 1 Description of the semistatic test system

Each third day the fish were transferred to a new aquarium containing
the initial concentration of the test compound. Samples of three fish and
50 ml of water were taken every third day and everyday respectively. The
water samples were kept in glass flasks with glass stoppers (completely
filled) at +4 °C and the fish samples were wrapped in aluminum foil and
kept at -20 °C until analysis.

Two sets of experiments were carried out.

1. To investigate the bioconcentration of 3CPH and lindane, filtered
    natural humic water sampled in April 1982 was used. The test chemical
    was added to all the aquariums twentyfour hours before the fishes were
    introduced. For 3CPH 10 ml of stock solution was added to each
    aquarium giving a final concentration of 21.2 µg/l. For lindane 20 µl
    of stock solution (2.33 µg/µl) was added to each aquarium giving a
    final concentration of 4.7 µg/l.

2. In the second experiment lindane and the PCB isomer (3CB) were used.
    The humus water sampled in October 1983 was filtered (0.45 µm) and
    allowed to equilibrate for 3 months in 5 x 2, 10 l glass bottles (in

    darkness, 10 ^{0}C) after addition of the test chemical. For lindane
    and 3CB 25 µl (1.83 µg/µl) and 100 µl respectively of stock solution
    were added to separate glass bottles giving a final concentration of
    4.6 µg/l lindane and 10.34 µg/l for the PCB isomer. After 3 months
    storage these mixtures were used in the bioconcentration tests. In
    the aquariums without humic water the chemicals were added twentyfour
    hours before the exposure started.

    When the test compounds were added the waters were strongly mixed.
    After adding the standard the pH was adjusted to 7.5 with 0.1 N NaOH.
    The aquariums were covered with plastic foil, not entirely tightened,
    and kept in natural light without air flushing at 10$\pm$ 1 ^{0}C.

3. ANALYTICAL PROCEDURE (11)
3.1 2,4,6-trichlorophenol - water samples
    The analysis of the chlorinated phenols were based on direct acetyl-
ation of the compound in the water phase with acetic anhydride. Dibromo-
phenol was used as internal standard. The derivatized chlorinated phenols
were analysed using a gas chromatograph (GC) equipped with an electron
capture detector (ECD).

3.2 2,4,6-trichlorophenol - fish
    The fishes were extracted with cyclohexane/methanol (1:1). The
cyclohexane phase was separated and the phenol isolated by use of an ion
exchange column (12). The phenol was acetylated before the GC analysis.

3.3 Lindane and PCB - water samples
    The samples were extracted with cyclohexane. The extract was treated
with equal amount of concentrated sulphuric acid and the chlorinated hydrocarbons were analysed by GC.

3.4 Lindane and PCB - fish
    The fishes were extracted with cyclohexane/methanol (1:1). The cyclo-
hexane phase was separated and treated with concentrated sulphuric acid
before the GC analysis.

4. CALCULATION OF BIOCONCENTRATION FACTOR

The bioconcentraction factor is defined as:

$$BCF = \frac{C_{F_{st}}}{C_{W_{st}}}$$

where

$C_{F_{st}}$ = concentration of the test component in fish (µg/g wet weight) at "steady state"

$C_{W_{st}}$ = concentration of the test component in water (µg/g) at "steady state"

"Steady state" for a component is reached when the amount of the component in the fish is at constant level during several days. The water
concentration ($C_{W_{st}}$) is the average concentration of the test component
during the time of "steady state". If there is no "steady state" obtained,
$C_{W_{st}}$ is the average value of all the water analysis during the accumulation time.

5. RESULTS AND DISCUSSION

The results of the bioconcentration of the 2,4,6-trichlorophenol is
presented in Fig. 2 where the wet weight concentration of the test compound
in the fish is plotted as a function of exposure time. The concentrations
of the test compound in the aquariums containing fish are also presented in
the figure. The concentrations of the test compounds were found to be
similar in both the control aquariums not containing fish, and no change in
concentration could be observed throughout the experiment. These results
have therefore not been included in the figure. The same was true for the
other test compounds. The results from the control aquariums and the
bioconcentration aquariums show that the semistatic test system is well
suited for these bioconcentration studies.

The bioconcentration factors (BCF) for each experiment are summarized
in table I. No steady-state was observed for 2,4',5-trichlorobiphenyl.
However, the bioconcentration curves seem to reach a plateau, and the fish
concentrations after 14 days exposure have been used for calculation of
the BCF.

In the first bioconcentration study, when the test compounds had been
equilibrated with the diluted springtime humus water for only 24 hours
before the addition of the fish, the presence of humus reduced the uptake
of both the 2,4,6-trichlorophenol and the lindane. The bioconcentration
factors for these compounds were about 18 percent lower from the humus
water than from the lakewater.

The results from a study of the recoveries of some chlorinated hydrocarbons, with the same aquatic springtime humus, indicated complexation
between the test compounds and the humus with time, since the recoveries
decreased with increasing contact time. After two months equilibration
time between the aquatic humus and the chlorinated compounds, only about
20 % of the initial test compound concentrations were recovered (7).

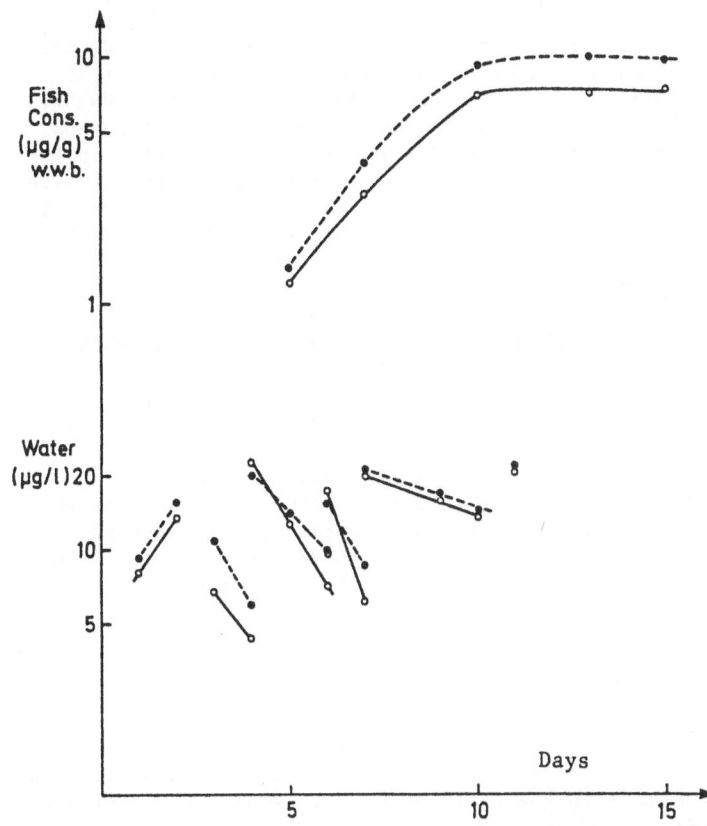

Fig. 2 Bioconcentration of 2,4,6-trichlorophenol in salmon fry. Fish
(wet weight) and water concentrations from ——— diluted humic
water (April 1982) and - - - - lake water experiments.

TABLE I

Bioconcentration factors for the test compounds in salmon
fry at steady state

Test chemical	Water quality	BCF
2,4,6-Trichlorophenol	Humus, springtime	560
"	Lake, "	690
Lindane	Humus, "	570
"	Lake, "	690
"	Humus, autumn	270
"	Lake, "	260
2,4',5-Trichlorobiphenyl[*]	Humus, "	4600
"	Lake, "	6800

[*] Steady state was not obtained for 2,4',5-trichlorobiphenyl.
The fish concentration after 14 days exposure was used for
the BCF calculation

Based on the above results an additional bioconcentration study was performed where a longer contact time between the humus and the test compounds had been allowed. This time humus collected in October 1983 was used. Lindane and 2,4',5-tricnlorobiphenyl were used as test compounds and a contact time of three months were chosen. However, contrary to the above mentioned results the lindane and trichlorobiphenyl recoveries were not reduced after three months equilibration period. A constant recovery of about 90 % indicated that a difference exists in complexation ability between aquatic humus taken from the same source at different times of the year. The same complexation variety has also been observed by others (13).

For 2,4',5-trichlorobiphenyl a reduction in uptake could be observed in the humus aquarium compared to the lake water aquarium, resulting in a 30 percent lower bioconcentration factor from the humus water. For lindane on the other hand no effect of the humus on the uptake could be observed.

In conclusion this investigation has shown that humus might reduce the bioavailability of organic micropollutants towards fish and that the effect is different for different compounds. It was also found that natural humus water taken from the same source at different times of the year might react differently towards the same compound. The experiments will be repeated with humus samples from the same source giving stronger complexation with the organic micropollutants.

REFERENCES

(1) Benes P, Gjessing E T, Steinnes E. Water Res. 10 (1976) 711-716.
(2) Hart B T, Davies S H R. In: International Conference on Management and Control of Heavy Metals in the Environment, (London: CEP Consultants) (1979) pp. 466-471.
(3) Boehm P D, Quinn J G. Geochim. Cosmochim. Acta. 37 (1973) 2459-2477.
(4) Ogner G, Schnitzer M. Geochim. Cosmochim. Acta 34 (1970) 921-318.
(5) Florence T M, Batley G E. CRC Critical Reviews in Anal. Chem. 9 (1980) 219-296.
(6) Gjessing E T, Berglind L. Arch. Hydrobiol. 92 (1981) 24-30.
(7) Carlberg G E, Martinsen K. Sci. Total Environ. 25 (1982) 245-254.
(8) Winner R W. Aquatic Toxicol. 5 (1984) 267-274.
(9) Andrew R W, Biesinger K E, Glass G E. Water Res. 11 (1977) 309-315.
(10) Sedlácek J, Källqvist T, Gjessing E T. In: Christman R F, Gjessing E T (Eds) Aquatic and terrestrial humic materials. Ann Arbor Science, Ann Arbor (1983) p 495-516.
(11) Carlberg G E, Martinsen K, Kringstad A, Gjessing E, Grande M, Källqvist T, Skåre J U. Submitted to Arch. Environ. Contam. Toxicol.
(12) Renberg L. Anal. Chem. 49 (1974) 459-461.
(13) Johnsen S, Krane J, Carlberg G E, Aamot E, Schou L. Presented at the 4th European Symposium on Organic Micropollutants in the Aquatic Environment, Vienna Oct. 1985.

MIXTURES OF HYDROPHOBIC CHEMICALS IN AQUEOUS ENVIRONMENTS: AQUEOUS SOLUBILITY AND BIOCONCENTRATION BY FISH OF AROCLOR 1254

Antoon Opperhuizen and Ronald P. Jongeneel

Laboratory of Environmental and Toxicological Chemistry,
University of Amsterdam, The Netherlands

SUMMARY

The aqueous concentrations of individual PCB congeners are influenced significantly by the presence of co-solutes. For saturated Aroclor 1254 solutions it was found that almost all aqueous concentrations of PCB components were below the saturation levels of the individual congeners.

The relative PCB mixture composition in water closely resembled the original Aroclor 1254 composition. In addition, the relative composition of PCB congeners in water was almost independent of the experimental temperatures. The solubility of Aroclor 1254 at $22^{\circ}C$ was 37 $\mu g/l$ and increased to 119 $\mu g/l$ at $65^{\circ}C$.

In addition it was found that all congeners of the Aroclor 1254 mixture were accumulated by fish. The calculated uptake rate constants were in excellent agreement with values found for individual congeners in previous studies. That however, after prolonged exposure as well as after elimination, variations in the relative PCB mixture composition in fish are found, must be explained by the observed differences in elimination rate constants between the various PCB congeners. This is because no significant differences in uptake rates constants between the various PCB congeners are found.

INTRODUCTION

PCB's as well as many other aromatic hydrocarbons are usually released into the environment as chemical mixtures. Hence, until a few years ago many papers reported on the physico-chemical properties, toxicological effects and bioaccumulation of whole mixtures. However, since analytical techniques and methodologies have improved, scientific interest has shifted towards the properties and environmental fate of single compounds.

Since capillary GC techniques often enable separation of complex mixtures into individual components, it has been proposed many times that one or more mixture components should be employed as reference chemicals by which the whole mixture should be characterized[1,2]. Unfortunately, however, it is not known whether or not such reference chemicals represent the properties of other mixture components.

Accurate data on the influence of co-solutes on the aqueous solubility of hydrophobic solutes, for instance, are very scarce[3]. Generally two solubility concepts are employed. In one concept, which is based essentially on statistics[4], it is assumed that the mixture composition in aqueous solutions is highly dependent on the composition of the original mixture. If so, close resemblence between these compositions will be observed, so that the aqueous solubility of the whole mixture can be expressed by one single value.

The second concept, which has a thermodynamic background, assumes that solubility is an equilibrium partitioning between solution and pure solute, which cannot be influenced by the presence of co-solutes. The solubility is only dependent on the solute's physico-chemical properties, its molecular structure and on the properties of water. Thus addition of co-solutes or co-solvents can only enhance the aqueous concentrations, i.e. creating super saturated solutions, but not the solubility.

The question whether mixture of hydrophobic chemicals behave as physico-chemical units or as a summation of individual components is not only of interest for aqueous solubility, but also for the bioavailability and accumulation by fish or other species. Although in many studies it has been shown that the uptake and elimination kinetics depend on the structure of the single solute[5] bioaccumulation or bioconcentration factors are still reported for PCB's[6] or other mixtures.

In the present study the aqueous solubulity of Aroclor 1254 is investigated, as this PCB mixture is mainly composed of congeners with low aqueous solubilities, for which the presence of co-solutes may be important. In addition the bioaccumulation by fish after aqueous exposure of Aroclor 1254 is studied, in relation to the bioavailability.

Since PCB congener contributing to Aroclor 1254 are extremely hydrophobic and very slowly metabolizable due to their high degree of chlorination, high bioconcentration factors are expected, which however, are not similar for all congeners.

By employing an accelerated test system the uptake and elimination kinetics of several PCB congeners is studied, and the results are compared to literature data.

EXPERIMENTAL

Aroclor 1254 (Monsanto, USA) was used without previous purification. Individual PCB congeners, which were used for the identification and quantification of Aroclor 1254 components, were available from previous experiments (7,8).

Water samples (200 ml) and fish samples (3 fishes each sample) were extracted with redistilled toluene by the method reported elsewhere (8). Final analysis of the samples was on a Packard-Becker 428 capillary gaschromatograph, equipped with a SGE on-column injector, and a [63]Ni electron capture detector (300°C), which was connected to a Shimadzu C-R2 AX computing integrator. A 25 m * 0.22 mm CpSil 5 CB, WCOT column was used. Nitrogen was employed as carrier gas at a rate of 0.57 ml/min. The flow rate of the detector make up gas was 19.5 ml/min. The following temperature program was applied:
- injection at 75°C, no initial hold
- temperature increase up to 175°C with a rate of 33°C/min.
- 175°C for 8 min.
- temperature increase up to 235°C with a rate of 3°C/min.

Mirex was added as internal standard to water and fish samples just before injection into the gaschromatograph. For the calculation of the concentrations of individual PCB congeners, mean response factors as reported by Mullin et al. (1984)9, and Onuska and Comba (1990)[10] were employed, together with experimental data for individual PCB's and Mirex.

The standard GC program provided 62 peaks, of which 41 could be assigned. From these assigned compounds, 17 were selected for quantification of total Aroclor 1254 samples (fig. 1). The 17 congeners contributed 70.8% to the original Aroclor 1254. Hence, the summed total of the 17 congeners concentrations of fish and water samples was divided by 0.708 to obtain an estimate of the Aroclor 1254 concentration of these samples.

Aqueous solutions of PCB's were prepared by application of a continuous flow through saturation system in which 2.0 l water was circulated with a flow rate of 4.0 l/hour[8]. Saturation of water was achieved by employment of chromosorb W-AW (100-120 mesh) which was impregnated with 1-5% Aroclor 1254, or with other test chemicals. To remove contaminated chromosorb particles from the solutions, water was filtered using two Millipore cellulose filters (0.22 μm).

To test bioconcentration, one year old male guppies (Poecilia reticulata) were exposed during 16 days to an aqueous solution of Aroclor 1254 in a static test system.[11] The test conditions were similar to those reported previously.[8] During the exposure period, water was sampled daily and fishes were sampled every third day. After 16 days, the remaining fishes were transferred to clean water to investigate the elimination kinetics.[8]

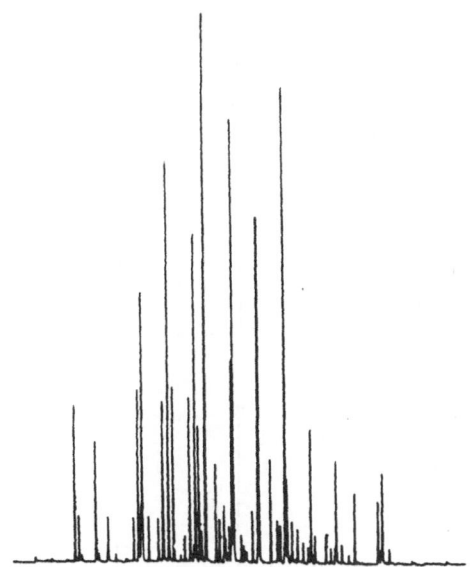

Figure 1 . Standard Aroclor 1254 gas chromatogram

RESULTS AND DISCUSSION

Aroclor 1254 was selected for the investigation of the fate of mixtures of hydrophobic chemicals in aqueous environments, because it is mainly composed of PCB congeners which have aqueous solubilities comparable or even lower than that of 2,2',5,5' tetrachlorobiphenyl.

It has been shown that solubility data of individual PCB congeners below 20-30 μg/l, obtained from different studies, normally show large scattering, which may be due to the application of inaccurate saturation methods in various studies.[12] In the present study a water saturation method comparable to a generator column[13], was exployed, which had previously provided good results for individual congeners.[12]

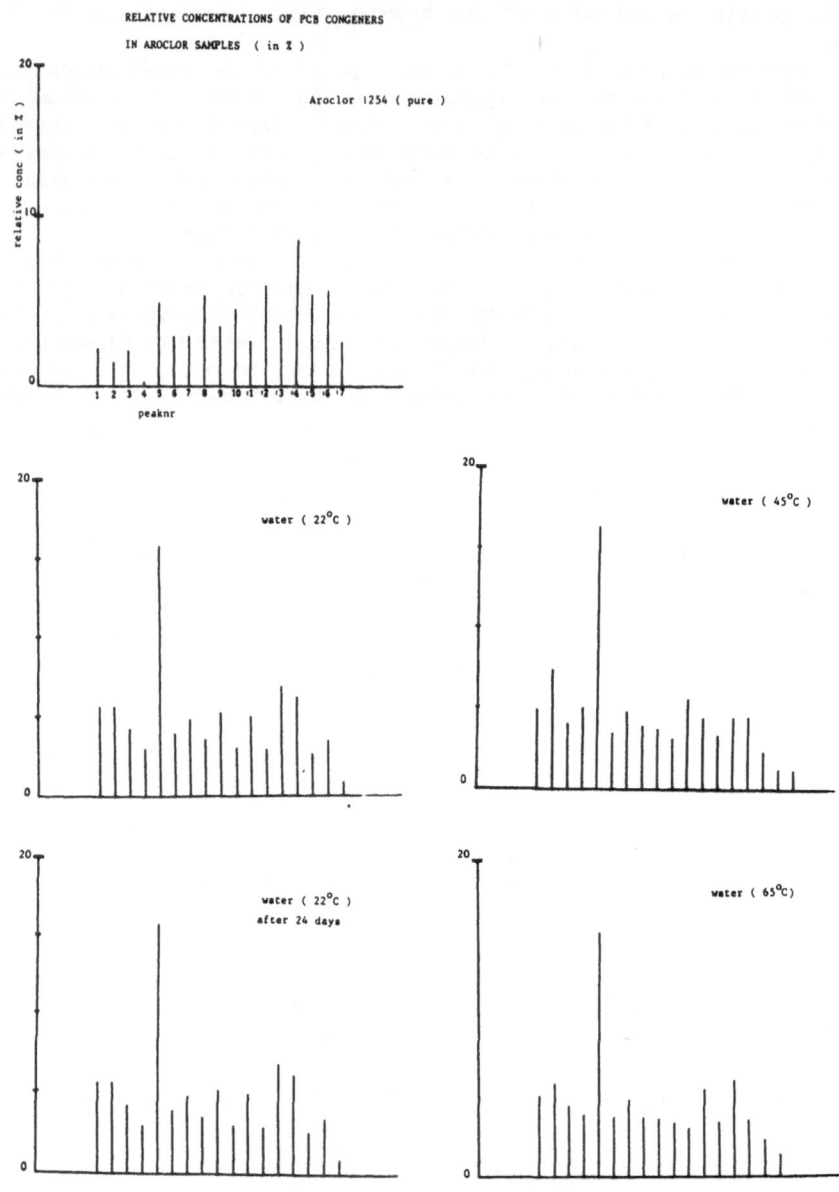

Figure 2. Relative concentrations of 17 PCB congeners (No 1 to 17) in Aroclor 1254 samples , expressed as percentage of the total mixture concentration.

TABLE 1. Aqueous solubilities of Aroclor 1254

solubility (μg/l)	water quality	ref.
56	distilled	14
41–43	distilled	15
40–44	distilled	16
70	distilled	17
12	distilled	18
45±10	distilled	19
300	distilled	20
24.7–28.1	seawater	21
300–1500	seawater	22
300–3000	freshwater	22
37±4 (T = 22°C)	distilled	this study
38±6 (T = 22°C, 24 days)	distilled	this study
68±14 (T = 45°C)	distilled	this study
119±12 (T = 65°C)	distilled	this study

In table 1, experimental solubilities of Aroclor 1254 are listed, together with values obtained from other studies. The data reported in the table are all measured after one day's circulation in the saturating system. The experimental value of 37 $\pm$ μg/l in good agreement with literature data.

As shown in table 2, the aqueous saturation level is neither influenced by the amount of contaminated chromosorb, nor by the percentage impregnation of the chromosorb. In addition it can be seen that the solubility of Aroclor 1254 increased at higher temperatures, up to 119 μg/l at 65°C (table 1). This increasing aqueous saturation concentration is in agreement with the generally considered endothermicity of the solution of hydrophobic chemicals.

In figure 2, the concentrations of the 17 congeners which are used for quantification of the PCB mixture are shown as percentage of the total PCB mixture (C_i/ Σ C_i). As shown, the PCB mixture compositions of the water samples at various temperatures are almost similar. This, however, is not in agreement with predictions based on the thermodynamics of solution, since the solubility of a solute is highly dependent on the experimental temperature relative to the melting point of the solute. Thus, it may be expected that variation of the experimental temperature would have a totally different influence on PCB congeners with either low or high melting points. Hence, since there is no significant influence of the system's temperature on the aqueous mixture composition, it may be concluded that solution theory of equilibrated aqueous solution at saturation level is inappropriate.

TABLE 2. The relationship between percentage impregnation and amount of impregnated chromosorb on the measured aqueous solubility of Aroclor 1254

amount of chromosorb (g)	impregnation (%)			
	0.1	1.0	2.5	5.0
1	33.4	35.6	35.3	34.8
3	38.2	36.7	37.0	39.1
5	37.2	38.1	39.0	36.8

TABLE 3. Aqueous concentrations of several PCB congeners in Aroclor 1254 saturated water and individual aqueous solubilities[12].

PCB Congener	concentration in aqueous Aroclor1254 solution ($\mu g/l$)	aqueous solubility ($\mu g/l$)
2,2',3,5, tetra	1.6	170
2,2',4,5' "	2.2	16.4
2,2',5,5' "	2.2	26.5
2,3,4,4' "	5.5	35.8
2,2',3,4,5' penta	1.0	23
2,2',3,4',5 "	1.1	9.8
2,2',3,3',4,4' hexa	0.6	0.44
2,2',4,4',5,5' "	1.3	1.1
2,2',3,3',4,4',6 hepta	1.6	2.17

The aqueous concentrations of several Aroclor 1254 components are listed in table 3. As is shown, only congeners with aqueous solubilities close to 1 $\mu g/l$ have solubilities close to the aqueous saturation concentration if Aroclor 1254 is added to the water. This is in contrast to the more soluble congeners like for instance 2,2',5,5' tetrachlorobiphenyl, which have aqueous saturation concentrations far below the aqueous solubilities of the pure solutes.

As shown in figure 2, the PCB mixture composition in water does not completely resemble the composition of the original Aroclor 1254. However, the similarity between the GC pattern of the Aroclor 1254 sample and aqueous solutions are obvious, especially for the higher chlorinated congeners. For the lower chlorinated congeners however, a slight increase of the relative concentrations, as percentages of the whole mixture, is found. In addition, it is shown in table 4, that the relative concentrations of PCB congeners with retention times shorter than that of 2,2',5,5' tetrachlorobiphenyl (peak nr. 1 in figure 2) are significantly enhanced in aqueous solution compared to the original mixture composition.In addition, from figure 2 it may be clear that the mixture composition in water does not change significantly if the contact time between aqueous solution and pure Aroclor 1254 is prolonged to 24 days. These observations are in agreement with the observed solubilities of some mono, di, tri and tetra PCB's in Aroclor 1242 and other PCB mixtures, reported as by Lee et al.(1979). This phenomenon cannot simply be explained from statistics as has sometimes been suggested[4], since from a thermodynamic point of view a solute's solubility is determined only by the properties of the pure solute and the solvent.

TABL 4. Relative concentrations of PCB congeners with retention times shorter than that of 2,2',5,5' tetrachlorobiphenyl in aqueous Aroclor 1254 samples.

Aroclor 1254 (pure)	< 1.0 %
water (22°C)	9.1 %
water (45°C)	8.7 %
water (65°C)	8.9 %

Additionally, since the amount of impregnated chromosorb, as well as the loading of the chromosorb, did not influence the aqueous saturation concentration, it may be clear that the low congeneric concentration also cannot be explained by a insufficient amount of Aroclor 1254 in the water-saturating system. Therefore, it must be assumed that aggregates of PCB's are present in aqueous Aroclor 1254 solutions, which prohibit thermodynamic equilibration of the solutes between pure solute and solution. The solubility of the less soluble congeners seems to be almost uninfluenced, while the aqueous concentrations of the more soluble congeners are probably strongly influenced by the presence of such mixture aggregates.

Although these clusters seem to be rather stable over a period of more than 24 days and the composition is similar at various temperatures it is unlikely that they persist indefinitely. The data of Lee et al., for instance, show that the a congener pattern of PCB mixtures (Aroclor 1242) in water still changing 150 days after addition of the PCB mixtures. Hence, achieving a final PCB composition in an aqueous solution of Aroclor 1254 may take several months or even years. For shorter contamination periods however, the composition of the aqueous solutions seems to be highly dependent on the composition of the original mixture.

For the investigation of the bioavailability for and the bioconcentration by guppies (Poecilia reticulata) of Aroclor 1254, from an initially saturated aqueous solution, a 16 days exposure period was employed, following by a 70 days clearance period. A static test system was used, in order to rapidly achieve a steady state situation between PCB congener concentrations in fish and water. The method was similar to those previously used for the testing of single PCB's[7,23] as well as other hydrophobic chemicals[8].

It was found for all PCB congeners that steady states were achieved in the concentrations in fish and water within the 16 days exposure period, so that from the ratio between the concentrations in fish and water, at the end of the exposure period the bioconcentration factor could be calculated.

For the calculations of the uptake and elimination rate constants, a first order kinetic model was employed, which is usually assumed to be accurate for the bioconcentration of hydrophobic chemicals[23]:

$$\text{water} \underset{k_2}{\overset{k_1}{\rightleftarrows}} \text{fish}$$

k_1, uptake rate constant $(ml\ g^{-1}\ d^{-1})$
k_2, elimination rate constant $(ml\ g^{-1}\ d^{-1})$

The range of the calculated uptake rate constant of several Aroclor 1254 components (table 5) is in agreement with experimental data of single PCB congeners, as well as with predictions made by a membrane permeation model for hydrophobic chemicals.

TABLE 5. Uptake and elimination rate constant of some PCB congeners measured after aqueous exposure of Aroclor 1254 to fish

PCB Congener	k_1	k_2	$K_C = k_1/k_2$
2,3',4,4' tetra	630	15 10^{-3}	4.2 10^4
3,3',4,4' "	530	9.5 10^{-3}	5.6 10^4
2,2',4,5,5' penta	380	8.1 10^{-3}	4.7 10^4
2,3',4,4',5 "	600	10.1 10^{-3}	5.9 10^4
2,2',3,4,4',5 hexa	640	4.9 10^{-3}	13.1 10^4
2,2',4,4',5,5' "	710	6.9 10^{-3}	10.2 10^4

In figure 3 the k_1 and k_2 values of the congeners listed in table 5 are combined with literature data and are plotted versus octan-1-ol/water partition coefficient. As shown, the data obtained from exposure of fish to a mixture are comparable to data from tests with single solutes.

As shown in table 5, the elimination rate constants are obviously not similar for the various solute structures so that due to the constancy of k_1, the bioconcentration factors K_C, i.e. k_1/k_2, will also vary. Hence, the differences between the relative compositions of the initial aqueous solution and in fish after several days of exposure must be explained by the differences in elimination rate constants, and not by differences in the uptake rate constants.

The alteration in PCB mixture composition in fish during the elimination period due to the various k_2 values, is illustrated in figure 4, in which the concentration of three PCB congeners relative to the concentration of 2,2',4,4',5,5' hexachlorobiphenyl in several fish samples during the elimination period are shown. From this figure it is clear that the PCB pattern in this fishes, sampled 50 days after exposure had stopped, does not closely resemble the original PCB pattern.

Hence, even for this very simple test system, matching of a PCB pattern found in fish to the Aroclor 1254 pattern will be very diffucult if neither the exposure nor the elimination period is known. This holds true especially for the lower chlorinated congeners with the faster elimination rates.

Since the highly chlorinated congeners all have very low elimination rate constants and the differences in k_2 values of these compounds are small, the best matching of PCB patterns found in biota with industrial PCB mixtures can be expected for the highly chlorinated biphenyl congeners. However, since even for these highly chlorinated compounds differences in elimination rate constants are observed, significant variation of the relative PCB composition in fish can occur after prolonged exposure or elimination.

CONCLUSIONS

After addition of a mixture of extremely hydrophobic chemicals to water, in a water saturating system which provided good aqueous solubility values for single solutes, aqueous solutions were created in which the individual solutes solubilities were influenced significantly by the presence of co-solutes. This was not caused by the presence of contaminated particles in the system, as Millipore filters were placed before the sampling point. This observation cannot be explained by the theory of the solubility of single non-electrolyte solutes. The uptake and elimination kinetics by fish of individual PCB congeners was however,

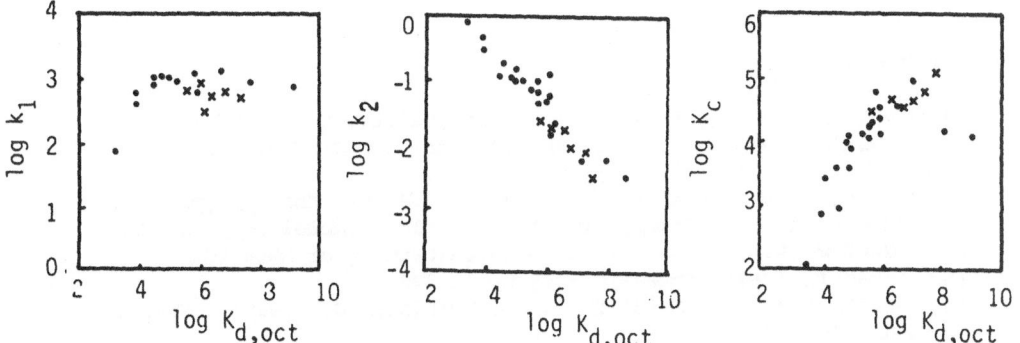

• data obtained from ref.5,8 and 23

✱ data presented in table 5

Figure 3. Relationships between uptake rate constant (k₁), elimination rate constants (k₂), bioconcentration factors (K_C) and Octan-1-ol/water partition coefficients K_d,oct, of several PCB congeners

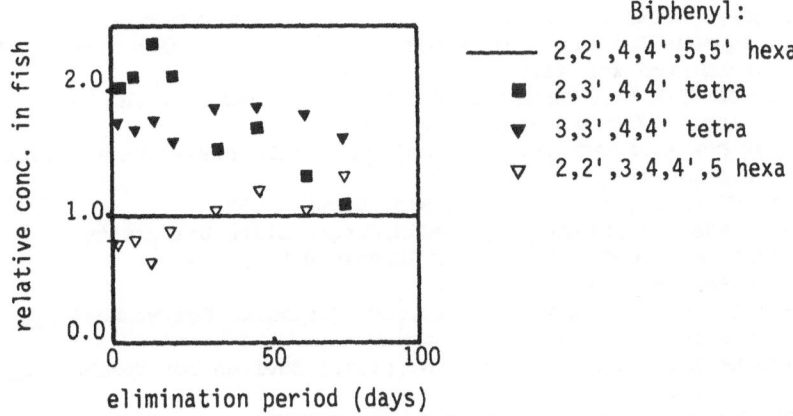

Biphenyl:

——— 2,2',4,4',5,5' hexa

■ 2,3',4,4' tetra

▼ 3,3',4,4' tetra

▽ 2,2',3,4,4',5 hexa

Figure 4. The concentrations in fish during the elimination period of three PCB congeners, relative to the concentration of 2,2',4,4',5,5' hexachlorobiphenyl

not influenced by the presence of co-solutes.

Thus, whereas mixtures of hydrophobic solutes in water have significantly different properties than the consisting components, the biotic fate of almost unmetabolizable hydrophobic chemicals is uninfluenced.

ACKNOWLEDGEMENT

Thanks are due to John Parsons for his useful comments and advice.

REFERENCES

(1) GEYER,H., (1984), Ecotoxicol.Environ.Safety, 8 , pp129-151.
(2) BROWN,M.P.,WERENER,M.B.,SLOAN,R.J. and SIMPSON,K.W.,(1985), Environ.Sci.Technol.,19,pp468-476
(3) LEINONEN,P.J. and MACKAY,D.,(1973),Can.J.Chem.Eng.,51,pp230-233
(4) KALMAZ,E.V. and KALMAZ,G.D.,(1979),Ecologic.Model.,6,pp223-251
(5) BRUGGEMAN,W.A.,(1983),Ph.D.Thesis,University of Amsterdam
(6) MATSUO,M.,(1980),Chemosphere,9,pp671-675
(7) BRUGGEMAN,W.A.,VANDERSTEEN,J. and HUTZINGER,O. (1982),J.Chromatogr. 238,pp335-346
(8) OPPERHUIZEN,A.,VANDERVELDE,E.W.,GOBAS,F.A.P.C.,LIEM,A.K.D., VANDERSTEEN,J.M.D. and HUTZINGER,O., Relationship between Bioconcentration in Fish and Steric Factors of Hydrophobic Chemicals , in press
(9) MULLIN,M.D.,POCHINI,C.M.,McCRINDLE,S.,ROMKES,M.,SAFE,S.H. and SAFE,L.M.,(1984),Environ.Sci.Technol.,18,pp468-476
(10)ONUSKA,F.I. and COMBA.M.,(1980), in 'Hydrocarbons and Halogenated Hydrocarbons in the Aquatic Environment' , edited by, Afgan,B. and MacKay,D. ,Plenum Press, pp285-302
(11)BANERJEE,S.,SUGATT,R.H. and O'GARDY,D.P.,(1984),Environ.Sci.Technol., 18,pp79-83
(12)OPPERHUIZEN,A.,GOBAS,F.A.P.C.,VANDERSTEEN,J.M.D. and HUTZINGER,O., Aqueous Solubility of PCB's Related to Molecular Structure, submitted for publication.
(13)MILLER,M.M.,GHODBANE,S.,WASIK,S.P.,TEWARI,Y.B. and MARTIRE,D.E. (1984), J.Chem.Eng.Data.,29,pp184-190
(14)HAQUE,R.,SCHMEDDING,D.W. and FREED,V.H.,(1974),Environ.Sci.Technol., 8,pp139-142
(15)NELSON,N.,(1972),Environ.Res.,5,pp249-362
(16)WIESE,C.S.,(1975),M.S.Thesis,Oregon State University
(17)LEE,M.S.,CHIAN,E.S.K. and GRIFFIN,R.A.,(1979), Water Res.,13,pp1249-1258
(18)MACKAY,D. and WOLKOFF,A.W.,(1973),Environ.Sci.Technol.,7, pp611-614
(19)LAWRENCE,J. and TOSINE,H.M.,(1976),Environ.Sci.Technol.,10, pp381-383
(20)EISENREICH,S.J.,LOONEY,B.B. and THORNTON,J.D.,(1981), Environ.Sci.Technol.,15,pp30-38
(21)WIESE,C.S. and GRIFFIN,D.A.,(1978),Bull.Environ.Contam.Toxicol. 19,pp403-411.
(22)ZITKO,V.,(1971),Bull.Environ.Contam.Toxicol.,5,pp279-285
(23)OPPERHUIZEN,A. ,Bioconcentration of Hydrophobic Chemicals by Fish, in press
(24)SPACIE,A. and HAMELINK,J.C.,(1982), Environ.Toxicol.Chem.,1, pp321-327

COMPARATIVE ANALYSIS, AT THE MOLECULAR LEVEL, OF THE INTERACTIONS BETWEEN ORGANIC AND INORGANIC MERCURY COMPOUNDS AND MEMBRANE LIPIDIC BILAYERS

A. BOUDOU*, J. FAUCON**, J.P. DESMAZES** and D. GEORGESCAULD**

* Laboratoire d'Ecologie fondamentale et Ecotoxicologie,
Université de Bordeaux I, 33405, Talence Cédex (France)
** Centre de Recherche Paul Pascal, CNRS,
Domaine universitaire, 33405, Talence Cédex (France)

Summary

In the context of a biophysical approach of the interactions between mercury compounds (CH_3HgCl - $HgCl_2$) and biological membranes, we did research works concerning :
- accessibility of organic and inorganic mercury to the hydrophobic core of membranes by looking at the fluorescence quenching of pyrene introduced in the lipidic bilayers.
- transmembrane flux of mercury using the technique of B.L.M. (bimolecular lipid membrane).
- effects of methylmercury and inorganic mercury on the dynamic of membrane lipids, by stady-state fluorescence polarization spectroscopy of diphenyl-hexatriene (DPH), embedded in the lipidic core of liposomes.
Each study was related to the chemical form and speciation of mercury, the abiotic parameters (temperature, pH, salinity) and the different species of lipids participating to the composition of model membranes (acidic and zwitterionic phospholipids).

1. INTRODUCTION

Lors de tout processus de contamination, les molécules exogènes pénètrent dans les organismes en traversant les "barrières biologiques" qui séparent le milieu intérieur et le milieu aquatique environnant : barrières cutanée, respiratoire et digestive. Ensuite, leur transport dans le système circulatoire, leur accumulation dans les organes et les cellules-cibles ainsi que les mécanismes d'excrétion reposent sur une succession d'interactions avec les membranes cellulaires.

Si chaque "barrière biologique" possède des spécificités structurales et fonctionnelles au sein d'un même individu et parmi les différentes espèces, une composante unitaire est toujours présente dans chacune d'elles : la membrane plasmique.

Le mercure et ses dérivés ont fait l'objet de plusieurs recherches, essentiellement biochimiques, visant à analyser la fixation du métal sur les constituants protéiques des membranes, grâce à leurs propriétés thioloprives. Ainsi, une action inhibitrice sur différentes activités enzymatiques et systèmes de transports a été mise en évidence (1,2,3,4) ; de même, le mercure agit sur la structure de la membrane cellulaire (résistance aux chocs osmotiques, "déformabilité",...) (5,6,7).

Par contre, très peu de travaux ont été consacrés à l'étude des interactions entre les dérivés du mercure et la composante lipidique des membranes plasmiques, malgré l'importance de cette bicouche hydrophobe à l'égard de la structure et des propriétés des barrières cellulaires (8).

Dans le cadre d'un programme de recherche en écotoxicologie expérimentale, ayant pour objectif principal l'étude des processus de bioaccumulation et de transfert des dérivés du mercure dans les systèmes aquatiques continentaux (9,10,11), nous avons orienté nos travaux vers une approche, au niveau moléculaire, des interactions entre deux dérivés du mercure - $HgCl_2$ et CH_3HgCl - et la composante phospholipidique des membranes biologiques. Trois étapes successives ont été élaborées :
- analyse de l'accessibilité des dérivés minéral et organique du mercure à la zone hydrophobe intra-membranaire.
- étude des flux transmembranaires du métal, flux global et flux liés aux espèces mercurielles ionisées.
- mise en évidence et quantification de l'action du mercure sur la microfluïdité des bicouches lipidiques.

Chacune d'elles a reposé sur la prise en compte de différents paramètres, relatifs à la physico-chimie du milieu (température, pH, concentration en chlore) et à la nature des phospholipides constitutifs des modèles de membrane (naturels ou synthétiques, neutres ou chargés).

2. ANALYSE DE L'ACCESSIBILITE DES DERIVES DU MERCURE A LA ZONE HYDROPHOBE INTRA-MEMBRANAIRE :

La quantification du degré d'accessibilité du mercure minéral ($HgCl_2$) et organique (CH_3HgCl) à la zone hydrophobe intra-membranaire a été réalisée sur des modèles de membrane (dispersions lipidiques et liposomes), en relation avec deux conditions de pH (5.0 et 9.0) et différents phospholipides (phosphatidylcholine, PC, et phosphatidylsérine, PS).

Certains marqueurs de fluorescence, comme le pyrène ou le benzo-(a)-pyrène, sont des molécules hydrophobes qui, en présence de modèles lipidiques de membrane ou de membranes biologiques, se localisent préférentiellement dans la bicouche phospholipidique. Très souvent utilisés pour mesurer l'état de fluïdité du système, ils permettent également de détecter la présence de substances étrangères, dans le micro-environnement de la membrane qu'ils occupent. Cette propriété est dûe à un effet d'inhibition ou "effet d'atome lourd", résultant de processus collisionnels entre le fluorophore et l'inhibiteur ; elle engendre une diminution de l'émission de fluorescence du marqueur, désignée par le terme de "quenching" (12,13). Ainsi, la mesure comparative des intensités des signaux de fluorescence, en absence ou en présence de mercure (Fo/F), selon les caractéristiques physico-chimiques du milieu et la nature des phospholipides, permet d'analyser l'accessibilité du métal à la zone hydrophobe intra-membranaire.

Nos recherches ont reposé sur l'utilisation du pyrène, introduit dans un très faible rapport de masse (pyrène/lipide - 1/500) à l'intérieur des bicouches lipidiques.

Les résultats obtenus (illustration figure 1) montrent que l'accessibilité du mercure à la zone hydrophobe intra-membranaire est fortement influencée par la forme chimique du métal ($HgCl_2$ - CH_3HgCl), par le pH du milieu et par la présence ou non de charges électriques négatives sur les têtes polaires des phospholipides (PS : une charge négative à pH acide - deux charges à pH alcalin).

Le pH du milieu agit à la fois sur la spéciation chimique des deux dérivés du mercure et sur les charges portées par les molécules de phosphatidylsérine. L'analyse détaillée des résultats (14)) attribue un rôle primordial aux espèces mercurielles cationiques - $HgCl^+$, $HgOH^+$, Hg^{++}, CH_3Hg^+ -, notamment par le biais d'attractions électrostatiques à l'interface membrane modèle - milieu aqueux. Bien que quantitativement mineures, d'après les diagrammes de spéciation chimique correspondant à nos conditions expérimentales (10,15), elles peuvent exercer une action importante par le jeu de déplacements d'équilibres.

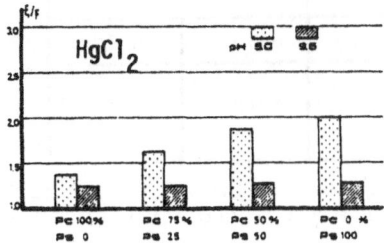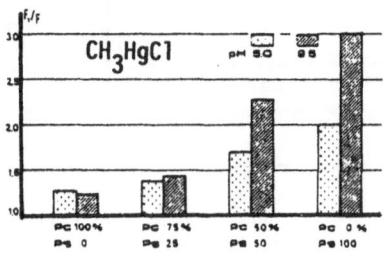

Figure 1 : Analyse comparée du quenching du pyrène (Fo/F) par HgCl$_2$ et CH$_3$HgCl, en relation avec le pH du milieu et la nature des phospholipides (PC : phosphatidylcholine - PS : phosphatidylsérine).

A ce stade de nos recherches, nous nous sommes confrontés à plusieurs interrogations, relatives aux mécanismes responsables de ces différents processus :
- les charges négatives portées par les têtes polaires des phospholipides assurent-elles uniquement une adsorption des espèces mercurielles cationiques ou bien ces espèces traversent-elles la bicouche lipidique ?
- quelle est la part des espèces mercurielles neutres, quantitativement dominantes dans les diagrammes théoriques de spéciation chimique, à l'égard du flux global de mercure au travers des membranes lipidiques ?
- le "quenching" de l'émission de fluorescence du pyrène par le mercure, attribué à des inhibitions collisionnelles, ne peut-il pas traduire également la présence du métal adsorbé au niveau des têtes polaires (transferts d'énergie entre le métal et le marqueur)?
Afin de répondre aux deux premières questions, nous avons orienté nos travaux vers une mesure des flux transmembranaires du mercure, au moyen de bicouches lipidiques planes.

3. ETUDE DES FLUX TRANSMEMBRANAIRES DU MERCURE MINERAL ET ORGANIQUE

Nous avons mis en place un programme de recherche reposant sur l'utilisation de bicouches lipidiques planes ou BLM ("black membranes"). L'association, dans les mêmes conditions de contrainte, de la mesure des flux électriques et de celle du flux global du mercure traversant la bicouche est particulièrement bien adaptée à l'élucidation des problèmes de perméabilité. Ainsi, il est possible de préciser si le flux observé est dû à une seule espèce mercurielle, s'il existe un contre-transport ou encore si le transport est réalisé sous forme d'entités électriquement neutres.
Parmi l'ensemble des résultats obtenus (16), nous citerons :
- avec des BLM réalisées avec le mélange lipidique "Phosphatidylcholine-Cholestérol", les flux globaux de mercure, pour HgCl$_2$ et CH$_3$HgCl, sont plus faibles à pH alcalin (9.0) qu'à pH acide (5.0) (tableau 1), les différences observées étant beaucoup plus marquées pour le dérivé minéral. Notons cependant que pour des conditions de pH opposées, les flux transmembranaires du mercure minéral et du méthylmercure sont similaires, ce résultat nuançant très fortement les affirmations rencontrées dans la littérature, attribuant au méthyl-mercure une forte capacité de bioaccumulation à cause de sa liposolubilité.

	HgCl$_2$		CH$_3$HgCl	
	pH 5.0	pH 9.0	pH 5.0	pH 9.0
Flux global de mercure x$_1$10^{10} (mol.cm^{-2}.s^{-1})	0.75	0.08	1.12	0.72

Tableau 1 : Flux global de mercure pour le système : Tampon (pH 5.0 ou 9.0) + HgCl$_2$ ou CH$_3$HgCl (5.10^{-4}M) / BLM (PC - Cholestérol) / Tampon (pH 5.0 ou 9.0).

- la présence au sein de la bicouche de phospholipides chargés électriquement (PS) ou de cholestérol ne modifie pas significativement les flux transmembranaires des deux dérivés mercuriels étudiés.
- toutes les mesures électrochimiques réalisées révèlent que le transport du mercure minéral ou organique, au travers des BLM, repose sur des espèces chimiques neutres. Les formes chlorées - HgCl$_2$ et CH$_3$HgCl - dominantes à pH acide, sont beaucoup plus aptes à traverser les bicouches lipidiques que les formes hydroxylées. Ce résultat est beaucoup plus marqué pour le chlorure mercurique et il est en accord avec les mesures réalisées par GUTKNECHT, J. (17). La prise en compte du paramètre "concentration en chlore" dans le milieu - gamme de 0 à 0.5 mol.NaCl.l^{-1} - nous a permis de confirmer ce résultat (tableau 2 et figure 2) : la présence des seules espèces anioniques - HgCl$_3$- et HgCl$_4$$^{--}$ - se traduit par une très brusque diminution des valeurs des flux globaux du mercure au travers des BLM.

HgCl$_2$ (M)	NaCl (mM) - c -				
10^{-3}	0	1	10	100	500
	0.20	0.22	0.22	0.14	0.08

Tableau 2 : Variation du flux unidirectionnel du mercure (J$_{Hg}$ x 10^9 mol. cm^{-2}.s^{-1}) en fonction de la concentration en NaCl, pour le système : tampon pH 5.0 + NaCl (c) + HgCl$_2$ / BLM (PC - Cholestérol) / tampon pH 5.0 + NaCl (c) + HgCl$_2$.

- les calculs des coefficients de perméabilité nous ont permis de mettre en évidence l'importance des couches d'eau non agitées ("unstirred layers") à l'égard du transport du mercure. Cela expliquerait notamment l'absence de modifications des flux du métal au travers des BLM lorsque la nature des phospholipides varie (PC, PS, Cholestérol) alors que l'accessibilité du mercure minéral ou organique à la zone hydrophobe intra-membranaire est fortement perturbée.

Nos résultats, tant au niveau des mesures d'accessibilité du mercure à la zone hydrophobe intra-membranaire qu'à celui des flux au travers des BLM, nous ont amené à rechercher une incidence du mercure sur les propriétés des bicouches phospholipidiques.

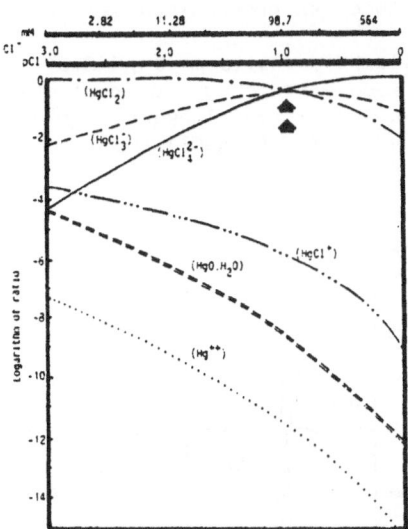

<u>Figure 2</u> : Diagramme de spéciation chimique du mercure minéral ($HgCl_2$), à pH 5.0, en fonction de la concentration en chlore (15).

4. ACTION DES DERIVES DU MERCURE SUR LA DYNAMIQUE DES PHOSPHOLIPIDES MEMBRANAIRES

Plusieurs auteurs, à partir de mesures indirectes de perméabilité (18) ou encore grâce à des études de complexation (19), ont émis l'hypothèse d'une perturbation des propriétés membranaires par le mercure, sans toutefois apporter de preuves expérimentales.

Nous avons élaboré un programme de recherche sur ce sujet, reposant dans une première phase, sur la technique de polarisation de fluorescence.

La mesure du taux de polarisation de la lumière émise par une molécule fluorescente donne des renseignements sur la molécule elle-même (géométrie, dimensions, durée de vie à l'état excité,...) mais également sur certaines caractéristiques de son environnement (température, viscosité,...).

C'est ainsi que l'introduction du diphénylhexatriène (DPH), marqueur fluorescent, dans la zone hydrophobe des bicouches lipidiques, permet d'analyser les diagrammes de phase des phospholipides, dont la température de transition est située dans une gamme allant de +5 à +60°C.

Plusieurs expérimentations ont été conduites, reposant sur l'utilisation de deux phospholipides - diphosphatidylglycérol (DPPG), lipide acide, et diphosphatidylcholine (DPPC), lipide zwitérion. Nous avons analysé comparativement l'incidence du mercure minéral ($HgCl_2$) et organique (CH_3HgCl) sur leur température de transition, en relation avec différentes conditions physico-chimiques (pH, concentration en chlore), agissant à la fois sur certaines propriétés des modèles de membrane et sur la spéciation chimique des dérivés mercuriels.

Les mesures de la température de transition de la DPPC ne révèlent pas d'effet significatif du mercure - $HgCl_2$ et CH_3HgCl, 500 mg.l^{-1} - sur ce paramètre (figure 3).

Par contre, les deux dérivés entrainent un déplacement de la température de transition du DPPG, à pH 5.0 (figure 4). Cet effet est plus marqué pour le méthylmercure que pour le mercure minéral ; il est dépendant de la concentration initiale du métal dans le milieu (figure 5). Les déplacements observés, traduisant une stabilisation des systèmes lipidiques, sont toutefois de faible amplitude, en comparaison par exemple avec un effet du calcium.

Des expériences similaires, réalisées à pH 8.5, révèlent une absence d'action des deux dérivés du mercure sur la température de transition du DPPG. Un résultat identique est observé lorsque, à pH 5.0, la concentration du chlore dans le milieu est de 0.5 M (NaCl) (20).

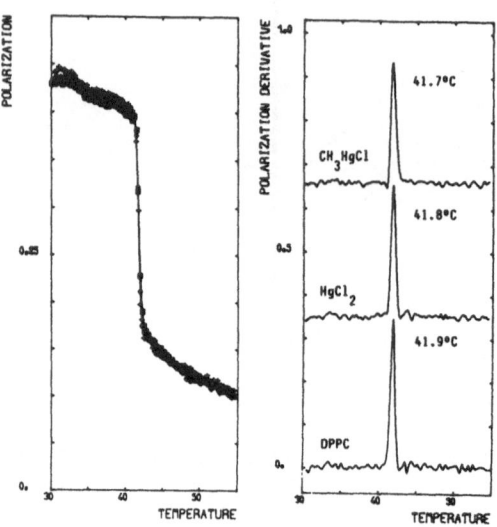

Figure 3 : Effet du $HgCl_2$ et CH_3HgCl (500 ppm) sur la polarisation de fluorescence du DPH dans les liposomes de DPPC (tampon CH_3COOH-NaOH, 50 mM ; pH 5.0).

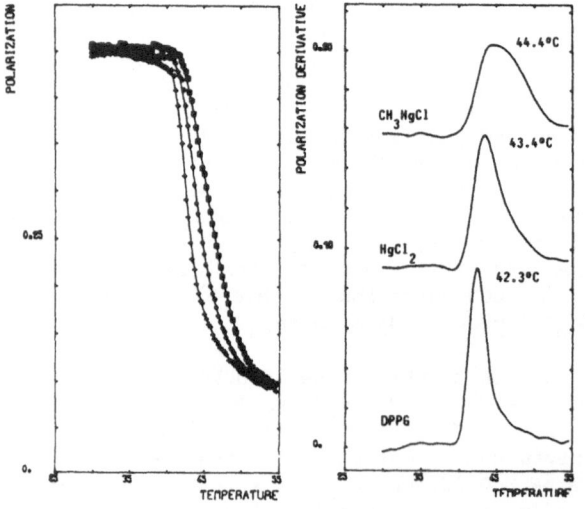

Figure 4 : Effet du $HgCl_2$ et CH_3HgCl (500 ppm) sur la polarisation de fluorescence du DPH dans des liposomes de DPPG (tampon CH_3COOH-NaOH, 50 mM ; pH 5.0).

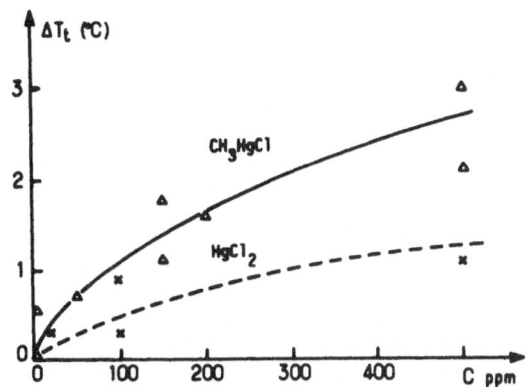

Figure 5 : Augmentation de la température de transition (T_t) du DPPG en fonction de la concentration du mercure dans le milieu (Cppm - $HgCl_2$ et CH_3HgCl) - (tampon acétate, pH 5.0)

Ces résultats sont en accord avec ceux obtenus lors de l'analyse comparative de l'accessibilité des deux dérivés du mercure à la zone hydrophobe intra-membranaire, révélant le rôle primordial joué par les charges négatives portées par les têtes polaires des phospholipides et par les espèces chimiques cationiques du mercure.

En effet, DPPC est une molécule zwitérion ; DPPG possède une charge négative. A pH 8.5, les diagrammes de spéciation ne font apparaître que des espèces neutres ($Hg(OH)_2$ et CH_3HgOH) ; de même, lorsque la concentration en chlore est de 0.5 M, à pH 5.0, seules des espèces anioniques ou neutres sont présentes. Par contre, à pH 5.0, sans ajout de chlore, des espèces cationiques mineures apparaissent et le déplacement de la température de transition n'est décelé que pour le phospholipide acide (DPPG).

Des recherches sont actuellement menées pour diversifier cette approche, grâce notamment à l'utilisation d'autres phospholipides, pris isolément ou en mélange. Une analyse plus précise des interactions entre les dérivés du mercure et les têtes polaires des phospholipides est également abordée, grâce à la technique des monocouches lipidiques, permettant la mesure simultanée du potentiel et de la pression de surface.

5. CONCLUSION

Les recherches que nous menons sur les interactions entre les dérivés du mercure et les barrières membranaires s'inscrivent dans le cadre d'une approche écotoxicologique des processus de bioaccumulation et de transfert de ces contaminants au sein des systèmes aquatiques continentaux.

Une démarche de complexification progressive a été retenue, tant au niveau du support biologique - modèles lipidiques de membrane, extraits membranaires, cellules en culture, épithélium - qu'à celui du nombre des paramètres pris en compte - nature chimique du contaminant, concentration, température du milieu, pH, concentration en chlore -, afin d'accroître la représentativité de nos recherches.

REFERENCES

(1) STORM, D.R. and GUNSALUS, R.P. (1974). Methylmercury is a potent inhibitor of membrane adenyl-cyclase. Nature 250, 778-779.

(2) ARHEM, P. (1980). Effects of some heavy metals ions on the ionic currents of myelinated fibers from Xenopus laevis. Journal of Physiology 306, 219-231.

(3)   KLIP, A. et al. (1980). Interaction of the sugar carrier of intestinal brush-border membranes with $HgCl_2$. Biochimica Biophysica Acta 548, 100-114.

(4)   ROTHSTEIN, A. (1981). Mercurials and red cell membranes, in "The function of red blood cells : erythrocyte pathobiology", Liss A.R. edit., pp. 105-131.

(5)   TANAKA, R. and NAKAI, K. (1977). Hemolysis and morphological changes in rat erythrocytes with mercurials. Japan Journal of Pharmacology 27(3), 413-419.

(6)   MELL, H.C. and REED, T.A. (1981). Biophysical responses of rat-cell membrane systems to very low concentrations of inorganic mercury. Cell biophysic 3(3), 233-251.

(7)   RALSTON, G.B. and CRISP, E.A. (1981). The action of organic mercurials on the erythrocyte membrane. Biochimica Biophysica Acta 649, 98-104.

(8)   BOUDOU, A. et al. (1983). Ecotoxicological role of the membrane barriers in transport and bioaccumulation of mercury compounds, in "Aquatic toxicology", J.O. Nriagu edit., Wiley New-York, pp. 118-136.

(9)   BOUDOU, A. and RIBEYRE, F. (1983). Processes of contamination of aquatic biocenoses by mercury compounds : an experimental ecotoxicological approach, in "Aquatic toxicology", J.O. Nriagu edit., Wiley New-York, pp. 73-116.

(10) BOUDOU, A. (1982). Recherches en écotoxicologie expérimentale sur les processus de bioaccumulation et de transfert des dérivés du mercure dans les systèmes aquatiques continentaux. Thèse Doctorat d'Etat n°748, Université de Bordeaux I, France, 297 p.

(11) RIBEYRE, F. (1985). Problems and methodologies in Ecotoxicology : biological models and experimental plans. Ecotoxicology and Environmental Safety 9, 346-363.

(12) DEUMIE, M. et al. (1981). Mécanismes spécifiques d'inhibition de la fluorescence dans les membranes d'érythrocytes : étude du pyrène et du BaP. Journal de Chimie Physique 78(5), 475-483.

(13) LAKOWICZ, J.R. and ANDERSON, C.J. (1980). Permeability of lipid bilayers to methylmercuric chloride : quantification by fluorescence quenching of a carbazole-labeled phospholipid. Chemico-Biological Interactions 30, 309-323.

(14) BOUDOU, A. et al. (1982). Fluorescence quenching study of mercury compounds and liposome interactions : effect of charged lipid and pH. Ecotoxicology and Environmental Safety 6(4), 379-387.

(15) HANNE, H.C.H. and KROONTJE, W. (1973). The simultaneous effects of pH and chloride concentrations upon mercury as a pollutant. Soil Sciences Society America Proceedings 37, 838-843.

(16) BIENVENUE, E. et al. (1982). Transfert of mercury compounds across bimolecular lipid membranes : effet of lipid composition, pH and chloride concentration. Chemico-Biological Interactions 48, 91-101.

(17) GUTKNECHT, J. (1981). Inorganic mercury transport through lipid bilayers membranes. Journal of membrane biology 61, 61-66.

(18) NAKADA, S. et al. (1978). Change in permeability of liposomes caused by methylmercury and inorganic mercury. Chemico-biological interactions 22, 15-23.

(19) REICHERT, W.L. and MALINS, D.C. (1974). Interactions of mercurials with Salmon serum lipoproteins. Nature 247, 569-570 .

(20) BOUDOU, A. et al. (1985). Effects of mercury compounds ($HgCl_2$ and CH_3HgCl) on the dynamic of membrane lipids : polarization fluorescence study. 5th International Conference "Heavy metals in the environment", Athens.

EXPERIMENTAL SYSTEMIC STUDY OF THE ACCUMULATION AND TRANSFER PROCESSES OF ORGANIC AND INORGANIC MERCURY IN FRESHWATER ENVIRONMENT - METHODOLOGY AND RESULTS

F. RIBEYRE, A. BOUDOU, R. MAURY and P. ENGRAND
Laboratoire d'Ecologie fondamentale et Ecotoxicologie
Université de Bordeaux I - 33405 Talence Cédex (France)

Summary

In our experimental study on processes of bioaccumulation and transfer of mercury compounds within continental aquatic systems we considered three basic ecotoxicological poles : biotic factors ; abiotic factors ; contamination modalities.
From 1974 to 1982 we worked with linear biological models - experimental trophic chains. Since 1982 we have been working on the development of a systemic approach, using interactive models. In this way we hope to study the effect of several ecotoxicological parameters on the processes involved when aquatic systems are contaminated.
The experimental method we decided upon was based on the use of factorial designs, which enable us to analyse the action of the various parameters being considered (for example : temperature, photoperiod, pH) and also the effects of their interactions on the different compartments of the experimental model.
The contamination levels, which were estimated from the mercury concentration and content, and also the effects on the organisms, were studied for different experimental periods. The equipment currently at our disposal makes it possible to use 512 experimental units simultaneously.
The significance of results from an approach such as ours depend to a very great extent on the devising and setting up of the experimental model, the definition of research protocols and the choice of method for the analysis of the data obtained. In fact results reflect the action of all the ecotoxicological factors (experimental parameters, measured factors, unknown factors) and the various sources of error (measurements, proportions,...). In order to define the actions and interactions of the parameters studied a very detailed analysis of all the stages of the experimental process is clearly necessary, from the definition of aims and objectives right up to the formulation of hypotheses.

1. INTRODUCTION

Depuis 1974, les recherches réalisées dans notre laboratoire ont pour objectif premier l'étude de la bioaccumulation et des transferts des dérivés du mercure au sein de modèles écotoxicologiques expérimentaux.
Ces modèles sont basés sur la prise en compte de paramètres ou "facteurs contrôlés" choisis parmi l'ensemble des facteurs abiotiques, biotiques et de contamination, qui sont les trois pôles fondamentaux de l'Ecotoxicologie. Leur intégration est effectuée progressivement, afin de préserver un niveau satisfaisant de maîtrise du modèle étudié et

d'atteindre une meilleure mise en évidence et une plus grande compréhension des mécanismes écotoxicologiques.

Cette complexification croissante des modèles expérimentaux nécessite d'une part, une continuité dans les programmes de recherche, notamment à l'égard des micropolluants étudiés, et d'autre part, de ne prendre en considération que des facteurs ou paramètres supplémentaires, lorsque les étapes antérieures ont été franchies de façon satisfaisante.

La démarche expérimentale ne se résumant pas uniquement à la réalisation des expériences, une place importante est accordée à la réflexion sur l'élaboration et la mise en oeuvre des autres étapes, notamment la conception des protocoles et l'analyse des résultats. Ainsi, le degré d'adéquation entre les objectifs fixés et les résultats obtenus s'accroît avec le niveau de cohérence de ces différentes étapes.

Dans le cadre de cette publication, seule une présentation synthétique de notre démarche expérimentale est proposée ; toutefois, afin d'illustrer les fondements et les aboutissants de nos préoccupations méthodologiques, certains résultats seront communiqués à titre d'exemples.

2. DESCRIPTION DE LA METHODOLOGIE

2.1. Constitution des modèles écotoxicologiques

2.1.1. Structure biologique

Les recherches conduites dans notre laboratoire, de 1974 à 1982, ont reposé sur un modèle biologique linéaire - chaîne trophique expérimentale (figure 1). Le contrôle des apports alimentaires permet de quantifier les transferts du mercure entre les différents maillons et d'apprécier ainsi les participations respectives des deux voies de contamination : directe et trophique. De nombreux résultats, relatifs à chaque niveau du modèle ainsi qu'à l'ensemble de la chaîne, en relation avec différents paramètres écotoxicologiques, ont été publiés (article de synthèse : 1).

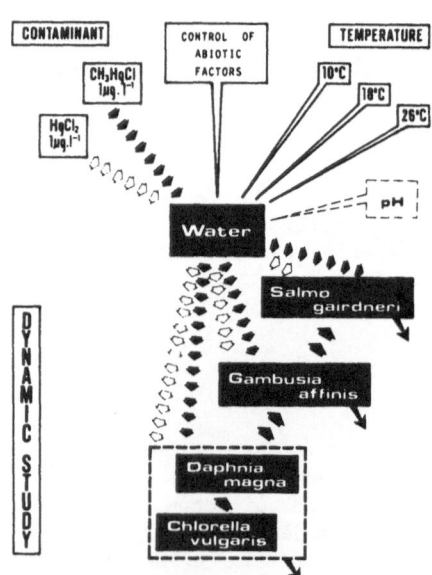

Figure 1 : Structure de la chaîne trophique expérimentale et des principaux paramètres pris en compte (1).

2.1.2. Paramètres_écotoxicologiques

Les modalités des paramètres retenus - biotiques, abiotiques et de contamination - sont définies en fonction des valeurs rencontrées en milieu naturel, des contraintes expérimentales et analytiques et de la variabilité des "réponses" observées. L'analyse de l'action de ces facteurs contrôlés sur les mécanismes étudiés, ainsi que leurs interactions, implique de les considérer simultanément. Ainsi, le nombre de "cas expérimentaux distincts" croît très vite avec le nombre de paramètres retenus et celui de leurs modalités (figure 2), ce qui limite rapidement la diversité des facteurs écotoxicologiques pouvant être pris en compte lors de la définition des protocoles expérimentaux. Notons cependant que l'utilisation de plans factoriels d'expérience est particulièrement bien adaptée à ce type de problématique.

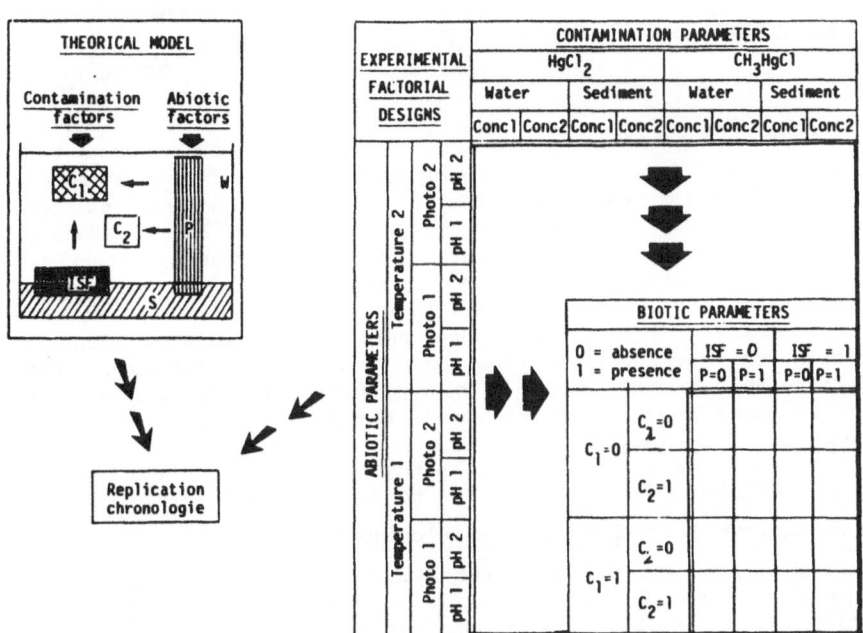

Figure 2 : Modèle écotoxicologique interactif et plan factoriel d'expérience (2).

2.1.3. Paramètres_expérimentaux

Si l'on veut accroître les connaissances sur le fonctionnement et l'évolution du système expérimental, afin de révéler au mieux les processus écotoxicologiques étudiés, il est nécessaire d'adjoindre aux paramètres écotoxicologiques, des paramètres expérimentaux, tels que la chronologie des observations, les répétitions...

- chronologie des observations :

La mise en évidence de certains mécanismes de transfert ne peut être effectuée que si "l'état du système" est analysé après plusieurs durées d'expérimentation. Une telle approche vise à apprécier les bilans des échanges du contaminant au sein du modèle, pour chaque "niveau d'analyse" et en fonction des paramètres écotoxicologiques pris en compte. Elle permet de dégager les tendances des processus de bioaccumulation et de transfert (illustration figure 3). Cela contribue à enrichir les hypothèses interprétatives proposées, par exemple sur les "capacités relatives d'accumulation, l'accessibilité aux sites de fixation, les équilibres "contamination - décontamination",...

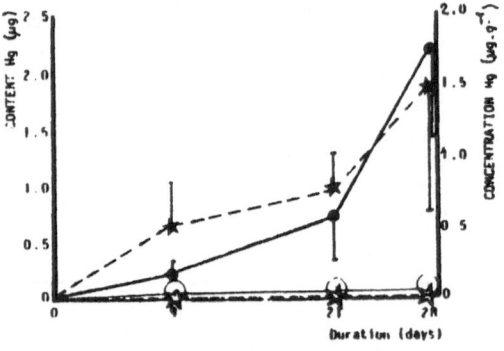

Figure 3 : Dynamique de la bioaccumulation du mercure - HgCl$_2$
et CH$_3$HgCl - chez <u>Elodea densa</u> -
Intervalles de confiance à P=0,05

Teneur   ●— CH$_3$HgCl

       ○— HgCl$_2$

Concentration   -★- CH$_3$HgCl

       -☆- HgCl$_2$

- <u>nombre de répétitions</u> :

Toutes les "réponses écotoxicologiques" enregistrées sont fonction de l'action de trois types de facteurs : ignorés, mesurés et contrôlés (paramètres). Si on admet l'hypothèse selon laquelle les facteurs ont une action déterminée, et une seule, sur les réponses observées, les fluctuations existant entre les différentes répétitions (conditions expérimentales similaires, de même durée et indépendantes) sont d'autant plus grandes que les facteurs non pris en compte dans l'analyse des résultats ont des effets importants. Parmi ces facteurs "ignorés", volontairement ou non, citons l'hétérogénéité au sein des compartiments (intra et inter unités expérimentales), les erreurs de mesure et de dosage, les décalages chronologiques,...

L'appréciation de la variabilité des résultats conduit à la définition du nombre de répétitions, pour chaque condition expérimentale.

2.2. <u>Analyse des modèles écotoxicologiques</u>

Nous distinguerons les "critères d'analyse" et les "niveaux d'analyse". Ils permettent de décrire le système à l'instant "t" et sont définis en fonction des objectifs fixés et des possibilités matérielles et humaines disponibles. L'information ainsi collectée est d'autant plus significiante que les critères et les niveaux retenus sont nombreux, précis et indépendants.

2.2.1. <u>Critères_d'analyse</u>

Parmi l'ensemble des critères, peuvent être séparés ceux relatifs à la contamination "sensu-stricto" et ceux décrivant la structure de certains éléments du système : biométrie, physico-chimie des milieux,... Ces derniers sont utilisés soit comme liens entre les critères du premier groupe, soit comme support complémentaire au cours de la phase d'interprétation des résultats.

L'appréciation des niveaux de contamination, ainsi que l'analyse détaillée des transferts, repose sur deux critères écotoxicologiques fortement complémentaires, bien que liés : concentration et teneur du contaminant. La prise en compte de ces deux critères présente d'autant plus d'intérêt que les valeurs des variables qui les lient (poids ou volume) peuvent se modifier au cours du temps ou changer en fonction des compartiments analysés et des conditions expérimentales retenues. A titre d'exemple, la figure 4 traduit l'évolution comparée de la concentration et de la teneur en mercure ($HgCl_2$ et CH_3HgCl), dans les reins de <u>Salmo gairdneri</u>, en phase de décontamination, après contamination par voie directe (5) : pour le mercure organique, les tendances sont similaires ; par contre, pour le chlorure mercurique, la teneur du métal augmente très nettement dans les reins alors que la concentration diminue.

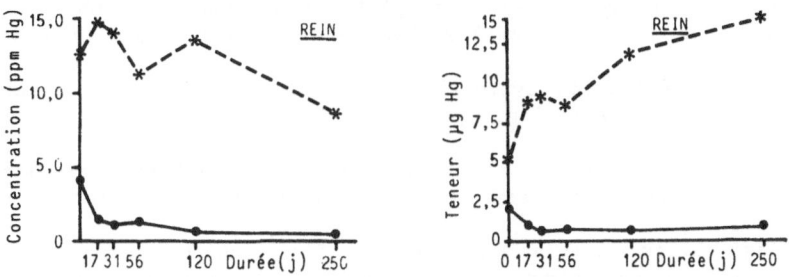

Figure 4 : Evolution comparée ($HgCl_2$ --- et CH_3HgCl ——) des concentrations et des teneurs en mercure chez <u>Salmo gairdneri</u> en phase de décontamination après contamination par voie directe (5).

2.2.2. Niveaux d'analyse

L'accès à tous les niveaux du système, du plus fin (moléculaire) jusqu'au plus large (système global) serait idéal, les observations collectées à chacun de ces niveaux étant très étroitement complémentaires. Cependant, nous nous heurtons une nouvelle fois aux moyens d'investigation disponibles, qui limitent le nombre d'échantillons pouvant être prélevés et analysés dans les différents compartiments du modèle.

Les choix que nous avons effectués visent à analyser les transferts des contaminants entre les principaux compartiments et sous-compartiments des écosystèmes expérimentaux (exemple figure 5).

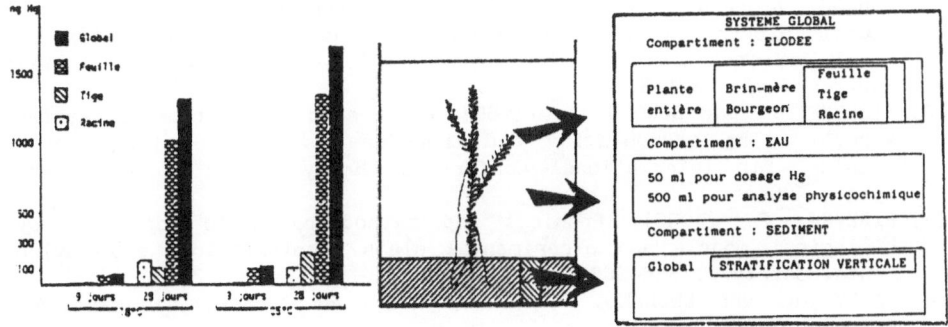

Figure 5 : Ecosystème expérimental à trois compartiments : "eau, sédiment, <u>Elodea densa</u>" - Représentation schématique des principaux critères et niveaux d'analyse utilisés (4).

2.2.3. Analyse des données et interprétation des résultats

Le nombre très important de données obtenues lors de telles expérimentations, lié d'une part à la structure du modèle écotoxicologique et d'autre part aux nombreux critères et niveaux d'analyse retenus, nécessite de procéder à un traitement méthodique de l'ensemble des observations. Il repose sur une analyse progressive des résultats, allant de l'utilisation de paramètres descriptifs, pour chaque variable, jusqu'à l'élaboration de modèles explicatifs. Cette méthode a pour objectifs essentiels :

a) la caractérisation synthétique des "états" du système et de leur évolution (valeurs centrales, paramètres de dispersion, analyse des correspondances,...),

b) la mise en évidence de liens entre les variables (corrélations, régressions,...) ou entre les observations et les variables (analyses multivariées...),

c) l'estimation, avec une probabilité fixée, de l'action des paramètres écotoxicologiques retenus, ainsi que de leurs interactions, à l'égard des processus de bioaccumulation et de transfert (analyse de variance, tests paramétriques et non paramétriques,...),

d) la confrontation des hypothèses avancées avec les résultats expérimentaux, au moyen de la modélisation mathématique et de la simulation (modèles à compartiments multiples,...).

3. CONCLUSION

La démarche expérimentale présentée ne doit pas être considérée comme une succession d'étapes isolées mais comme un ensemble de préoccupations, de méthodes, d'outils et d'opérations étroitement imbriqués. Les choix effectués à chacune de ces étapes sont modulés ou imposés par l'ensemble des travaux antérieurs, afin d'optimiser les coûts globaux des expérimentations et d'améliorer l'adéquation entre les résultats et les objectifs de recherche.

REFERENCES

(1) BOUDOU, A. and RIBEYRE, F. (1983). Processes of contamination of aquatic biocenoses by mercury compounds : an experimental ecotoxicological approach, in "Aquatic* toxicology", J.O. Nriagu edit., Wiley New-York, pp. 73-116.

(2) RIBEYRE, F. and BOUDOU, A. (1985). First step on experimental systemic study on the accumulation and transfer processes - Methodology and results. 5th International Conference "Heavy metals in the environment", Athens.

(3) RIBEYRE, F. (1985). Problems and methodologies in Ecotoxicology : biological models and experimental plans. Ecotoxicology and Environmental Safety 9, 346-363.

(4) MAURY, R. and ENGRAND, P. (1985). Thèse de Doctorat, Université de Bordeaux I (in press).

(5) RIBEYRE, F. and BOUDOU, A. (1984). Etude expérimentale des processus de décontamination chez Salmo gairdneri, après contamination par voie directe avec deux dérivés du mercure ($HgCl_2$ et CH_3HgCl) - Analyse des transferts aux niveaux "organisme" et "organes". Environmental Pollution (A) 35, 203-228.

BIOACCUMULATION BY FISH IN RELATIONSHIP TO THE OXYGEN CONCENTRATION IN WATER

S. MARCA SCHRAP   and   ANTOON OPPERHUIZEN

Laboratory of Environmental and Toxicological Chemistry, University of Amsterdam, Nieuwe Achtergracht 166, 1018 WV. Amsterdam, The Netherlands

Direct uptake from ambient water is often the most important intake route of hydrophobic chemicals for aquatic organisms (1). Due to the low aqueous oxygen concentrations, fish for instance, must pass large quantities of water across their gills to satisfy their oxygen demand (2).

The exchange of hydrophobic chemicals between water and gills is considered to be an equilibrium partitioning, which is reached almost instantaneously. Therefore, it has been proposed previously that the rates of uptake of hydrophobic chemicals by fish are proportional to the ventilation volume passing the gills. Since, in addition, the ventilation volume is inversely proportional to the aqueous oxygen concentration, it may be expected that lowering of the oxygen concentration in water results in increasing uptake rates of pollutants by fish (1,3).

In the present experiment the uptake rate of 2,2',5,5' tetrachlorobiphenyl by guppies (Poecilia reticulata) at three different oxygen concentrations (2.5, 4.0 and 6.0 ppm) has been measured. For the simultaneous regulation of oxygen concentration and addition of test chemical to water, a water circulation system has been employed in which the test chemical was added via impregnated chromosorb, while the oxygen concentration was regulated by a combination of nitrogen and air inlets. Throughout the experiments no significant variation of the aqueous concentrations of either the test chemical or oxygen was observed.

Water and fish (two each sample) were sampled daily, and after extraction with hexane, analyzed at a gas chromatograph, which was equipped with an electron capture detector. From these data, uptake rates of 2,2',5,5' tetrachlorobiphenyl by fish under various oxygen regimes have been calculated. As can be seen from figure 1, no significant differences between the uptake rates by fish were found for the three different aqueous oxygen concentrations.

Hence, it can be concluded that the ventilation volume passing the gills, does not influence the uptake rate constant, but the uptake efficiency.

REFERENCES
(1) BRUGGEMAN, W.A., MARTRON, L.B.J.M., KOOIMAN, D, and HUTZINGER, O. (1981), Chemosphere, 10, pp 811-832
(2) NORSTROM, R.J., McKINNON, A.E. and DEFREITAS, A.S.W., (1976), J. Fish. Res. Board Can., 33, pp 248-267
(3) NEELY, W.B., (1979), Environ. Sci. Technol.,13, pp 1506-1510.

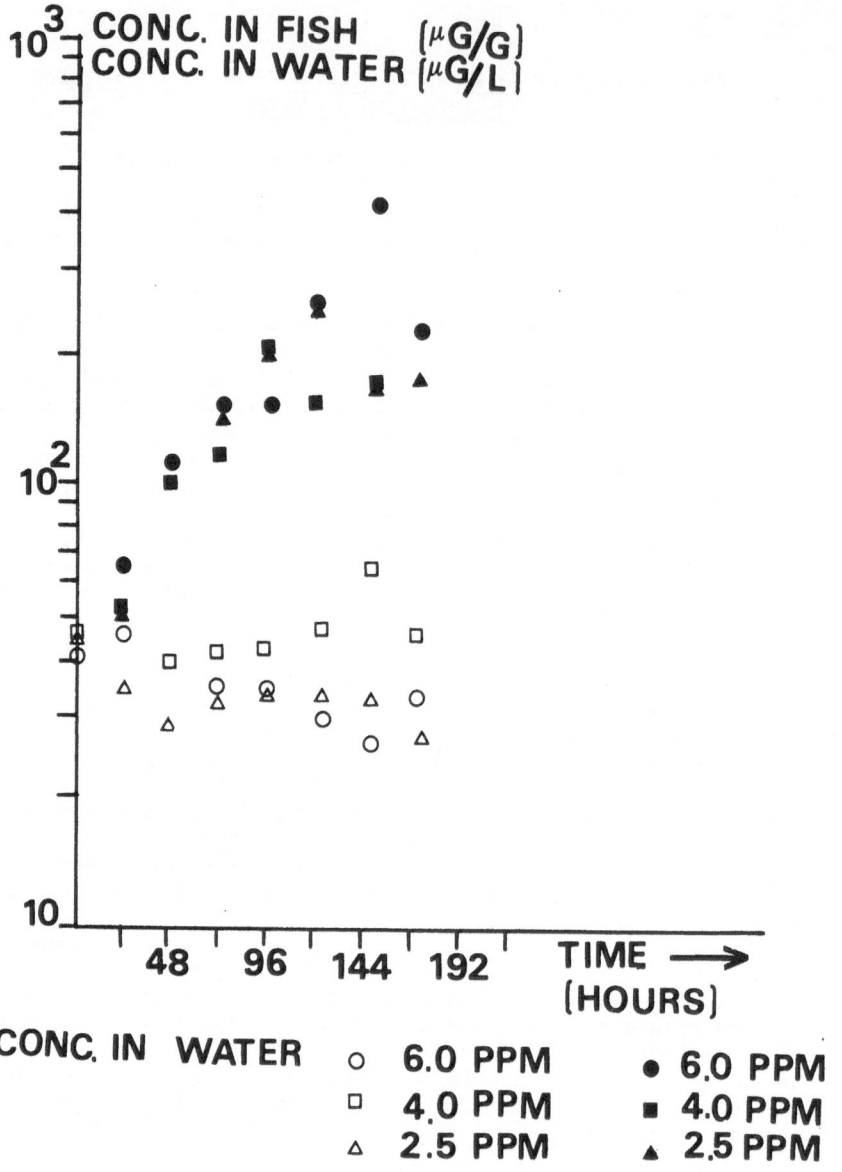

Figure 1. Concentrations of 2,2',5,5' tetrachlorobiphenyl in fish and water during exposure, under three different oxygen regimes

TRANSFORMATION IN THE AQUATIC ENVIRONMENT

Biodegradation of chlorinated compounds

Review of microbial transformations

Biotransformation of nitroaromatic compounds

Aquatic photochemistry : future developments

New reactions involved in non-aerobic degradation of aromatic compounds

Total biodegradation of chlorinated biphenyls by a pseudomonas strain

Possibilities of predicting the production of lipophilic volatile organohalides chlorination of sea water and fresh water samples in laboratory conditions

Orchard spraying with organophosphorus pesticides : Occurrence and fate of parent compounds and phosphorus containing hydrolytic products in adjacent ponds

Microbial degradation of nitrilotriacetate (NTA) in groundwater: Laboratory studies under aerobic and anaerobic conditions

Characterization and biodegradability of lignin containing wastewater from forest industry

BIODEGRADATION OF CHLORINATED COMPOUNDS

GOSSE SCHRAA and ALEXANDER J.B. ZEHNDER
Department of Microbiology, Agricultural University,
H.v.Suchtelenweg 4, 6703 CT Wageningen, The Netherlands

Summary

In this paper the important aspects of biodegradation of chlorinated compounds are discussed; namely, the mechanisms of biodegradation, the change in the chemical characteristics of molecules with chlorine substituents and, some factors affecting biodegradation. The existing knowledge about biodegradation of a number of chlorinated aliphatic hydrocarbons and chlorinated monoaromatic hydrocarbons is summarized and discussed in more detail for some selected compounds. Special attention is given to the role of molecular oxygen in the breakdown of these chemicals and to the way in which the ring is cleaved in aromatic compounds. In addition, some general remarks are made about biodegradation of xenobiotic compounds.

1. INTRODUCTION

Halogenated compounds, and among them specifically the chlorinated hydrocarbons, have been produced synthetically on a large scale during the last few decades (Table I). They are formed as products or by-products in a number of industrial processes and are used as solvents, refrigerants, fire retardants, insulators, lubricants and pesticides. But, significant amounts of chlorinated hydrocarbons are also produced during chlorination of drinking water and wastewater (25). Although over 200 halogenated hydrocarbons have been found to be produced naturally (59), most of the chlorinated compounds being released into or produced in the environment are of anthropogenic origin. In recent years these man-made chemicals have led to the contamination of air, water and soil (53). Some of these cases have attracted a lot of attention since a number of the xenobiotics found as pollutants were originally produced as deleterious agents against certain forms of life. Other xenobiotics have, in the meantime, proven harmful to man and animals.

To assess the hazard of a xenobioticum to the biota, introduced accidentally or purposely into the environment, information should be available concerning the following aspects: (i) the acute toxicity of the compound; (ii) its long term effects, especially when present in low concentrations; and (iii) its fate in the environment. Since the scope of this paper is limited to the discussion of the biodegradation of chlorinated aliphatic and aromatic hydrocarbons, we will not go into the toxicological aspects but focus specifically on the biological mechanisms which lead to the transformation or elimination of these chemicals in the environment.

Table I. Present production and uses of a selected number of
chlorinated aliphatic and aromatic hydrocarbons (49)

Name	World production 1,000 tons/year	Major uses
chloromethane	400	intermediate
dichloromethane (methylene dichloride)	500	solvent
trichloromethane (chloroform)	250	intermediate, solvent
tetrachloromethane	1,000	intermediate, fumigant
chloroethene (vinyl chloride)	10,000	monomer for plastics
trichloroethene (tri)	600	solvent
tetrachloroethene (per)	1,100	solvent
chloroethane	400	solvent, intermediate
1,2-dichloroethane	13,000	intermediate
chlorobenzene	600	solvent, intermediate
1,2-dichlorobenzene	80	solvent, intermediate
1,4-dichlorobenzene	80	deodorizer, insect repellent
1,2,3-trichlorobenzene	1	intermediate
1,2,4-trichlorobenzene	30	intermediate
1,3,5-trichlorobenzene	1	intermediate
1,2,4,5-tetrachlorobenzene	5	intermediate
chlorotoluene	10?	intermediate

2. FACTORS AFFECTING BIODEGRADATION

2.1. Mechanism of biodegradation

Biodegradation is a process in which an organic compound is
structurally changed via enzymatic reactions. The intermediates and/or
end products formed undergo a change in physical and/or chemical
properties, whereby they may also become more or less toxic. The
bioconversion process can be complete or incomplete. In the first
case, the organic compound is metabolized to inorganic end products, a
reaction which is called mineralization. In case the conversion reaction is incomplete, at least one of the end products will consist of a
partially degraded organic molecule (Fig. 1).

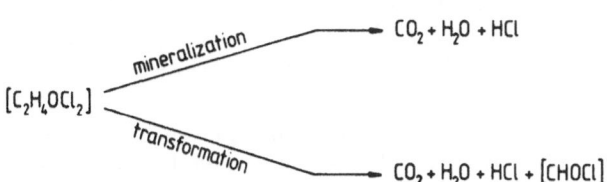

$[C_2H_4OCl_2]$ — mineralization → $CO_2 + H_2O + HCl$

$[C_2H_4OCl_2]$ — transformation → $CO_2 + H_2O + HCl + [CHOCl]$

Fig. 1 Biodegradation of halogenated organic compounds.

The latter process is termed transformation. Both transformation and mineralization reactions can support growth of an organism. In addition, a number of reactions are known in which a compound is metabolized without an obvious benefit to the organism. These kinds of reactions are called cometabolic processes. Horvath (28) defined cometabolism as "any oxidation of substances without utilization of the energy derived from the oxidation to support microbial growth". Although cometabolism is thought to play an important role in nature, its existence is only demonstrated in laboratory studies (1). A synthetic chemical can either be used as additional or only carbon and energy source or metabolized without an obvious benefit to the organism. Whether it is (co)metabolized or not depends on various factors, the most important of which include: structure and concentration of the chemical, kind and species of organisms involved, and environmental conditions.

2.2. Characteristics of halogen substituents

The presence of halogen substituents adds two new aspects to the metabolism of aliphatic and aromatic hydrocarbons: (i) the catabolic enzymes have to overcome the steric and negative inductive effects (-I) of the halogen atoms, and (ii) at some stage of the degradation, removal of the halogen atom is required. An example of the first aspect is shown in Table II. The large size and the -I-effect of a chlorine substituent influence the dioxygenation of benzoate and 2-,3- and 4-chlorobenzoate (38). The steric effect of the chlorine atom at the ortho position prevents the substrate binding to the enzyme. The effect has less influence when the chlorine is present at the meta and para position. The reason for the much faster oxygenation of 3-chlorobenzoate than of 4-chlorobenzoate has to be sought in the negative inductive effect and the positive mesomeric effect (+M) of the substituent. Only when the chlorine is in the meta position, will the -I-effect be partly compensated for by the +M-effect of the chlorine atom.

Tabel II. Relative rates of dioxygenation of substituted benzoates by whole cells of Alcaligenes eutrophus B9(38)

Substrate	Product	Relative rate of dioxygenation
Benzoate	DHB*	1000
2-Chlorobenzoate	-	0
3-Chlorobenzoate	3-Chloro-DHB	50
"	5-Chloro-DHB	100
4-Chlorobenzoate	4-Chloro-DHB	1

*DHB = Dihydrodihydroxybenzoate

The strong withdrawal of electrons from the nucleus may also place a constraint on the ability of the benzene ring to be cleaved by oxygen. The higher the number of halogen atoms, the more deactivated the nucleus will become. This may be the basis of an explanation for the resistance of multihalogenated aromatic compounds to biological oxidation (65).

Removal of a halogen atom, the second aspect, may chemically proceed via two mechanisms: nucleophilic substitution and ß-elimination (Fig. 2).

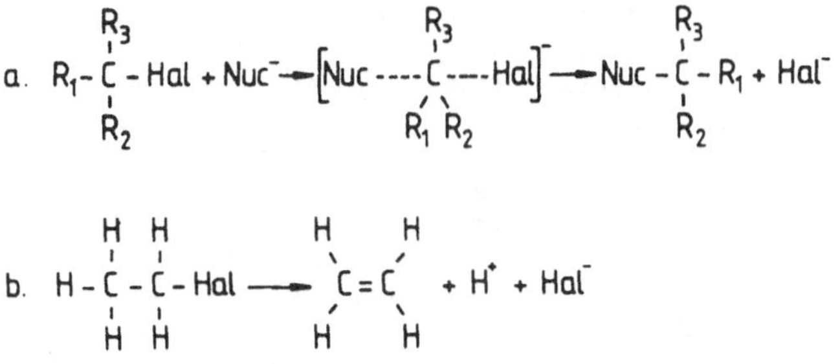

$$\text{a.} \quad R_1-\underset{\underset{R_2}{|}}{\overset{\overset{R_3}{|}}{C}}-Hal + Nuc^- \longrightarrow \left[Nuc ---- \underset{\underset{R_1\,R_2}{\diagup\,\diagdown}}{\overset{\overset{R_3}{|}}{C}} ---Hal \right]^- \longrightarrow Nuc -\underset{\underset{R_2}{|}}{\overset{\overset{R_3}{|}}{C}}- R_1 + Hal^-$$

$$\text{b.} \quad H-\underset{\underset{H}{|}}{\overset{\overset{H}{|}}{C}}-\underset{\underset{H}{|}}{\overset{\overset{H}{|}}{C}}-Hal \longrightarrow \underset{\diagdown\,H}{\overset{H\,\diagup}{C}}=\underset{\diagdown\,H}{\overset{\diagup\,H}{C}} + H^\bullet + Hal^-$$

Fig. 2. Dehalogenation mechanisms:
a. Nucleophilic substitution (S_N reactions)
b. β-elimination

Aliphatic hydrocarbons with a halogen substituent are relatively susceptible to both nucleophilic displacement and to ß-elimination. Halogenated aromatic compounds on the other hand, are considered to be chemically inert to nucleophilic substitution reactions. Only at high temperatures and under very basic conditions will nucleophilic substitution take place. In addition, an unsaturated ring structure as found in aromatic hydrocarbons is unfavorable for ß-elimination reactions.

Microorganisms, however, have evolved enzymes which make a dehalogenation possible before ring cleavage. One such dehalogenation reaction has been described by Engesser et al. (19). Pseudomonas sp. B13 dehalogenates 2-fluorobenzoate through regioselective dioxygenation by means of a highly adapted enzyme. Suflita et al. (57) demonstrated a reductive dehalogenation of halogenated benzoates by a methanogenic bacterial consortium. The reaction required strictly anaerobic conditions and was independent of the number of halogen substituents. Wood (65) has stated that under aerobic conditions the substitution reaction of a halide by a hydroxyl group on the intact aromatic ring may occur quite frequently through the action of either mixed function oxygenases or dioxygenases. In most of the investigated cases of dehalogenation, however, the halogen is removed only after cleavage of the ring (e.g., 26, 29, 51).

2.3. Limits to biodegradation

Despite the demonstrated biodegradation of a number of chlorinated compounds in laboratory experiments (11, 13, 51), many of these compounds are not readily metabolized in soil and water. Their persistence may be attributed to chemical, physical or environmental factors (1). One of the most important causes is certainly the chemical structure of the compound which will affect biodegradation in three ways. Firstly, no suitable enzyme system exists (yet) for a catabolic process. This might be due to the fact that these synthetic chemicals have only been around for a few decades and consequently the microorganisms had not enough time to get "acquainted" with these molecules. In case the synthetic molecule has no natural chemical analog, existing enzyme systems cannot be modified for its degradation and biodegradation will not occur. The chemical will remain biologically inert until it is able to induce the synthesis of a suitable enzyme system in an organism. Secondly, the molecule may contain substituents which affect the mode and rate of reaction (see previous section). Thirdly, the chemical structure gives the molecule physical properties which minimize its susceptibility to microorganisms (e.g., extremely low solubility in water).

Optimal environmental conditions exist for the biodegradation of every organic compound. Apart from the presence of essential growth factors, favorable temperatures, water activity and pH, the concentration of the compound and the presence of a suitable electron acceptor are of decisive importance. It seems that for some organic compounds the concentration not only influences the rate of mineralization, but also the type of biodegradation observed (2). Isopropyl N-phenyl-carbamate, for instance, is mineralized in samples of fresh water at concentrations of 400 µg/l, but is cometabolized at 1.0 mg/l (63). Other compounds may be toxic at higher concentrations and no longer metabolized at lower concentrations, i.e., threshold concentrations below which no further transformation occurs. Threshold concentrations seem to exist especially for those compounds which support growth (2, 46).

The presence or absence of a specific electron acceptor will also determine the success of biodegradation. Some compounds seem to be degraded in the presence of molecular oxygen (aerobic conditions) and others only in its absence (anaerobic conditions). Because of the vital role of molecular oxygen (dioxygen) in the degradation of some organic compounds, those reactions in which molecular oxygen is directly or indirectly involved will be briefly summarized. Oxygen serves two different functions in the degradation of organic compounds, as a terminal acceptor of electrons which are released during oxidation of organic carbon, and as a reactant in the primary attack on a substrate. In the absence of oxygen, the first function may be transferred to other oxidized compounds such as nitrate, oxidized metal ions, sulfate, carbon dioxide or some organic carbon compounds. However, there is no equivalent to oxygen which can substitute as a reactant in the primary transformation of some types of organic compounds (e.g., benzene). This will be discussed in Chapter Four. The exact role of alternative electron acceptors like nitrate, sulfate, carbon dioxide, etc. during anaerobic degradation processes is not yet fully elucidated.

In summary we can say that the three factors influencing the biodegradation of a chlorinated compound in a natural habitat are: (i) the chemical structure of the compound, (ii) the prevailing environmental conditions, and (iii) the indigenous microbial population (30).

3. BIODEGRADATION OF CHLORINATED ALIPHATIC COMPOUNDS

Most of the research on the degradation of halogenated aliphatic compounds has focused on chlorinated C1 and C2 hydrocarbons (Table III). These compounds represent a significant portion of the known environmental pollutants (49). Although some information exists on the degradation of higher alkanes, e.g., cometabolic degradation of 1-chloroheptane and 1,9-dichlorononane reported by Omori and Alexander (47), we will limit this review to the chlorinated C1 and C2 hydrocarbons.

3.1. Anaerobic conditions

The important and often first step in the degradation of chlorinated aliphatic hydrocarbons is dechlorination. This reaction is as-sumed to be a reductive dechlorination with the necessary electrons obtained from a reduced organic compound (41).

Table III Degradation of chlorinated C1 and C2 hydrocarbons by microorganisms

Compound	Conditions			
	Anaerobic	Reference	Aerobic	Reference
dichloromethane	n.d.		+	13,35,43,52,55
trichloromethane				
(chloroform)	+	9,11,48	-	11
tetrachloromethane	+	9,10	n.d.	
1,2-dichloroethane	+	9	+	31,65
1,1,1-trichloroethane	+	9,48	-	8
1,1,2,2-tetrachloroethane	+	9	n.d.	
trichloroethylene (TRI)	+	9,37,48	-	11
tetrachloroethylene (PER)	+	9,48,62	-	11

+   degraded or transformed
-     no degradation observed
n.d. not determined

Reductive dehalogenation has been reported to be responsible for the degradation of chloroform, carbon tetrachloride, 1,2-dichloroethane, 1,1,1-trichloroethane, 1,1,2,2-tetrachloroethane, trichloroethylene and tetrachloroethylene under methanogenic conditions with acetate added as the main carbon source (9). The extent of degradation was tested with chloroform and carbon tetrachloride. Carbon-14 measure-ments indicated that these chemicals were almost completely oxidized to carbon dioxide by the methanogenic microbial population. The same authors (10) also investigated the degradation of chloroform, carbon tetrachloride and 1,1,1-trichloroethane in the absence of molecular oxygen but in the presence of nitrate as a terminal electron acceptor and with ethanol as the main carbon source. Only carbon tetrachloride was degraded, while at the same time production of chloroform was observed. This observation suggested reductive dechlorination as a possible mechanism. Some carbon dioxide, presumably originating from carbon tetrachloride, was also detected.
Using natural sediments from wetlands, Parsons and Lages (48) demon-strated the anaerobic dechlorination of chloroform, 1,1,1-trichloro-ethane, trichloroethylene and tetrachloroethylene to less chlorinated analog compounds. With carbon-13 experiments, Kleopfer et al. (37) confirmed the biological degradation of trichloroethylene to di-chloroethylene in soils in the absence of oxygen. 1,2-Dichloroethylene was the only endproduct observed. In a study by Vogel and McCarty (62) it was shown that tetrachloroethylene was degraded by reductive de-chlorination to trichloroethylene, dichloroethylene, and vinyl chlo-ride, with the concomitant production of carbon dioxide under methano-genic conditions. The authors hypothesized that after reductive de-chlorination, vinyl chloride could be converted partially to carbon dioxide.
The described experiments have all been performed in the presence of an extra carbon source, many times higher in concentration than the compounds examined. Anaerobic microorganisms, able to grow on chlo-rinated aliphatic compounds as a sole carbon and energy source have

not yet been isolated.

3.2. Aerobic conditions

In the presence of dioxygen, only a few chlorinated aliphatic hydrocarbons have been found to be biodegradable.
Rittmann and McCarty (52) and Klecka (35) observed degradation of dichloromethane by a mixed microbial population. In more detailed studies with isolated bacteria growing on dichloromethane as a sole source of carbon and energy, more insight has been obtained as to the degradation pathway (13, 55). Both Hyphomicrobium DM2 and Pseudomonas sp. strain DM1 used by Stucki et al. (55) and Brunner et al. (13), respectively, were thought to convert dichloromethane via a nucleophilic substitution reaction with a halidohydrolase to monochloromethanol. This intermediate would then be spontaneously decomposed to formaldehyde which is assimilated by the serine pathway. In a study by LaPat-Polasko et al. (43), dichloromethane was shown to be degraded preferentially over acetate by a Pseudomonas sp. strain LP.

1,2-Dichloroethylene has been shown to be biodegradable and even to serve as carbon and energy source for different bacteria (31, 56). A dichloroethylene degrading bacterium isolated by Janssen et al. (13) was classified as a strain of Xanthobacter autotrophicus. This organism was also able to degrade a number of halogenated aliphatics with up to four carbon atoms.

No degradation of trichloromethane, trichloroethylene and tetrachloroethylene has been observed in the presence of molecular oxygen (8), although all three aliphatic compounds are degradable under anaerobic conditions.

4. BIODEGRADATION OF CHLORINATED AROMATIC COMPOUNDS

Halogen substitution in aromatic hydrocarbons generally results in a compound which is less susceptible to microbial degradation (23), see also Section 2.2. Nevertheless, microbial degradation of halogenated aromatic hydrocarbons has been observed and microorganisms have been isolated which are able to grow on a number of these compounds. Gibson and Subramanian (23) suggest that the catabolic reactions may occur via two types of mechanisms; namely by a cometabolic type of process or by an ordinary mineralization process. They hypothesize that in the first reaction type, an existing enzyme system with low specificity can also transform halogenated analogs. This frequently results in a partial degradation, since not all enzymes of the enzyme cascade have the same low specificity.

For complete mineralization of a halogenated aromatic compound, however, a highly specific enzyme system is needed. A number of hypotheses exist concerning the evolution of such a specific enzyme system not originally present in nature (22, 50, 64). It goes beyond the scope of this paper to discuss these theories here. The interested reader should refer to the previous mentioned literature.

In Table IV, a number of chlorinated monoaromatic compounds are given together with their potential for degradation by microorganisms. The degradation of several of these hydrocarbons will be discussed in more detail in the following sections. Since these steps are substantially different from the metabolic reactions occurring during the breakdown of non-chlorinated aromatic compounds, we will restrict ourselves to chlorine elimination and ring cleavage.

Table IV. Degradation of chlorinated monoaromatic hydrocarbons
        by microorganisms

Compound	Conditions			
	Anaerobic	Reference	Aerobic	Reference
chlorobenzene	n.d.		+	4,5,51
1,2-,1,3-,1,4-dichlorobenzene	-	42	+	4,5,42
1,2,3-,1,3,5-trichlorobenzene	n.d.		+	4,5,44
1,2,4-trichlorobenzene	+	60	+	4,5,44
2-,4-chlorobenzoate	n.d.		+	15,26,34, 45,58
3-chlorobenzoate	+	54	+	15,26,32, 29,58
2,4-,3,4-,3,5-dichlorobenzoate	n.d.		+	15,26,27, 58,61
2-,3-,4-chlorophenol	+	12	+	3,40
2,3-,2,5-,2,6-dichlorophenol	-	3,12	+	3,33
2,4-,3,4-,3,5-dichlorophenol	+	3,12	+	3,7,33
2,4,5-,2,4,6-trichlorophenol	-	3	+	3,33
pentachlorophenol	-	3	+	3,14,18, 33,36
3-chlorotoluene	n.d.		+	24,61

+   degraded or transformed
-       no degradation observed
n.d. not determined

4.1. Anaerobic conditions

     Apart from dechlorination, degradation processes of chlorinated
aromatic compounds also require a step for the cleavage of the aromatic ring. Under aerobic conditions this cleavage is preceded by a
hydroxylation in which molecular oxygen is incorporated by means of an
oxygenase. Anaerobic bacteria cannot carry out oxygenative reactions;
thus, they are forced to use a different enzyme system for ring cleavage. Under anaerobic conditions, non-halogenated rings are first de-
aromatized. Ring cleavage subsequently occurs by a hydrolytic step
(20). The basic requirement for anaerobic ring cleavage, however,
appears to be the presence of oxygen in a ring substituent. Due to its
electronic structure in the ring, benzene is a very stable compound.
Through the presence of oxygen in a ring substituent, e.g., a hydroxyl
or carboxyl group, the benzene ring will be slightly destabilized and
more susceptible to hydrogenation. Upon hydrogenation, the oxygen of a
water molecule (instead of molecular oxygen) can be introduced and,
as a consequence, the ring may be opened (21).In contrast to phenol
and benzoate, benzene has not been observed to undergo reduction or
ring cleavage under anaerobic conditions. The effect of a halogen atom
as a sole substituent is unknown. It is of interest to mention here
the results from experiments done by Tsuchiya and Yamaha (60). In a
batch culture of Staphylococcus epidermis, 1,2,4-trichlorobenzene
underwent reductive dechlorination leading to the formation of monochlorobenzene. No further degradation was observed once all the 1,2,4-
trichlorobenzene was transformed to monochlorobenzene. Ring cleavage,
however, did occur in experiments where, in addition to a chlorine,

the ring contained a carboxyl or hydroxyl group. 3-Chlorobenzoate (54) and mono- and dichlorophenols (12) first underwent dechlorination and then, as normal benzoates or phenols were mineralized to carbon dioxide and methane. A methanogenic consortium, from which seven bacteria were isolated, was responsible for the degradation of 3-chlorobenzoate. The dechlorination to benzoate was performed by only one bacterium. The observation that this bacterium was only capable of growth on pyruvate and not on benzoate, illustrates the complexity of a microbial community responsible for total degradation of a chlorinated compound. Little or no information exists on the degradation of chlorinated aromatic hydrocarbons in anaerobic environments where nitrate or sulfate is present as an electron acceptor.

4.2. Aerobic conditions

Formation of catechols
    The majority of the chlorinated monoaromatic hydrocarbons which have been examined up to now are degraded by microorganisms that leave the carbon-chlorine bond intact until chlorocatechols are produced and the ring is cleaved by oxygenases. Chlorinated benzenes, with or without an additional carboxyl- or phenol group as a substituent, may be converted to chlorocatechols via different steps and enzymes (Fig. 3).

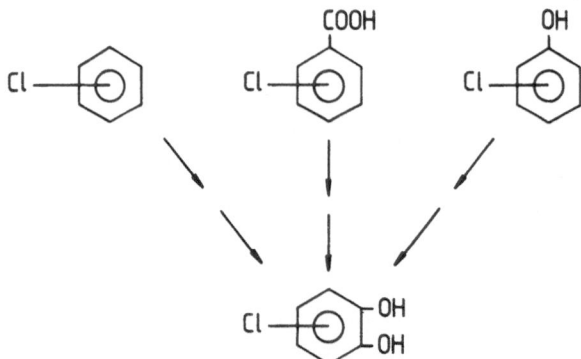

Fig. 3. Schematic reactions for the formation of chlorocatechols from chlorinated benzenes,-benzoates and -phenols.

Either monooxygenases or dioxygenases catalyze the initial reactions in which the nucleus is hydroxylated in the presence of molecular oxygen and a hydrogen donor (NADH or NADPH). Monooxygenases will initially form chlorophenols which will then undergo a second hydroxylation to chlorocatechols. This pathway has been postulated by Ballschmiter and Scholz (4) for the degradation of a number of di- and trichlorobenzenes by different Pseudomonas species. Dioxygenases, on the other hand, have been found to catalyze the first steps in the degradation of chlorobenzene (51), and 3-chloro-, 4-chloro-, and 3,5-

dichlorobenzoate (26). The dioxygenases form cis-dihydrodiols which subsequently undergo dehydrogenation, and in case of the benzoates, also decarboxylation, to yield chlorocatechols. A dioxygenase was found to also be involved in the degradation of 3-chlorotoluene (24).

Ring cleavage

Chlorocatechols may undergo ortho- or meta-cleavage which is achieved by dioxygenase catalyzed reactions in which both atoms of molecular oxygen are incorporated (Fig. 4).

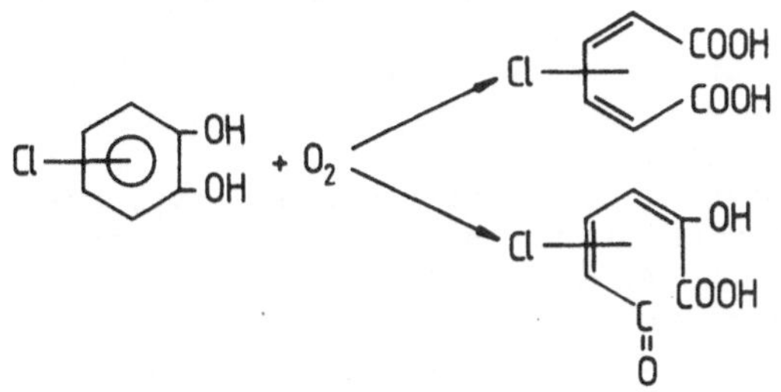

Fig. 4. Ortho- and meta-pathway of ring cleavage of a chlorocatechol.

The chlorocatechols represent critical metabolites during bio-degradation. The negative inductive effect of the chlorine atoms (5) decreases the nucleophilicity of the catechol and this, plus the steric effect, hinder an electrophilic attack by dioxygenases which are normally involved in the ortho-cleavage of non-halogenated catechols. The relative activity of such a dioxygenase for 3-chloro-catechol has been found to be less than 1% in comparison with that for catechol (16). As a result of the low activity of this enzyme system, the chlorocatechols and their dark colored autoxidation products accu-mulate in the culture medium. Thus, enzymes with a high activity are necessary to avoid the accumulation of chlorocatechols. Pseudomonas sp. B13 is one organism in which such an enzyme for ring cleavage has been demonstrated (17).

During the degradation of 3-chlorocatechol, meta-cleavage may occur. Bacteria that can degrade non-halogenated compounds such as benzene and toluene mostly use the meta-cleavage pathway. A chlori-nated compound which is present may subsequently be degraded by a cometabolic kind of reaction leading to the formation of acyl halides, which irreversibly bind to the enzyme. This inactivation of a key enzyme is lethal for the microorganisms (6).

The 4-chlorocatechol isomer, however, has been demonstrated to be completely degraded via ortho-cleavage in Pseudomonas sp. WR912 (26). The degradation of 4-chlorocatechol will also come to an end in case meta-cleavage takes place. As a result, a semialdehyde will accumulate as a dead end product (39).

The inefficient ortho-cleavage activity for chlorinated catechols by bacteria which normally act on non-chlorinated aromatics and the accumulation of chlorocatechols and suicide products in the case of meta-cleavage activity are thought to be the main reasons for the difficulties encountered in the biodegradation of chlorinated aromatic compounds. However, several of these compounds have been shown to serve as sole carbon and energy sources. The isolation of bacteria able to use e.g., chlorobenzene (51) and 1,2- and 1,4-dichlorobenzene and 1,2,4-trichlorobenzene (Schraa et al, in prep.) as sole carbon and energy sources indicates that a degradation pathway must exist which leads to the total mineralization of these chemicals. Some chlorinated compounds, e.g., 3-chlorobenzoate (32) and 4-chlorobenzoate (34, 45), have also been shown to undergo dechlorination before ring cleavage. The mechanisms of these dechlorinations remain unknown. The operation of an oxygenase is inconsistent with the data for the dechlorination of 4-chlorobenzoate to 4-hydroxybenzoate, suggesting that halidohydrolysis is the most likely mechanism (45). A halidohydrolase or a monooxygenase has been suggested to be involved in the production of chlorophenol intermediates in the mineralization of 1,2,3- and 1,2,4-trichlorobenzene in soil (44). However, conclusive evidence has not been presented.

The resistance of halogenated aromatic compounds to microbial degradation is dependent on the number and position of the halogen atoms on the ring (65). An increase in the number of halogen atoms would result in a greater resistance to microbial attack. The widely used pesticide pentachlorophenol, however, has been shown to be extensively degraded (36), and utilized as a sole carbon and energy source by several bacteria (14, 18). The biochemical pathways of its biodegradation remain to be elucidated.

5. CONCLUDING REMARKS

The chlorinated and aromatic hydrocarbons dealt with in this chapter are man-made and have only recently been brought into the environment. Nevertheless, the experiments performed to date show that many of these compounds can be degraded by microorganisms in a controlled environment in the laboratory. This indicates that microorganisms have apparently evolved the ability to degrade these compounds, either via existing enzyme systems (cometabolism) or via newly evolved ones (complete mineralization). The genes for the degradation of halogenated aromatic compounds have often been found to be plasmid-associated. Although not discussed in this review, the evidence is strong that microorganisms which have the genetic information for complete degradation of a xenobiotic compound may have obtained it by gene transfer among bacteria in mixed microbial communities (50). Plasmids are found to be exchanged by conjugation (64). This "natural" genetic engineering may produce an organism which has a complete enzyme system for the biodegradation of a novel xenobiotic compound. The degradation of a xenobiotic compound is not necessarily dependent on one microorganism; a consortium of microorganisms which are interdependent can have the same result.

The observation that compounds which are degradable in laboratory experiments but are persistent in nature may be the result of limitations based on, e.g., concentration and availability of the compound, environmental conditions (electron acceptors, nutrients, additional carbon sources, etc.) and active bacteria present. All these factors must lay within certain limits to allow the compound to be degraded. These limits, however, may vary from one compound to an other.

Important questions to which this review has not addressed itself or has just pointed to include: What is the role of plasmids? What are the mechanisms of the enzymes? How do degradation pathways evolve? What are the potential applications of laboratory findings for the removal of xenobiotic compounds from contaminated environments? etc.

Only a concerted research effort will allow us to significantly increase our knowledge and as a result explore the potential abilities of microorganisms. Subsequently this may lead to a use of their capacity to deal successfully with the pollution of our environment by xenobiotic compounds.

ACKNOWLEDGMENT
We thank Patricia J. Colberg and Daniel A. Kluepfel for critical reading of the manuscript.

REFERENCES
1. ALEXANDER, M. 1981. Science 211:132-138.
2. ALEXANDER, M. 1985. Environ. Sci. Technol. 18:106-111.
3. BAKER, M.D. and MAYFIELD, C.I. 1980. Water Air Soil Pollution 13:411-424.
4. BALLSCHMITER, K. and SCHOLZ, Ch. 1980. Chemosphere 9:457-467.
5. BALLSCHMITER, K., UNGLERT, C., HEINZMANN, P. 1977. Angew. Chem. Int. Ed. Engl. 16:645.
6. BARTELS, I., KNACKMUSS, H.-J., REINEKE, W. 1984. Appl. Environ. Microbiol. 47:500-505.
7. BELTRAME, P., BELTRAME, P.L., CARNITI, P., PITES, D. 1982. Water Res. 16:429-433.
8. BOUWER, E.J. and McCARTY, P.L. in: Proceedings of the ASCE Environ. Eng. Div. Specialty Conf. 196, July 8-10, 1981.
9. BOUWER, E.J. and McCARTY, P.L. 1983a. Appl. Environ. Microbiol 45:1286-1294.
10. BOUWER, E.J. and McCARTY, P.L. 1983b. Appl. Environ. Microbiol. 45:1295-1299.
11. BOUWER, E.J., RITTMANN, B.E., McCARTY, P.L. 1981. Environ. Sci. Technol. 15:596-599.
12. BOYD, S.A. and SHELTON, D.R. 1984. Appl. Environ. Microbiol. 47:272-277.
13. BRUNNER, W., STAUB, D., LEISINGER, T. 1980. Appl. Environ. Microbiol. 40:950-958.
14. CHU, J.P. and KIRSCH, E.J. 1972. Appl. Microbiol. 23:1033-1035.
15. DIGERONIMO, M.J., NIKAIDO, M., ALEXANDER, M. 1979. Appl. Environ. Microbiol. 37:619-625.
16. DORN, E. and KNACKMUSS, H.-J. 1978. Biochem. J. 174:85-94.
17. DORN, E., HELLWIG, M., REINEKE, W., KNACKMUSS, H.-J. 1974. Arch. Microbiol. 99:61-70.
18. EDGEHILL, R.U. and FINN, R.K. 1982. Eur. J. Appl. Microbiol. Biotechn. 16:179-184.
19. ENGESSER, K.-H., SCHMIDT, E., KNACKMUSS, H.-J. 1980. Appl. Environ. Microbiol. 39:68-73.
20. EVANS, W.C. 1977. Nature 270:17-22.
21. FEWSON, C.A. in Leisinger et al. (eds.), Microbial Degradation of Xenobiotics and Recalcitrant Compounds, 141-179, Academic Press, London, 1981.
22. GHOSAL, D., YOU, I.-S., CHATTERJEE, D.K., CHAKRABARTY, A.M. 1985. Science 228:135-142.
23. GIBSON, D.T. and SUBRAMANIAN, V. in D.T. Gibson (ed.), Microbial Degradation of Organic Compounds, 180-252, Marcel Dekker, Inc. New York, 1984.
24. GIBSON, D.T., KOCH, J.R., SCHULEL, C.L., KALLIO, R.E. 1968. Biochemistry 7:3795-3802.
25. GLAZE, W.H. and HENDERSON, J.E. 1975. J. Water Pollut. Control Fed. 47:2511-2515.

26. HARTMANN, J., REINEKE, W., KNACKMUSS, H.-J. 1979. Appl. Environ. Microbiol. 37:421-428.
27. HOROWITZ, A., SUFLITA, J.M., TIEDJE, J.M. 1983. Appl. Environ. Microbiol. 45:1459-1465.
28. HORVATH, R.S. 1972. Bact. Rev. 36:146-155.
29. HORVATH, R.S. and ALEXANDER, M. 1970. Appl. Microbiol.20:254-258.
30. HUTZINGER, O. and VEERKAMP, W. in Leisinger et al. (eds.), Microbial Degradation of Xenobiotics and Recalcitrant Compounds, 3-45, Academic Press, London, 1981.
31. JANSSEN, D.B., SCHEPER, A., DIJKHUIZEN, L., WITHOLT, B. 1985. Appl. Environ. Microbiol. 49:673-677.
32. JOHNSTON, H.W., BRIGGS, G.G., ALEXANDER, M. 1972. Soil Biol. Biochem. 4:187-190.
33. KARNS, J.S., KILBANE, J.J., DUTTAGUPTA, S., CHAKRABARTY, A.M. 1983. Appl. Environ. Microbiol. 46:1176-1181.
34. KLAGES, U. and LINGENS, F. 1979. FEMS Microbiol. Lett. 6:201-203.
35. KLECKA, G.M. 1982. Appl. Environ. Microbiol. 44:701-707.
36. KLECKA, G.M. and MAIER, W.J. 1985. Appl. Environ. Microbiol. 49:46-53.
37. KLEOPFER, R.D., EASLEY, D.M., HAAS, B.B.Jr., DEIHL, T.G., JACKSON, D.E., WURREY, C.J. 1985. Environ. Sci. Technol. 19:277-280.
38. KNACKMUSS, H.-J. 1975. Chemiker Zeitung 99:213-219.
39. KNACKMUSS, H.-J. in Leisinger et al. (eds.), Microbial Degradation of Xenobiotics and Recalcitrant Compounds, 189-212, Academic Press, London, 1981.
40. KNACKMUSS, H.-J. and HELLWIG, M. 1978. Arch. Microbiol 117:1-7.
41. KOBAYASHI, H. and RITTMANN, B.E. 1982. Environ. Sci. Technol 16:170A-183A.
42. KUHN, E.P., COLBERG, P.J., SCHNOOR, J.L., WANNER, O., ZEHNDER, A.J.B., SCHWARZENBACH, R.P. Environ. Sci. Technol., in press.
43. LAPAT-POLASKO, L.T., McCARTY, P.L., ZEHNDER, A.J.B. 1984. Appl. Environ. Microbiol. 47:825-830.
44. MARINUCCI, A.C. and BARTHA, R. 1979. Appl. Environ. Microbiol. 38:811-817.
45. MARKS, T.S., SMITH, A.R.W., QUIRK, A.V. 1984. Appl. Environ. Microbiol. 48:1020-1025.
46. McCARTY, P.L., REINHARD, M., RITTMANN, B.E. 1981. Environ. Sci. Technol. 15:40-51.
47. OMORI, T. and ALEXANDER, M. 1978. Appl. Environ. Microbiol 35:867-871.
48. PARSONS, F. AND LAGES, G.B. 1985. J. Am. Water Works Assoc. 77:52-59.
49. PEARSON, C.R., in O. Hutzinger (ed.), The Handbook of Environmental Chemistry: Anthropogenic Compounds, Vol. 3, Part B, 69-116, Springer-Verlag, Berlin, 1982.
50. REINEKE, W. AND KNACKMUSS, H.-J. 1979. Nature 277:385-386.
51. REINEKE, W. and KNACKMUSS, H.-J. 1984. Appl. Environ. Microbiol. 47:395-402.
52. RITTMANN, B.E. and McCARTY, P.L. 1980. Appl. Environ. Micriobiol. 39:1225-1226.
53. SCHNEIDER, M.J. Persistent Poisons: Chemical Pollutants in the Environment, The New York Academy of Sciences, N.Y., 1979.
54. SHELTON, D.R. and TIEDJE, J.M. 1984. Appl. Environ. Microbiol. 48:840-848.

55. STUCKI, G., GäLLI, R., EBERSOLD, H.-R., LEISINGER, T. 1981. Arch. Microbiol. 130:366-371.
56. STUCKI, G., KREBSER, U., LEISINGER, T. 1983. Experientia 39:1271-1273.
57. SUFLITA, J.M., HOROWITZ, A., SHELTON, D.R., TIEDJE, J.M. 1982. Science 218:115-116.
58. SUFLITA, J.M., ROBINSON, J.A., TIEDJE, J.M. 1983. Appl. Environ. Microbiol. 45:1466-1473.
59. SIUDA, J.F. and DEBERNARDIS, J.F. 1973. Lloydia 36:107-143.
60. TSUCHIYA, T. and YAMAHA, T. 1984. Agric. Biol. Chem. 48:1545-1550.
61. VANDENBERGH, P.A., OLSEN, R.H. and COLARUOTOLO, J.F. 1981. Appl. Environ. Microbiol. 42:737-739.
62. VOGEL, T.M. and McCARTY, P.L. 1985. Appl. Environ. Microbiol. 49:1080-1083.
63. WANG, Y.S., SUBBA-RAO, R.V., ALEXANDER, M. 1984. Appl. Environ. Microbiol. 47:1195-2000.
64. WILLIAMS, P.A. 1982. Phil. Trans. R. Soc. Lond. B. 297:631-639.
65. WOOD, J.M. 1982. Environ. Sci. Technol. 16:291A-297A.

REVIEW OF MICROBIAL TRANSFORMATIONS

H.A. PAINTER
Water Research Centre, Medmenham, UK.

Summary

The various ways in which organic chemicals can be changed by the action of micro-organisms are reviewed. All organic compounds are capable of releasing sufficient energy for growth on degradation, but the necessary enzyme systems are often not available for the conversion of xenobiotic compounds to mineral products. Microbial enzymes having broad substrate specificity are capable of relatively slowly effecting single, simple structural changes, or a series of changes in xenobiotic chemicals, resulting in transformation products which are sufficiently recalcitrant for them to persist in the environment. Three broad classes of reaction are identified – transformation, cleavage and conjugation – and examples of many types of reaction are quoted. Difficulties in predicting events likely to occur in the environment from results of tests with monocultures with single or multi-substrates are discussed. A large body of data on transformations are available and it is suggested that a register be set up and maintained recording the known reactions for application to environmental problems.

1. INTRODUCTION

This review was prepared in the context of COST 64b and COST 641, which have produced an extensive list (CICLOPS) of organic compounds, and their concentrations, found in the aquatic environment – rivers, lakes, effluents etc. It is little concerned with detailed enzymology of microbial transformations, but rather with what products are known, or likely, to be formed. The work was undertaken not just as a didactic exercise to help identify unknown ´peaks´ or decide about further compounds to seek, but for the practical objectives of aiding interpretation of data so far collected and assisting in tracing sources of pollution thus helping to avoid harmful effects on the environment.

Essentially the interest lies in the transformations of xenobiotic compounds brought about by micro-organisms and in this context Alexander (1) has brought together a valuable list of reactions together with examples of chemicals of environmental concern most of which were pesticides. The present paper, which is far from comprehensive, owes a great deal to Alexander´s work.

It may be mentioned, in passing, that computer searches of the literature lead to many false trails because of linguistic difficulties and , presumably, because of the mis-use and/or non-use of key words in the original papers. In particular, the words – fate, transport, persistence, removal, behaviour, pathway, transformation, degradation – can be used in the physical sense, leading the unsuspecting reader to papers quite unconnected with microbial transformations.

At the outset it is necessary to distinguish between biodegradation and microbial transformations. The definitions recently given by Nielson et al (2) are useful for this purpose, since they do not involve the contentious word "cometabolism". <u>Biodegradation</u> implies drastic changes in the molecule leading to release of energy and microbial growth, and nearly always resulting finally in mineralisation. <u>Transformations</u>, however, are considered to be relatively small changes in molecular structure, not involving growth and not necessarily leading to further changes in structure.

2. ORIGINS OF CHEMICALS IN THE AQUATIC ENVIRONMENT

It is worthwhile briefly to consider the sources from which organic chemicals reach the environment. First, there is the obvious source - the direct discharge by domestic and industrial users of chemicals to sewers and surface waters. A very wide range of compounds are discharged and will include surfactants and additives, solvents, plasticisers, dyestuffs, explosives, flame-retardants etc, as well as the compounds from which they are synthesised. and by-products formed. Another important source of xenobiotics is the pharmaceutical industry and it should be noted that many metabolites (ie transformation products) of drugs are excreted by man and thus reach the environment. Marten (3) discussed a number of these metabolites and lists seven commonly observed initial reactions by which the body (liver) attempts to detoxify the consumed drug. Richardson and Bowron (4) have investigated the fate of some 200 pharmaceutical chemicals in the aquatic environment.

The second source is waste-water treatment during which microbial transformations of discharged xenobiotics may occur. The introduction of xenobiotics to sewage treatment is particularly important. Often it is here that micro-organisms are first challenged by new chemicals and in a sense, waste water treatment acts as a first line of defence of the environment against toxic effects of xenobiotics. However, the residence time of the liquid wastes in the aerobic process (activated sludge) is relatively low (3-12 h) giving little opportunity for microbial action although in anaerobic treatment of sludge (containing insoluble and adsorbed xenobiotics) the retention time is 20-25 d. A third source of chemicals reaching natural waters is the soil, in which both aerobic and anaerobic microbial processes take place. Pesticides and their transformation products will form the majority of chemicals leached out of soil, but the practice of applying sewage sludge to soil will increase the numbers of non-pesticide xenobiotics present.

Finally, reactions occur within the aquatic environment itself, giving rise to derivatives; transformations take place not only by microbial agencies, but also within aquatic animals and plants, and in bottom deposits in which both aerobic and anaerobic conditions exist.

It should be noted that dynamic situations exist in bodies of water; both cell synthesis and cell lysis occur simultaneously so that at any given time simple, biodegradable compounds such as acetate, glucose, amino-acids will be present, albeit at extremely low concentrations. Also, since biochemical reactions do not go to completion and since complex chemicals are transformed only slowly, traces of readily degradable compounds, such as aniline, higher fatty acids, theophylline, will usually be found in environmental samples and sometimes even in drinking water (5).

3. TYPES OF MICROBIAL METABOLISM

When a chemical is biodegraded to mineral products, growth of new cells, of pure or mixed cultures, normally takes place, as in laboratory cultures, organic intermediates may or may not be found in the culture medium. In sewage treatment, the concentration of substrate is considerably lower than in normal laboratory cultures so that the specific growth rate of the micro-organisms is much lower than their maximum specific growth rate. As a result a large proportion of the population is thought to be non-viable. The substrate is degraded and growth occurs but it is not clear what the role of the non-viable cells is. Sometimes growth can occur on a degradable moiety of a molecule, leaving the remainder intact; for example, chloridazon is converted to a derivative of catechol, which allows growth, and 5-amino,4-chloro-3(2H)- pyridazinone, which persists (6), and activated sludge converts about 50% of ampillicin to carbon dioxide, presumably leaving a penicillin residue (7) intact.

Microbial transformations of compounds which are not biodegraded, as defined above, take place as a result of what has been defined as "co-metabolism" in which no energy is released for growth and the product accumulates. Other terms have been coined to describe other reactions, for example, analogue, dead-end, suicide, fortuitous and concurrent metabolism, which are self-explanatory. However, the need for a term such as cometabolism has been questioned by a number of authors, especially since microbial transformations occur in single substrate cultures without the need for a second, growth, substrate. Dagley (8) has suggested that cometabolism is equivalent to detoxication in mammals and in some fungi. It is to be noted that ´cometabolic´ products can be less or even more toxic, to a given species, than the original compound, while transformation in mammals normally results in detoxication (or detoxification). There are exceptions, however, since mammals can ´activate´ dosed precursors of active drugs (3).

More importantly, the question of whether and when growth (mineralisation) or cometabolism (transformation) occurs in natural waters, in which oligotrophic bacteria probably play an important role, is complex. Experimentally, studies have to be made with radio-labelled compounds. Alexander (9) has recently discussed factors relating to this problem, namely threshold concentration of xenobiotic, apparent inhibition, carbon assimilation and kinetics of degradation. He concludes that the behaviour depends on the nature of the chemical; some are mineralised at rates proportional to their concentration, while others are mineralised at trace levels but may be only transformed (cometabolised) at higher concentrations. Yet other usually mineralisable molecules are not converted to carbon dioxide below individual threshold concentrations. He further concludes that at ´trace´ concentrations of xenobiotics, micro-organisms may not be able to produce the acclimatised populations needed for enhanced biodegradation. In other papers (10,11) he has shown how to predict the minimum concentrations of chemical needed to support microbial growth.

4. TYPES OF EXPERIMENTATION; SOURCES OF INFORMATION
There is a large and growing literature on the biodegradation and microbial transformation of xenobiotic compounds; a number of useful reviews have been written over the past ten years in journals(1,8,12-15) and in books (16-18).

Most experiments have been made with single pure stock cultures or with pure cultures isolated from the pertinent environment (soil, sewage, river water etc) and some have been done with known mixed cultures. Fewer attempts have been made to identify transformation products in laboratory experiments using natural communities of organisms, or microcosms (eg river water, sediments) but work in this field is increasing. Even fewer attempts have been reported of the identification of transformation products in the environment.

While pure culture experiments can give clues as to what products are formed, it cannot be firmly concluded from such results that these products will be formed in nature, nor that, if they are, they will be formed in the same proportions as found in vitro. The purpose of mono-culture studies is usually to establish metabolic pathways by which the selected species degrades/transforms the chemical and how new abilities are acquired; mixed cultures would not convincingly do this, if at all. Also, a single species may transform a xenobiotic into a product which is stable in that culture because further attack was blocked, but had other species been present, as they probably would be in nature, additional transformations or even mineralisation might have occurred. For example, various mono-chlorobiphenyls are converted in mono-culture by some species to the corresponding mono-chlorobenzoic acids which accumulated. In other monocultures, these mono-chlorobenzoic acids were converted, after periods of acclimatisation, to intermediates identified as 3-and 4-chlorocatechols (19) while with other strains the possibility of chlorocatechols was excluded, the products being thought to be monochlorohydro-quinones or monochloropyrogallols (2). On the other hand, mixed cultures, in the form of activated sludge grown on sewage, were able to mineralise 2-and 3-chlorobenzoic acids, although some sludges needed much longer acclimatisation than others (7): this does not necessarily mean that these chloro-compounds will not appear in the environment.

Some chemicals can be biodegraded or transformed only by communities of micro-organisms (see for example Slater and Lovatt in (9)).

It is thought by some workers that it is imperative that transformations be experimentally established in natural ecosystems or in models of them, although it has been conceded (2) that virtually all transformations carried out by a consortium of species or by a single species in the presence of a co-substrate have also been demonstrated for a single species-single substrate system. Nielson et al (2) discussed factors which affected the interpretation of results obtained from mono- and multi-substrated experiments. They concluded that, since the presence of a second substrate could sometimes (a) lead to a higher cell density, (b) cause accumulation of toxic metabolites, (c) cause inhibition by unsubstituted analogues of the xenobiotic, or (d) alter the regulation of catabolic enzymes at low substrate conditions, results of experiments using more than one substrate do not necessarily have greater environmental significance than those utilising only single substrates.

5. EXCLUDED TRANSFORMATIONS

Reported in the literature under the heading of "microbial transformations" or "bioconversions" are many reactions in which products are formed in high yield by mono-cultures grown or maintained under special conditions often unrelated to those in the environment. These reactions, used for preparative purposes, often produces enantiomorphs difficult to prepare by purely chemical reactions. Examples are acetamido-cinnamic acid $\longrightarrow$ L-phenylalanine, α-allenic alcohols $\longrightarrow$ α-allenic acids, 2-butanol $\longrightarrow$ methyl ethyl ketone and squalene $\longrightarrow$ squalene-12-one. Such reactions are not thought to be representative of what would occur in the environment and are excluded from further consideration. The reader is referred to the Handbook of Microbiology (20) and such journals as Applied Microbiology and Technology for details of 'bioconversions' of hydrocarbons, steroids, alkaloids, terpenes, pesticides and antibiotics.

6. HOW TRANSFORMATIONS OCCUR

From thermodynamic considerations it is concluded that virtually all organic compounds could provide sufficient energy, if degraded, to support growth - the question is whether the relevant enzyme system exists and is available. Biodegradable chemicals are broken down, normally after entry into the cell, by a series of enzymic reactions, the first steps are usually hydrolytic rather than oxidative. The enzymic attack continues until the substrate is completely mineralised. For example, the side chains of benzenoid compounds are manipulated in a series of steps, by ω- and then β- oxidation and hydroxylation of the ring then occurs to form catechol, protocatechuic acid or gentisic acid (or simple derivatives of them), after which ring cleavage takes place. Ring fission products are then converted by further sequential reactions to form simple compounds like pyruvate and fumarate, which are then used in anabolic processes for cell synthesis.

Hydrolytic enzymes employed in the initial steps often have fairly wide substrate-specificity and can sometimes act, fortuitously if rather slowly, on xenobiotics especially if there is some similarity of structure with the normal substrate. The facility to degrade the xenobiotic may increase if the cell learns to modify the structure of the enzyme accordingly; alternatively, the initial step(s) proceed very slowly or may have to await the emergence of a mutant species which can synthesize the necessary enzyme system.

If the cell is able to adapt at each enzymic step so that at some stage an intermediate product is formed, which is a recognised member of a known metabolic pathway, complete degradation, or mineralisation, will result. However, if at some stage the succeeding step does not occur because the appropriate enzyme is not present or does so much more slowly than the preceding step, the transformation product will accumulate and persist, providing that it is excreted by the organism. In mixed cultures, and in some mono-cultures, a number of different initial transformation steps may take place simultaneously and sequentially, so that a variety of transformation products can be formed. Also, halogenated aromatic compounds may be dehalogenated early in the pathway and this facilitates subsequent cleavage of the ring, but if ring-fission occurs before dehalogenation the halogenated aliphatic produce is less easy to transform. So-called "dead-end" metabolites can thus be formed, eg fluoro-acetate from 3-fluorocatechol, and fluoro-citrate which are recalcitrant and can even be strongly toxic to other pathways.

Some general indications, rather than rules, can be gleaned from the literature as to the ease with which organic compounds can be degraded, eg aliphatic hydrocarbons (21), cyclo-aliphatic (22) and substituted aromatics. In the last case, degradation is hindered by a high degree of side-chain branching, and by groups such as halogen, nitro, sulphonic, nitrile and trifluoro-methyl. Aromatic ring cleavage normally requires the presence of two hydroxyl groups but 1-hydroxy,2-naphthoic acid and 5-chloro-salicylic acid can be directly oxidised to aliphatic products.

Xenobiotics may be transformed via different pathways in different species, so yielding different products. Occasionally abiotic reactions - photo, hydrolytic, molecular rearrangements - intervene to help in the general attack on the xenobiotic. Some compounds can be transformed by only relatively few species which do not seem to be widely distributed. Thus, this would explain the uneven appearance of some products and partly explains why biodegradability tests give disparate results from time to time with a given source of inoculum, and from one source of inoculum to another for some problematic chemicals.

7. TYPES OF TRANSFORMATIONS

Table I summarises some of the transformation reactions recorded in the literature together with examples; in Table II several types of cleavage reactions are given. The classification of these reactions is as given by Alexander (1). All the steps listed have been established in various microbial systems, as indicated, and in most cases the transformations have seemed to be biological and not abiotic. Some of the sequences, such as hydroxylation, ester hydrolysis and dehalogenation, are common, while for others only a few, or single, examples could be found. The initial steps can take place in any order and products can undergo transformation; for example, Aldicarb can be converted to a sulphoxide, sulphone, oxime and nitrile, as well as to oximes and nitriles of sulphoxides and sulphones (23). The products listed in the Tables were detected using whole cells (and not enzyme preparations) so that they are extra-cellular and will appear in the environment.

A minority of the reactions listed occur only under anaerobic conditions; for instance, the reductive dehalogenation of carbon tetrachloride and 1,1,1- trichloroethane requires at least anoxic conditions, while the reduction of nitro groups probably require anaerobic conditions. Some transformations need a preliminary step before they take place- the aerobic β-oxidation step may first need an ω-oxidation.

In addition to the transformations listed in Table I, Alexander (1) records others which take place in soil and/or natural waters. Triple bond ($RC\equiv CH \rightarrow RCH=CH_2$) reduction is exemplified by the conversion of buturon, while DDT is an example of a xenobiotic being reduced ($Ar_2C=CH_2 \rightarrow Ar_2CH.CH_3$) or hydrated ($Ar_2C=CH_2 \rightarrow Ar_2.CH_2OH$) at a double bond. Parathion and ethylenethiourea are examples of the replacement of $>S$ by $>O$, while phorate exemplifies the reduction of sulphoxide to sulphide. Additional cleavage reactions not listed in Table II involve nitrogen, sulphur and phosphorus (1). The thioalkyl group attached to P in Hinosan and Kitazin P is replaced by hydroxyl in mixed cultures. The hydroxyl group also replaces the amino group attached to S in Oryzalin, while the -S-S- bond is reduced to -SH in Thiram, both in soil.

Ring cleavage. Cleavage of the aromatic ring normally leads to growth and mineralisation. However, in some circumstances aliphatic products are formed which are recalcitrant or only very slowly transformed. Examples, in Table III, are some halogenated benzene compounds which form acylhalides or halo-aliphatic acids, which accumulate. In multiring systems, the less heavily substituted ring is cleaved, leaving the more heavily substituted ring(s) intact, but containing extra hydroxyl and/or carboxyl groups.

Other reactions. A few other reactions not fitting in with those already listed are worth mentioning. Pentachloronitrobenzene has been reported (24) to be transformed by a protozoan, via the mediation of glutathione, to pentachlorothioanisole, as well as pentachloroaniline. The anisole and the derived 2,3,5,6-tetrachloro,1,4-dithiomethylbenzene have been identified in mussels in coastal waters (25), but their origin was not traced.

In an aerobic soil, a new organophosphorus carbamate insecticide, containing RN-S-NR^1, split at the two S-N linkages to give RNH and R'NH (thiophosphoramide) (26).

In the presence of nitrate, 3,4-dichloroaniline, formed from phenylurea and acylanilide pesticides, was transformed by a minor pathway not only to the azo and azo-amino derivatives but also to 3,4,3',4'-tetrachlorodiphenyl (27,28).

Hydroxylations of aromatic compounds can sometimes lead to

intra-molecular rearrangements, as well as to the expected hydroxy product. For example, 2,4-dichloro-phenoxyacetic acid was hydroxylated in a fungal culture in the normal 5-position, but in addition 4-hydroxy,2,5-dichlorophenoxyacetic acid was also detected (29).

The piscicide, squoxin(1,1´-methylene bis(2-naphthol)) was observed to degrade in river water to a series of hydroxy- and keto-products, of which 1-hydroxy,2-naphthol was the most stable (30).

Conjugations. Besides effecting degradative reactions, micro-organisms can add simple groups to xenobiotics and also form dimers, polymers and N-ring compounds. The known reactions, listed in Table IV, appear to be limited to methylation, ether formation, N-acylation, nitration and nitrosation: Alexander (1) cites other examples. Polymerisation of aromatic amines to coloured products has often been catalysed by extra-cellular fungal enzymes; halo-,alkyl- and alkoxy-anilines were converted to dimers-pentamers by laccase from a soil fungus (31).

8. CONCLUSIONS

Transformations of chemicals are carried out by enzymes in or on microbial cells and can be considered as attempts either to degrade the chemical to obtain energy for growth, leading finally to mineralisation, or to render the xenobiotic chemical less toxic (detoxication, "cometabolism"). Products formed by the latter reactions are likely to accumulate and persist in the environment.

The transformations by micro-organisms of xenobiotic chemicals discharged to the environment are numerous and varied; most, if not all, classes of chemicals have been found to be involved. The accumulated and growing body of information on individual xenobiotics is sufficient to make it worthwhile to collect, compile and keep up to date a register, similar to CICLOPS, of transformation, cleavage and conjugation reactions, together with the products. Such a register would aid the identification of more compounds present in natural waters and help prevent the occurrence of harmful effects to the environment.

REFERENCES

(1) ALEXANDER, M. (1981). Science, **211**, 132-138.
(2) NIELSON, A.H., ALLARD, A-S and REMBERGER, M. (1985) In the Handbook of Environmental Chemistry, Vol. 2, Part C (Hutzinger, O. ed.) 28-86. Berlin:Springer-Verlag.
(3) MARTEN, T.R. (1985). Chem. in Britain, August, 745-748.
(4) RICHARDSON, M.L. and BOWRON, J.M. (1985). J. Pharm. Pharmacol. **37**, 1-12.
(5) CRATHORNE, B. et al (1984). Env. Sci. Tech. **18**, 797-802.
(6) EBERSPACHER, J. and LINGENS, F. (1981). In Microbial degradation of xenobiotic and recalcitrant compounds. FEMS Symp. No. 12 (Leisinger, T. et al) 271-285, Academic Press, London.
(7) PAINTER, H.A. and KING, E.F. (unpublished).
(8) DAGLEY, S. In Microbial degradation of organic compounds (Gibson, D.T. ed.) pp.1-10, New York:Marcel Dekker, Inc.
(9) ALEXANDER, M. (1985). Env. Sci. Tech. **18**, 106-111.
(10) SCHMIDT, S.K., SHULER, M.L. and ALEXANDER, M. J. Theor. Biol. (in press).
(11) SIMKISS, S. and ALEXANDER, M. (1984). Appl. Environ. Microbiol. **47**, 1299-1306.
(12) DAGLEY, S. (1975). Essays Biochem. **11**, 81.

(13) KOBAYASHI, H. and RITTMAN, B.E. (1982).   Env. Sci. Tech. **16**, 172A-183A.
(14) LAL, R. and SAXENA, D.M. (1982). Microbiol. Rev. **46**, 95-127.
(15) ATLAS, R.M. (1981) Microbiol. Rev. **45**, 180-209.
(16) LEISINGER, T. et al. (1981).   Microbial degradation of xenobiotics and recalcitrant compounds.   FEMS Symp. No. 12 London: Academic Press.
(17) CHAKRABARTY, A.M. (1982).   Biodegradation and detoxicification of environmental pollutants. CRC Press, Boca Raton.
(18) GIBSON, D.T. (1984).   Microbial degradation of organic compounds. Marcel Dekker, New York.
(19) KNACKMUSS, H-J. In ref. 16, pp 190-212.
(20) LASKIN, A.I. and LECHEVALIER, A.H. (1984). Handbook of Microbiology, 2nd Ed. Vol. VII. Microbial transformation. CRC Press, Boca Raton.
(21) BRITTON, L.N. In Ref. 18, pp. 89-130.
(22) TRUDGILL, P.W. IN REF. 18, PP. 131-180.
(23) MILES, C.J. and DELFINO, J.J. (1985).   J. Agric. Food Chem. **33**, 455-460.
(24) MURPHY, S.E. et al. (1982). Chemosphere **11**, 33-39.
(25) DE VOS, N. et al, (1979). Sci. Tot. Envir. **13**, 225-233.
(26) JOHNSON, D.B. and COX, B.L. (1985).   J. Agric. Food Chem. **33**, 255-259.
(27) MINARD, R.D. et al. J. Agric. Food Chem. **25**, 841.
(28) CORKE, C.T. et al. J. Agric. Food Chem. **27**, 644.
(29) FAULKNER, J.K. and WOODCOCK, D. (1965). J. Chem. Soc. 1187.
(30) OLIVER, J.E. et al. (1983). J. Agric. Food Chem. **31**, 1178-1183.
(31) HOFF, T. et al (1985). Appl. Envir. Microbiol. **49**, 1040-1045.
(32) BOUWER, E.J. and McCARTY, P.L. (1983). Appl. Envir. Microbiol. **45**, 1295-1299.
(33) PARSONS, F. and LAGE, G.B. (1985). J. Amer. Wat. Wks. Assn. **77**, 5, 52-59.
(34) CASTRO, C.E. et al (1983). J. Agric. Food Chem. **31**, 1184-1187.
(35) NOVICK, N.J. and ALEXANDER, M. (1985). Appl. Envir. Microbiol. **49**, 737-743.
(36) EVANS, W.C. et al, (1971). Biochem. J. **122**, 543.
(37) SUZUKI, T. (1983). J. Pesticide Sci. **8**, 419.
(38) TABAK, H.H. and BARTH, E.F. (1978). J. Wat. Pollut. Control Fed. **50**, 552-558.
(39) GROSSENBACHER, H. et al (1984). Appl. Envir. Microbiol. **48**, 451-453.
(40) SCHILLING, R. et al (1985). Chemosphere, **14**, 267-270.
(41) ROSENBERG, A. (1984). Bull. Envir. Contam. Toxicol. **32**, 383-390.
(42) LEE, A-H. et al. (1976). J. Envir. Qual., **5**, 482-486.
(43) CERNIGLIA, C.E. et al. (1984). Appl. Envir. Microbiol. **47**, 111-118.
(44) WONG, L.K. et al. (1983). Appl. Envir. Microbiol. **46**, 1239-1242.
(45) ISHIDA, M. (1972).   In Environmental Toxicology of Pesticides (Matsumura, F et al. eds.) pp 281-306, Academic Press, New York.
(46) CAPLAN, J.A. et al. (1984). J> Agric. Food Chem. **32**, 166-171.
(47) COATS, J.R. et al. (1976). Envir. Health Persp. **18**, 167-179.
(48) FURUKAWA, K. et al. (1983). Appl. Envir. Microbiol. **46**, 140-145.
(49) RAJAGOPAL, B.S. and RAO, V.R. (1984).   Canad. J. Microbiol. **30**, 1458-1466.
(50) MIYAZAKI, S., BOUSH, G.M. and MATSUMARA, F. (1969). Appl. Microbiol. **18**, 972.
(51) LEIDNER, H. et al. (1980). Xenobiotica, **10**, 47-56.
(52) STEBER, J. and WIERICH, P. (1985).   Appl. Envir. Microbiol. **49**, 530-537.

(53) GILLAN, F.T. et al. (1983). Appl. Envir. Microbiol. **45**, 1423-1428.

(54) CATELANI, D. et al. (1977). Appl. Envir. Microbiol. **34**, 351-354.

(55) WOROBEY, B.L. (1984). Chemosphere, **13**, 1103-1111.

(56) BULL, D. and IVIE, G.W. (1982). J. Agric. Food Chem. **30**, 150-155.

(57) KODAMA, K. et al. (1973). Agr. Biol. Chem. **37**, 45.

(58) MUIR, D.C.G. and YARECHEWSKI, A.L. (1982). J. Agric. Food Chem. **30**, 1028-1032.

(59) HALLAS, L.E. and ALEXANDER, M. (1983). Appl. Envir. Microbiol. **45**, 1234-1241.

(60) McCORMICK, N.G., FEEHENY, F.E. and LEVINSON, H.S. (1976). Appl. Envir. Microbiol. **31**, 949-958.

(61) SMITH, A.E. and CULLIMORE, D.R. (1974). Canad. J. Microbiol. **20**, 773.

(62) OHKAWA, H., KIKUCHI, R. and MIYAMOTO, J. (1980). J. Pestic. Sci. Jap. **5**, 11-22.

(63) PARIS, D.F., WOLFE, N.L. and STEEN, W.C. (1984). Appl. Envir. Microbiol. **47**, 7-11.

(64) HIGNITE, C. and AZARNOFF, D.L. (1977). Life Sciences, **20**, 337-342.

(65) ZHANG, L-Z, et al, (1984). J. Agric. Food Chem. **32**, 1207-1211.

(66) HALES, S.G. et al. (1982). Appl. Envir. Microbiol. **44**, 790-800.

(67) STEPHANOU, E. and GIGER, W. (1982). Env. Sci. Tech. **16**, 800-805.

(68) SARIASLANI, F.S. and ROSAZZA, J.P. (1983). Appl. Envir. Microbiol. **46**, 468-474.

(69) ECKENRODE, F.M. and ROSAZZA, J.P. (1983). J. Nat. Prod. (Lloydia) (USA) (1983) **46**, 884-893.

(70) KHAN, S.U. and IVARSON, K.C. (1982). J. ENVIR. SCI. HEALTH, **17**, 737-749.

(71) YANG, Y-S., MADSEN, E.L. and ALEXANDER, M. (1985). J. Agric. Food Chem. **33**, 495-499.

(72) SEWELL, G.J. et al. (1984). Appl. Microbiol. Biotech. **19**, 247-241.

(73) MATSUMURA, F., GOTOH, Y. and BOUSH, G.M. (1971). Science, **173**, 49-51.

(74) BOURQUIN, A.W. (1977). Appl. Envir. Microbiol. **33**, 356-362.

(75) SZETO, S.Y. and SUNDARAM, K.M.S. (1982). J. Agric. Food Chem. **30**, 1032-1035.

(76) LIEBERMAN, M.T. and ALEXANDER, M. (1983). J. Agric. Food Chem. **31**, 265-267.

(77) KU, Y. and ALVAREZ, G.H. (1982). Appl. Envir. Microbiol. **43**, 619-622.

(78) HEITKAMP, M.A. et al. (1984). Envir. Sci. Tech. **18**, 434-439.

(79) SAEGER, V.W. and THOMPSON, Q.E. (1980). Envir. Sci. Tech. **14**, 705-709.

(80) FURUKAWA, K. et al. (1983). Appl. Envir. Microbiol. **46**, 140-145.

(81) BAXTER, R.M. and SUTHERLAND, D.A. (1984). Envir. Sci. Tech. **18**, 608-610.

(82) MORRIS, C.M. and BARNSLEY, E.A. (1982). Canad. J. Microbiol. **28**, 73-79.

(83) DAGLEY, S. and JOHNSON, P.A. (1963). Biochim. Biophys. Acta. **78**, 377.

(84) ALLARD, A-S., REMBERGER, M. and NEILSON, A.H. (1985). Appl. Envir. Microbiol. **49**, 279-288.

(85) MIYAZAKI, T., YAMAGISHI, T. and MATSUMOTO, M. (1984). Bull. Envir. Contam. Toxicol. **32**, 227-232.

(86) WATANABE, I., KASHIMOTO, T. and TATSUKAWA, R. (1983). Bull. Envir. Contam. Toxicol. **31**, 48-52.

(87) SUBBA-RAO, R.V. and ALEXANDER, M. (1977).   Appl. Envir. Microbiol.
    33, 101-108.
(88) YONEGAMA, K. and MATSUMURA, F. (1984°).   Arch. Envir. Contam. Toxicol.
    13, 501-507.
(89) CLARK, A.M. et al. (1984). Appl. Envir. Microbiol. **47**, 537-539.
(90) SYLVESTRE, M. et al. (1982). Appl. Envir. Microbiol. **44**, 871-877.
(91) KEARNEY, P.C. et al (1977). J. Agric. Food Chem. **25**, 1177-1181.
(92) BARTHA, R. and PROMER, D. (1970). Adv. Appl. Microbiol. **13**, 317.

Table I   Transformation reactions

Type	Reaction	Example[+]	Ref.
Dehalogenation	$R.CH_x.Cl_n \longrightarrow R.CH_{x+1}$ $.Cl_{n-1}$	CCl_4 (G) 1,1,1,-trichloro-ethane (E) chloropicrin (M)	32,33 34
	$R.CH_2Cl \longrightarrow RCH_2OH$	Propachlor (G)	35
	$ArCl \longrightarrow ArH$	Chlorophenoxyacetate (M)	36
	$ArCl \longrightarrow ArOH$	Pentachlorophenol (S)	37
Deamination	$ArNH_2 \longrightarrow ArOH$	Benzidine (G) Chlorotriazine (M)	38,39
	$ArNH_2 \longrightarrow ArH$	Metribuzin (M)	40
Decarboxylation	$ArCOOH \longrightarrow ArH$	Fenac (G) Bifenox (E)	41,42
Methyl oxidation	$ArCH_3 \longrightarrow ArCH2OH$ $\longrightarrow ArCOOH$	Methylnaphthalenes (M) Methylbenzanthracene (M) Pentachlorobenzylalcohol (S)	43 44 45
Hydroxylation and ketone formation	$ArH \longrightarrow ArOH$	Fenvalerate (E) Phenothiazine (E)	46 47
	$ArH_2 \longrightarrow Ar(OH)_2$ $RCH_2R' \longrightarrow R.CHOH.R'$ $\longrightarrow R.CO.R$	Pentachlorobiphenyl (M) Carbofuran (S) Dichlorobenzilic acid (M)	48 49 50
β-oxidation	$Ar(CH_2)_n CH_2CH_2COOH$ $\longrightarrow Ar(CH_2)_n COOH$	ABS (G) Alcoholethoxylates (G) Phytol (M)	51,52 53
N-oxidation	$Me_3C.CH_2R \longrightarrow Me_3C.COOH$ $ArNNAr \longrightarrow ArN(-O)NAr$	2,2-dimethylheptane (M) Tetrachloroazobenzene (S)	54 55
S-oxidation	$RSR' \longrightarrow RS(O)R'$ $\longrightarrow RS(O_2)R'$	Phenothiazine (E) RH0994*(S) Aldicarb (W) Dibenzothiophene (M)	47,56 23 57
Nitro metabolism	$RNO_2 \longrightarrow RNH_2$	Niclosamide (W) Dinitrobenzene (G) Trinitrotoluene (M)	58 59 60
Oxime metabolism	$RCH=NOH \longrightarrow RC\equiv N$	Aldicarb (W)	23
Nitrile/amide metabolism	$RC\equiv N \longrightarrow RCONH_2$ $\longrightarrow R.COOH$	Fenvalerate (E) Dichlobenil (W) Sumicidin (W)	64 61 62

+   Reaction demonstrated in model ecosystem (E), Sewage (G), microbial
    culture (M), soil (S) or natural waters (W).
*   O[4-[(4-chlorophenyl)-thio]phenyl-O-ethyl S-propylphosphorothioate.

Table II Cleavage reactions

Substrate	Reaction	Example	Ref.
Ester	RCOOR´—>RCOOH	Chlorophenoxyacetic esters (W)	63
		Clofibrate, aspirin (W)	64
		Bifenox (W)	42
		Fenvalerate (W) Dinocap (S)	46
		Deltamethrin (S)	65
Ether	$ROCH_2R´$	Polyethylene glycols (G)	52
		Alkyltriethoxysulphate (M)	66
		Nonylphenolethoxylate (G)	67
	ArOR—>ArOH	7-ethoxycoumarin (M)	68
		Dihydrovindoline (M)	69
C-N	$RNH.Alk—>RNH_2$	Prometryn (S)	70
	$RNAlk_2—>RNH_2$	Monuron (S)	71
	R(R´)NR"—>RR´NH	Amitryptiline (M)	72
Peptide,	$RNHC(O)R´—>RNH_2$	Monuron (S) Niclosamide (W)	71,58
carbamate	R(R´)NC(O)R"—>R(R´)NH	Propachlor (G,W)	35
=NOC(O)R	RCH=NOC(O)R —>RCH=NOH	Aldicarb (W)	23
C-S	R-S-R´—>ROH, R´SH	Metribuzin (S)	40
Hg-O	$RCOOHgR´—>Hg(R´)_2$	Phenylmercuric acetate (M)	73
C-O-P	$(AlkO)_2P(S)R$ —>AlkO(HO)P(S)R	Malathion (W)	74
	ROP(S)(R)R´	Chlorpyriphos (W)	75
		Dichlovos (G)	76
	—>ROH+HOP(S)(R)R´	Tricresylphosphate (G)	77
		RHO994* (S)	56
Sulphate ester	$RCH_2OS(O_2)OH$ —>RCH_2OH	Alkyltriethoxysulphate (M)	66
C-C	$ArCH(CH_3)_2$ —>$ArCH_2CH_3$	Isopropylphenyl diphenyl phosphate (E)	78

* See Table I

Table III   Ring cleavage reactions

Reaction	Example	Ref.
(1) $Ar(Hal)_h-CH_2-Ar$ $\longrightarrow Ar(Hal)_nCH_2.COOH$	2,6-Dichloro dipenyl methane (W) 2,3,4 Trichlorodiphenyl methane (W)	79 79
(2) $Ar(Hal)_n-Ar(Hal)$ $\longrightarrow Ar(Hal)_nCOOH$	Tri-and Tetrachlorobiphenyls (M)	80,81
(3) Substituted condensed ring system $\longrightarrow$ hydroxylated, substitutedsimpler ring system	Chloronaphthalenes (M) $\longrightarrow$ 2-hydroxy, 3-chloro and 2-hydroxy 5-chloro benzoic acids	82
	4-hydroxy-2quinolinic acid (M) $\longrightarrow$ 5-carboxyethyl, 4-hydroxy, 2-picolinic acid	83
	Chloridazon (S) $\longrightarrow$ 5-amino-4-chloro-3(2H)pyridazinone	6
(4) R(Hal)ArCOOH $\longrightarrow$ halo-catechol $\longrightarrow$ acyl halide or 2-halo-muconic acid or 5-halo, 2-hydroxy muconic acid	3-chloro-benzoic acid (M) 2-fluoro-benzioc acid (M) 4-chloro-catechol (M)	19 19 19

Table IV Conjugation and other reactions

Reaction type	Reaction	Example	Ref.
Methylation	$ArOH \longrightarrow Ar\ OCH_3$	Chloroguaiacols (M) Triclosan (W) 7-hydroxycoumarin (M)	86 87 68
	$ArOH \longrightarrow Ar(OCH_3)_2$	Pentachlorophenol (S) Tetrabromobisphenol (E)	37 88
Ether formation	$R\ CH_2R \longrightarrow R(R)CH\ O\ CH(R)R$	Diphenylmethane (M)	89
N-Acylation	$Ar\ NH_2 \longrightarrow ArNHCOCH_3$	Nitraniline (G) Methylene bis (2-chloroaniline) (M)	59 90
	$ArNH_2 \longrightarrow ArN(OH)COCH_3$	Primaquine (M)	91
Nitration	$ArH \longrightarrow ArNO_2$ (via epoxide, $+NO_3^-$)	4-Chlorobiphenyl (M)	92
N-Nitrosation	$R_2NH \longrightarrow R_2N.NO$ ($\mp NO_2^-$)	Atrazine (S)	93
Dimerization	$2ArNH_2 \longrightarrow ArN=NAr$	Linuron (S) Propanil (S)	94
	$2ArH \longrightarrow ArCH_2\ Ar$	Primaquine (M)	91
N-heterocyle formation	$\longrightarrow$ Benzimidazole $\longrightarrow$ Quinoline	Nitraniline (G) Nitrobenzene (G)	59 59

BIOTRANSFORMATION OF NITROAROMATIC COMPOUNDS

J. Zeyer
Swiss Federal Institute for Water
Resources and Water Pollution Control (EAWAG)
6047 Kastanienbaum, Switzerland

SUMMARY

Nitroaromatic compounds are used as dyes, pesticides, explosives and solvents. They are ubiquitous pollutants in the environment. Their microbial mineralization is generally slow and proceeds through: (i) an initial reduction to an aminoaromatic compound (aniline), followed by an oxidative removal of the amino-substituent as ammonium, or (ii) an initial oxidative liberation of the nitro-substituent as nitrite. Both pathways eventually lead to catechol, which is a central meta-bolite in the aerobic degradation of most aromatic compounds and rapidly degradable.

Environmental conditions and compound specific parameters largely determine the pathway and the rate of the metabolism of nitroaro-matics. Oxygen limitations as often encountered in soils and lakes and electron withdrawing substituents on the aromatic ring favour the reduction to anilines. Further mineralization of anilines, however, requires oxygen and is therefore prevented in anaerobic environments. Consequently, anilines accumulate temporarily and polymerize slowly to persistent compounds as a result of enzymatic and non-enzymatic pro-cesses.

1.INTRODUCTION

Nitroaromatic compounds are frequently used as pesticides (16), explo-sives (24) and dyes (4). In chemical industry, they also serve as solvents and as precursors for the production of aminoaromatic derivatives (4). Some nitroaromatics are believed to be formed in waste water plants through biological processes (7, 32). As a consequence, nitroaromatics are ubi-quitous in waste waters (7, 18), rivers (14, 39), ground waters (26), pesticide treated soils (8, 16) and in the air (21). Some nitroaromatics such as 2-nitrophenol and 2,4-dinitrophenol are listed as priority pollu-tants by the Environmental Protection Agency (18). Nevertheless, their importance as pollutants in aquatic systems is often underestimated since they are rather polar and therefore not detectable by the commonly used "stripping method" (14).

Nitroaromatics are slowly degraded by microorganisms in the environ-ment, but most of the metabolic steps involved in the degradation are poorly documented. Two major catabolic pathways are thought to be of domi-nant importance (Fig. 1). One pathway involves an initial reduction of the nitroaromatic compound to an aniline intermediate (3, 5, 8, 10, 24), followed by a liberation of the amino-substituent as an ammonium ion (23, 28). Since many anilines are reactive, they are often not degraded but modified (e.g. hydroxylated, acetylated (5, 15)) or converted to very persistent azocompounds or polymers by enzymatic or non-enzymatic processes

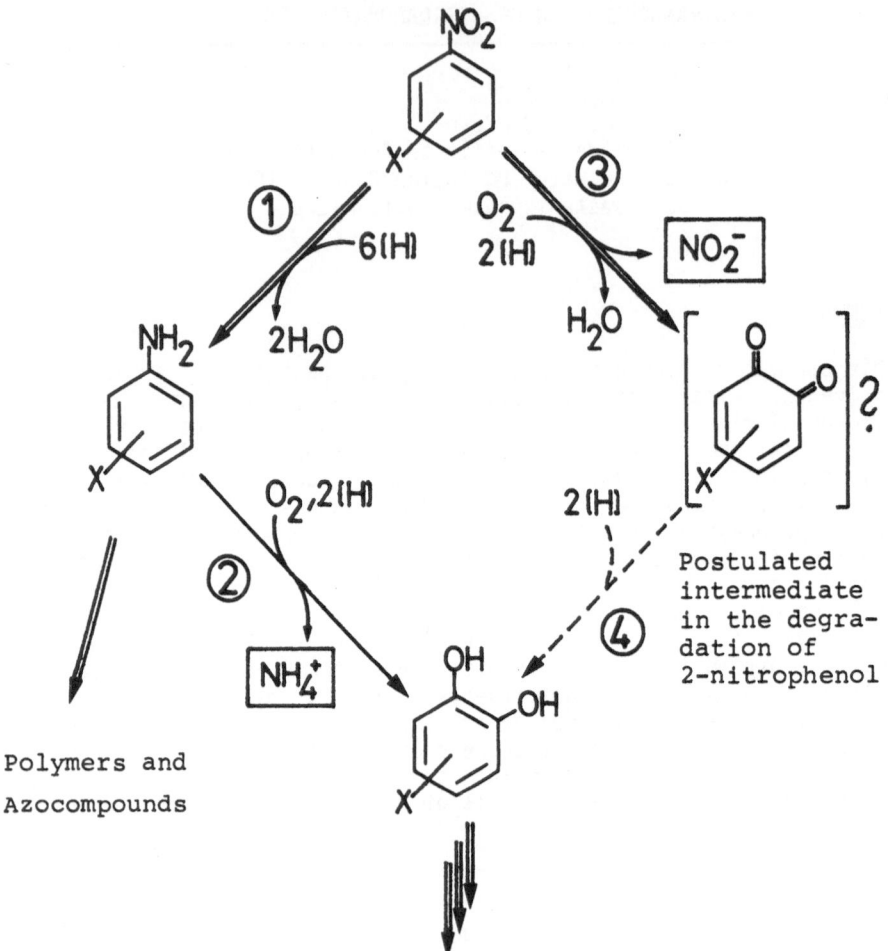

Figure 1: Microbial degradation of nitroaromatic compounds

Metabolic steps: ⟹ Steps demonstrated with whole cells
and enzyme extracts

⟶ Steps demonstrated with whole cells

---⟶ Steps postulated

Enzymes: ① Nitroreductase (aerobic and anaerobic)

② Aniline-Oxygenase (aerobic)

③ Nitrophenol-Oxygenase (aerobic)

④ Chinonereductase ?

(12, 17, 25, 27). The other pathway is believed to involve the direct removal of the nitro-substituent as nitrite (3, 9) and the formation of an intermediate such as a chinone (31). Both metabolic routes lead to catechol, which is the central intermediate in the microbial degradation of most aromatic compounds (22). It is noteworthy that both pathways involve a removal of the nitro-substituent from the aromatic core before the ring cleavage occurs. Catechol is quickly degraded to physiological compounds through the well established ortho- or meta-cleavage pathways (22).

The microbial degradation of nitroaromatics to catechol probably involves a series of reductions and oxidations, which are generally catalyzed by enzymes such as reductases and oxygenases. Knackmuss and coworkers (6, 22, 29) studied the activities of oxygenases on a number of substituted benzoates and catechols. They found that the oxygenation of aromatic compounds is an electrophilic reaction and that ring substituents with a high electron withdrawing effect (i.e. high Hammett constant δ, ref. 11) deactivate the ring system and therefore lower the reaction rate of oxygenases. In some cases, however, turnover rates of substituted aromatics are also impaired by steric effects (29).

This paper reports experiments performed with pure microbial strains to clarify some steps involved in the conversion of nitroaromatics to catechol. In particular the effects of compound-specific characteristics on the enzymatic degradation will be discussed. The paper focuses on the metabolism of substituted nitrophenols and anilines commonly used by the chemical industry for the manufacture of dyes and pesticides (4, 16). Some of the results were previously published (36, 37).

2. MATERIALS AND METHODS

The microbial strains, media and instruments as well as the microbiological, biochemical and analytical methods mentioned in this report have been described previously (36, 37, 38).

3. RESULTS AND DISCUSSION

3.1. Conversion of nitroaromatics to catechol and ammonium ions through aniline intermediates

A large number of nitroaromatics are chemically or enzymatically reduced to anilines in soil (8, 34) waste water (10) or microbial cultures (3, 5, 15). The action of microbial nitroreductases was demonstrated in enzyme extracts under aerobic and anaerobic conditions (24, 33, Fig. 1). Most nitroreductases have a broad substrate specificity. Substituents with a high electron withdrawing effect such as halogens and low redox potentials facilitate the reduction of nitroaromatics (8, 24, 34).

The microbial degradation of the non-substituted aniline to catechol by an aniline-oxygenase is well established (28). The substituted and, in particular, the halogenated anilines, however, were considered to be rather persistent compounds (5, 15). Some reports described a partial microbial metabolism of chlorinated anilines in presence of additional nutrients (28, 30). Only recently, two microorganisms were isolated which rapidly mineralize halogenated anilines as a sole source of energy, carbon and nitrogen (23, 37). One organism (a Pseudomonas sp.) was derived in Knackmuss' laboratory using a genetical recombination method (23), whereas we isolated an organism (a Moraxella sp.) directly from soil (37).

The Moraxella sp. grows rapidly (generation time: 3-6 hours) on aniline and a number of halogenated anilines but no growth was detectable on 4-iodo-, 4-methyl-, 4-methoxy- and 3,4-dichloroaniline (Table 1). The aniline-oxygenase which catalyzes the conversion to catechol has a broad substrate specificity and is well induced by a number of anilines. The enzyme requires oxygen and releases ammonium ions in stoichiometric amounts (36). In contrast to data obtained with an aniline-oxygenase of a Pseudomonas sp. (28), the activity of the aniline-oxygenase of this Moraxella sp. correlates more with size than with the electron withdrawing effect of the substituent. Anilines with substituents larger than 0.2 nm were poorly converted. Since these kinetic studies were not done with enzyme extracts but with suspensions of whole cells, the reported turnover rates (Table 1) might have partially been impaired by uptake rates.

Table 1: Degradation of anilines by a Moraxella sp.

Substrate NH_2 / X, X	Degradation as sole source of carbon and nitrogen 1)	Characteristics of substituent X		Characteristics of aniline-oxygenase 2)	
		Hammett constant δ	Size (nm)	Induction by anilines 3)	Activity on anilines 4)
4-H	+	0	0.120	100 5)	100 5)
4-F	+	+ 0.06	0.135	250	100
3-Cl	+	+ 0.37	0.180	100	100
4-Cl	+	+ 0.23	0.180	150	70
4-Br	+	+ 0.23	0.195	150	70
4-J	−	+ 0.18	0.215	130	20
4-CH3	−	− 0.17	0.200	75	20
4-OCH3	−	− 0.27		30	30
3,4-diCl	−	+ 0.60		30	< 1

1) + : > 85% degradation within 2 days
   − : < 10% degradation within 5 days
2) Assayed by measuring the rate of oxygen consumption after addition of a particular substrate to a suspension of whole cells (details see ref. 37)
3) Aniline-oxygenase induced by the aniline listed in the first column. Activity determined on 4-chloroaniline.
4) Aniline-oxygenase induced by 4-chloroaniline, Activity determined on the anilines listed in the first column.
5) Rates in % relative to aniline.

All attempts, to isolate an aniline-oxygenase and to demonstrate its activity in an enzyme extract failed. It is likely that the aniline-oxygenase consists of a complex of several proteins (like the benzoate-oxygenase, [35]) and is therefore rapidly inactivated upon cell breakage. In absence of oxygen, anilines are generally not mineralized but only modified (1, 13). To our knowledge, only the degradation of 2-carboxyaniline (anthranilic acid) has been reported under denitrifying conditions (2).

3.2. Conversion of nitroaromatics to catechol and nitrite

The aerobic microbial degradation of some nitroalkanes (19, 20) and nitroaromatics (in particular nitrophenols and nitrobenzoates, [3, 9]) is often not accompanied by an accumulation of ammonium ions but rather by a liberation of nitrite. The microbial enzymes releasing nitrite from nitroalkanes were purified and well characterized (19, 20). Oxygenases which liberate nitrite from nitroaromatics, however, were only demonstrated in enzyme extracts from bacteria growing on 4-nitrophenol and 2-nitrophenol, respectively, as the sole sources of energy, carbon and nitrogen (31, 36). The mechanism of the nitrite liberation from nitroaromatics and, in particular, the existence of an intermediate, are still controversial. The oxygenase active on 4-nitrophenol had a narrow substrate specificity, was membrane bound and converted 4-nitrophenol to nitrite and hydroquinone. Mechanistic studies suggested 1,4-benzoquinone to be an intermediate but analytical evidence was not given and the enzyme was not be purified (31).

The oxygenase active on 2-nitrophenol was not membrane-bound but soluble. It converted 2-nitrophenol to catechol and had a broad substrate specificty. Even compounds such as 2-nitro-4-methylphenol and 2-nitro-4-chlorophenol induced the enzyme and were rapidly converted to the corresponding catechols (38, Table 2). These two nitrophenols, however, did not serve as growth-substrates because the catechols which were produced cannot be mineralized (22). The turnover rates of the nitrophenols correlated better with the electron withdrawing effect (Hammett constant) than with the size of substituents. For nitro-, carboxy- and formyl-substituents, a corelation which is only based on Hammett constants is somewhat questionable, since these substituents are charged and/or have mesomeric effects on the aromatic ring. A 40-fold purification of the enzyme revealed that it had a molecular weight of about 62'000, a pH-optimum of 7.5 and a Km for 2-nitrophenol of 7 μM (38). The enzyme requires oxygen and releases nitrite in stoichiometric amounts. Ortho-chinone is likely to be an intermediate but a definite proof is still lacking (Fig. 1).

Table 2: Degradation of 2-nitrophenols by a <u>Pseudomonas</u> sp.

The substrate is a benzene ring with NO_2 at position 2, OH adjacent, and X at the para position (relative to OH).

Substrate X	Degradation as sole source of carbon and nitrogen [1]	Characteristics of substituent X		Characteristics of nitrophenol-oxygenase [2]	
		Hammett constant δ (para)	Size (nm)	Induction by nitrophenols [3]	Activity on nitrophenols [4]
H	+	0	0.120	100 [5]	100 [5]
CH_3	-	- 0.17	0.200	100	83
Cl	-	+ 0.23	0.180	100	20
COOH	-	+ 0.31		< 1	2
CHO	-	+ 1.03		< 1	8
NO_2	-	+ 1.24		< 1	< 1

1) See legend Table 1
2) Assayed by measuring the rate of the nitrophenol disappearance in an enzyme extract. The disappearance was determined spectrophotometrically (details see ref. 36, 38).
3) Nitrophenol-oxygenase induced by the nitrophenol listed in the first column. Activity determined on 2-nitrophenol.
4) Nitrophenol-oxygenase induced by 2-nitrophenol. Activity determined on the nitrophenols listed in the first column.
5) Rates in % relative to 2-nitrophenol.

4. REFERENCES

1. Bollag J.-M., Russel S. (1976), Microbial Ecology 3, 65 - 73.
2. Braun K., Gibson D. T. (1984), Appl. Environ. Microbiol. 48, 102 - 107
3. Cartwright N. J., Cain R. B. (1959), Biochem. J. 71, 248 - 261
4. Color Index, The Society of Dyers and Colourists, GB, and The American Assoc. of Textile Chemists and Colorists, USA, 2nd Edit., 1956
5. Corbett M. D., Corbett B. R. (1981), Appl. Environ. Microbiol. 41, 942 - 949
6. Dorn E., Knackmuss H.-J. (1978), Biochem. J. 174, 85 - 94
7. Giger W., Schaffner C. (1981), In "Advances in the Identification and Analysis of Organic Pollutants in Water", Vol. 1, Ann Arbor Science
8. Golab T., Althaus W. A., Wooten H. L. (1979), J. Agric. Food Chem. 27, 163 - 179
9. Gundersen K., Jensen H. L. (1956), Acta Agr. Scand. 6, 100 - 114
10. Hallas L. E., Alexander M. (1983), Appl. Environ. Microbiol. 45, 1234 - 1241
11. Hansch C., Leo A. (1979), "Substituent Constants for Correlation Analysis in Chemistry and Biology", Wiley-Interscience
12. Hoff T., Liu S.-Y., Bollag J.-M. (1985), Appl. Environ. Microbiol. 49, 1040 - 1045

13. Horowitz A., Shelton D. R., Cornell C. P., Tiedje J. M. (1982),
    Developments in Industrial Microbiology 23, 435 - 444
14. Karrenbrock F., Haberer K. (1984), Gewässerschutz, Wasser, Abwasser 72,
    139 - 160
15. Kaufman D. D., Plimmer J. R., Klingebiel U. I. (1973), J. Agric. Food
    Chem. 21, 127 - 132
16. Kearney P. C., Kaufman D. D. (1975), "Herbicides: Chemistry,
    Degradation, and Mode of Action", Marcel Dekker
17. Kearney P. C., Plimmer J. R., Guardia F. B. (1969), J. Agric. Food
    Chem. 17, 1418 - 1419
18. Keith H. L., Telliard W. A. (1979), Environ. Sci. Technol. 13, 416- 423
19. Kido T., Hashizume K., Soda K. (1978), J. Bacteriol. 133, 53 - 58
20. Kido T., Soda K., Suzuki T., Asada K. (1976), J. Biol. Chem. 251,
    6994 - 7000 ·
21. Kinouchi T., Ohnishi Y. (1983), Appl. Environ. Microbiol. 46, 596 - 604
22. Knackmuss H.-J. (1981), in "Microbial Degradation of Xenobiotics and
    Recalcitrant Compounds", FEMS-Symposium, 189 - 212, Academic Press
23. Latorre J., Reineke W., Knackmuss H.-J. (1984), Arch. Microbiol. 140,
    159 - 165
24. McCormick N. G., Feeherry F. E., Levinson H. S. (1976), Appl. Environ.
    Microbiol. 31, 949 - 958
25. Parris G. E. (1980), Residue Reviews 76, 1 - 30
26. Piet G. J., Smeenk J. G. M. M. (1985), In "Ground Water Quality",
    122 - 144, Wiley Interscience
27. Pillai P., Helling C. S., Dragun J. (1982), Chemosphere 11, 299 - 317
28. Reber H., Helm V., Karanth N. G. K. (1979), Europ. J. Appl. Microbiol.
    Biotechnol. 7, 181 - 189
29. Reineke W., Knackmuss H.-J. (1978), Biochim. Biophys. Acta 542,
    412 - 423
30. Schukat B., Janke D., Krebs D., Fritsche W. (1983), Current Microbiol.
    9, 81 - 86
31. Spain J. C., Wyss O., Gibson D. T. (1979), Biochem. Biophys. Res.
    Commun. 88, 634 - 641
32. Sylvestre M., Masse R., Messier F., Fauteux J., Bisaillon J.,
    Beaudet R. (1982), Appl. Environ. Microbiol. 44, 871 - 877
33. Villanueva J. R. (1964), J. Biol. Chem. 239, 773 - 776
34. Willis G. H., Wander R. C., Southwick L. M. (1974), J. Environ. Quality
    3, 262 - 265
35. Yamaguchi M., Fujisawa H. (1982), J. Biol. Chem. 257, 12497 - 12502
36. Zeyer J., Kearney P. C. (1984), J. Agric. Food Chem. 32, 238 - 242
37. Zeyer J., Wasserfallen A., Timmis K. N. (1985), Appl. Environ.
    Microbiol. 50, 447 - 453
38. Zeyer J., Kocher H.-P. (in preparation)
39. Zoeteman B. C. J., Harmsen K., Linders J. B. H. J. (1980), Chemosphere
    9, 231 - 249

AQUATIC PHOTOCHEMISTRY : FUTURE DEVELOPMENTS

P.C.M. VAN NOORT
National Institute of Public Health and Environmental Hygiene
The Netherlands

Summary
An account is given of the present knowledge about the occurrence and
quantitative description of the various photochemical reactions in the
aquatic environment. The lack of knowledge about quantitative aspects
is specified. An indication is given of some photochemical reactions
expected to occur, but not yet identified, in the aquatic environment.

1. INTRODUCTION
        The environmental effects of the introduction of organic
micropollutants into the aquatic environment are determined by the various
transport and transformation processes. Solar irradiation is one of the
causes for transformation. Initially, work was done to demonstrate that
solar irradiance could induce relevant transformations. The relevance of
sunlight induced processes also appears from the following. The photon flux
of midday summer sunlight of a wavelength range of 300-350 nm is about 10^{15}
photons/cm^2s. If each photon induces the transformation of one molecule,
and if they are absorbed within the first 100 cm, then the (overall)
transformation rate will be 1 µ mol/1.min. Not all of these photons are
effective,nor are they effective in the same way. Therefore, nowadays
research is being done to investigate the fate of photons or, in other
words, to identify the various photochemical pathways to derive such
quantitative expressions that combination of the quantitative description
of the aquatic environment with relevant molecular properties of
(potential) micropollutants allows a quantitative assessment of the
transformation rates. The structure of the transformation products may be
deduced from the type of photochemical process. In other words, a complete
understanding of the various processes allows the transformation forecast
of new compounds. Such knowledge can only be obtained through the study of
relevant model substrates. Additional information may be obtained from
studies on the transformation of specific compounds such as priority
pollutants if these studies enclude identification of the transformation
products.
        This review will give an account of the present knowledge about the
various sunlight induced processes taking place in surface waters. The lack
of knowledge about these processes will be specified. Finally, specific
areas to be developed will be given on the basis of some selected
transformation product studies. The various processes will be treated based
on the following.
        The transformation pathway depends on the micro-environment of the
organic micropollutant. In the aquatic environment relevant
micro-environments are : the surface film, particulate matter, biota, the
dissolved organic matter, and the water itself. The various sunlight
induced transformation processes for organic micropollutants in the
dissolved phase in aquatic environments are given in Scheme 1.

SCHEME 1. THE AQUATIC PHOTOCHEMICAL PROCESSES FOR DISSOLVED COMPOUNDS

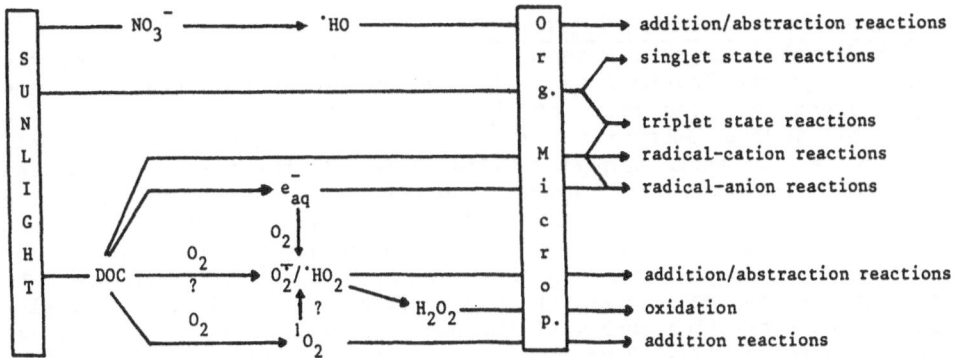

2. HYDROXYL RADICALS

The formation of the highly reactive ·HO radical in natural waters seems to be caused by the presence of NO_3^- (1) which on irradiation yields O^- (2). This reactive radical has been detected in seawater (3). For some U.S. eutrophic surface waters the near-surface ·HO steady-state concentrations, $[·HO]_{ss}$, were found to be $0.15 - 1.8 \times 10^{-17}$ M (4). In filtered German surface waters $[·HO]_{ss}$ was found to be approximately 50×10^{-17} M (1). For the Swiss Greifensee $[·HO]_{ss}$ was found to be 30×10^{-17} M (5). These values were obtained using cumene (4), benzene (1), or 1-chlorobutane (5) as HO scavengers. The $[·HO]_{ss}$ value for the Swiss Greifensee may be an upper limit since the hydrated electron released in natural waters (this will be discussed later) will very rapidly be taken up by 1-chlorobutane resulting in an additional disappearance pathway for this substrate. The two orders of magnitude difference between the various values for $[·HO]_{ss}$ are difficult to account for, since the authors did not specify which part of the incident light was being absorbed by NO_3-. From these $[·HO]_{ss}$ values and rate constants for the reaction of ·HO with organic compounds (6) near-surface disappearance rates for organic micropollutants dissolved in natural waters may be estimated.

Alkylperoxy radicals were also detected in some U.S. eutrophic surface waters. Their near-surface steady-state concentrations amounted $0.45 - 9.5 \times 10^{-9}$ M (4). Their origin is not clear. Perhaps they are formed by reaction of alkyl radicals, formed from organic material and ·HO, with oxygen.

3. DIRECT PHOTOTRANSFORMATION

The environmental factors influencing the transformation reactions by direct light absorbtion are the solar irradiance intensities and the light attenuation characteristics of the surface waters. These factors will not be treated here. The interested reader should consult reference (7) for a discussion and references. The relevant properties of the organic micropollutant are its (UV) absorbance spectrum and the quantum yield for transformation. The micropollutant may react from two different excited states. The singlet excited state is directly obtained through light absorbtion. The triplet excited state is reached from the singlet excited state through electron spin inversion. In general these two excited states have different reactivities. Most singlet excited states are sufficiently short-lived to allow only unimolecular decay or reaction with H_2O to occur in the aquatic environment. Triplet excited states are relatively long-lived. They may therefore react with various compounds present in natural waters.

4. INTERACTION WITH PHOTO-EXCITED DISSOLVED ORGANIC MATERIAL

The dissolved organic material (DOC) in natural waters absorbs the major part of the incident sunlight. The triplet excited states of the various light absorbing parts of the DOC may transfer their excitation energy to various substrates in natural waters. As a result these substances are in their triplet excited state. If the triplet excitation energy of the donor (the DOC) exceeds the triplet excitation energy of the acceptor by more than 10 kJ/mole, the energy transfer will proceed at a diffusion controlled rate. If the energy transfer is endothermic, the transfer rate will decrease rapidly by several orders of magnitude. The efficiency of the energy transfer is therefore determined by the triplet lifetime and excitation energy of the donor. From a study on the triplet energy transfer to (Z)-1,3-pentadiene in natural waters, Zepp obtained an average lifetime of about 2 µs and a triplet excitation energy of 250 kJ/mol (8). From flash-spectroscopic measurements on natural waters a lifetime of 5 µs and an excitation energy of 240 kJ/mole was obtained (9) on excitation with λ355 nm. Lower lying triplet states may also be present in surface waters since, as will be given later, the quantum yield for singlet oxygen formation decreases steadily on increasing wavelength below 550 nm.

Hydroquinone moieties are present in the DOC of natural waters. It has been found from the irradation of hydroquinone itself in the presence of Aroclor 1254 in water-alcohol mixtures that loss of Cl^- from the PCBs occurs as a result of electron transfer from the excited state of the hydroquinone anion to PCBs (10).
Under similar conditions methoxychlor was found to be dechlorinated via a similar mechanism (11).

$$HQ \quad \xrightarrow{h\upsilon} \quad {}^1HQ \longrightarrow H^+ + {}^1Q^-$$

$$ {}^1Q^- + ArCl \longrightarrow Q\cdot + ArCl^-\cdot$$

$$ArCl^-\cdot \longrightarrow Cl^- + Ar\cdot \longrightarrow \quad products$$

It is not clear if this process will occur at a relevant rate in surface waters because of the short lifetime of singlet excited states and the low concentration of dissolved organic micropollutants.

Another possible reaction may occur through interaction of the excited states of DOC with organic micropollutants : electron transfer from the pollutant to the excited state of the DOC. For instance, irradiation of atrazine in aqueous fulvic acid solution resulted in the dealkylation of the primary photoproduct of atrazine (12). This proces may be rationalized through an electron transfer mechanism :

$$R-NH-CH_2-CH_3 + DOC^* \longrightarrow DOC^-\cdot + R-NH^+\cdot-CH_2-CH_3$$

$$\xrightarrow{-H^+} R-NH=\overset{\cdot}{C}H-CH_3 \xrightarrow[H_2O]{O_2} R-NH_2$$

The transformation of aniline (13) and 3,4-dichloroaniline (14) into azobenzenes may also be explained in this way :

$$R-NH_2 + DOC^* \xrightarrow{} DOC^- \cdot + R-NH_2^+ \cdot \xrightarrow{-H^+} R-NH \cdot \longrightarrow \text{products}$$

The acceptor properties of the DOC are not yet known. They might be quantified on the basis of the reduction potential and the excitation energy. The relevant substrate property is its oxidation potential.

5. THE HYDRATED ELECTRON

The excited states of DOC may also release an electron to the aqueous environment. Recently spectroscopic evidence for this process was obtained. The lifetime of the hydrated electron in a natural water was found to be 0.5 µs at pH7, the quantum yield for formation at $\lambda355nm$ was 0.1, the near-surface steady-state concentration in natural waters was estimated to be $10^{-13}M$ (9). The hydrated electron adds rapidly, in many cases at a diffusion controlled rate, to many compounds (6) to form radical-anions.

6. SUPEROXIDE

The hydrated electron in natural waters is expected to react with dissolved oxygen to form superoxide at a diffusion controlled rate $(1.88 \times 10^{10}$ $M^{-1}s^{-1}$ (6). The rate of formation of superoxide can be estimated to be 1800 µ M/h from $[e_{aq}^-] = 10^{-13}M$ and $[O_2] = 250$ µM.
Reports on measurements of superoxide formation in natural waters are scarce. For a Canadian natural water the rate of superoxide formation at a light intensity of 250 mWcm^{-2} was found to be 5 µM/h (15) from the formation of NO_2^- from added hydroxylamine. Light intensity for 40° N midday summer sunlight is 110 mWcm^{-2}. For this water the rate of superoxide formation is therefore 3 orders of magnitude lower than expected. From the relative rates of formation at $\lambda366$ nm of singlet oxygen and superoxide in a Dutch natural water (16) and average rates of formation per mg/1 DOC for singlet oxygen at 110 mW/cm^2 (17) the rate of superoxide formation can be estimated to be about 5 µM/h.
Differences between observed and estimated rates of superoxide production may well be the result of the very low number of data on e_{aq}^- - production, but may also be the result of a poor detection technique for superoxide since the kinetics and mechanism of the reaction of hydroxylamine with superoxide in natural waters are not sufficiently known (18). Steady-state concentrations of superoxide are not known. They can be calculated in the following way. Under steady-state conditions of a reactive intermediate the rate of its formation is equal to its rate of disappearance which is equal to the product of the rate constant for disappearance and the steady-state concentration. The steady-state concentration is therefore equal to the quotient of the rate of formation and the (first order) rate constant for disappearance. Upper limits can be calculated on the assumption of decay of superoxide via disproportionation only, to give H_2O_2 and O_2, from the disproportionation rate (19) and rates of formation.
For the (model) natural water with $[e_{aq}^-] = 10^{-13}$ M the steady-state concentration at pH7 is about $10^{-8}M$. Actual steady-state concentrations may very well be lower since the rate of H_2O_2 production in the presence of superoxide dismutase was found to be twice that without superoxide dismutase added (20), suggesting ways of decay other than

disproportionation. The actual steady-state concentrations will not be very much lower (50%) if superoxide is the sole source of H_2O_2. The reactivity of superoxide in water is low because of its high aqueous solvation energy (21). However, in water part of the superoxide is protonated (21). The reactivity of $\cdot HO_2$ towards some organic compounds was found to be greater than the reactivity of superoxide (22). It may therefore well be that $\cdot HO_2$ is the reactive form in natural waters.

7. HYDROGEN PEROXIDE

The formation of H_2O_2 in natural waters was reported by Zika (20). He found the rate of accumulation to depend on total organic carbon. For some surface waters with TOC contents of about 15 mg/l the accumulation rate was about 3 µM/h in June midday sunlight at latitude 25°44'N. The addition of superoxide dismutase resulted in a rate twice that of the control. This means that at least 50% of the H_2O_2 originates from superoxide. Also, Crosby reported the sunlight induced formation of H_2O_2 in natural waters (23). His data for the half-life in one natural water and for the concentration after 5 h of irradiation suggest a rate of formation of about 1 µM/h. It is interesting to note that the observed rates of formation are close to the rates of formation of superoxide.

8. SINGLET OXYGEN

The first report on the formation of singlet oxygen in natural waters appeared already in 1977 (24). The singlet oxygen was trapped by 2,5-dimethylfuran (DMF). The rate of disappearance of DMF was taken as a measure for the rate of formation of singlet oxygen. Recently, the use of furfuryl alcohol (FFA) as an alternative for DMF was published (25). The advantage of FFA over DMF is that FFA is well soluble in water and does not volatilise as DMF does. The formation of singlet oxygen is the result of excitation energy transfer from the triplet excited state of dissolved organic material in natural waters (24, 26). The rates of singlet oxygen formation in some (optically matched) American natural waters and solutions of humic and fulvic acid differed by less than a factor of 1.5 (27). At low concentrations of aquatic humus the production rates are proportional to the aquatic humus concentrations (26).
The quantum efficiency of light induced formation of singlet oxygen in natural waters was found to decrease with increasing wavelength, at 550 nm virtually no singlet oxygen was formed (28). Also for some Dutch natural waters a fair correlation between singlet oxygen production rates were determined, together with quantum yields at 4 wavelengths (30). Steady-state concentrations of singlet oxygen in natural waters were calculated from production rates and the rate of decay of singlet oxygen in pure water. A compilation of data is given in ref. (30). They are in the range of $1.3 - 5.8 \times 10^{-14}$ M per mg/L DOC for the Swiss and American waters. These values are given for an unspecified temperature, and may be in error if the lifetime of singlet oxygen in natural waters differs substantially from that in pure water. However, this seems not to be a serious problem. The ratios of the rate constants for decay of singlet temperatures between 4 and 20°C were found to be independent of the temperature and only 50% lower than the one for pure water (31). The rate of disappearance of organic micropollutants in natural waters via reaction with singlet oxygen may be calculated from singlet oxygen steady-state concentrations and the rate constant for that reaction. For a variety of organic compounds the rate constants are compiled in ref.(32). Unfortunately, for many compounds only the rate constant for singlet oxygen deactivation is known which includes, but is not necessary equal to, the chemical reaction.

9. BIOTA MEDIATED REACTIONS

From the irradiation of 22 non-ionic compounds in water containing algae it appeared that for some of them the rate of disappearance was fast relative to the irradiation in pure water (33). The algae induced transformation did not occur via a metabolic pathway. There was no difference between algae, dead or alive. On the basis of kinetic data it was concluded that the compounds are transformed on association with the algae. No product studies were performed. Therefore, the type of transformation reaction can not be established. For instance, does energy or electron transfer occur, or are reactive intermediates, e.g. singlet oxygen, formed ?

10. PARICLE ASSOCIATED COMPOUNDS

Part of the incident solar irradiance is being absorbed by particulate matter in the aquatic environment. A technique for determination of light attenuation in turbid waters is given in ref. (38). Using an elegant technique to determine which part of the incident light is absorbed by particulate matter, Zepp found from product and kinetic studies on the photolysis of an aromatic ketone and DDE that the photochemical reaction takes place in a micro-environment less polar than water and that rates did not differ substantially from those in pure water (34).

The rate of photolysis of methoxychlor as a saturated (0.12) ppm solution in water was found to decrease on increasing amount of added sediments (36). The authors rationalized this dependence in terms of increasing light attenuation on increasing (suspended) sediment concentration. They did however not consider the sorption of methoxychlor by the suspended sediments. Also the type of sorption was not considered. The method employed to sorb compounds may affect the results of photolysis (34).

The sorption site may also influence photochemical behaviour. Increasing equilibration times prior to irradiation of DDE in water containing suspended sediments result in decreasing overall rates of disappearance (35). The kinetic data could be explained by assuming that the DDE is moving slowly to sediment sites not available for photochemical reaction.

The photolysis rate of hexachlorocyclopentadiene in sediment water systems was found to be independent of the sediment concentration (37). The authors did not specify their method for preparing the solution. It may well be that because of the short lifetime (half-lives of less than 10 min under their conditions) the hexachlorocyclopentadiene is transformed before transport to unreactive sites occurs.

All these results demonstrate that photo-induced transformation may occur. It is not clear which procsses occurred. Are they the result of direct light absorption, or of energy transfer, or of other processes ?

11. SURFACE FILMS

The photochemical behaviour of compounds in surface films may differ from that in the water phase. Concentrations of organic compounds are higher, thus making competition with dissolved ogygen more important. Further, the micro-environment is less polar and concentrations of potential hydrogen atom donors are higher. At least some of the organic constituents seem to be formed as a result of the presence of photochemically formed singlet oxygen (39).

The surface film is also the interface for air-sea interactions. A very rough estimate of the balance for some airborne radicals ($\cdot HO_2$ and $CH_3O_2\cdot$) of the flux from the air and the disappearance via radical-radical reactions indicated for the upper layers of the sea (20-200 μm) that the rates are of comparable magnitude (40). Concentrations will therefore be smaller than 10^{-10}-10^{-11} M, which values would hold for the situation without disappearance by reaction. For $\cdot HO_2$ they are therefore smaller than estimated for the photo-induced formation (Chapter 6). The influx of $\cdot HO$ is much smaller than that for $\cdot HO_2$ (40). Therefore it may be concluded that the air is not a dominant source for these radicals.

Studies on a model surface film comprising n-pentadecane and anthraquinone floating on seawater demonstrated the sunlight induced production of ketones, γ-diketones and 1-alkenes. The initial step was proposed to be hydrogen atom abstraction by the photoexcited anthraquinone (41). It still remains to be established if natural sensitizers in surface films may abstract hydrogen atoms.

12. ASSOCIATION WITH DISSOLVED ORGANIC MATERIAL

Hydrophobic micropollutants in surface water appear to be partly associated with the dissolved organic material (DOC) (42, 43, 44, 45). The degree of association is a function of the hydrophobicity, or K_{ow} (42, 44, 45). It is also determined by the type of dissolved organic material (44) and by other compounds in surface waters (43).

At present no studies on the photochemical transformation of DOC-associated organic micropollutants are known. Some of the photochemical reactions induced by DOC for dissolved compounds (e.g. energy transfer) may occur more efficiently for DOC-associated compounds as competition by dissolved oxygen will be less in the latter case.

13. CONCLUSION

A lot of knowledge has been gained on processes involving dissolved organic micropollutants. The formation of singlet oxygen is quantitatively well documented. Data on the formation of $\cdot HO$, the hydrated electron, and superoxide are scarce. More data, and especially those aiming at establishing rates of formation and/or steady-state concentrations in terms of aquatic environmental properties, are certainly needed. Some quantitative information on environmental properties for energy transfer from DOC has appeared. Data relevant to "high energy" acceptors are still lacking. The electron donor and acceptor properties of DOC are not available in quantitative terms.

The present knowledge, if there is at all, about processes involving organic micropollutants associated with biota, particles, DOC, or the surface layer allows the conclusion that in these micro-environments sunlight induced transformations may occur. The type of these processes, specified in the same way as for dissolved organic micropollutants, still has to be established quantitatively.

The herefore mentioned areas to be developped are those based on a priori assumptions of possible reaction pathways. They apply to the initial step in the sunlight induced transformation. The study of specific, e.g. priority, pollutants including the identification of products in natural waters should be performed to test for completeness of the a priori assumed pathways.

An interesting example is the formation of 2,4-dichlorophenol from the 2-butoxyethyl ester of 2,4-D in river water (46). The product formation and the transformation rate can not be explained in terms of direct light absorption. A transformation via ipso substitution by $\cdot HO$ seems not very likely. Perhaps electron transfer from photoexcited DOC to the 2,4-D ester is the initiating step.

The sunlight induced transformations of methoxychlor in some river waters are up to about 100 times as fast as in distilled water. In one case DMDE, formed by dehydrohalogenation, accounted completely for methoxychlor disappearance at 30% conversion. Singlet oxygen formation or triplet energy transfer could not explain the rates and product formation (47). Also in this case the initiating step may be the uptake of an electron by methoxychlor. Electron transfer from photoexcited N,N-diethylaniline to DDT resulted in the formation of DDE (among others) by dehalogenation (48).

In natural creek water chlorsulfuron was found to photolyse 4 times faster than in distilled water. The products are difficult to explain in terms of the presence of hydroxyl radicals or singlet oxygen. Other processes were thought to cause the phototransformation (49).

In addition to these types of research investigations have to be carried out to establish the fate of intermediates formed in the initial transformation steps. The eventual addition of radical intermediates to the DOC could play a role in this respect.

REFERENCES

(1) H. Russi, D. Kotzias, and F. Korte, Chemosphere, 11 (1982) 1041.
(2) M. Daniels, R.V. Meyers, and E.V. Belardo, J. Phys.Chem., 72 (1968)
(3) O. Zafiriou, M. McFarland, and R.H. Bromund, Science, 207 (1980) 637.
(4) T. Mill, D.G. Hendry, and H. Richardson, Science, 207 (1980) 886.
(5) J. Hoigné, W. Haag, and H. Bader, paper presented at the 189th National Meeting of the A.C.S., Miami, Florida, 1985.
(6) M. Anbar and P. Neta, Int. J. appl. Radiat. Isotopes, 18 (1967) 493.
(7) O.C. Zafiriou, J. Joussot-Dubien, R.G. Zepp, and R.G. Zika, Environ. Sci. Technol., 18 (1984) 358A.
(8) R.G. Zepp, P.F. Schlotzhauer, and R. Merritt Sink, Environ. Sci. Technol., 19 (1985) 74.
(9) A. Fischer, D. Kliger, J. Winterle, and T. Mill, paper presented at the 189th A.C.S. National Meeting, Miami, Florida, 1985.
(10) S.K. Chaudhary, R.H. Mitchell, and P.R. West, Chemosphere, 13 (1984) 1113.
(11) S.K. Chaudhary, R.H. Mitchell, P.R. West, and M.J. Ashwood-Smith, Chemosphere, 14 (1985) 27.
(12) S.U. Khan and M. Schnitzer, J. Environ. Sci. Health, B13 (1978) 299.
(13) R.G. Zepp, G.L. Baughman, and P.F. Schlotzhauer, Chemosphere, 10 (1981) 109.
(14) G.C. Miller, R. Zisook, and R.G. Zepp, J. Agric. Food Chem., 18 (1980) 1053.
(15) R.M. Baxter and J.H. Carey, Nature, 306 (1983) 575.
(16) P.C.M. van Noort, E. Wondergem, and P. Breugem, unpublished results.
(17) W.R. Haag, J. Hoigné, E. Gassman, and A.M. Braun, Chemosphere, 13 (1984) 641.
(18) B.H. Bielski, R.L. Arudi, D.E. Cabelli, and W. Bors, Anal.Biochem., 142 (1984) 207.
(19) B.H.J. Bielski and A.O. Allen, J. Phys. chem., 81 (1977) 1048.
(20) W.J. Cooper and R.G. Zika, Science, 220 (1983) 711.
(21) D.T. Sawyer and J.S. Valentine, Acc. Chem. Res., 14 (1981) 393.
(22) J.M. Gebicki and B.H.J. Bielski, J. Amer. Chem. Soc., 103 (1981) 7020.
(23) W.M. Draper and D.G. Crosby, Arch. Environ. Contam. Toxicol., 12 (1983) 121.
(24) R.G. Zepp, N.L. Wolfe, G.L. Baughman, and R.C. Hollis, Nature, 267 (1977) 422.
(25) W.R. Haag, J. Hoigné, E. Gassman, and A.M. Braun, Chemosphere, 13 (1984) 631.
(26) R.G. Zepp, G.L. Baughman, and P.F. Schlotzhauer, Chemosphere, 10

(1981) 119.

(27) R.G. Zepp, G.L. Baughman, and P.F. Schlotzhauer, Chemosphere, 10 (1981) 109.

(28) R.G. Zepp and G.L. Baughman, in O. Hutzinger, L.H. van Lelyveld, and B.C.J. Zoeteman (Eds), Aquatic Pollutants : Transformation and Biological Effects, Pergamon, Oxford, 1978, p. 237.

(29) C.J.M. Wolff, M.T.H. Halmans, and H.B. van der Heijde, Chemosphere, 10 (1981) 59.

(30) W.R. Haag, J. Hoigné, E. Gassman, and A.M. Braun, Chemosphere, 13 (1984) 641.

(31) P.C.M. van Noort and E. Wondergem, unpublished results.

(32) F. Wilkinson and J.G. Brummer, J. Phys. Chem. Ref. Data, 10 (1981) 809.

(33) R.G. Zepp and P.F. Schlotzhauer, Environ. Sci. Technol., 17 (1983) 462.

(34) G.C. Miller and R.G. Zepp, Environ. Sci. Technol., 13 (1979) 860.

(35) R.G. Zepp and P.F. Schlotzhauer, Chemosphere, 10 (1981) 453.

(36) B.G. Oliver, E.G. Cosgrove, and J.H. Carey, Environ. Sci. Technol., 13 (1979) 1075.

(37) N.L. Wolfe, R.G. Zepp, P. Schlotzhauer, and M. Sink, Chemosphere, 11 (1982) 91.

(38) G.C. Miller and R.G. Zepp, Water Res., 13 (1979) 453.

(39) A. Momzikoff et al., C.R. Acad. Sci. Ser II, 297 (1983) 261.

(40) A.M. Thompson and O.C. Zafiriou, J. Geophys. Res., 88 (1983) 6696.

(41) M.G. Ehrhardt and G. Petrick, paper presented at the 189th National Meeting of the A.C.S., Miami, Florida 1985.

(42) E.M. Perdue and N.L. Wolfe, Environ. Sci. Technol., 16 (1982) 847.

(43) C.W. Carter and I.H. Suffet, Environ. Sci. Technol., 16 (1982) 735.

(44) C.W. Carter and I.H. Suffet, Org. Geochem., 8 (1985) 145.

(45) J.F. McCarthy and B.D. Jimenez, Org. Geochem., 8 (1985) 141.

(46) R.G. Zepp, N.L. Wolfe, J.A. Gordon, and G.L. Baughman, Environ. Sci. Technol., 9 (1975) 1144.

(47) R.G. Zepp, N. Lee Wolfe, J.A. Gordon, and R.C. Fincher, J. Agric. Food Chem., 24 (1976) 727.

(48) L.L. Miller and R.S. Narang, Science, 169 (1970) 368.

(49) M. Hermann, D. Kotzias, and F. Korte, Chemosphere, 14 (1985) 3.

NEW REACTIONS INVOLVED IN NON-AEROBIC DEGRADATION
OF AROMATIC COMPOUNDS

B. SCHINK
Fakultät für Biologie, Universität Konstanz, D-7750 Konstanz,
W.-Germany

Summary

A considerable amount of natural and synthetic aromatic compounds is
subject to anaerobic transformations, either in sediments, water
-logged soils, or gastro-intestinal tracts of animals and man. Our
knowledge on the types of anaerobic transformations of these compounds
has grown remarkably in the recent years, and many reactions have been
described which occur either preferentially or exclusively in the
absence of molecular oxygen. Mononuclear aromatics are subject to
demethoxylation, decarboxylation, dehydroxylation, deamination, and
dehalogenation, and most of these reactions are important steps in the
mineralization of the respective compounds under anaerobic conditions.
Hydroxyl and carboxyl substituents largely determine the pathway how
the ring is cleaved to an open-chained derivative. Azo linkages between aromatic ring structures are cleaved in the absence of oxygen
much more efficiently than in its presence. The various new reactions
of aromatic ring modification and cleavage under anaerobic conditions
are discussed. They give hope to developing new technical devices in
the future which could help to degrade a broad variety of aromatic
pollutants which so far proved to be resistant to the "classical"
aerobic degradation mechanisms.

1. INTRODUCTION: ENVIROMENTAL IMPORTANCE OF ANAEROBIC DEGRADATION PROCESSES
    Increasing amounts of synthetic compounds are being released annually
into our natural environment, either directly as various biocides or fer-
tilizers, or indirectly as waste materials, synthesis byproducts, and products of incomplete degradation. Among these are various aromatic compounds, especially halogen derivatives in biocides, nitrogen derivatives
in synthetic dyestuffs, and   phenolic compounds. The fate of these compounds in the natural environment, especially their biodegradation, has so
far only been regarded as a process occuring in the presence of molecular
oxygen; e.g., the legislative regulations on testing the biodegradability
of new chemical products only require degradability tests by activated
sludge in aerated test devices. Degradation and modification of synthetic
compounds in the absence of oxygen has been largely disregarded in the
past, basically for three major reasons: i) anoxic environments were considered to be only of marginal importance to the global cycling of organic
matter; ii) the degradative capacity of anaerobic bacteria appeared to be
always inferior to those of aerobic ones; and iii) the principles and
limits of anaerobic degradation processes were barely understood due to a
lack of suited methods for culturing and studying fastiduous strict anaerobes. After the development of such methods, mainly by Robert Hungate and

his school (20), a new and fascinating world was opened to scientific research.

Increasing interest in microbial ecology in the last two decades revealed that anaerobic metabolism may quite significantly contribute to the global degradation processes. The still progressing eutrophication of our lakes, rivers and oceans converted their sediments and hypolimnic water bodies into oxygen-deprived environments which often undergo complete anaerobiosis, at least for parts of the year. Moreover, it turned out that also many apparently oxygen-saturated environments such as oxygenated waters, soil particles or even the human skin are inhabited by strictly anaerobic, oxygen-sensitive bacteria which thrive in these habitats in close connection with aerobes and profit from their respiratory activity as a means of protecting them from toxic reduced oxygen species (29). Thus, also an apparently aerobic environment may contain innumerable anoxic microniches in which a given substrate is subject to degradation processes much different from those prevailing in the surroundings (34). The gastro-intestinal tract of higher animals represents another anaerobic environment which is maintained in an oxygenic atmosphere by a well-organized co-operation of aerobic and anaerobic partners. Nonetheless, our knowledge on the extent of cooperation of the bacterial flora and the host organism in the digestive tract is small, especially with respect to the metabolism of synthetic or otherwise unusual constituents of the host's diet.

After consumption of oxygen in an oxygen-limited environment, nitrate and sulfate can take over its role as a terminal acceptor of the electrons released during substrate oxidation. Finally, carbon dioxide is reduced to methane which normally escapes back into the oxygenic world. However, oxygen also serves an important function as a reaction partner in the primary attack on substrate molecules, e.g. by means of mono- or dioxygenase reactions. Unlike the role of oxygen as electron acceptor, its function as a reaction partner in oxygenase reactions cannot be taken over by any other oxidized compound. Therefore, degradation processes which in the presence of oxygen involve oxygenase reactions have to proceed in its absence in an entirely different manner.

Degradation of aromatic compounds is so intimately associated with the action of oxygenases in the biochemist's mind that anaerobic degradation of these compounds was for long times considered to be impossible although reliable data on methanogenic degradation have been reported already 50 years ago (40). Today we know that a broad variety of aromatic compounds is anaerobically degraded (13,16,17,21), and the variety of bacteria and of metabolic pathways involved appears to be at least as diverse as those active in aerobic degradation (34). In the present survey, the author wants to briefly review the various types of anaerobic transformation of organic compounds which have mainly been described in the recent past. It will be shown that the degradation potential of anaerobic microbial communities is at least equivalent to that of aerobic bacteria, and this potential opens up new perspectives, both from the technological and the environmental point of view.

2. REACTIONS OF RING SUBSTITUENTS

2.1. DEMETHOXYLATION

Plant tissues contain a broad variety of phenolic compounds among which the precursors of lignin biosynthesis may be the most important ones. These phenylpropane derivatives mostly carry one or more methoxyl groups on the aromatic ring in meta- or paraposition. In the presence of oxygen, the phenylmethyl ether linkage is transformed into an unstable halfacetal bond by a monooxygenase reaction, and the methyl carbon is released as formaldehyde. In the absence of oxygen, several homoacetogenic bacteria cleave the methyl group off and ferment it to acetate, and leave the corresponding phenol derivative unchanged (1, 6 ,14 ,15). The mechanism of this demethoxylation reaction is not known yet. Methanol does not appear as a free intermediate (41); it is likely that the methyl group is directly transferred to a cobamide carrier. The result of this anaerobic ether cleavage reaction is the formation of a phenol group which is by far more reactive than its methylated precursor. Phenols tend to react with a broad variety of other functional groups, especially if traces of oxygen allow the formation of a radical structure. In any case, accumulating phenolic compounds appear to be highly toxic to aerobic as well as anaerobic microorganisms.

2.2. DECARBOXYLATION

The carboxylic group plays an essential role in anaerobic benzoate degradation (12). Benzoate is activated to the CoA derivative right after entering the cell, and this activation is prerequisite to the subsequent ring saturation and cleavage reactions. It is evident that decarboxylation of the ring structure is beneficial only if other functional groups, e.g. hydroxyl or carboxyl groups, are present and allow further degradation; decarboxylation of benzoate would lead to benzene which to our knowledge today is not degraded in the absence of oxygen (33). Hydroxylated benzoates undergo decarboxylation if the position of the hydroxyl group allows this reaction. Gallic acid and 2.4.6-trihydroxy-benzoic acid are thus transformed to the respective trihydroxybenzenes prior to ring cleavage by Pelobacter acidigallici (30,35). 2.4- and 2.6-dihydroxybenzoate are decarboxylated in a similar way to resorcinol by other recently isolated strains of anaerobes which also ferment resorcinol but not 3.5-dihydroxybenzoate (42). 4-hydroxybenzoate and protocatechuate are transformed to the respective phenols by enrichment cultures, and phthalic acid is converted to benzoate in a similar manner (43). In all cases, the bacteria involved were highly specialized to the transformation of the respective substrate they were enriched with. It is too early to speculate if they benefit to any extent from the free energy change of the decarboxylation reaction (about -20 kJ per mol); it has been shown recently that decarboxylation of dicarboxylic acids can establish an ion gradient across the membrane which might be used at least for transport purposes (11).

2.3. DEHYDROXYLATION

The reductive removal of hydroxyl function from phenol compounds has been reported repeatedly (3, 7 ,28 ,31), however, the bacteria involved could never be identified. It was shown recently with fairly homogeneous enrichment cultures that both catechol and hydroquinone are degraded via reductive dehydroxylation to phenol (39). Both reactions are considered to be important steps in the anaerobic degradation of these compounds (see below). 3-hydroxybenzoate is degraded by a defined culture of strict anaerobes via reductive dehydroxylation to benzoate (43). Whereas in this case the dehydroxylating organism also opens and degrades the benzoate ring, it is unclear in the former cases to which extent the dehydroxylating bacteria participate in the further ring degradation. Indications exist that the dehydroxylation reaction mainly serves as an electron sink for these bacteria, but the substrate concomitantly oxidized is unknown. The dehydroxylation reaction could well proceed via an intermediate dihydrodiol formation as suggested (3), however, this hypothesis has still to be corroborated by further experimental evidence. At present, such a reaction sequence appears to be energetically quite unfavourable. A special kind of hydroxyl group transformation is the conversion of pyrogallol to phloroglucinol during fermentative degradation of gallic acid and pyrogallol by Pelobacter acidigallici (30). This 2.5-transhydroxylation reaction could proceed via an resorcinol epoxide intermediate as suggested for the backwards reaction catalyzed by Fusarium solani (44), but so far any experimental evidence for such a reaction is lacking.

2.4. DEHALOGENATION

Anaerobic degradation of 3.5-dichlorobenzoate proceeds via reductive dechlorination to 3-chlorobenzoate and benzoate, and subsequent benzoate degradation (37). The reductive removal of a chlorine substituent to form hydrochloric acid and the respective benzoate derivate was a new and revolutionizing discovery. Aerobic chlorine elimination would always lead to the corresponding hydroxybenzoates if possible at all (23). The bacteria involved in anaerobic dehalogenation are highly sensitive to oxygen and exhibit a well-pronounced preference for certain halobenzoate and halophenol isomers (8, 9,19,38) which is different from that exhibited by aerobic haloaromatic-degrading bacteria. The various bacteria active in anaerobic 3-chlorobenzoate degradation have been isolated and characterized (36). The dehalogenating bacterium only catalyzes the reductive dechlorination and does not participate in the further ring degradation. The substrate oxidized during halobenzoate reduction is still unknown. This fact makes the cultivation of this also morphologically quite unusual bacterium very difficult and explains why the biochemistry of reductive dehalogenation has not yet been studied in detail. Basically, reductive dechlorination is a reaction chemically similar to the reductive dehydroxylation reaction mentioned above, and both could function as electron sinks at a redox potential (E_o') of about +100 mV (39). The release of electrons at this redox potential could be advantageous for an anaerobic bacterium which lacks a hydrogenase enzyme and cannot participate in interspecies hydrogen transfer.

2.5 METABOLISM OF NITROGEN SUBSTITUENTS

Nitroaromatic compounds are readily reduced to the corresponding amino derivatives, preferentially in the absence of oxygen (25,26). Azo bridges between two aromatic nuclei are reductively cleaved to two amino residues by activated sludge and pure cultures of aerobic bacteria (24,45). The same reaction occurs in the absence of oxygen with anoxic sewage sludge, however, the bacteria catalyzing this reduction have not yet been characterized (43).

The further degradation of aminoaromatic compounds in the absence of oxygen is less clear. Aniline appears to be recalcitrant in the absence of oxygen (18) and efforts in the author's laboratory to enrich for aniline -degrading anaerobes failed. Anthranilic acid is degraded by denitrifying anaerobes, and at least part of the amino nitrogen appears finally as ammonia (10). It is not known yet whether ammonia is released by a reductive reaction analogous to reductive dehydroxylation or dehalogenation, or whether a hydrolytic or nitrite-dependent reaction occurs. Salicylic acid appeared as an intermediate of anthranilate degradation by sediment samples (2,4) suggesting a hydrolytic ammonia elimination. In methanogenic enrichment cultures, benzoate was found as an intermediate of anthranilate degradation (43). Further studies on the anaerobic degradation of aminoaromatics are needed to elucidate the degradation mechanisms and to explain the observed recalcitrance of some aniline derivatives, especially with respect to the enormous amounts of these compounds formed during incomplete azo dye degradation.

2.6. DEGRADATION OF ALKYL BENZENES

Anaerobic degradation of toluene, xylene, or ethylbenzene by defined laboratory cultures has so far never been reported, and it has been argued that these compounds as well as benzene pose extreme difficulties to degradation in the absence of oxygen (33,34; and literature quoted therein). Nonetheless, degradation and transformation of these compounds in anoxic sediment column reactors has been reported recently (Kuhn et al.; Grbic -Galic and Reinhardt; posters at the Gordon Conference on Applied and Environmental Microbiology, New London, New Hampshire, 1985). On the basis of the identified degradation intermediates, both hydroxylation of the ring system and hydroxylation of the methyl group of e.g. toluene appears to be possible; in the case of ethylbenzene degradation also a desaturative attack on the side chain is discussed. These considerations pose extreme problems with respect to both the kinetics and the mechanism of such a reaction, and the experimental evidence of these reactions is still rather weak. Other authors find evidence for a reductive demethylation of cresol to phenol (47), a reaction which is hard to understand as well. Further research will show if aromatic hydrocarbons are degraded in the absence of oxygen, and if other partners (e.g. nitrogen compounds or heavy metals) participate in the initial activation reactions.

3. RING CLEAVAGE REACTIONS

3.1. BENZOATE DEGRADATION

The mechanism of anaerobic benzoate degradation as studied until now appears to be basically similar, no matter if nitrate-reducing or phototrophic bacteria or methanogenic mixed cultures are concerned (12). Benzoate is activated to the CoA derivative right after entering the cell, and reduced in a one-step reaction with 6 electrons to form cyclohexyl-CoA. The further oxidative cleavage to a C-7 dicarboxylic acid derivative is a reaction sequence analogous to the ß-oxidation of fatty acids, which does not pose any basic problems. It was shown with a defined mixed methanogenic culture that benzoate is stoichiometrically transformed to 3 acetate residues and perhaps one formate, together with a balance of molecular hydrogen which is used by the methanogenic component to form methane (27). Propionate and valerate which were reported in the elder literature (22) as primary products of benzoate degradation by methanogenic enrichment cultures (see 12) are probably formed secondarily from acetate and C-1 compounds by contaminating bacteria similar to Pelobacter propionicus (32). At present, the benzoate degradation scheme as suggested by Evans (12) is in good agreement with the results of recent studies on more defined cultures (13,46) although enzymatic studies on the degradation pathway are still lacking. Especially the initial ring saturation reaction is a challenge to our imagination since 6 electrons have to be transferred in one step to overcome the energy barrier posed by the mesomeric π -electron structure.

3.2. DEGRADATION OF PHENOLS

Phenol, catechol, resorcinol, hydroquinone, pyrogallol and phloroglucinol all were reported to be completely degraded in the absence of molecular oxygen. Since they lack a carboxylic group which plays a central role in ring activation and subsequent ring saturation, it is hard to believe that they are degraded in a manner as analogous to benzoate degradation as the current literature suggests (2, 3,12,46).

Methanogenic degradation of phenol and hydroxybenzoates was studied extensively in the author's laboratory. Basic differences in their degradability were observed already in the initial enrichment cultures. Benzoate was always degraded by far faster than its hydroxy derivatives or the noncarboxylated phenols. Moreover, none of the enrichment cultures obtained with phenols as substrates ever degraded benzoate, and benzoate enrichments never attacked noncarboxylated phenols either. Among the phenol enrichments, both the adaptation periods for the respective compound and the degradation rates in adapted cultures showed a well-expressed preference for the respective isomers which was nearly exactly contrary to that of aerobic degradation: resorcinol was degraded the fastest, followed by phenol, and catechol and hydroquinone caused the most difficulties in anaerobic degradation (39). This is understandable since the latter two both undergo reductive dehydroxylation to phenol before ring cleavage (see above), and the bacteria catalyzing this reaction probably get only a very small share of the free energy available from this total degradation process. The preference of resorcinol over phenol suggests that the former is easier accessible to ring cleavage than the latter. Defined and pure cultures of resorcinol degraders were isolated which also degraded 1.3-dioxocyclohexane (42). The two hydroxyl groups of resorcinol exert a strong inductive effect on the π-electron system of the ring, strong enough to form the tautomer 1.3-dioxocyclohexene in well detectable quantities. The now isolated double bond in the ring could be reduced with one electron pair to form 1.3-cyclohexane as intermediate which is subject to hydrolytic or thiolytic cleavage. Thus, full ring saturation is not necessary at the beginning because the dioxo-tautomer is polarized sufficiently for a selective reduction reaction. A similar hypothetical pathway can be assumed for anaerobic degradation of phloroglucinol (35). In this case, the equilibrium between the trihydroxybenzene and the tautomeric trioxocyclohexane structure is by far more on the side of the latter than with the resorcinol couple. Whether the polarization of the π-electron system by one hydroxyl substituent is sufficient to allow a similar pathway also for phenol degradation has still to be examined (5,39). Nonetheless, the strong inductive effects of hydroxy groups on the mesomeric π-electron system have to be respected as important determinats for the degradation pathways used.

3.3. DEGRADATION OF HYDROXYBENZOATES

It is evident from the above considerations that both carboxyl and hydroxyl groups determine the degradation pathway of an anaerobic compound. The question is not easy to answer which one of both is more important if both are present. Most of the double- and triple-hydroxylated benzoic acid derivatives are first decarboxylated to the corresponding phenols and further metabolized as those. Only in the case of 3.5-dihydroxybenzoate, the carboxylic group appears to stay at the ring and thus determine a degradation pathway different from that of resorcinol (42). Among the monohydroxybenzoates, both alternatives exist: 3-hydroxybenzoate is dehydroxylated to benzoate, 4-hydroxybenzoate is decarboxylated to phenol (43). At the present stage it cannot be ruled out that both alternative pathways are even used by different bacterial isolates obtained with the same compound. After nearly 5 years of systematic study on the degradation of hydroxybenzoates in our lab, we are amazed by the variety of degradation pathways existing for the various isomers. It also appears that the anaerobic enrichment cultures and isolated are by far more specialized to the degradation of their respective substrates than this is known for their aerobic counterparts: Only few of them grow with non-aromatic substrates, and most are confined to only a small number of closely related aromatic compounds and their degradation derivatives.

4. CONCLUSIONS

The degradative capacity of anaerobic microbial communities has seen a vivid expansion in the last 10 to 20 years. Many compounds which are difficult to degrade in the presence of oxygen proved to be degradable in its absence, and in several cases anaerobic degradation was even more efficient than the aerobic process (34). Especially the recently discovered reductive dehydroxylation and dehalogenation reactions make a broad variety of aromatic compounds accessible to anaerobic decomposition which so far were considered to withstand microbial degradation. These new discoveries are encouraging, both for the environmental biologist and the biotechnologist: There is hope that a broad variety of aromatic pollutants is degraded during the passage of e.g. river water through anoxic sediments before entering ground water layers. And many difficult-to-degrade byproducts and waste materials of chemical production processes may be suited for anaerobic degradation in special anaerobic wastewater treatment devices.

However, it has to be emphasized that anaerobic degradation systems have some disadvantages compared to aerobic ones. Anaerobes grow comparably slow and have low growth yields. For efficient substrate degradation, they usually depend on a complex network of cooperation with partner organisms to exchange metabolites and maintain suited environmental conditions (pH, redox potential, hydrogen partial pressure). These apparent shortcomings for a technological application of anaerobic degradation processes have been overcome recently by the development of suited reactor designs which retain the active biomass in the system and thus maintain a stable complex population of active anaerobic bacteria (34). This development has just started, and we have reasons to assume that more refined devices than those existing today will do their job even more efficiently in the future.

On the other hand, we should realize that some kinds of synthetic chemical compounds, e.g. anilines, polyanellated phenols, or alkylphenols with complex side chains still appear to be not accessible to anaerobic decomposition, and that halogenated aromatics are only very slowly degraded. Accumulation of these compounds which are formed from azo dyes, commercial detergent components and biocides could cause severe environmental problems in anoxic sediments of rivers and lakes which serve directly or indirectly as drinking water ressources. These considerations urge us again to postulate complete biodegradability both in the presence and absence of oxygen for any synthetic chemical compound which enters our environment at any significant rate. We cannot afford any longer to transform the sediments of our freshwater resources into cemeteries for the products of a too inventive chemical industry.

ACKNOWLEDGEMENT

The author is indebted to Prof. Dr. N. Pfennig for generous support, and to Andreas Tschech for helpful discussions and criticism. Experimental work in the author's lab was financially supported by the Deutsche Forschungsgemeinschaft and the Fakultät für Biologie der Universität Konstanz.

REFERENCES

(1) BACHE, R. and PFENNIG, N. (1981). Selective isolation of Acetobacterium woodii on methoxylated aromatic acids and determination of growth yields. Arch. Microbiol. 130, 255-261.

(2) BALBA, M. and EVANS, W.C. (1977). The methanogenic fermentation of aromatic substrates. Biochem. Soc. Trans. 5, 302-304.

(3) BALBA, M.T. and EVANS, W.C. (1980). The methanogenic biodegradation of catechol by a microbial consortium: evidence for the production of phenol through cis-benzenediol. Biochem. Soc. Trans. 8, 452-454.

(4) BALBA, M.T. and EVANS, W.C. (1980). Methanogenic fermentation of the naturally occurring aromatic amino acids by a microbial consortium. Biochem. Soc. Trans. 8, 625-627.

(5) BAKKER, G. (1977). Anaerobic degradation of aromatic compounds in the presence of nitrate. FEMS Microbiol. Lett. 1, 103-108.

(6) BARIK, S., BRULLA, W.J. and BRYANT, M.P. (1985). PA-1, a versatile anaerobe obtained in pure culture, catabolizes benzenoids and other compounds in syntrophy with hydrogenotrophs, and P-2 plus Wolinella sp. degrades benzenoids. Appl. Environ. Microbiol. 50, 304-310.

(7) BOOTH, A.N. and WILLIAMS, R.T. (1963). Dehydroxylation of catechol acids by intestinal contents. Biochem. J. 88, 66P-67P.

(8) BOYD, S.A. and SHELTON, D.R. (1984). Anaerobic biodegradation of chlorophenols in fresh and acclimated sludge. Appl. Environ. Microbiol. 47, 272-277.

(9) BOYD, S.A., SHELTON, D.R., BERRY, D. and TIEDJE, J.M. (1983). Anaerobic biodegradation of phenolic compounds in digested sludge. Appl. Environ. Microbiol. 46, 50-54.

(10) BRAUN, K. and GIBSON, D.T. (1984). Anaerobic degradation of 2-aminobenzoate (anthranilic acid) by denitrifying bacteria. Appl. Environ. Microbiol. 48, 102-107.

(11) DIMROTH, P. (1982). Decarboxylation and transport. Bioscience Reports 2, 849-860.

(12) EVANS, W.C. (1977). Biochemistry of the bacterial catabolism of aromatic compounds in anaerobic environments. Nature (London) 270, 17-22.

(13) FERRY, J.G. and WOLFE, R.S. (1976). Anaerobic degradation of benzoate to methane by a microbial consortium. Arch. Microbiol. 107, 33-40.

(14) FRAZER, A.C. and YOUNG, L.Y. (1985). A gram-negative anaerobic bacterium that utilizes O-methyl substituents of aromatic acids. Appl. Environ. Microbiol. 49, 1345-1347.

(15) GRBIĆ-GALIĆ, D. and YOUNG, L.Y. (1985). Methane fermentation of ferulate and benzoate: anaerobic fermentation pathways. Appl. Environ. Microbiol. 50, 292-297.

(16) HEALY, J.B., jr and YOUNG, L.Y. (1978). Catechol and phenol degradation by a methanogenic population of bacteria. Appl. Environ. Microbiol. 35, 216-218.

(17) HEALY, J.B. and YOUNG, L.Y. (1979). Anaerobic biodegradation of eleven aromatic compounds to methane. Appl. Environ. Microbiol. 38, 84-89.

(18) HOROWITZ, A., SHELTON, D.R., CORNELL, C.P. and TIEDJE, J.M. (1982). Anaerobic degradation of aromatic compounds in sediments and digested sludge. Dev. Ind. Microbiol. 23, 435-444.

(19) HOROWITZ, A., SUFLITA, J.M. and TIEDJE, J.M. (1983). Reductive dehalogenations of halobenzoates by anaerobic lake sediment microorganisms. Appl. Environ. Microbiol. 45, 1459-1465.

(20) HUNGATE, R.E. (1969). A roll tube method for cultivation of strict anaerobes. In: NORRIS, J.R. and RIBBONS, D.W.,(eds). Methods in Microbiology. Vol. 3B, 117-132.

(21) KAISER, J.P. and HANSELMANN, K.W. (1982). Fermentative metabolism of substituted monoaromatic compounds by a bacterial community from anaerobic sediments. Arch. Microbiol. 133, 185-194.

(22) KEITH, C.L., BRIDGES, R.L., FINA, L.R., IVERSON, K.L. and CLORAN, J.A. (1978). The anaerobic decomposition of benzoic acid during methane formation. IV. Dearomatization of the ring and volatile fatty acids formed on ring rupture. Arch. Microbiol. 118, 173-176.

(23) KLAGES, U., MARKUS, A. and LINGENS, F. (1981). Degradation of 4-chlorophenylacetic acid by a Pseudomonas species. J. Bacteriol. 146, 64-68.

(24) KULLA, H.G. (1981). Aerobic bacterial degradation of azo dyes. In: LEISINGER, Th., COOK, A.M., HÜTTER, R. and NÜESCH, J. (eds). Academic Press, London, New York, pp. 387-399.

(25) LIN, D., THOMSON, K. and ANDERSON, A.C. (1984). Identification of nitroso compounds from biotransformation of 2.4-dinitrotoluene. Appl. Environ. Microbiol. 47, 1295-1298.

(26) MCCORMICK, N.G., FEEHERRY, F.E. and LEVINSON, H.S. (1976). Microbial transformation of 2,4,6-trinitrotoluene and other nitroaromatic compounds. Appl. Environ. Microbiol. 31, 949-958.

(27) MOUNTFORT, D.O. and BRYANT, M.P. (1982). Isolation and characterization of an anaerobic syntrophic benzoate-degrading bacterium from sewage sludge. Arch. Microbiol. 133, 249-256.

(28) PEPPERCORN, M. and GOLDMAN, P. (1971). Caffeic acid metabolism by bacteria of the human gastrointestinal tract. J. Bacteriol. 108, 996-1000.

(29) PFENNIG, N. (1979). Formation of oxygen and microbial processes establishing and maintaining anaerobic environments. In: SHILO, M., (ed). Dahlem Konferenzen, Berlin, pp. 137-148.

(30) SAMAIN, E., DUBOURGIER, H.C. and ALBAGAC, G., unpublished manuscript.

(31) SCHELINE, R.S. (1973). Metabolism of foreign compounds by gastro-intestinal microorganisms. Pharmacol. Rev. 25, 452-523.

(32) SCHINK, B. (1984). Fermentation of 2.3-butanediol by Pelobacter carbinolicus sp. nov. and Pelobacter propionicus, sp. nov., and evidence for propionate formation from C2 compounds. Arch. Microbiol. 137, 33-41.

(33) SCHINK, B. (1985). Degradation of unsaturated hydrocarbons by methanogenic enrichment cultures. FEMS Microbiol. Ecol. 31, 69-77.

(34) SCHINK, B. (1985). Principles and limits of anaerobic degradation - environmental and technological aspects. In: ZEHNDER, A.J.B., (ed). Environmental Microbiology of Anaerobes. John Wiley and Sons, New York (in press).

(35) SCHINK, B. and PFENNIG, N. (1982). Fermentation of trihydroxybenzenes by Pelobacter acidigallici gen. nov. sp. nov., a new strictly anaerobic, non-sporeforming bacterium. Arch. Microbiol. 133, 195-201.

(36) SHELTON, D.R. and TIEDJE, J.M. (1984). Isolation and partial characterization of bacteria in an anaerobic consortium that mineralizes 3-chlorobenzoic acid. Appl. Environ. Microbiol. 48, 840-848.

(37) SUFLITA, J.M., HOROWITZ, A., SHELTON, D.R. and TIEDJE, J.M. (1982). Dehalogenation: a novel pathway for the anaerobic biodegradation of haloaromatic compounds. Science (Wash.) 218, 1115-1117.

(38) SUFLITA, J.M., ROBINSON, J.A. and TIEDJE, J.M. (1983). Kinetics of microbial dehalogenation of haloaromatic substrates in methanogenic

environments. Appl. Environ. Microbiol. 45, 1466-1473.

(39) SZEWZYK, U., BRENNER, R. and SCHINK, B. (1985). Methanogenic degradation of hydroquinone and catechol via reductive dehydroxylation to phenol. FEMS Microbiol. Ecol. 31, 79-87.

(40) TARVIN, D. and BUSWELL, A.M. (1934). The methane formation of organic acids and carbohydrates. J. Am. Chem. Soc. 56, 1751-1755.

(41) TSCHECH, A. and PFENNIG, N. (1984). Growth yield increase linked to caffeate reduction in Acetobacterium woodii. Arch. Microbiol. 137, 163-167.

(42) TSCHECH, A. and SCHINK, B. (1985). Fermentative degradation of resorcinol and resorcylic acids. Arch. Microbiol. (in press).

(43) TSCHECH, A. and SCHINK, B., manuscript in preparation.

(44) WALKER, J.R.L. and TAYLOR, B.G. (1983). Metabolism of phloroglucinol by Fusarium solani. Arch. Microbiol. 134, 123-126.

(45) WUHRMANN, K., MECHSNER, Kl. and KAPPELER, Th. (1980). Investigation on rate-determining factors in the microbial reduction of Azo-dyes. Eur. J. Appl. Microbiol. Biotechnol. 9, 325-338.

(46) YOUNG, L.Y. (1984). Anaerobic degradation of aromatic compounds. In: GIBSON, D.T. (ed). Microbial degradation of organic compounds. Marcel Dekker, New York, pp. 487-523.

(47) YOUNG, L.Y. and RIVERA, M.D. (1985). Methanogenic degradation of four phenolic compounds. Water Res. 19, 1325-1332.

TOTAL BIODEGRADATION OF CHLORINATED BIPHENYLS BY A PSEUDOMONAS STRAIN

J.R. Parsons and T.H.M. Sijm
Laboratory of Environmental and Toxicological Chemistry
University of Amsterdam
Nieuwe Achtergracht 166
1018 WV Amsterdam, The Netherlands

O. Hutzinger
Chair of Ecological Chemistry and Geochemistry
University of Bayreuth
Postfach 3008
D-8580 Bayreuth, Germany

SUMMARY

A biphenyl-degrading Pseudomonas strain has been isolated. The
strain is also able to degrade both chlorinated biphenyls and
chlorinated benzoic acids. These latter compounds have previously
been identified as the end products of the metabolism of
chlorinated biphenyls by other bacterial strains.
The strain described here is capable of degradation of chlorinated
biphenyls beyond the chlorinated benzoic acid stage.

INTRODUCTION

Bacteria and other microorganisms play an important role in the
degradation of organic materials, releasing carbon dioxide and other
nutrients which can be used by plants. Xenobiotic organic compounds
present in the environment may also be subjected to biodegradation by
this process (1).
Chlorinated biphenyls (CB) are known to be degraded by a variety of
species of bacteria (2). Metabolism of these compounds usually proceeds
via initial dioxygenation of the least chlorinated ring, followed by
degradation of this ring to yield a chlorinated benzoic acid (CBA); see
for example, the pathway proposed by Furukawa et al. (3). None of the
bacterial strains capable of degrading chlorinated biphenyls isolated
so far have been shown to be able to degrade these compounds beyond the
chlorobenzoic acid stage (2). Total degradation or mineralisation
(degradation to CO_2) of chlorinated biphenyls has been observed in
mixed cultures obtained from river sediments (4).
Sylvestre et al. found that in a mixed culture, consisting of a
strain able to grow on 4-CB and a Pseudomonas strain able to grown on
4-CBA, 4-CB was degraded almost completely, with only traces of 4-CBA
and no hydroxylated derivatives of 4-CB being detected (5).
We are in this laboratory engaged in a study of the cometabolism
of chlorinated aromatic compounds in aqueous solution. Cometabolism is
the process whereby microorganisms metabolise a compound on which they
cannot grow, while utilising another compound as growth substrate (1).
We have isolated a bacterial strain which can grow on both biphenyl and
benzoic acid, and can cometabolise chlorinated derivatives of these
compounds

<u>EXPERIMENTAL</u>

The bacterial strain described here was isolated by incubating samples of soil in biphenyl medium (see below). This strain was subsequently identified as a <u>Pseudomonas</u> species. Cultures were grown in medium of the following composition (per 1).

0.61 g benzoic acid (5m$\underline{M}$)
4.03 g $Na_2HPO_4.12H_2O$
2.14 g KH_2PO_4
1.00 g $(NH_4)_2SO_4$
0.20 g $MgSO_4.7H_2O$
0.07 g $Ca(NO_3)_2.4H_2O$
0.03 g iron(III) ammonium citrate
1   ml trace elements solution (6)

The medium used for cultures growing on biphenyl or 4-CB was similar, except that ca. 10 mg/l of one of these compounds replaced the benzoic acid. Experiments in batch culture were carried out in 300 ml erlenmeyer flasks containing 100 ml medium, to which the required amount of CBA or CB had been added as a solution in water or acetone, respectively. In order to limit losses of CB due to volatilisation, these compounds were only added once growth was visible (usually 24 h after innoculation from a biphenyl-grown culture) and the flasks were kept sealed. Experiments in chemostat culture were carried out in a system as described previously (7).

The extracellular fluid was analysed for CBA after filtration (0.2 μm nitro-cellulose filter) by RP-HPLC using a Zorbax ODS column (15 cm x 2.7 mm) and 55/45 10 m$\underline{M}$ H_3PO_4/MeOH as mobile phase). Analysis for chlorinated biphenyls were carried out as follows. A sample of culture was extracted with an equal volume of hexane. The hexane layer was removed after centrifugation and reduced in volume under a nitrogen stream to ca. 1 ml. This extract was then passed through a column containing ca. 1 g 100-120 mesh silica-H_2SO_4 (ca. 40% w/w) above ca. 1 g silica - 1N NaOH (ca. 33% w/w) to remove polar and hydrolysable compounds. The clean sample was then analysed by GC-ECD (2% Dexil 300 GC on Chromosorb 750, 160-180 mesh, 1 m x 2 mm). Recoveries of CB's were greater than 95%. Controls were carried out for all the degradation experiments under identical conditions in the absence of bacteria. All solvents were distilled before use.

<u>RESULTS AND DISCUSSION</u>

The <u>Pseudomonas</u> strain described here grows on either biphenyl or 4-CB. In both cases a greenish yellow colour becomes visible. In the case of growth on biphenyl this colour disappears after a few days, but the yellow colour does not disappear from 4-CB-grown cultures. The formation of yellow metabolites of biphenyl is consistent with earlier reports of the degradation of these compounds, where the yellow colour was assigned to the products of aromatic ring cleavage by the <u>meta</u> route (2,3). Although this strain grew well on biphenyl and 4-CB, we used a compound with a higher aqueous solubility as growth substrate in the cometabolism experiments. Benzoic acid, a compound previously identified as a microbial metabolite of biphenyl (2) was used. A transient yellow colour also appeared during growth on benzoic acid.

The cometabolic degradation of 4-CB and 2,5-CB was investigated in batch culture. As is clear from the data in Table I, both these compounds were degraded readily. As it was necessary to carry out these experiments

in sealed systems it is possible the growth of bacteria was limited by
the amount of oxygen present in the flasks. However, this did not appear
to have any effect on the degradation of the CB's.

TABLE I. Degradation of Chlorinated Biphenyls and Benzoic Acids in
Batch Culture.

Compound	Incubation time (h)	Initial conc. (g/l)	Final conc. (g/l) Culture	Control
4 CB	24	1.5×10^{-3}	$-^{a)}$	1.8×10^{-3}
2,5-CB	24	1.2×10^{-4}	0.4×10^{-6}	1.2×10^{-4}
4-CBA	96	5.2×10^{-2}	4.5×10^{-2}	5.4×10^{-2}
2,5-CBA	96	3.6×10^{-2}	3.6×10^{-2}	3.5×10^{-2}

$^{a)}$Not detected

Since the Pseudomonas strain grew well on benzoic acid, we also
investigated whether it was capable of degrading chlorinated benzoic
acids, the expected metabolites of chlorinated biphenyls (2). The results
obtained for the degradation of 4-CBA and 2,5-CBA in batch culture are
given in Table I and indicate that 4-CBA was degraded under these conditions, but that no metabolism of 2,5-CBA had taken place. The use of a
sealed system was not necessary for these compounds.

The use of a chemostat continuous culture system in the investigation
of the biodegradation of chlorobenzoic acids and PCB mixtures has been
described previously (7,8). The major advantage that such a system offers
is the possibility of long term exposure of a growing culture to a
xenobiotic compound, which may result in the induction of degradation
otherwise not observed in batch culture. Consequently a chemostat may be
a better model system than batch culture for the prediction of biodegradation in the environment.

The results of the addition of 2,5-CBA to a chemostat culture of
Pseudomonas sp. at an initial concentration of 33.2 mg/l are shown in
figure 1, together with those from a similar experiment in the absence
of bacteria. No indication of possible degradation was seen up 24 h
after the start of the experiment. In contrast, continuous exposure to
2,5-CBA as a component of the medium did appear to lead to degradation
of this compound: the equilibrium concentration in the culture being
approximately half that in the medium (fig. 2).

We therefore conclude that the biphenyl-degrading Pseudomonas strain
isolated by us is capable of cometabolic degradation of both chlorinated
benzoic acids and biphenyls and thus of total degradation of a number of
the latter compounds. We intend to use chemostat cultures of this strain
to study the cometabolism of the more highly chlorinated biphenyls and
other aromatic compounds.

ACKNOWLEDGEMENT
The authors are indebted to Dr. O.M. Neijssel for fruitful discussions
and for his help in the characterisation of the strain used in this work.

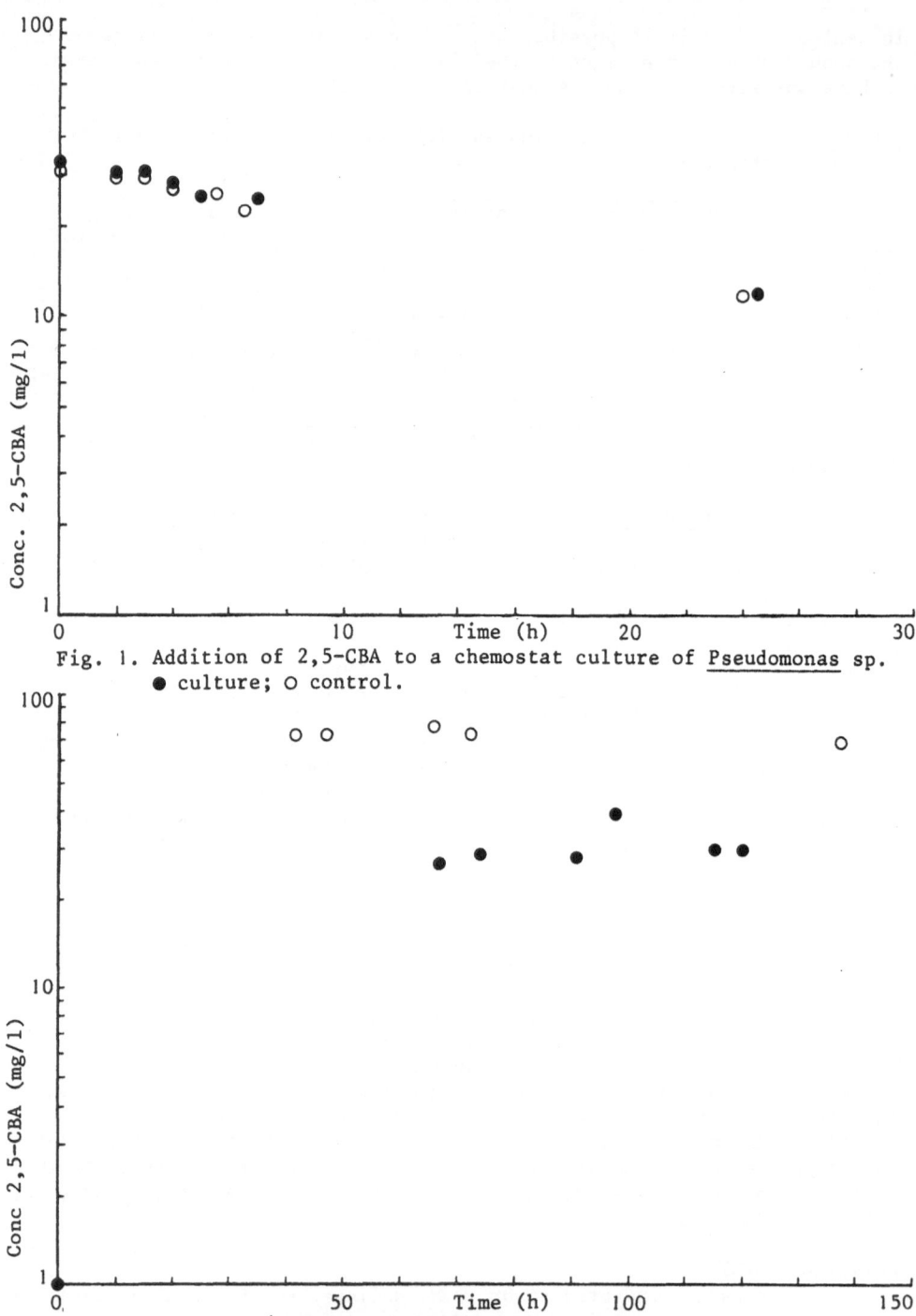

Fig. 1. Addition of 2,5-CBA to a chemostat culture of <u>Pseudomonas</u> sp.
● culture; ○ control.

Fig. 2. Continuous exposure of a chemostat culture to 2,5-CBA.
● culture (medium concentration 60.7 mg/1);
○ control (medium concentration 71.7 mg/1).

REFERENCES

(1) ALEXANDER, M., Science, 211, 132-138 (1981).
(2) PARSONS, J.R., VEERKAMP, W. and HUTZINGER O., Toxicol. Environ.
    Chem., 6, 327-350 (1983).
(3) FURUKAWA, K., TOMIZUKA, K. and KAMIBAYASHI, A., Appl. Environ.
    Microbiol., 38, 301-310 (1978).
(4) KONG, H.-L. and SAYLER, G.S., Appl. Environ. Microbiol., 46, 666-672
    (1983).
(5) SYLVESTRE, M., MASSE, R., AYOTTE, C., MESSIER, F. and FAUTEUX, J.
    Appl. Microbiol. Technol., 21, 192-195 (1985).
(6) PFENNING, N. and LIPPERT, K.D., Arch. Microbiol., 55, 245-256 (1966).
(7) VEERKAMP, W., PEL, R. and HUTZINGER O., Chemosphere, 12, 1337-1343
    (1983).
(8) LIU, D., Bull. Environ. Contam. Toxicol., 29, 200-207 (1982).

POSSIBILITIES OF PREDICTING THE PRODUCTION OF LIPOPHILIC VOLATILE ORGANOHALIDES CHLORINATION OF SEA WATER AND FRESH WATER SAMPLES IN LABORATORY CONDITIONS

M. PICER, V. HOCENSKI and N. PICER

Center for Marine Research Zagreb, "Rudjer Bošković" Institute,
Zagreb, Yugoslavia

Summary

Investigation of the relationship between the concentration of the organic matter in the samples of river, waste and ground water and the samples of seawater on one side and the production of lipophilic volatile organohalogen materials during the process of their chlorination at the other side is described. The production of the halomethanes and other ECD response materials during the chlorination of water samples in laboratory conditions is compared with the concentrations of organic materials in water samples. From linear correlation coefficients possibilities of predicting amounts of organohalides produced during the chlorination process by using very simple methods for the organic materials estimation in investigated water samples are obtained.

1. INTRODUCTION

Water chlorination leads to the generation of various halogenated degradation products of water organic matter (1,2,3). Besides more than 2000 organic compounds have been reported to be present in natural waters, ground water, drinking water and industrial effluents (4). Isolation and identification of so many organic compounds is a very difficult task. However, by using a relatively simple and unexpensive FID and ECD fingerprint method it is also possible to obtain some useful data about such a complex mixture (5). Relationship between the concentrations of organic matter in the samples of the Sava river and the production of lipophilic volatile organohalogen materials during the process of their chlorination by using ECD fingerprint method is described in our paper presented in Oslo in 1983 (6). These investigations have been extended to other rivers in Croatia and to the ground waters and seawater as well. The aim of this paper is to find our possibilities in predicting the amounts of organohalides produced during the chlorination of water samples, where the concentrations of organic matter were estimated by using very simple methods of analysis.

2. METHODOLOGY

For estimation of organic materials in water samples several spectrophotometric methods are used (6). Lignin matter is estimated by using four methods (L_A, L_B, L_C, L_D) and humic materials are estimated by using two methods of analysis (H_A, H_B). For estimation of total organics in the water the UV spectrophotometric method is used (UV). In some samples stan-

dard BOD, COD and O_2 measurements are also taken (6). The amounts of halomethanes and other ECD response volatile lipophilic materials produced
during chlorination of water samples are obtained by substracting the
amounts of the ECD response materials in the water samples before chlorination from the amounts of ECD response materials in water samples after
the chlorination process (6).

3. RESULTS AND DISCUSSION

Table 1 presents high significant linear correlation coefficients obtained after comparing concentrations of humic, lignin and "total" organic
matter burden in river water and waste water samples heavily polluted by
ECD response lipophilic volatile organic materials produced during chlorination in laboratory conditions.

TABLE 1. Correlation between produced ECD response materials and organic
matter burden in chlorinated waste water and heavily polluted
river water samples.

Fraction of ECD response materials	More specific definition of ECD response materials	Organic matter and method of their determination in water sample	Number of pairs	Linear correlation coefficient		
				r	r_1	r_2
LVOH fraction	Total LVOH	HL_A	20	0.954	0.909	0.977
		L_A	20	0.959	0.919	0.980
		L_B	20	0.987	0.974	0.994
	Δ LVOH	COD	16	0.998	0.995	0.995
	Rest	BOD	9	0.993	0.995	0.995
TCB fraction	Rest	BOD	7	0.980	0.947	0.993
	RRT = 3.4-3.1	L_B	11	0.970	0.952	0.988
DDT I fraction	Total DDT I	L_B	15	0.983	0.970	0.990
	Δ DDT I	UV	15	0.978	0.961	0.988
	Rest	BOD	7	0.999	0.995	0.995
	RRT = 11.8-11.1	H_A	8	0.997	0.993	0.995
		L_A	8	0.985	0.964	0.994
		L_B	8	0.986	0.966	0.994
	RRT = 2.9-2.8	H_A	8	0.974	0.938	0.989
		L_B	8	0.987	0.969	0.995
	RRT = 2.2-2.1	L_B	12	0.983	0.967	0.991
		BOD	6	0.971	0.910	0.991
DDT II fraction	RRT = 5.7-3.0	H_A	10	0.961	0.919	0.981
		L_A	10	0.955	0.907	0.978
	RRT = 2.9-1.8	L_C	6	0.984	0.950	0.995
		UV	6	0.972	0.914	0.991

Table 2. Correlation between produced ECD response materials and organic matter burden in chlorinated relatively clean river water, ground water and seawater samples.

Sort of water samples	Fraction of ECD response materials	More specific definition of ECD response materials	Organic matter and method of their determination in water samples	Number of pairs	Linear correlation coefficients		
					r	r_1	r_2
River water samples		$CHCl_2Br$	H_A	10	0.935	0.820	0.978
			L_A	10	0.948	0.850	0.938
			L_D	10	0.878	0.680	0.958
	LVOH Fraction	$CHClBr_2$	L_B	8	0.847	0.540	0.956
		Δ LVOH Rest	H_A	10	0.964	0.870	0.990
			L_A	10	0.964	0.870	0.990
			L_D	10	0.898	0.670	0.971
Ground water samples	LVOH Fraction	$CHCl_2Br$	L_A	13	-0.707	-0.400	-0.870
			L_C	13	-0.744	-0.420	-0.890
			UV	13	-0.712	-0.400	-0.870
	TCB Fraction	Total TCB	UV	13	-0.818	-0.600	-0.923
	DDT I Fraction	RRT = 3.5	L_D	13	0.903	0.770	0.960
	DDT II Fraction	Total	L_C	5	0.986	0.890	0.998
Sea water samples	LVOH Fraction	Total	UV	4	0.971	0.572	0.995
		Δ LVOH Rest	H_A	6	0.893	0.691	0.963
	DDT I Fraction	RT = 7.1-6.8	H_A	4	-0.943	-0.291	-0.995
		RT = 5.4-5.2	H_B	5	0.938	0.658	0.976
		Δ DDT I Rest	L_B	5	-0.774	-0.300	-0.939

In the first and the second column fractions of ECD response materials produced during water chlorinaiton are given. For instance "total" LVOH fraction means that all peaks of this fraction are calculated as the concentration of $CHCl_3$. $\Delta_{"rest"}$ means all other peaks besides the identified peaks or separately calculated peaks with various retention times. In the third column the method for estimation of organic matter burden of water samples is designated. In the fourth and fifth column number of pairs and linear correlation coefficients are designated. The first r is linear correlation coefficient, r_1 and r_2 are minimum and maximum values of linear correlation coefficient at 0.95 confidence limit. From the LVOH fraction very high linear correlation coefficient (higher than 0.95) is obtained when concentration of all peaks which appeard in LVOH fraction are compared with the concentrations of humic matter (estimated by using Method A) and those of lignin matter (estimated by using methods A and B). Such high positive linear correlation coefficients are also obtained for Δ_{Rest} LVOH fraction when comparing with COD and BOD parameters.

On Fig. 1 the highest correlation coefficients and also all the coefficients higher than ± 0.7 are presented. Here appeard some halomethanes ($CHCl_3$ and $CHCl_2Br$) and there was significant negative correlation coefficient between chloroform production and oxygen concentrations in the examined heavily polluted river and waste water samples. Only two very high values of correlation coefficients appeard during the investigation of TCB fraction. Correlation coefficients higher than 0.7 show larger number of samples as it can be seen from the Fig. 1. The highest number of very high positive linear correlation coefficients found during the investigation of DDT I Fraction. High correlation coefficients are obtained not only for total ECD response materials but also for some groups of peaks which appeard at several relative retention times to p,p, DDT. DDT II fraction obtained after alumina cleaning of DDT I fraction (7) shows lower number of high significant correlation coefficients in comparison with DDT I fraction (Table 1 and Fig. 2).

The highest correlation coefficients obtained during the investigation of slightly polluted river water, ground water and seawater samples are presented on Table 2 and in Fig. 2. As it can be seen high significant linear correlation coefficients for these samples are found in significantly lower number in comparison with heavily polluted river water and waste water samples.

4. ACKNOWLEDGEMENT

The authors express their gratitude to the Self-managing Community of Interest for the Scientific Research of S.R. Croatia for the financial support.

REFERENCES

(1) JOLLEY, R.L. (Editor) 1976. Water Chlorination, Environmental Impact and Health Effects. Ann Arbor Sci., Ann Arbor, Vol. 1.

(2) JOLLEY, R.L., GORCHEV, H., and HAMILTON, D.H. Jr. (Editors) 1978. Water Chlorination, Environmental Impact and Health Effects. Ann Arbor Sci., Ann Arbor, Vol. 2.

(3) JOLLEY, R.L., BRUNGS, W.A., CUMMING, R.B., and JACOBS, V.A. (Editors) 1980. Water Chlorination, Environmental Impact and Health Effects. Ann Arbor Sci., Ann Arbor, Vol. 3.

(4) ELLIS, D.D., JONE, C.M., LARSON, R.A., SCHAEFFER, D.J. (1982). Orga-

nic Constituents of Mutagenic Secondary Effluents from Wastewater
Treatment Plants, Arch. Environm. Contam. Toxicol. 11, 373.

(5) PICER, M. (1981). The use of ECD and FID Fingerprint Techniques for
     the Evaluation of River Water Purification Contaminated with Organic
     Pollutants. Proc. of the Second European Symposium "Analysis of
     Organic Micropollutants in Water", Killarney, Eds. A. Bjørseth and
     G. Angeletti, 48.
(6) PICER, M., HOCENSKI, V., and PICER, N. (1983). The Relationship Bet-
     ween the Concentration cf Organic Matter in Natural Waters and the
     production of Lipophilic Volatile Organohalogen Compounds during
     their Chlorination, Proc. of the third European Symposium "Analysis
     of Organic Micropollutants in Water", Oslo, Eds. A. Bjørseth and
     G. Angeletti, 301.
(7) PICER, N., and PICER, M. (1980). Evaluation of Macroreticular Resins
     for the Determination of Low Concentrations of Chlorinated Hydrocar-
     bons in Sea Water and Tap Water. J. Chromat. 193, 357.

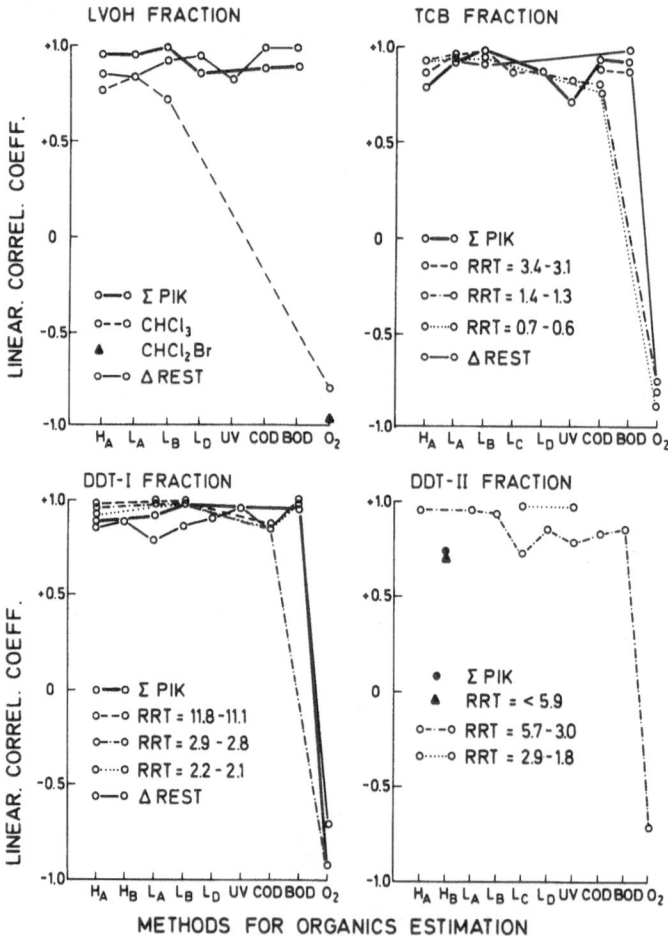

Fig. 1. Relationship between linear correlation coefficients and the methods
        used to estimate concentrations of organic materials in heavy polluted
        river water and waste water samples.

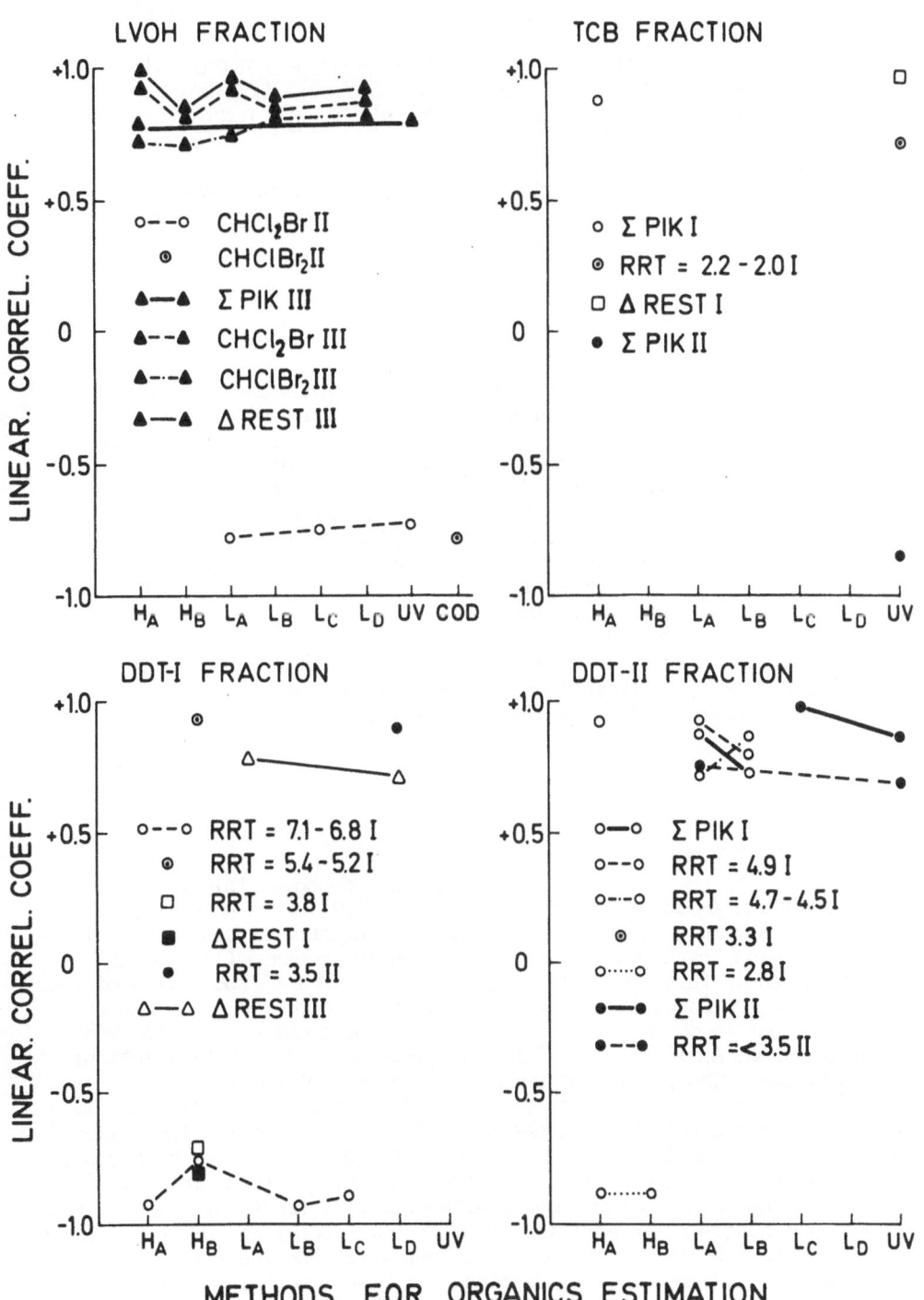

Fig. 2 Relationship between linear correlation coefficients and the methods used to estimate concentrations of organic materials in relatively clean river water (I), ground water (II) and marine water (III) samples.

ORCHARD SPRAYING WITH ORGANOPHOSPHORUS PESTICIDES:
OCCURRENCE AND FATE OF PARENT COMPOUNDS AND PHOSPHORUS
CONTAINING HYDROLYTIC PRODUCTS IN ADJACENT PONDS

V. Drevenkar, Z. Fröbe, B. Štengl and B. Tkalčević
Institute for Medical Research and Occupational Health,
Zagreb, Yugoslavia

Summary

A two-year control of the water from four ponds in two orchards
regularly treated with triesters of phosphorothioic and -dithioic
acids showed, during and immediately after the spraying season, a
considerable increase in the concentration of dialkyl phosphorothioates and phosphorodithioates. Intact organophosphorus pesticides
were also determined in some of the analysed water samples showing
that the compounds were not completely degraded before reaching the
pond water by leaching and washover of the contaminated soil. The
degradation rate of phosphorodithioic acid triesters to more stable
specific diesters in the pond water - related silt model systems
and translocation of parent compounds and arising degradation products from contaminated silt to the water was dependent on the physical and chemical properties of the pesticide as well as on the
characteristics of the water and silt.

1. INTRODUCTION

The occurrence of dimethyl and diethyl phosphorodithioates (DMDTP
and DEDTP) and phosphorothioates (DMTP and DETP) in the water from three
ponds adjacent to an apple orchard was a consequence of seasonal application of organophosphorus pesticides (OP) in the orchard (1). In the
continuation of this work a two-year control - in 1983 and 1984 - of these
OP pesticide residues in three ponds adjacent to orchard KS and in one
pond adjacent to orchard DS was undertaken. The orchards were sprayed
regularly from April to July with about 1500 dm^3ha^{-1} of 0.1% solutions of
OP pesticides. The pesticides were all triesters of phosphorothioic and
-dithioic acids: Chlorpyrifos, Diazinon, Demeton-S-methyl, Azinphos-methyl,
Methidathion and Phosalone. Orchard KS (320 ha) was treated four times in
1983 and six times in 1984 and orchard DS (93 ha) four times each year.
The water samples were collected before the last treatment in July and
again one and three to four months after the last treatment, i.e. in September and November. In addition to DMTP, DETP, DMDTP and DEDTP in samples
collected in 1984 the intact OP pesticides were also identified. Their
presence was indicated by a screening test based on cholinesterase inhibition (2).

For the better understanding of the results of field experiments the
persistence of two applied phosphorodithioic acid triesters Phosalone and
Methidathion, their degradation to more stable specific diesters - DETP
and DEDTP for the former and DMTP and DMDTP for the latter pesticide -

and the possibility of translocation of parent pesticides and arising degradation products from the contaminated silt to the pond water were investigated on model systems by addition of the two pesticides to pond water or to related silt.

2. EXPERIMENTAL

2.1. Analysis of pond water

Dialkylphosphorus anions and OP pesticides were accumulated from 10 and 100 cm^3 pond water samples, respectively, by C_{18} reversed-phase (Sep-Pak C_{18} cartridge) enrichment procedure and analysed by gas chromatography using an alkali flame ionization detector (3).

The inhibitory effect on cholinesterase activity of the concentrated samples, after passing 100 cm^3 of pond water through a Sep-Pak C_{18} cartridge, was determined by a test for monitoring traces of cholinesterase inhibitors (2). In this experiment the methanolic eluate was mixed with a phosphate buffer, pH 7.4, methanol was removed under a stream of nitrogen and the remaining buffer solution of organophosphates added to a solution of cholinesterase.

2.2. The behaviour of OP pesticides in pond water - related silt model systems

The persistence of Phosalone and Methidathion and their degradation to corresponding dialkyl phosphorodithioates and dialkyl phosphorothioates in the water and silt of Ponds KS and Pond DS were examined by comparing their behaviour in three different systems. In systems A and B 1 dm^3 of pond water and 100 g of air dried related silt were placed in a glass 10 cm in diameter; system C consisted only of 1 dm^3 of pond water. An amount of 4-5 mg of each pesticide was added in a methanolic solution either to the pond water in systems A and C or was mixed with dry silt for system B. The water was air-bubbled in all systems for 19-24 days at room temperature (20 °C) and the concentrations of pesticides and their degradation products were determined during this period by analysing 20 cm^3 water samples every 3-10 days.

3. RESULTS AND DISCUSSION

The results of analysis of OP pesticides and residues in the pond water are listed in Table I. The values are given only for one pond adjacent to orchard KS because the results for all three ponds were very similar. Owing to the more intensive treatment of the orchard KS with OP pesticides in 1984 the concentrations determined in the samples collected in that year were higher than in 1983. The highest concentrations in the water from Pond KS were observed immediately before the end of the spraying season in July 1984. In the water from Pond DS residue concentrations were equally high in July, and one month after the end of the spraying, in September 1984.

As OP pesticides are specific inhibitors of cholinesterase activity, a cholinesterase assay was applied as a preliminary indicator of the presence of intact pesticides in pond water samples (2). None of the water samples added directly to the enzyme solution influenced its activity but the concentrated samples from Pond KS water collected in July 1984 and from Pond DS water collected in July and September 1984 inhibited the enzyme activity by 18, 35 and 41%, respectively. The highest inhibition of enzyme activity determined with Pond DS water was accompanied by the highest concentrations of parent OP pesticides and specific residues.

Traces of intact Phosalone were detected in the Pond DS water even four months after its application in the orchard. This confirmed a greater stability and longer residual life of arylsubstituted organophosphates like Phosalone compared to alkylphosphates like Demeton-S-methyl (4).

The results of laboratory experiments shown in Figs. 1 and 2 illustrate the differences in the behaviour of two OP pesticides in two pond water - related silt systems depending on their chemical and physical properties as well as on the type of water and silt. In Pond DS water both, Phosalone and Methidathion concentrations, decreased faster than in the water from Pond KS, whether the silt was added or not. Phosalone (Fig. 1) was mostly degraded to DEDTP which explains a much higher concentration of this residue than of DETP as determined by pond water analysis (Table I). The addition of Phosalone to the silt in both KS and DS systems resulted also in the increase of predominantly DEDTP concentration in water indicating the adsorption of Phosalone on the silt. The tendency of adsorption was also suggested by a much faster decrease of Phosalone concentration in the water of Pond KS in the presence of related silt than in the silt absence (Fig. 1, Pond KS A and C). After 24 days Phosalone was not detected in the water and only about 42% of it was degraded to DEDTP in the system A. In the system C over 70% of the initially added Phosalone was still present in the water while about 20% was degraded to DEDTP in the same period.

Compared to Phosalone, Methidathion is more soluble in water (240 mg dm^{-3} and 10 mg dm^{-3} at 25 °C, respectively) and as an 0,0-dimethyl derivative has a lower octanol-water partition coefficient than Phosalone (4). In our experiments (Fig. 2) no significant adsorption of Methidathion on the silt was observed. It was degraded in a similar extent to DMDTP and DMTP in all investigated systems. After addition to the silt about 80% of intact compound was released to the water in the Pond KS water - silt system in the period of 19 days. In the Pond DS water - silt system this percentage was lower probably because of faster degradation of the released Methidathion in the Pond DS water than in the Pond KS water.

4. CONCLUSION

The treatment of orchards with OP pesticides increases the concentration not only of specific dialkylphosphorus residues but also of parent compounds in the water from adjacent ponds during and immediately after the spraying season. The increase is mostly a consequence of leaching and washover and is less due to a direct contamination from aerial spray. Despite their ability to undergo degradation or chemical alteration in the environment shortly after application OP pesticides are not completely degraded before reaching the pond water and some of them can be detected in the water as late as four months after application in the orchard. Their degradation in the water-related silt system, adsorption on the silt and translocation from the silt to the water depend on the chemical and physical properties of the pesticide as well as on the type of water and silt.

References

(1) Drevenkar, V., Fröbe, Z., Štengl, B., and Tkalčević, B., Species and Persistence of Pollutants in the Pond Water from an Orchard Area Treated with Organophosphorus Pesticides. In "Analysis of Organic Micropollutants in water", Proceedings of the Third European Symposium, Oslo 1983, A Bjørseth and G. Angeletti (eds), D. Reidel Publishing Comp., Dordrecth, Boston, Lancaster, 289-293.

(2) Drevenkar, V., Vasilić, Ž., and Štefanac, Z., Mikrochim. Acta (Wien) II (1981) 45-56.

(3) Drevenkar, V., Fröbe, Z., Štengl, B., and Tkalčević, B., Mikrochim. Acta I (1985) 143-156.

(4) Freed, V.H., Schmedding, D., Kohnert, R., and Haque, R., Pestic.Biochem.Physiol. 10 (1979) 203-211.

Table I - Concentration of dialkylphosphorus anions and of some parent compounds in the water from two ponds adjacent to orchards treated with organophosphorus pesticides

Month of sampling	Compound	Concentration, μg dm^{-3}			
		POND KS		POND DS	
		1983	1984	1983	1984
JULY	DMTP	1.83	2.26	<1.00	<1.00
	DMDTP	1.81	9.79	ND	2.23
	DETP	ND	4.14	ND	1.94
	DEDTP	ND	19.16	ND	22.75
	Phosalone	–	ND	–	ND
	Methidathion	–	ND	–	18.9
	Demeton-S-methyl	–	1.9	–	4.0
SEPTEMBER	DMTP	1.99	2.69	3.88	4.02
	DMDTP	3.02	8.52	5.58	20.65
	DETP)	ND	1.91	5.60	5.93
	DEDTP	ND	ND	36.74	31.45
	Phosalone	–	ND	–	5.82
	Methidathion	–	ND	–	3.50
	Demeton-S-methyl	–	0.17	–	2.34
NOVEMBER	DMTP	2.85	2.35	3.81	<1.00
	DMDTP	<1.00	3.54	7.52	2.83
	DETP	1.02	1.19	2.33	<1.00
	DEDTP	1.78	ND	5.91	1.00
	Phosalone	–	ND	–	0.49
	Methidathion	–	ND	–	ND
	Demeton-S-methyl	–	ND	–	ND

ND=not detected; -=not measured

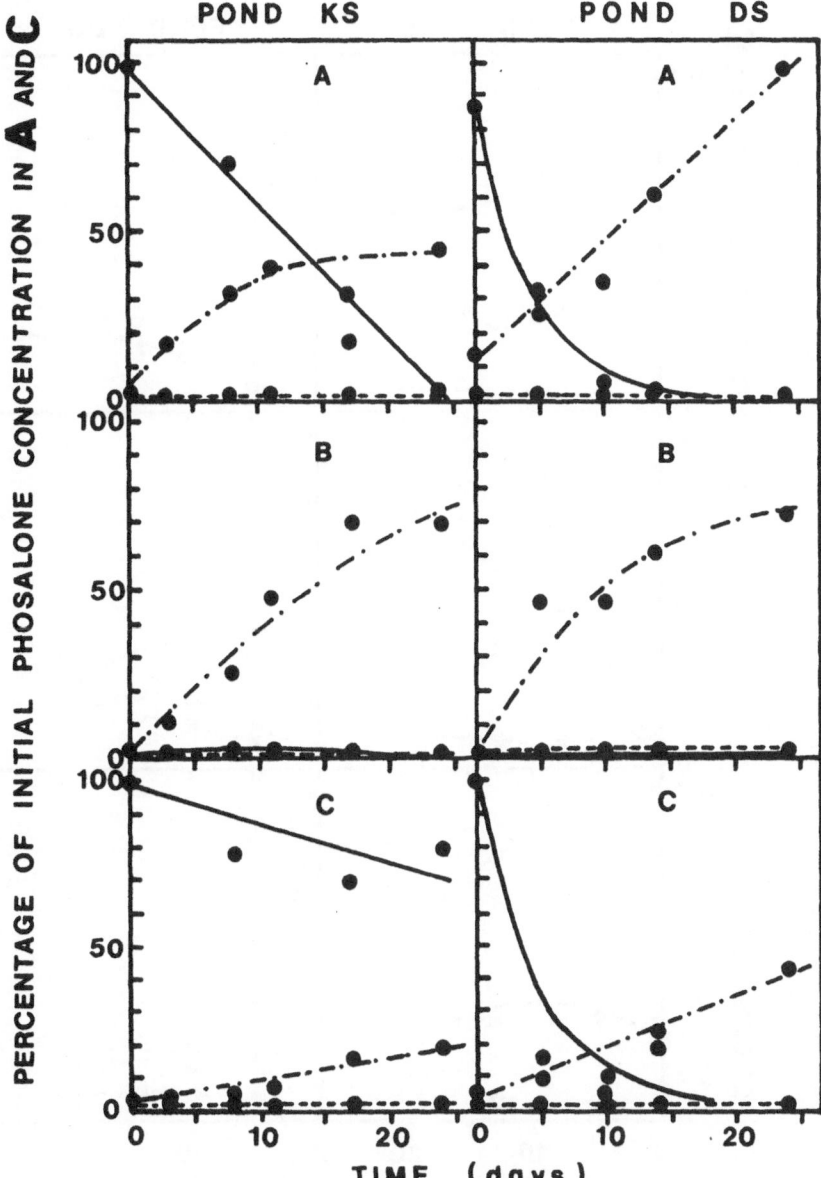

Fig. 1 Change in the concentration of Phosalone and its degradation
products DEDTP and DETP in the air-bubbled water from Ponds KS
and DS with (A and B) and without (C) related silt added

A and C: Phosalone added to the water (4.8 mg dm^{-3})

B　　: Phosalone added to the silt (4.8 mg/100 g)

—— Phosalone　　—·— DEDTP　　--- DETP

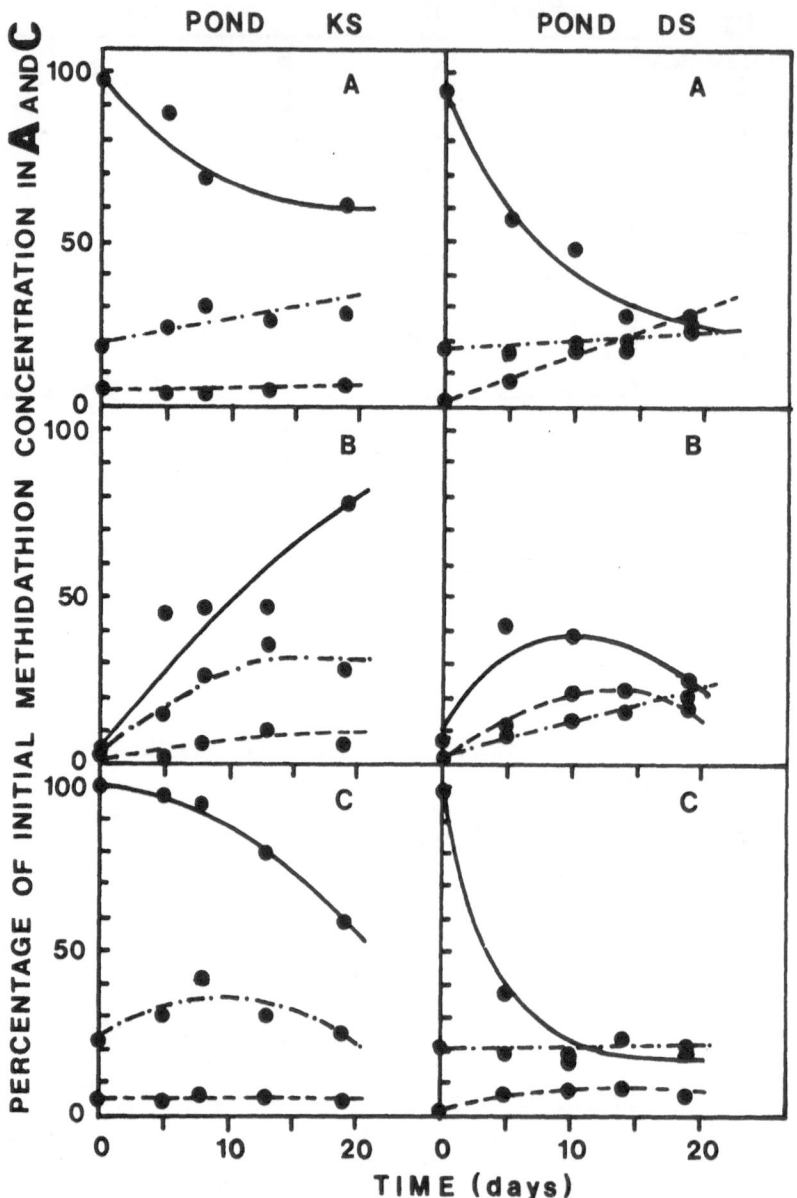

Fig. 2 Change in the concentration of Methidathion and its degradation products DMDTP and DMTP in the air-bubbled water from Ponds KS and DS with (A and B) and without (C) related silt added

A and C: Methidathion added to the water (4.2 mg dm^{-3})

B : Methidathion added to the silt (4.2 mg/100 g)

—— Methidathion   —·— DMDTP   ——— DMTP

MICROBIAL DEGRADATION OF NITRILOTRIACETATE (NTA) IN GROUNDWATER: LABORATORY STUDIES UNDER AEROBIC AND ANAEROBIC CONDITIONS

E. KUHN, M. VAN LOOSDRECHT, R.P. SCHWARZENBACH, W. GIGER

Swiss Federal Institute for Water Resources and
Water Pollution Control (EAWAG)
CH-8600 Dübendorf, Switzerland

ABSTRACT

Plexiglas columns filled with aquifer material where continously fed with a mineral salt medium supplemented with NTA. Different parameters (electron acceptor, temperature, concentration of NTA) were tested for their influence on the microbial degradation of NTA. Under aerobic conditions, more than 90% of NTA was removed during the column passage starting from initial concentrations of 105, 160 and 600 $\mu g/l$. At a smaller concentration (11 $\mu g/l$) the degradation rate decreased. A change from aerobic to denitrifying conditions resulted in a partial breakthrough of NTA at the column effluent. After 5 days of adaption, however, again more than 90% of NTA was degraded. The degradation of NTA was determined over the temperature range of 5 to 20 °C. The NTA concentration in the column effluent increased temporarily when the temperature was decreased, but after adaption times of 4-20 days, high degradation rates were again obtained. [^{14}C] NTA was mineralized to $^{14}CO_2$ under aerobic ($\geqslant$ 68%) and denitrifying conditions ($\geqslant$ 85%).

1. INTRODUCTION

The addition of phosphate to detergents is limited or will be banned in many countries, and NTA is increasingly used as a substitute.

The environmental fate of NTA under various conditions is well described (1, 2), but the behaviour of NTA during river water/groundwater infiltration is still poorly understood. A field study (see Poster by Schaffner et al., this volume) in the lower Glatt Valley, Switzerland, showed that NTA (concentration in river: 8-83 $\mu g/l$) is eliminated within the first seven meters of infiltration to residual concentrations of 0.5 $\mu g/l$. To evaluate the degradation of NTA under aerobic and denitrifying conditions at various concentrations and incubation temperatures, laboratory studies using aquifer columns were conducted.

2. METHODS

A) The apparatus used to study the microbial degradation of NTA is depicted in Fig. 1. Aquifer material taken from a river water/groundwater infiltration site was dry sieved, and the fraction 63-500 μm was filled into a plexiglas column. The reservoir containing the mineral salt medium and NTA was autoclaved. The connections to the column were sterilized by autoclaving or purging with a formaldehyd solution prior to each experiment. The bacterial filter between the pump and the column prevented a contamination of the reservoir. The column was operated in upflow mode and under saturated conditions. The experiments were conducted by step input variations.

B) Analysis: Samples (5 ml) were taken at the column influent and at the sampling ports (normally, after 1.8, 6.8 and 21.8 cm flow distance). NTA was determined using the method described in Reference 3.

3. RESULTS

3.1. Microbial degradation of NTA under aerobic conditions at different concentrations

In our columns, a retardation factor of 1.2 was determined for NTA indicating that NTA was only weakly sorbed by the aquifer material. After the initial addition of NTA to the column, NTA degradation could be observed immediately (Fig. 2). After 5 hours, about 10% of the NTA (influent concentration = 105 μg/l) was eliminated over the length of the column (21.8 cm); 13 days later, 97% of the NTA was removed.

Changing the influent concentration of NTA to a lower or higher level, resulted in a temporary small decrease in the overall elimination. Within a few days, however, the organisms had adapted to the new concentration (Table 1). The residual concentrations in the effluent were between 2 and 6 μg/l for all input concentrations.

Table 1   Microbial degradation of NTA under aerobic conditions.

Adaption time (days)	Concentration of NTA (μg/l) at the end of adaption time		%-removal
	influent	effluent	
13	105	3	97
4	11	2	82
5	160	2	99
9	600	6	99

3.2. Microbial degradation of NTA upon removal of molecular oxygen as electron acceptor

Removal of molecular oxygen from the inflowing medium led to denitrifying conditions. NTA degradation was inhibited temporarily and was then quickly resumed (Fig. 3). Twenty hours after oxygen removal, 30% of the NTA was eliminated over the length of the column (21.8 cm). Four days later, more than 90% was degraded.

Upon removal of nitrate from the inflowing medium, NTA degradation ceased immediately.

3.3. Microbial degradation of NTA at different temperatures

Lowering the temperature under aerobic conditions from 10 to 5 °C had a bigger effect on the degradation rate of NTA (factor 2.3) than lowering it from 20 to 10 °C (factor 1.6 / see table 2).

When the temperature was changed from 20 to 5 °C under denitrifying conditions, the degradation rate was decreased by a factor of 3.9. The removal of NTA after adaption times of 4-20 days were, however, at least 95%.

Table 2   Rates of degradation and removal efficiency of NTA at different temperatures under aerobic and denitrifying conditions.

Conditions (NTA conc)	Temp. (°C)	Rate of degradation k [h^{-1}] (assumption: first order kinetic)		%-removal of NTA after adaption
		immediately after temp. shift	after adaption of 4-20 days	
aerobic 600 µg/l	20		6.7	∿ 99
	10	4.2	5.1	∿ 99
	5	2.2	3.6	∿ 95
	20	6.7	8.9	∿ 99
denitrify- ing 650 µg/l	20		2.7	∿ 99
	5	0.7	1.0	∿ 99
	20	4.6	6.1	∿ 98

3.4. Evidence for mineralization

During a period of 12.5 hours [^{14}C] NTA was continously added to a NTA adapted aquifer column under aerobic conditions. Ten hours after the begin of the experiment, at 6.8 cm flow distance 68% of the ^{14}C could be recovered as $^{14}CO_2$ (see table 3). Only minor amounts of [^{14}C] NTA were not transformed (1.9%).

Table 3   Degradation of ^{14}C NTA under aerobic conditions.

flow distance (cm)	Water phase (% of inflowing ^{14}C)		
	^{14}C in NTA	^{14}C in biomass	$^{14}CO_2$
0	100%	-	-
1.8	2.9%	0.1%	67.5%
6.8	1.9%	0.01%	67.9%
21.8	1.5%	0.01%	62.0%

After 12.5 hours, the column was disassembled, and about 15% of the totally added ^{14}C was found to be incorporated into the biomass attached to the aquifer material.

The same experiment was repeated under denitrifying conditions. After 10 hours, 85% of the added ^{14}C was recovered as $^{14}CO_2$, but only 3% was incorporated into the biomass attached to the aquifer material.

4. SUMMARY AND CONCLUSIONS

1. Under saturated conditions, NTA was found to be mineralized at similar rates under aerobic and denitrifying conditions. Upon a change from aerobic to denitrifying conditions the NTA degradation was only temporarily inhibited.

2. Under both, aerobic and denitrifying conditions, there is evidence that NTA was used as sole source of carbon and energy. This might explain the rapid adaption of the microorganisms to higher concentrations.

3. The removal efficiency of NTA was concentration dependent. The degradation of NTA at concentrations of > 100 $\mu g/l$ amounted $\geqslant$ 97%. At a concentration of 11 $\mu g/l$ the overall degradation was only 82%.

4. Lowering the temperature resulted in a moderate decrease of the degradation rate of NTA. However, the microbial population adapted within a few days.

5. REFERENCES

1. Bernhardt H., et al., Studie über die aquatische Umweltverträglichkeit von Nitrilotriacetat., Verlag Hans Richarz, St. Augustin, (1984).
2. Perry R., et al., Environmental aspects of the use of NTA as a detergent builder, Water Res., 18, 255-276, (1984).
3. Schaffner C. and Giger W., Determination of nitrilotriacetic acid in water by high-resolution gas chromatography.
   J. Chromatography, 312, 413-421, (1984).
4. Kuhn E.P., et al., Microbial transformations of substituted benzenes during infiltration of river water to groundwater: Laboratory column studies. Environ. Sci. Technol., 19, 961-968, (1985).

ACKNOWLEDGMENT

We are grateful to J. Zeyer for reviewing the manuscript.

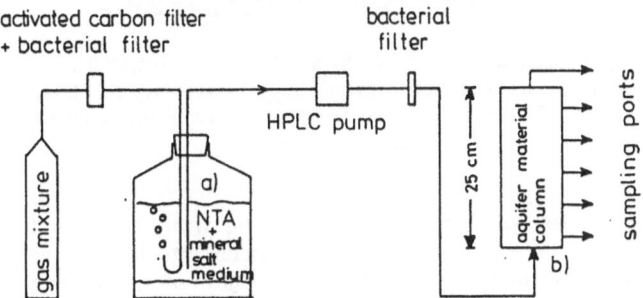

Fig. 1 Apparatus used to study the microbial degradation of NTA in an
aquifer column.
  a) A carbonate buffer system was established by stirring mineral
     salt medium (composition see Ref. 4) supplemented with NTA
     over solid $CaCO_3$ and by purging with air for aerobic condi-
     tions or a N_2/CO_2 (99.5/0.5, v/v) mixture for denitrifying
     conditions
  b) Flow rate in the aquifer column : 7.9 cm/h

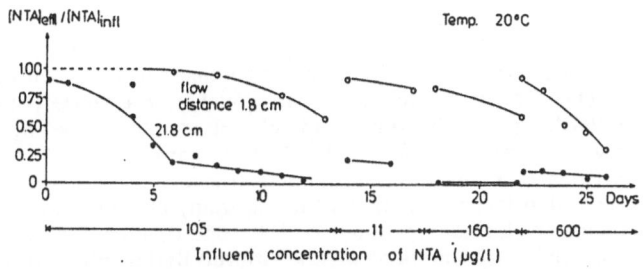

Fig. 2 Microbial degradation of NTA under aerobic conditions at different concentrations

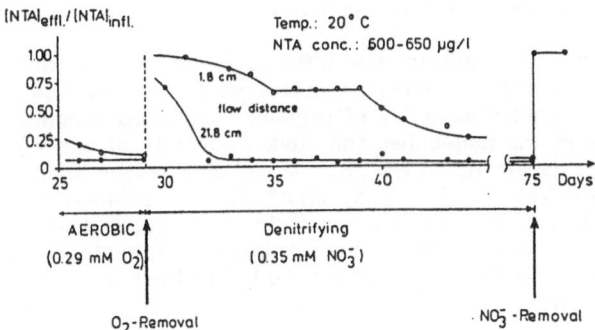

Fig. 3 Microbial degradation of NTA upon removal of molecular oxygen
with nitrate as electron acceptor

CHARACTERIZATION AND BIODEGRADABILITY OF LIGNIN CONTAINING WASTEWATER FROM FOREST INDUSTRY

J. PELLINEN* AND M.S. SALKINOJA-SALONEN
Department of General Microbiology, University of Helsinki
* Present address The Finnish Pulp and Paper Research Institute, Box 136
Helsinki, Finland

Summary. Lignin derived compounds originating from forest industry represent the major organic constituent in waste streams into Finnish recipient waters. Pulping and bleaching of pulp transforms the lignin of wood chips into a variety of aromatic polymer molecules widely differing in size (MW from 10^2 to 10^5) and properties. The aim of this study was to see, whether the different wastewater lignins also differ in their biodegradability. We monitored the molecular weight distributions of lignins in waste streams from different stages of pulping and different bleaching processes at four different pulp mills and also in the recipient water. Our method of analysis was high performance size exclusion chromatography (HPSEC), shown to give reliable results on the molecular weight distribution of lignins, provided that precautions are taken to avoid hydrogen bonding and lipophilic or electrostatic interaction. The wastewater lignins proved poorly degradable by bacteria if used as sole substrate.

1. INTRODUCTION.

Wastewaters from forest industry have a dominant role in the pollution of surface waters in Finland. In 1983 about 90 % of the BOD_7 released into the watercourse was from forest industry (1). The most important wastewaters are those from debarking of wood and from washing and bleaching of chemical pulp. The major organic constituent of these wastewaters is lignin or lignin-like material.

Lignin in the effluents is very heterogeneous material consisting of a large spectrum of molecules of different sizes containing many phenolic and acidic groups (2,3). From the annual production of about $5x10^6$ tons of chemical pulp (of which $3x10^6$ tons of bleached pulp) the amount of chlorinated lignin released by the forest industry can be estimated to be approximately 150 000 tons per year in Finland.

The high-molecular weight lignin has been seen as potential environmental problem because toxic or mutagenic compounds may be formed from it in the nature or in the chlorination of drinking water (4,5). Chlorolignin seems very stable in the aqueous environment (5,6). One harm caused by pulp mill effluents is that the absorption of light is increased due to its colour which may affect water ecosystem (5).

Biodegradation studies may give us new information on the fate of lignin in the environment and make possible to develope better ways to treat lignin containing wastewaters. Since the biodegradability of organic compounds depends on the size and chemical structure of the molecules, the biodegradability of wastewater lignins can be expected to vary depending on the process and chemicals used in the industry. This paper describes the results of a study undertaken to evaluate the molecular weight distribution (MWD) of various lignins and its effect on biodegradability using recently developed high-performance size exclusion chromatography (HPSEC) methods (7,8). The HPSEC method was used also to follow the changes in the MWD of lignin in recipient water systems.

2. EXPERIMENTAL

Chromatography. The chromatography of water-soluble samples was performed as described in ref. 7 using TSK PW type columns (Toyo Soda Manufacturing Co., Tokyo,Japan). The mobile phase was $NaHCO_3$ (0.025 M)-NaOH buffer (pH 10.5) with

0.5g/l of polyethylene glycol 6000 (Fluka AG, Buchs, Switzerland) and 0.5 g/l of ethylene glycol (E. Merck, Darmstadt, FRG). Poly(styrene sulfonate) standards were used for calibration (Pressure Chemicals Inc.,Pittsburgh, Pa.,USA).

The water-insoluble samples were chromatographed as described in ref. 8 using Ultrastyragel columns (Waters Associates Inc., Milford, Mass., USA) with tetrahydrofuran as mobile phase (Rathburn, Walkerburn, Peeblesshire, Scotland, HPLC-grade). The calibration was done with poly(styrene) standards (Pressure Chemicals Co.).
UV-absorption detector was used at 280 nm. The chromatograms were normalised to equal area with a microcomputer.

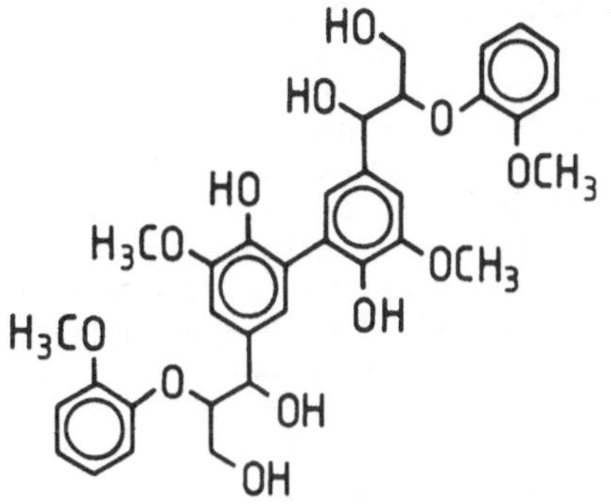

Fig. 1. Structure of the tetrameric lignin model compound used in this study, 1,1'-(4,4'-dihydroxy-5,5'-dimethoxy-3,3'-biphenylene)-bis- 2-(2-methoxyphenoxy)-1,3-propanediol (MW 638)

Samples. The bleaching liquors were all from bleaching of kraft pulp. The bleaching sequences were the following: mill 1 (D/C)EHED, mill 2 (D/C)EHED, mill 3 OCE (prebleaching). The effluent from the mill 4 was combined from two bleacheries with sequences (C/D)EHDED and (C/D)EHPD. The kraft lignin from birch (mill 2) was purified by precipitation at pH 3, filtered, washed, dissolved in 0.1 N NaOH and ultrafiltered with an Amicon UM 2 filter (Amicon Corp., Danvers, MA, USA) with a nominal cut-off of 1000 dalton. The filtrate was used in the experiments. Indulin AT-R softwood kraft lignin was obtained from Westvaco Corp., North Charleston, SC, USA. Humic acid was from Fluka (Buchs, Switzerland).

The tetrameric lignin model compound 1,1'-(4,4'-dihydroxy-5,5'-dimethoxy-3,3'-biphenylene)-bis- 2-(2-methoxyphenoxy)-1,3-propanediol (MW 638, Fig.1) was synthesized as described in the literature (9,10). It was characterized by MS and NMR spectra.

Analyses. Total organic carbon (TOC) was determined with Ionics Mod 1258 analyzer using acetic acid for calibration. Total organic chloride (TOCl) was determined by burning the sample in oxygen after adsorption to activated carbon and titrating the formed hydrochloric acid coulometrically.

3. RESULTS AND DISCUSSION

Tables 1 and 2 and Figs. 2 and 3 show data of the lignins of various waste streams from four kraft pulp mills. The bleaching liquor lignins from the four millshave an average molecular weight varying from Mn 300 to 1700 and Mw of 1350 to 4400. The smallest molecules were found in the chlorination stage effluent of bleaching and the biggest molecules in the oxygen stage effluent of OCE bleaching. It seems that kraft lignins are of smaller molecular weight than bleached lignins extracted with alkali (Table 1). These observations are well in accordance to what has been reported when other methods have been used (2,3,11,12).

The large difference in the molecular weight distributions of the different bleaching effluents can be seen from Figs. 2 and 3. The proportion of low molecular weight lignin in D/C stage liquor was not constant even in the same mill (Table 1) although the mill was in normal operation at both times and used the same type of wood (birch) and bleaching process and sampling was extended both times over 7 subsequent days to avoid the effects of temporary fluctuation. It seems probable that the peak at MW 100 in Fig. 3C corresponds to that in Fig. 3B. Both of these peaks as also that at MW 1000 in Fig. 3B were missing from the samples collected in April-May .

Figs. 3C and D together with Table 1 show how the MWD's and average molecular weights changed within the effluent treatment system of a mill and in the recipient watercourse. It can be seen that Mn is very sensitive to small changes of MWD. The real trend of the change of MW in this kind of material is better seen from the weight average molecular weights. The screening wastewater lignin had a higher Mw than the other samples including washing water lignin of the same mill. The Mw of discharged lignin decreased somewhat in the effluent treatment lagoon and in the recipient water compared to El-stage. This probably is caused by the lower pH of the treatment lagoon, as compared to that at El stage spent liquor, which has lead the largest molecules to precipitate in the clarifier or lagoon.

The different waste liquors were further characterized by chemical methods to see if the changes were reflected in the other parameters also. The results are presented in Table 2. Large changes were not seen but COD_{Cr}, colour and TOC values were somewhat lower in April-May than in August. The clarifier and lagoon lignin Mw's in August were similar to those in May inspite of the fact that Mw of El stage lignin was higher in May compared with August. An explanation for this may lie in the lower pH of the clarifier and lagoon in May, causing the largest lignin molecules to precipitate.

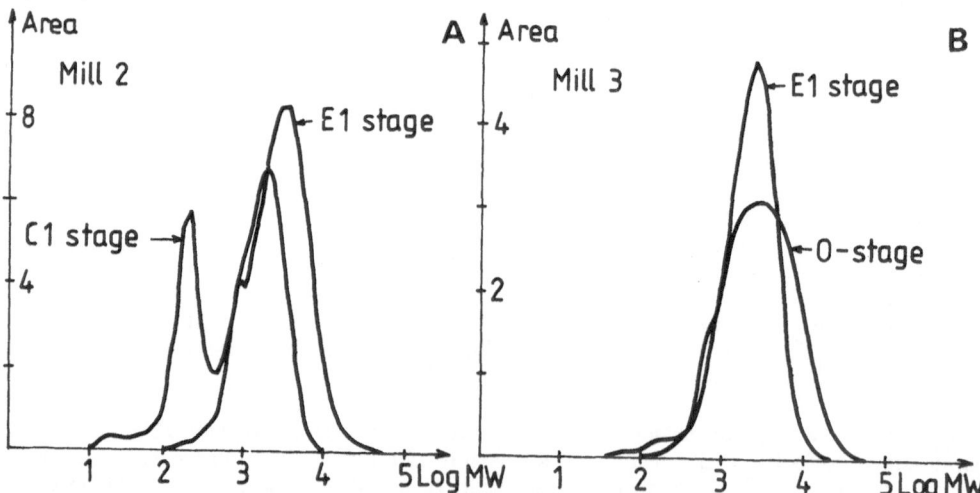

Fig. 2. Molecular weight distributions of kraft pulp bleaching liquors. A) D/C and El stages (mill 2), B) O and El stages (mill 3).

Fig. 3D and Table 1 show the effect of pulp mill effluents on molecular weight distribution of UV-absorbing material in inland recipient. The molecular weight of

TABLE 1. Average molecular weights of lignins from pulping and bleaching processes.

Sample	Mn	Mw	Mw/Mn
Mill 1: April-May 1984 [2]			
Washing of pulp	1140	2200	1.93
Screening of pulp	760	5560	7.30
Cl stage of bleaching	870	1770	2.05
El stage of bleaching	1350	4310	3.18
Clarifier discharge to lagoon	1240	3030	2.46
Lagoon discharge to sea	850	3090	3.62
Recipient sea, 1 km	1330	2370	1.78
August 1984 [3]			
Washing of pulp	610	2860	4.70
Screening of pulp	540	11300	21.04
Cl stage of bleaching	230	1420	6.13
El stage of bleaching	1150	3610	3.14
Clarifier discharge to lagoon	980	2950	3.00
Lagoon discharge to sea	910	2820	3.11
Recipient sea, 1 km	600	3610	3.93
Kraft lignin (hardwood, mill 2)	490	690	1.40
Kraft lignin (softwood) [1]	460	2330	5.07
D/C stage from D/CEHED (mill 2)	290	1350	4.64
El stage from D/CEHED (mill 2)	1740	3570	2.06
O stage from OCE (mill 3)	1690	4410	2.61
El stage from OCE (mill 3)	1180	2650	2.24
Mill 4:			
20 km upstream the mill	1630	2630	1.49
Discharge of the mill to recipient [4]	380	2410	6.31
10 km downstream the mill	1030	2340	2.26
Humic acid (reference)	1480	3390	2.29

1) Indulin AT-R
2) Samples collected between April 27th and May 3rd 1984
3) Samples collected between August 20th and 27th 1984
4) Bleaching sequences (C/D)EHDED and (C/D)EHPD

the discharged bleaching effluent lignin was somewhat smaller than that of natural lignins (humic matter) (mill 4 in Table 3). The increase in the UV-absorbance of the water downstream the mill was 1.7 fold as compared to the water sample upstream although upstream there was no pulp or paper industry nor any other source of industrial lignin pollution. Lower pH and ionic strength in the recipient water may increase apparent molecular weight and cause sedimentation of lignin (7).

The conclusion from the observations cited above is, that biodegradation of the lignins in the effluent treatment system and recipients of the industrial discharges, if occurred at all, was in any case not visible in the molecular weight distribution of such lignins. The amount of lignins found below the mill as compared to above the mill was not yet twofold, which makes it difficult to observe any such changes,

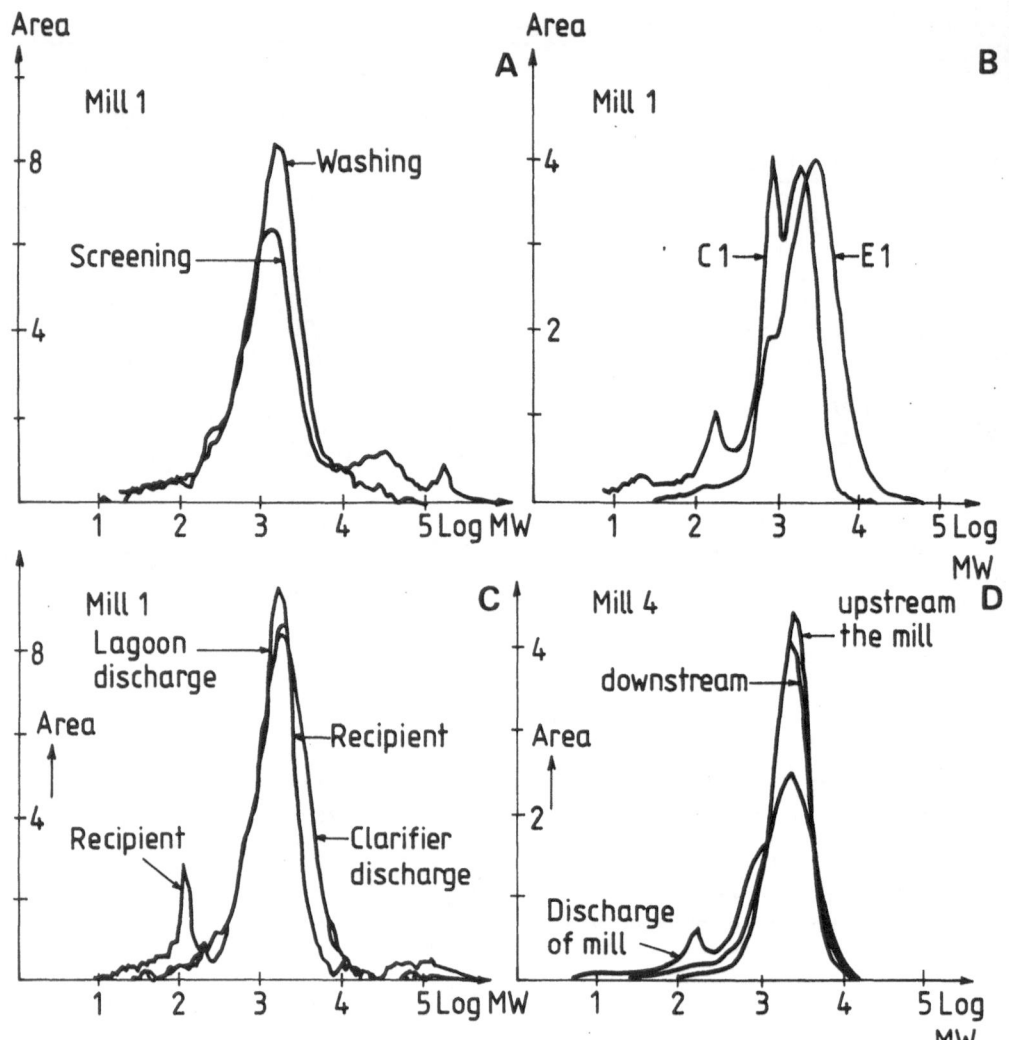

Fig. 3. Average molecular weight of discharged lignin from kraft pulping (mill 1).
Lignin discharged from the cookery (washing) and screening (A), bleach plant (B) and
effluent treatment system (C). Fig. (D) shows similar curves for natural lignin (humic
substances) upstream the mill, mixed effluent of the mill 4 and the mixture of both in
recipient lake sample taken downstream the mill.

especially because the difference of molecular weight distribution of the industrial
discharge lignin and that of natural humic matter was so small.

Laboratory experiments on biodegradation of industrial lignin. It is well known
that fungi are able to degrade natural lignin and also some industrial lignins (13), but
we studied bacteria because the present-day wastewater treatment systems have
conditions unfavourable for fungal colonization and thus rely almost exclusively on

the degradation activity of bacteria. We have earlier shown that bacteria were able to degrade completely synthetic lignin oligomer compounds of a molecular weight comparable to that of industrial lignins discussed above (10).

The results in Tables 3 and 4 and Fig. 4 show that very little degradation of bleaching waste lignin cr kraft lignins occurred when the bacterial culture able to degrade tetrameric model compounds was used. In some cases the area of the chromatogram of inoculated sample was larger than that of uninoculated control. This was possibly due to the introduction of new chromophoric groups which was shown to happen as the first step in the oxidation of the tetramers (10). At pH 8 only the area of the tetramer peak increased (indicating oxidation) but not that of lignin when they were incubated together (Table 3, Fig 4B). In our earlier experiments we found that the tetrameric model compound was completely degraded in 5 days when it was used as the sole source of carbon (10). Now we found that the tetramer was more slowly degraded in the precence of black liquor lignin than in the absence of it. The model compound (Fig. 1) used in the present study was not completely pure as can be seen from Fig. 4A.

One reason for the industrial lignins not being degraded although the model compound of the same size was, might be that in the industrial lignins there are very little β-O-4 ether linkages between aromatic units (2) which may be important for the degradation. On the other hand we have shown that the bacterial culture used was able to degrade a variety of interunit linkages of dimeric lignin model compounds (14). Another difference is that the aromatic structures are largely cleaved in bleaching (2,3,11).

4. CONCLUSIONS

The fractionation of spent bleaching liquors can be done by SEC if care is taken to avoid phenomena which can disturb fractionation (7,8). It was noted that the molecular weight distribution of bleaching liquor lignins was different during two sampling periods, under apparently identically process situations. This makes it difficult to study possible changes of lignin molecular weight during effluent treat-

TABLE 2.Characteristics of lignin containing discharges from bleached pine pulping

Stream analysed	pH	COD_{Cr} mg /l	BOD_7 mg/l	Colour mg Pt/l	TOC mg/l	TOCl mg/l
Mill 1: April-May 1984						
Washing of pulp	8.8	398	92	510	50	0.24
Screening of pulp	7.7	467	158	650	91	0.29
C1 stage from bleaching	2.1	1260	322	750	315	136.9
E1 stage from bleaching	11.0	2060	476	3770	432	86.5
Clarifier disch. to lagoon	5.5	491	181	580	119	
Lagoon discharge to sea [1]	6.3	473	147	630	89	
Recipient sample, 1km [1]	5.6	47	5	120	7	0.31
Recipient sample, 5km [1]	7.6	27		130	<2	0.22
August 1984						
Washing of pulp	9.9	787	133	650	66	0.13
Screening of pulp	7.5	813	172	950	117	0.29
C1 stage from bleaching	2.3	1050	283	820	296	207.0
E1 stage from bleaching	11.7	2340	514	4700	598	167.0
Clarifier disch. to lagoon	7.1	619	163	740	132	
Lagoon discharge to sea	6.9	387	86	490	83	5.7
Recipient sample	6.9	64	5	160	12	0.55

1) The recipient was the Gulf of Bothnia.

ment or in recipient water. The change in the average molecular weights in the lagoon treatment was small at both sampling times although there was large difference in the activity of the lagoon. COD_{Cr} and BOD_7 were more effectively removed in August than in the spring , probably because of higher lagoon temperature in August.

In the laboratory experiments a small amount of low-molecular weight lignin was degraded in the wastewater by added bacteria, although a lignin model compound of a MW close to that of wastewater lignins was almost completely degraded by the same bacteria. A too large MW should thus could not explain the recalcitrance of lignin. The bacteria we used had been shown to degrade most types of linkages which occur at least in kraft lignin so that the interunit structures should not alone be prohibitive for biodegradation. It seems possible that dimeric and tetrameric lignin model compounds are degraded in a different way what would be needed for the degradation of wastewater lignins.

TABLE 3. Degradation of industrial lignins by bacteria.

Substrate	Time (weeks)	pH	Peak area, % of control [1] with tetramer	without tetramer
El stage liquor from	2	7	100	
OCE bleaching	19	7	98	
kraft lignin (hardwood)	2	7	93	106
	19	7	94	60
kraft lignin (softwood)	4	8	99	139

[1] Detection by UV at 280 nm.

TABLE 4. Change in the average molecular weights on treatment with bacteria.

Sample	Inocul.	Time	pH	Mn	Mw	Mw/Mn
El stage liquor from	-	2	7	700	1640	2.35
OCE bleaching [1]	+			620	1610	2.61
	-	19	7	790	1650	2.13
	+			870	1630	1.89
kraft lignin (softwood) [2]	-	4	8	450	3510	7.84
	+			530	3660	6.94

[1] Mill 3. [2] Mill 2.

5. ACKNOWLEDGEMENTS

The authors thank Riitta Boeck and Jane Langshaw for performing much of the analytical work, Eira Toivanen (City of Helsinki Water Works) for analysis of organically bound chlorine, Reino Lammi and Sven-Ingvar Nord for providing the in-mill samples, the National Board of Waters, the Water Districts of Kokkola and Oulu and the Academy of Finland (to MSS) for financial support, Tellervo Kyläharakka for her continuous interest in this work and G. Brunow for synthesis of the lignin model compounds.

6. REFERENCES

(1) National Board of Waters, Finland (1983). Report 224 (in Finnish).
(2) K. Lindström and F. Österberg (1984). Characterization of the high molecular mass chlorinated matter in spent bleach liquors. Part 1. Holzforschung 38:201-212.

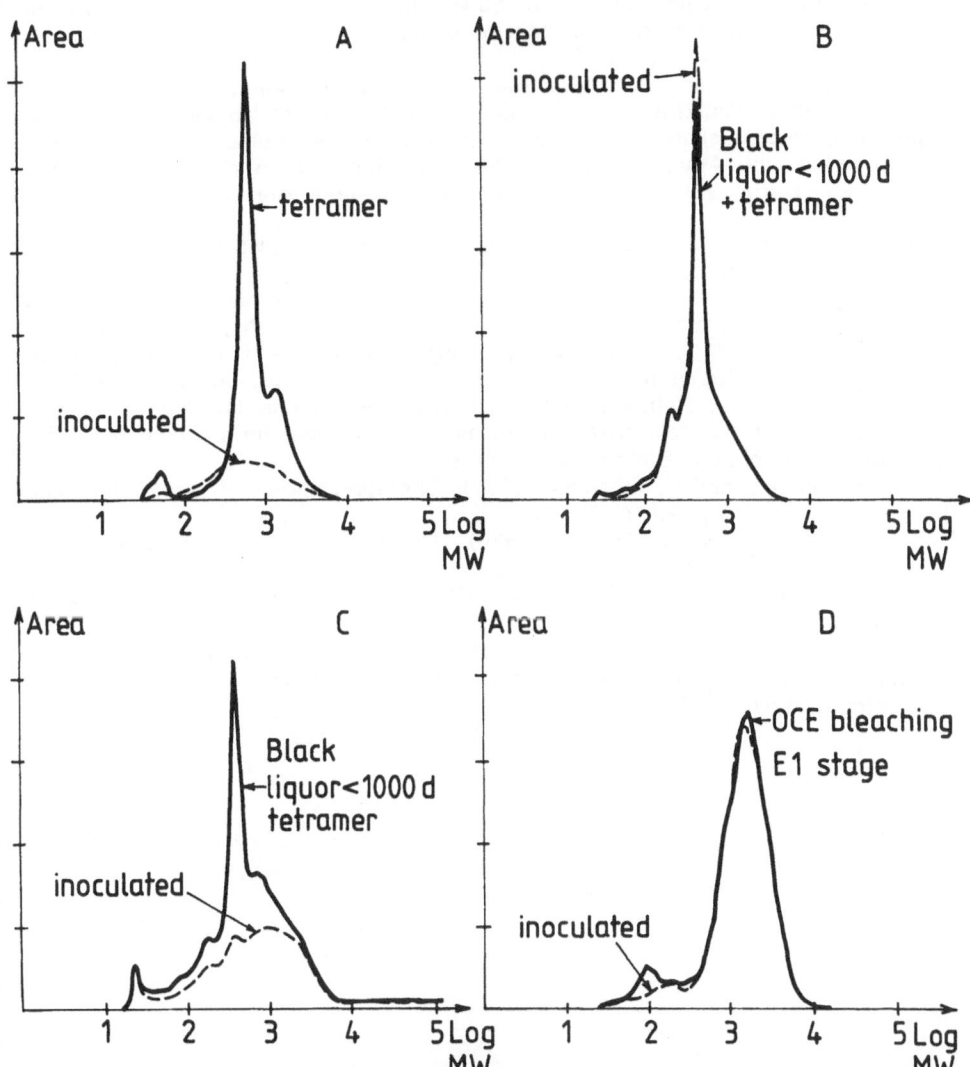

Fig. 4. Molecular weight distributions of lignins upon incubation with bacteria that were able to degrade a tetrameric lignin model compound (MW 638). (A) Tetrameric model compound (4 weeks), (B) black liquor lignin (ultrafiltrated) + tetramer model compound (2 weeks), (C) black liquor lignin (ultrafiltrated) + tetrameric model compound (19 weeks), (D) E1 stage bleaching liquor lignin from OCE sequence (19 weeks).

(3) F. Österberg and K. Lindström (1985). Characterization of the high molecular mass chlorinated matter in spent bleach liquors. Part II. Holzforschung 39:149-158.
(4) K.P. Kringstad and K. Lindström (1984). Spent liquors from pulp bleaching. Environ. Sci. Technol. 18:236A-248A.

(5) K.P. Kringstad, P.O. Ljungquist, F. de Sousa and L.M. Strömberg (1983). On the formation of mutagens in the chlorination of humic acid. Environ. Sci. Technol. 17:553-555.

(6) M. S. Salkinoja-Salonen, M.-L. Saxelin, J. Pere, T. Jaakkola, J. Saarikoski, R. Hakulinen and O. Koistinen (1981). Analysis of toxicity and biodegradability of organochlorine compounds released into the environment in bleaching effluents of kraft pulping. In: Advances in the Identification and Analysis of Organic Pollutants in Water, Vol 2 (ed. by L. Keith), p. 1131-1164. Ann Arbor Science, Ann Arbor, Mich.

(7) J. Pellinen and M. S. Salkinoja-Salonen (1985). Aqueous size exclusion chromatography of industrial lignins .J. Chromatography, 322:129-138.

(8) J. Pellinen and M.S. Salkinoja-Salonen (1985). High performance size exclusion chromatography of lignin and its derivatives. J. Chromatography, 328:299-308.

(9) J.C. Pew and W.J. Connors (1969). New structures from the enzymic dehydrogenation of lignin model p-hydroxy-α-carbinols. J. Org. Chem. 34:580-584.

(10)   J. Jokela, J. Pellinen, M.S. Salkinoja-Salonen and G. Brunow (1985). Biodegradation of two tetrameric lignin model compounds by a mixed bacterial culture. Appl. Microbiol. Biotechnol, in press.

(11)   K. Pfister and E. Sjöström (1979), Characterization of spent bleaching liquor. Part 3. Paperi ja Puu 61:367-370.

(12)   K. Pfister and E. Sjöström (1979). Characterization of spent bleaching liquors. part 5. Paperi ja Puu 61:525-528.

(13)   G. Sundman, T.K. Kirk and H.-m. Chang (1981). Fungal decolorization of kraft bleach plant effluent. Tappi 64:145-148.

(14)   J. Pellinen, E. Väisänen, M. Salkinoja-Salonen and G. Brunow (1984). Utilization of dimeric lignin model compounds by mixed bacterial cultures. Appl. Microbiol. Biotechnol. 20:77-82.

WATER TREATMENT PROCESSES

Formation and removal of organic micropollutants in drinking water treatment

Chemical and biological characterization of industrial effluents and applied mathematical modelling - A case study

Risk estimation as a method for modelling the fate and impact of industrial effluents in marine environments

Das NTA-Problem

Behaviour of nonionic surfactants in biological waste water treatment

A data base on behaviour and effects of organic pollutants in waste water treatment processes

PAH in water systems, effect of aquatic humus and chlorination

Properties of mutagenic compounds formed during chlorination of humic water

Behaviour of organic micropollutants during infiltration of river water into ground water : Results of a field study in the Glatt Valley, Switzerland

FORMATION AND REMOVAL OF ORGANIC MICROPOLLUTANTS IN DRINKING WATER TREATMENT

F. VAN HOOF
Study Center for Water Research
c/o Antwerp Waterworks
Mechelsesteenweg 64
2018 Antwerp - Belgium

Summary

A short review is given of the possibilities and restrictions of different unit operations in drinking water treatment in relation to micropollutant removal.

The currently used techniques perform poorly with regard to the removal of micropollutants with low k_{ow} partition coëfficiënts. A more thorough evaluation of the relation between physico-chemical characteristics of organic micropollutants and their removal in water treatment is needed. Results of such studies could be helpful in orienting future studies.

The formation of oxidation by-products is looked at in more detail with emphasis on ozonization, which produces a.o. low molecular weight aldehydes and short chain carboxylic acid. Experimental evidence suggests that these compounds can be removed easily through microbiological activity.

1. REMOVAL OF MICROPOLLUTANTS

The analysis of organic micropollutants in water has drawn considerable attention during the last decades. An overview of the litterature indicates that in relation to water treatment most attention has been paid to the presence of organochlorine compounds formed by chlorination. Although this effort seemed initially justified by reports concerning the carcinogenic character of some trihalomethanes, more recent findings related to the toxicity of different volatile organochlorine compounds do not seem to justify this one-sided approach. In a more general way, the methods used for the analysis of organic micropollutants have been to a large extent limited to compounds which are readily amenable to gas chromatography. However recent reports using other techniques indicate that a wide range of compounds, which until now had gone undetected are present in raw waters and drinking waters. (1)

Although the application of more advanced analytical techniques will open new frontiers for organic micropollutants analysis, the use of some of these techniques will remain expensive and therefore priorities will have to be set on what types of compounds will have to be analyzed for and to what extent, both for studies on raw water sources and on the influence of water treatment processes.

Ideally the planning of a search for micropollutants should start with a comprehensive study of the discharges into the raw water and of the manufacturing processes used in different industries, evaluating in qualitative and quantitative terms the possible presence of organic micropollutants. (2)

Such an approach, however, can only be used in those cases in which an integrated water quality management is applied. Due to the lack of the above mentioned information most studies are directed towards the presence of compounds meeting certain criteria e.g. widespread use, production in sufficient amounts and potential toxicological concern. Sometimes physico-chemical characteristics, mostly limited to volatility, are taken into account. A criterion which is seldom used is the potential removal of micropollutants during drinking water treatment. This parameter seems important since compounds passing through purification systems will reach the consumer in higher concentrations than those which are more easily retained.

Studies on major European rivers have recently produced usefull information in this field. A Dutch study on the rivers Rhine and Meuse (1979-1982) resulted in a list of 59 compounds which were detected in raw and drinking waters. From this list 14 substances were considered undesirable in drinking water or raw water used for drinking water production due to their carcinogenic or mutagenic properties. (3) A similar investigation conducted by ourselves on the river Meuse in Belgium (1982-1983) (4) resulted in a list of 17 compounds (alkanes excluded) which were identified in both surface and drinking water. Eight of them are suspected of mutagenic and/or carcinogenic activity. For many of the compounds adequate toxicological information is lacking. (Table 1)

Both studies indicate that the majority of compounds passing through purification systems have log k_{ow} values below 3,2. A similar observation was made by Noordsy et al. in a study on bank filtered water in the Netherlands. (5)

The limitations of water treatment installations for removal of hydrophilic micropollutants has been extensively demonstrated. The coagulation flocculation (-sedimentation) step, which is often applied to remove suspended materials only removes a limited number of micropollutants through coprecipitation or adsorption.

Compounds which do not show this behaviour are poorly removed. (Tables 2 and 3) (6,7) The subsequent filtration step removes flocculated particles but has little effect on the presence of dissolved micropollutants. If it has any effects in this context, they are usually linked to microbiological degradation. Other treatments e.g. the application of oxidants are strongly limited with respect to micropollutant removal, while slow sand filtration gives varying results (Table 4) (8) Although ozone performs better than chlorine or potassium permanganate in this context, many micropollutants remain unaffected under practical ozonization conditions (9).

Among the limited number of techniques available the use of granular activated carbon is generally considered the most effective barrier against organic micropollutants. Its use however has limitations due to displacement effects caused by competitive adsorption and reequilibration, causing a rapid breakthrough of moderately polar components. This phenomenon has been extensively documented for a.o. tributylphosphate, benzoic acid, benzaldehyde, dibutylphtalate and haloforms (10-12).

The limitations of each of the unit operations makes it desirable to combine several of them in order to minimize exposure to micropollutants. This practice however has strong limitations a.o. in terms of purification costs and seems to act as a counterincentive for raw water source protection. Moreover most drinking water facilities practice post-desinfection with chlorine, often cancelling to a certain extent the beneficial effects of previous treatment.

TABLE 1

COMPOUNDS DETECTED IN RIVER MEUSE AND DRINKING WATER
(1982 - 1983)

COMPOUND	MAXIMUM CONCENTRATION IN MEUSE WATER	M	C	log KOW
Xylene	++	-	-	3,06
Styrene	+	S	S	
Ethylbenzene	+	-	S	3,06
C_3 - benzene	++	-	-	3,58
benzaldehyde	+	-	S	-
dimethylpyridine	++	-	-	1,92
diphenylamine	++	-	-	3,02
N-butyltoluamide	+	-	-	-
dichlorobenzene	++	-	S	3,51
naphtalene	++	-	S	3,30
benzothiazole	+	-	-	-
diphenylsulphone	+	-	-	-
dimethylphtalate	+	S	S	2,29
diethylphtalate	+++	S	-	3,33
dibutylphtalate	++	-	-	5,41
dioctylphtalate	++	S	+	9,56
picolylacetate	+++			

+ < 0,1 µg l^{-1} M : mutagenicity

++ 0,1 - 1,0 µg l^{-1} C : carcinogenicity

+++ > 1 µg l^{-1} S : suspect.

TABLE 2

EFFICIENCY OF SEVERAL MINERAL COAGULANTS FOR MICRO-POLLUTANT REMOVAL FROM SURFACE WATER

Coagulant	$Al_2(SO_4)_3$	WAC	PCBA	$FeCl_3$	Fe $ClSO_4$
meq l^{-1} Fe or Al	0,168	0,089	0,099	0,177	0,174
flocculation aid	Anionic Polyelec-trolite	-	-	Na alginate	Na alginate
Dosage mg l^{-1}	0,15	-	-	0,10	0,10
Lindane	34	26	40	31	33
Aldrin	18	40	46	70	19
Dieldrin	6	18	47	47	66
pp' DDT	21	39	62	43	25
DDD	71	60	77	74	35
Diethylphtalate	15	10	20	17	15
Dibutylphtalate	30	25	30	25	30
Chloroform	24	12	20	19	23
CCl_4	13	24	20	14	6
Dichlorobromo methane	18	38	20	26	25

F. FIESSINGER (1978)

TABLE 3

REMOVAL OF POLAR LOW MOLECULAR WEIGHT COMPOUNDS BY COAGULATION

COMPOUND	REMOVAL - PERCENT	
	MEAN ($\bar{X}$)	STANDARDDEVIATION (s.d)
Benzoic acid	4,2	1,4
Octanoic acid	3,4	1,8
Salicylic acid	18,8	2,5
Phenol	2,3	1,6

* Begin concentration : 100 ,ug l^{-1}
SEMMENS and AYERS (1985)

TABLE 4

REMOVAL OF PESTICIDES IN TREATMENT PROCESSES

	COAGULATION	SLOW SAND FILTRATION	Cl_2	$KMnO_4$	ACTIVATED CARBON
ALDRIN	±		+	+	+
ENDRIN	±		-	-	+
DIELDRIN	±	±	-	-	+
HEPTACHLOR		±	-	+	
CHLORDANE					+
TOXAPHENE	-	-	-		
LINDANE	±	±	-	-	+
DDT	±	±	-	-	+
ENDOSULFAN	+	±			+
HEXACHLORO-BENZENE		-			
PARATHION	-		+		+
MALATHION		+	+		
DIMETHOATE		-	+		
METASYSTOX		+			
2,4 D		-	-		+
2,4 DICHLORO-PHENOL			+		+
MONURON		-			
DIURON		-			

2. FORMATION OF OXIDATION BY-PRODUCTS

The concern about organic micropollutants in drinking water is not limited to the presence of compounds in water sources and their removal. Organic micropollutants in raw waters are only exceptionally encountered in concentrations above 10 μg l^{-1}, most of the time the majority of micropollutants is present in concentrations below 1 μg l^{-1}. The formation of organic micropollutants during treatent which is mainly due to the application of oxidizing agents can produce by-products in quantities far above 1 μg l^{-1}. The consequences of chlorination practices are well documented as far as the formation of haloforms is concerned. Recent studies have pointed out that the spectrum of chlorination by-products is very complex, yielding structurally very different compounds. (13) This may explain why studies trying to link side effects of chlorination, e.g. mutagenicity, with the presence of priority pollutants, have failed to show positive correlations.

The formation of oxidation by-products is not restricted to chlorination practices. Studies on the influence of ozone and chlorine dioxide have demonstrated the formation of different non chorinated micropollutants : quinones, hydroquinones, mono- and dicarboxylic acids have been reported as by-products of chlorine dioxide application (14-17) while aldehydes and carboxylic acids have been frequently cited as degradation products for many organic molecules of ozonation. (18) Although the interest in the application of ozone and chlorine dioxide has been growing during the last years little is known about their potential to produce micropollutants under circumstances encountered during water treatment practice. This lack of knowledge is largely due to the absence of suitable techniques for the detection and quantification of low molecular weight polar and volatile products.

As a consequence data on the relation between the formation of oxidation by-products and widely known phenomena such as microbial aftergrowth caused by ozone application are lacking.

Recently methods were developed enabling the analysis of low molecular weight aldehydes, taking advantage of the formation of 2,4 dinitrophenylhydrazones for their isolation and assay by HPLC (19), and of short chain carboxylic acids through esterification with α , p - dibromoacetophenone (20). Both groups of compounds were shown to be formed through ozonation, their concentrations increasing with increasing ozone dose applied. Flocculation had little influence on the presence of low molecular weight aldehydes, while their removal through rapid double layer filtation was strongly depending on water temperature : significant removal bein only observed above 10°C. (Table 5) Under these circumstances the best aldehyde removals correlated strongly with high TOC removals, suggesting the implication of filtration and biodegradation in their removal. The evidence for the role of biodegradation is strenghtened by the fact that their concentrations are significantly lowered by storage of the filtered water in a covered reservoir, before activated carbon filtration. (Table 6)

The good correlation which are observed between AOC, low molecular weight aldehydes and acetic acid concentrations respectively suggest that the assay of these micropollutants can give valuable indications on the aftergrowth potential of ozonated waters. (Table 7)

TABLE 5

INFLUENCE OF RAPID DOUBLE LAYER FILTRATION ON LOW MOLE-
CULAR WEIGHT ALDEHYDES, TOC AND UV_{254}

	TOC (mg^{-1})	UV_{254} (m^{-1})	aldehydes (μ mole 1^{-1})	Temperature °C
1.	- 0,1	- 0,4	- 0,04	13,7
2.	- 0,1	- 0,3	- 0,08	13,7
3.	- 0,7	- 0,3	- 0,26	15,2
4.	- 1,5	- 0,3	- 0,32	20,5
5.	- 1,7	- 0,0	- 0,19	19,6
6.	- 0,1	- 0,3	- 0,0i	2,0
7.	- 0,6	- 1,4	+ 0,02	2,0
8.	- 0,6	- 4,5	+ 0,06	4,5
9.	- 0,1	- 3,3	+ 0,03	10

Δ TOC, Δ UV_{254} and Δ aldehydes represent the changes in TOC, UV_{254} absorption and aldehyde levels caused by rapid double layer filtration.

TABLE 6

INFLUENCE OF RAPID DUAL LAYER FILTRATION AND STORAGE OF
THE FILTERED WATER ON THE PRESENCE OF LOW MOLECULAR
WEIGHT ALDEHYDES

	before filtration	After filtration	After storage
1.	-	0,04	0,00
2.	0,06	-	0,00
3.	0,07	-	0,00
4.	-	0,06	0,04
5.	0,21	0,12	0,07
6.	0,14	0,20	0,13
7.	-	0,10	0,01

All results are expressed as μ mole aldehydes 1^{-1} (C_1 - C_3)

TABLE 7

RELATIONSHIP BETWEEN AOC CONTENT, LOW MOLECULAR WEIGHT ALDEHYDES AND ACETIC ACID AFTER OZONIZATION OF PRECHLORINATED FILTERED WATER

	AOC (μg C l^{-1})	Aldehydes (μ mole l^{-1})	Acetic acid (μ mole l^{-1})
Before ozone	22	0,45	0,30
Ozone (2 mg l^{-1})	56	0,60	0,40
Ozone (4 mg l^{-1})	91	0,77	0,53

3. CONCLUSIONS

There is growing evidence indicating that in the future the analysis of organic micropullutants in relation to drinking water production will go beyond the search of compounds which are amenable to gas chromatography as such.

The cost of analysis for micropollutants using sophisticated techniques and the wide variety of potential pollutants will force us to select research targets carefully.

The gaps which widely used treatment show both for micropollutant removal and generation of oxidation by-products necessitates further research in new treatment technique e.g. alternative adsorbents and combination of oxidants.

REFERENCES

1. B. CRATHORNE, M. FIELDING, C.P. STEEL a.d. C.D. WATTS
   Environ. Sci. Technol., 18, 797-802, 1984.

2. H. FISH and S. TORRANCE, Control of specific industrial pollution of water sources, in Proceedings 12th IWSA Congress, Kyoto, Japan, October 2-6, 1978.

3. P.C. NOORDAM, Toxicologische beoordeling van een aantal in het water van Rijn en/of Ysselmeer en/of Maas als in het hieruit bereide drinkwater aangetroffen xenobiotische organische verbindingen. KIWA 1983.

4. RIWA, Het kwaliteitsprofiel van de Maas, 19-30 september 1982.

5. A. NOORDSIJ, L.M. PUIJKER and M.A. VAN DER GAAG,
   The Quality of Drinking Water prepared from Bank-filtered river water in the Netherlands, Paper presented at the Int. Symp. on Organic Micropollutants in Drinking Water and Health. Amsterdam, June 11-14, 1985.

6. M.J. SEMMENS and K. AYERS, Removal by Coagulation of Trace Organics from Mississipi River Water JAWWA, 77, 79-84, 1985.

7. F. FIESSINGER, Coagulation et flocculation, in Proceedings 12th IWS Congress, Kyoto, Japan, October 2-6, 1978.

8. U. BAUER, VOM WASSER, 39, 161-187, 1972.

9. J. HOIGNE and H. BADER, Water Research, 17, 173-183, 1983.

10. I.H. SUFFET, L. BRENNER, J.T. COYLE and P.R. CAIRO Environ. Sci. Technol., 12, 12, 1315-1322, 1978.

11. K. ALBEN and E. SHPIRT, Distribution profiles of chloroform, weak organic acids, and PCB's on Granular Activated Carbon colums from Waterford New York.
Environ. Sci. Technol. 17, 4, 187-192, 1983.

12. B. HULTMAN, Elimination of Organic Micropollutants. Paper Presented at the IAWPRC Symposium. Micropollutants in the Environment, Brussel 22-25/10/1981.

13. E.W.B. DE LEER, J.S. SINNINGHE DAMSTE, C. ERKELENS AND L. DE GALAN, Environ. Sci. Technol., 19,6, 512-522, 1985

14. A.A. STEVENS, D.R. SEEGER AND C.J. SLOCUM, Products of chlorine dioxyde treatment of organic materials in water. Paper presented at the Workshop on Ozone/Chlorine Dioxyde Oxidation Products of Organic Materials, November 17-19, 1976, Cincinnati, Ohio.

15. J. HOIGNE and H. BADER, Vom Wasser 59, 253-267, 1982.

16. C. RAV-ACHA, Water Research, 18, 11, 1329-1341, 1984.

17. C.A. COLCLOUGH, J.D. JOHNSON, R.F. CHRISTMAN AND D.S. MILLINGTON, Organic Reaction Products of Chlorine Dioxide and Natural Aquatic Fulvic Acids, in Water Chlorination, Environmental Impact and Health Effects, 4, 1, 219-229, R.J. Jolley et al. Eds, Ann. Arbor, 1983.

18. J. MALLEVIALLE, Produits de réaction identifiés au cours de l'ozonation in l'Ozonation des eaux, W.J. Masschelein Ed., Paris 1980, p 78-99.

19. F. VAN HOOF, A. WITTOCX, E. VAN BUGGENHOUT and J. JANSSENS, Analytica Chimica Acta, 169, 419-424, 1985.

20. M.J. BARCELONA, H.M. LILJESTRAND and J.J. MORGAN, Anal. Chem., 52, 321-325, 1980.

CHEMICAL AND BIOLOGICAL CHARACTERIZATION OF INDUSTRIAL EFFLUENTS

AND APPLIED MATHEMATICAL MODELLING - A CASE STUDY

Å. Granmo, L. Renberg* and G. Sundström*
National Environmental Protection Board

Kristineberg Marine Biological Station, Pl.2130 S-450 34 Fiske-bäckskil, Sweden.
* Special Analytical Laboratory, Box 1302 S-171 25 Solna, Sweden.

Summary A case study is described for aquatic investigations around a petrochemical industrial centre at the west coast of Sweden. A biological/chemical characterization of the wastewaters were followed by a field program and a mathematical model was developed for predicting the environmental impact of the wastewaters. A summary of the results so far obtained is given.

Background

Stenungsund is situated on the Swedish west coast about 40 km north of Gothenburg (Fig.1). Here a centre for the Swedish petrochemical industry has existed for 20 years. Stenungsund is situated in a fjord system characterized by a rocky and broken coastline with a complex water current. The receiving water body for the discharge, Askeröfjorden has a relatively rapid exchange of water.

In March 1982 the Swedish government decided that an investigation of the effluents from 5 petrochemical industries and one power plant in Stenungsund should be carried out under the supervision of the National Environmental Protection Board, the County administration and the Municipality of Stenungsund. The investigation was divided into 4 separate projects: Air, water, epidemiology and technology. The present paper will focus on the water project.

A flow scheme shows the connection between the petrochemical industries and their products (Fig.2).

Chemical and biological characterization of effluents

A plan for the research program was set up during 1983 (1). Earlier, several investigations have been carried out in the ambient water body mainly by a fishery and bottom fauna sampling program. Due to difficulties in interpreting the results from this work the new program was primarily focused on the wastewaters and their characteristics (Fig.3). Representative water

samples were therefore taken from eight major wastewaters in
the area and were immediately frozen for later distribution to
the different investigators. The wastewater samples were screened
for toxicity using biological tests. They were chosen in accordance with experience gained from previous tests carried out
within the Environmental Protection Board (2),(3) and included
short-term toxicity tests with 20 species of marine microalgae,
higher plants (growth retardations of onion roots), crustaceans
(Nitocra spinipes) and fish (Gasterosteus aculeatus). In addition
an Ames' test was also performed.

        The strategy adopted for the chemical characterization
included both inorganic and organic analyses. The inorganic analyses included 18 metals and 6 anions. In order to get an overview of the large number of organic compounds present in the
different wastewaters, 73 compounds were selected from the US.
Environment Protection Agency list of priority pollutants as
being relevant for Scandinavian conditions and analyzed by gas
chromatography and mass spectrometry (GC/MS) at the Center for
Industrial Research (SI) in Oslo, Norway. Also, compounds known
to be used and produced in the industrial plants, as well as
byproducts, were included. In addition to the quantitative GC/MS
analyses all mass spectra records were stored to facilitate a
subsequent identification and semiquantitative determination of
unknown compounds.

        Biodegradation tests were considered to be very important
for the wastewater characterization. The test selected was a
modified OECD screening test(4). To simulate natural conditions
it was performed at two temperatures typical of seasonal variations in the marine environment (4°C and 15°C). Degradation was
followed by measurements of dissolved organic carbon (DOC) for 60
days. Some of the wastewaters were also analyzed for adsorbable
organohalogens (AOX). A stabilization test during two months was
also performed for those wastewaters which were assumed to contain the more persistant fractions as well as potential bioaccumulating components.The stabilization process was monitored by
DOC measurements. After the stabilization period the water samples were investigated for:
1. Algal toxicity (carbon-14 assimilation of Skeletonema costatum).
2. The contents of potential bioaccumulating compounds using
   reversed phase thin layer chromatography (5).
3. Chemical analyses of phenols and PAH.
4. Analyses of adsorbable organic halogens (AOX).
        Samples from the stabilized waters were frozen for later
analyses and tests. These samples were sent to SI in Oslo for a
renewed "priority pollutant" analysis and a repeated Ames' test
was performed. The biodegradability tests were performed at the
Water Quality Institute, Hørsholm, Denmark.

Mathematical modelling
        The next step in the stratified attempts for predicting the
potential environmental impact of the wastewaters was to establish suitable mathematical models. The purpose of such models
were twofold. First, a model for predicting environmental concentrations of specific organic pollutants and secondly, a dilu-

tion model to allow estimations on the ambient toxicity, based on the toxicity of the wastewater effluents. In both cases a hydrodynamic model of the Stenungsund area was a prerequisite, and such a model has now been developed by the Danish Hydraulic Institute (6).

Work with the dilution model is still in progress. However, a distribution and fate model (FATEMODEL) for predicting environmental concentrations of suspended matter, sediment and biota compartments in the water has just recently been established at the Danish Water Quality Institute (7), though the model has not yet been applied to the organic compounds of interest. The receiving water body id divided into six segments, each segment consisting of a two layer box. The model is a steady state model and is based on the assumption that within a box all compartments are in equilibrium with each other. The transformation processes taken into account are evaporation, hydrolysis and biodegradation, all approximated to be first order kinetic processes (Fig.4).

After the modelling work has been carried out the intention is to validate the model by analysing samples from different compartments for the most pertinent compounds.

It should be mentioned that there exist at least two other modelling concepts, which are of particular interest. One model similar to FATEMODEL is the EXAMS model developed by US Environmental Protection Agency (8) and the fugacity models developed by Mackay and co-workers (9, and references citied therein). We have applied some parts of the fugacity model for an initial ranking of the various compounds emitted from the different industrial activities (10). However, it would be of great interest to use these models not only for a ranking but to predict the actual concentrations in the receiving water bodies.

Field program

A more detailed knowledge of the wastewaters should give better possibilities to optimize a field program. Therefore only limited field activities were performed during the first years of the project. The strategy adopted for the field program was thus to:
1. search for specific components in the field which were identified in the wastewaters.
2. search for biological effects in the receiving water body.
3. validate the mathematical fate model.
4. find relevant methods which could be used in a future permanent control program for the area.

With reference to the commercial fishery in the area it was considered important to continue an already existing fishery program studying the abundance and size of fish. As a result of a fishery report (11) indicating decreasing stocks of plaice (Pleuronectes platessa) in the area, a detailed study on the production of young epibenthic fauna in shallow waters was performed. Also physiological investigations were recently carried out using the bottom-dwelling flounder (Platichtys flesus). After sampling in a gradient from the outlet the activity of the MFO-liver enzyme system, carbohydrate metabolism, osmo- and ion-regulation and hematology were studied. In order to establish the possible presence of bioaccumulating substances caged mussels (Mytilus edulis L.) were placed during 6 weeks in a gradient from one of the outlet pipes of one industry (A). The reason why this plant was selected was that high levels of p-nonylphenol (12000 μg/l) were identified in the undiluted wastewater. Meanwhile mussels were exposed in the laboratory to different of wastewater. A test with periphyton community colonization was also performed in the fjord.

As a minor part of the field investigations, a number of juvenile eiderducks were shot in the inner part of the fjord. Liver- and muscle samples were taken for chemical analyses and the MFO-activity in liver was investigated.

Recently a number of sediment samples from the fjord have been subjected to chemical analyses for the most frequent lipophilic organic compounds previously identified in the wastewaters.

Preliminary results

A considerable amount of raw data has been produced from the various tests. However, as results are still accumulating and the data are being continuously evaluated, only a brief summary of results obtained will be presented.

The results of the biological screening tests (1) were ranked in order of toxicity. From Table 1 it is obvious that for most of the tests the ranking order is similar, indicating that industry A and B has the most toxic effluents.

If the flow rates are also taken into consideration (Tab.2), the toxicity emmission rate (TER) (12) can be simply calculated (Tab.3.). It is evident that the same industries also cause the highest toxicologicalal impact on the environment. However, the assumption is made that the toxic effects of all the effluents are strictly additive. This assumption is not always valid.

The chemical characterization showed that the inorganic components were of minor importance. The organic "priority pollutants" found are listed in Table 4. As seen there are some compounds such as p-nonylphenol, phenol and phthalates which are present at higher concentrations than the others. From the mass spectra it was possible to identify also other compounds and for one of the industries (A), nonylphenol with short ethylene oxide chains were semiquantitatively determined to about the same concentrations as that of p-nonylphenol.

Results from the biodegradation tests (13) show that the most toxic wastewaters were not ready biodegradable especially at low temperature (Fig.5).

The effects on photosynthetic activity in Sceletonema costatum, which was investigated before and after an aerobic stabilization test (2 months), were less pronounced in the stabilized water, but the waters from industry A and B still were the most toxic. The screening for potential bioaccumulating compounds in these wastewaters show (Table 5) that components which may be bioaccumulated are present also after stabilization.

Only some results are available at present from the field investigations. These do not indicate any serious biological effects in the area.

Literature cited

(1) Granmo,Å.(1984). Översiktsplan samt biologiska tester av industriavloppsvatten från Stenungsund. National Environmental Protection Board, pm no. 1845, in Swedish.

(2) Ekelund,R.,E. Emanuelsson and Å. Granmo(1983). Comparison of methods for assessing effects of industrial wastewater on the mussel Mytilus edulis L.. Vatten, vol.39 275-285.

(3) Granmo,Å. and R. Ekelund(1983). Toxicity testing of industrial effluents with fish, crustaceans and molluscs. National Environmental Protection Board, pm no. 1733. 76pp. In Swedish.

(4) Nyholm,N.(1983). Methods for testing biodegradability of chemicals in water. Experiences from EEC's and OECD's ring test programmes.Report to the Danish Council of Technology. 99 pp.

(5) Renberg,L., G. Sundström and A-C. Rosen-Olofsson: The determination of partition coefficients of organic compounds in technical products and waste waters for the estimation of their bioaccumulation potential using reversed phase thin layer chromatography. Toxicol. Environm. Chemistry.In press.

(6) Dansk Hydraulisk Institut.(1985). Hydrografiske modellberegninger i forbindelse med beregninger af chemical fate for stoffer udledt till Stenungsund-recipienten. National Environmental Protection Board, report no. 3058. 73 pp. In Danish.

(7) Vandkvalitetsinstituttet Hørsholm.(1985). Beregninger af chemical fate for stoffer udledt till Stenungsund recipienten. Preliminary stenciled report. In Danish.

(8) U S Environmental Protection Agency.(1982). Exposure analyses modelling system (EXAMS): User manual and system documentation. EPA-600/3-82-023.

(9) Mackay, D. and S. Paterson.(1982). Fugacity revisited. Environm. Sci. Technol. Vol.16, pp.654A-660A.

(10) Sundström, G. and L. Renberg. (1984). Tillämpning av Mackay's fugacitetsmodell på några organiska ämnen i Stenungsundsrecipienten, stenciled report. In Swedish.

(11) Jacobsson, A.(1984). Årsrapport Vatten 1982-1983. Stenciled report. National Environmental Protection Board, 82-2651-84 Mkk.In Swedish.

(12) California Water Resources Control Board. (1972). Water quality control plan for ocean waters of California. State of California, Water Resources Control Board, 18 pp.

(13) Vandkvalitetsinstituttet Hørsholm. (1984). Bionedbrydelighedsundersøgelser af industrispildevand udledt til marint miljø. National Environmental Protection Board, report no. 3037, in Danish.

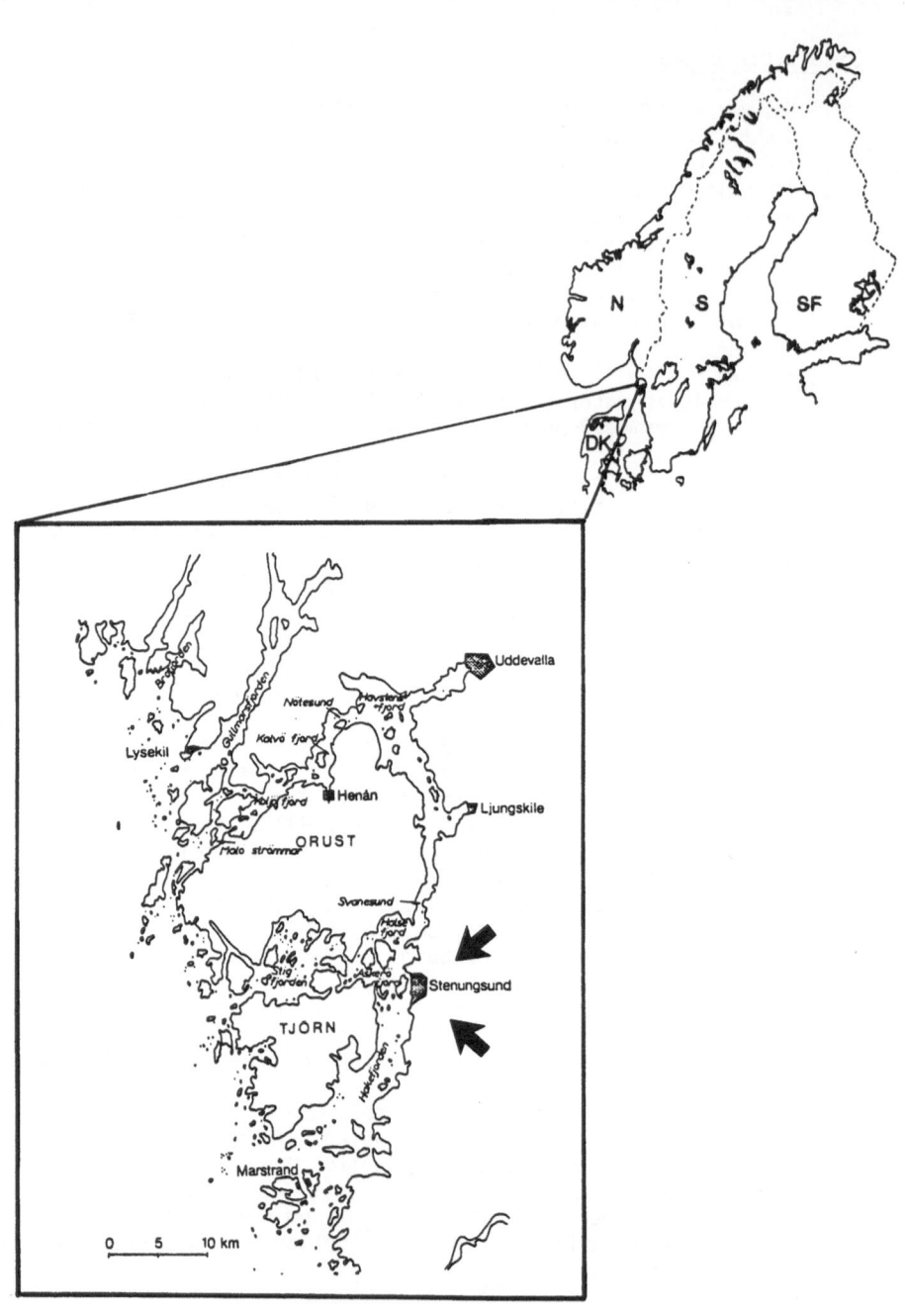

Fig. 1 Situation of the investigated area.

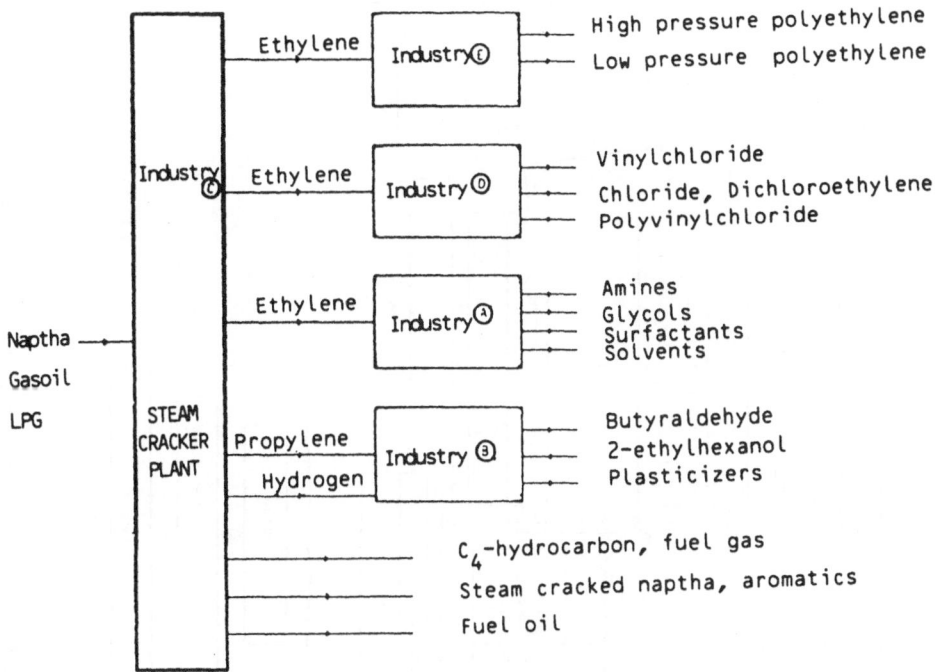

Fig. 2 The petrochemical industries in Stenungsund. Flow scheme for raw materials and products.

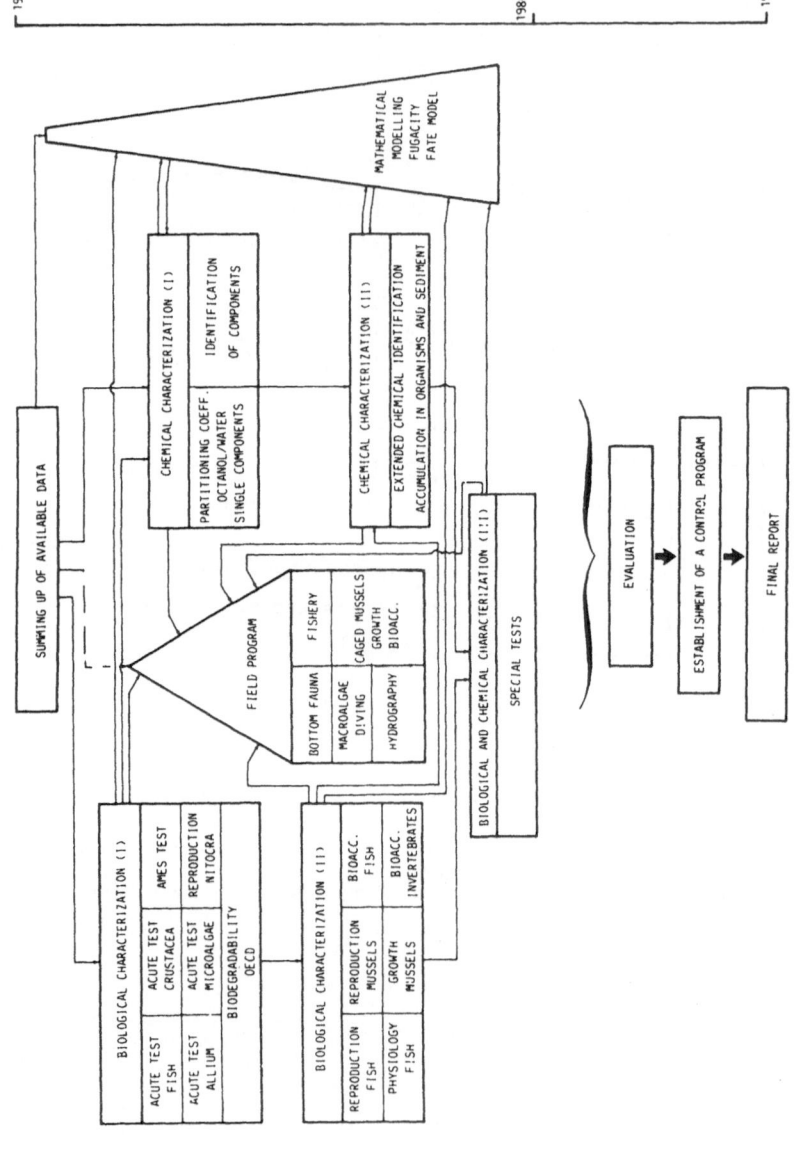

Fig. 3 Plan for the investigations in Stenungsund – the aquatic part.

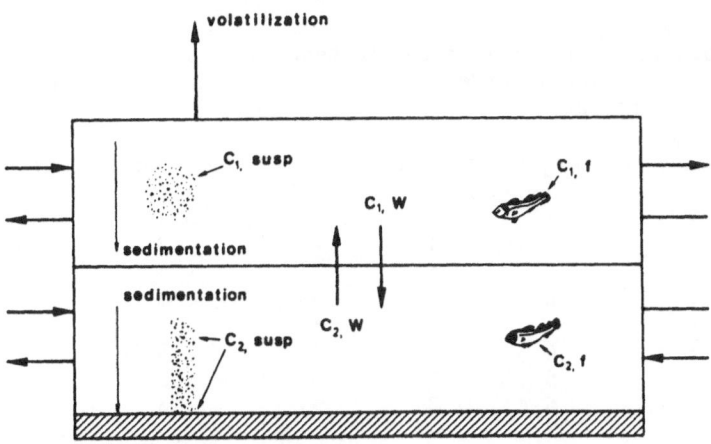

Fig. 4 The chemical fate model used in Stenungsund. C_W: concentration of dissolved compounds (µg/l). C_{susp}: concentration of particle-adsorbed compounds (µg/l). C_f: concentration of compound in fish (µg/kg). (From VKI, preliminary report, 1985).

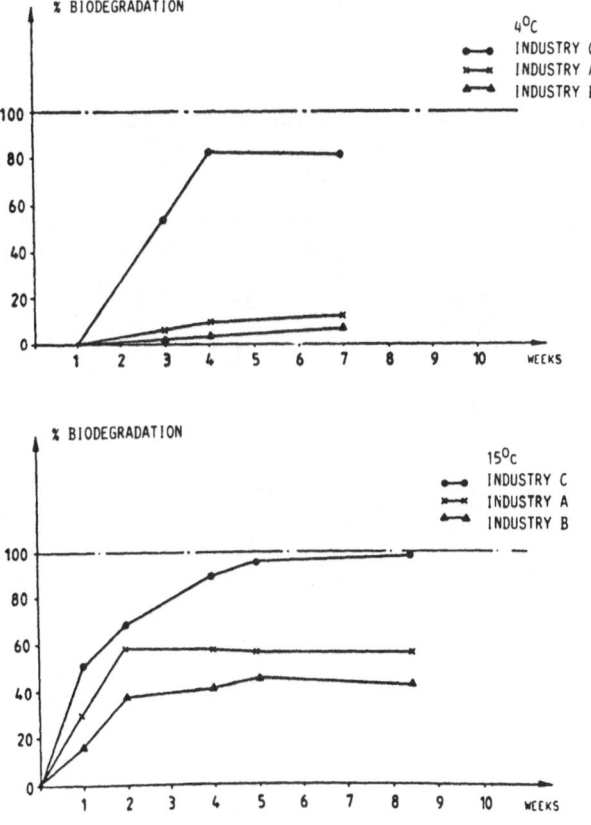

Fig. 5 Biodegradation of wastewaters in sea-water at $4^\circ C$ and $15^\circ C$ expressed as per cent of the initial DOC value. (From VKI, report, 1984.)

Table 1 Results from screening tests on wastewaters from Stenungsund ranked in order of toxicity. 1 = most toxic, 8 = least toxic effect.

Industry	Algae	Allium	Nitroca	Gasterosteus	Ames
A	1	3	1	1	1
B	3	1	2	2	4
C	4	6	6	3	4
D	2	2	4	3	2
E	4	7	6	3	4
F	4	4	6	3	4
G	5	5	3	3	3
H	5	8	5	3	4

Table 2 Flow rates and dilution factors for 8 different wastewaters from Stenungsund.

Industry	Flow rate m3/h	Dilution before discharge (times)	Estimated initial dilution (times)
A	8-10	125	100 - 200
B	10-15	11	300 - 500
C	160	1.5	100 - 300
D	70	250	1000
E	30	300	1500
F	3	-	?
G	150	-	?
H	2	via C	100 - 300

Table 3 Toxicity emission rates of the industries in Stenungsund.

TER = T.U x flow rate
T.U. = 1:EC 10
TER = (1:EC 10) x flow rate
EC 10 = conc. giving an effect for 10% of the test organisms.

Values are expressed as per cent of volume of wastewater. Flow
rate expressed as cubic meter per hour.
      The computation of TER values facilitates comparison of the
environmental impacts of large and small discharges of wastewaters having widely different toxicities.

Tests:

Allium test   (Effects measured as growth retardations of onion
                 roots during 3 days).
Nitocra 96 h LC test
Sticklebacks 96 h LC test.

                              TER.

Industry A   3845
Industry B   1967
Industry C   816
Industry D   581
Industry F   557
Industry E   63
Industry G   28
Industry H   7

Table 4 Priority pollutants (μg/l) in wastewaters from Stenungsund.

Industry A.
Ethylbenzene	3
M-/P-xylene	5
O-xylene	2
Naphtalene	2
Phenol	1100
O-cresol	3
P-nonylphenol	12022
Di-N-butylphtalate	46
Butylbenzylphtalate	32
Di-2 ethylhexylphtalate	855
Dioxane	3000
Chloroform	3

Industry B.
Di-2-ethylhexylphtalate	1430

Industry C.
Phenol	5
O-cresol	10
Di-2 ethylhexylphtalate	1
Dioxane	11
1,1,1-trichloroethane	4

Industry D
Phenol	7
M-/P- Cresol	10
P-nonylphenol	11
2,4,6 trichlorophenol	9
Di-2 ethylhexylphtalate	4
Dioxane	29
Chloroform	8
1,1,2- trichloroethane	1

Industry E.
Di-2-ethylhexylphthalate	3

Industry F..
Di-2-ethylhexylphthalate	3
Trichloroethene	5
Tetrachloroethene	16

Industry G.
Naphtalene	9
1-methylnaphtalene	1
Benzo(A) pyrene	3
O-cresol	2

Industry H.
Pyrene	1
Flouranthene	10
Crysene	1
Benzoflouranthene	7
Phenol	4
Dioxane	32

Table 5 Octanol/water partition coefficients (P ow) in the acidified/ neutral fraction from some wastewaters in Stenungsund. (From VKI, report, 1984.)

	LOG P (O/W)	
SAMPLE	BEFORE STABILIZATION	AFTER STABILIZATION
INDUSTRY A	6.65	
	4.95	
		4.65
		2.59
INDUSTRY B	6.65	
		3.05

RISK ESTIMATION AS A METHOD FOR MODELLING THE FATE AND IMPACT OF INDUSTRIAL EFFLUENTS IN MARINE ENVIRONMENTS

J. FOLKE
VKI, Water Quality Institute
11, Agern Allé, DK-2970, Hørsholm, DENMARK

SUMMARY.

Discharge of hazardous substances contained in industrial wastewater (emission) into a receiving water body will cause an immission, the size of which may be estimated as concentrations of substance in time and space in the receiving waters. Risk estimation of industrial effluents seeks by extrapolation to predict the nature and extent of the immision under given circumstances. The experimental basis for the extrapolation is hazard assessment results and *in situ* experiments and studies of the receiving waters. The ultimate goal is an evaluation of the environmental consequences of the situation. Possibly, the modelling of environmental fate, impact and effects could be evaluated by *(i)* identifying primary, affected areas of the receiving waters where biological effects can be directly recognized owing to the discharge, *e.g.* by biological survey, (ii) identifying secondary, affected areas where life conditions are expected to have deteriorated owing to the discharge, *e.g.* by extrapolations of laboratory experiments, and *(iii)* identifying areas of the receiving waters exposed to substances of the effluent, *e.g.* by chemical analyses in biota and sediment, and/or fate modelling of effluent substances. Unfortunately, it may in many instances turn out, that our scientific knowledge on functioning and vulnerability of natural ecosystems is too poor to predict long-term environmental consequences of many industrial effluents. Consequently, politicians often have to take their decisions based on an inclomplete, scientific knowledge.

1. INTRODUCTION.

Environmental impact assessment of industrial effluents is a complicated multidisciplinary process, which has recently been reviewed[1] with emphasis on the pulp and paper industry[2]. The concepts of environmental impact assessment have also been subject to presentation within a workshop (WP4) of the COST 641 action[3], a presentation dividing the impact assessment into three main phases, *i.e.* hazard assessment, risk determination, and societary evaluation. Furthermore, the process of hazard assessment was reviewed in presentations of two other workshops of the COST 641 action[4,5]; hazard assessment may be divided into subphases of hazard identification[4] and hazard analysis[5] and is concerned with the analysis of environmental threats. Following these presentations this paper will review the phase of risk determination of industrial effluents, a phase which is concerned with the measures of environmental threats in terms of quantified *estimates* [6], and therefore may be denoted risk estimation.

2. RISK ESTIMATION.

In general, risk estimation seeks to measure the likelihood of an adverse event (of some stated magnitude) occurring and the likelihood and nature of the consequences that follow; its methods are: revelation, intuition, and extra-polation. Revelation as a method is related to religion and supernatural inspiration, and the value of the outcome is closely related to one's degree of belief. Intuition is akin to revelation except that it is internal to the person, reaching a conclusion on the basis of less explicit information than ordinarily required to reach that conclusion. Extrapolation relies on individual or shared experience, and may, metaphorically speaking, be forward - extrapolating the likelihood of future events by past experience; backward -from unknown but imagined events and consequences to known events and consequences; or sideways - by analogy and transfer of experience from different but similar place, situation or thing[6].

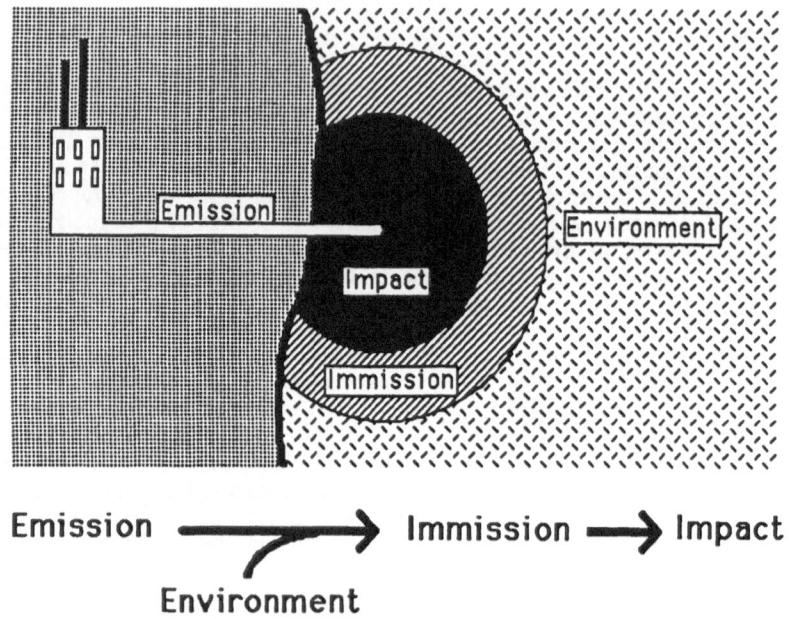

Fig. 1.   *Interactions between an emission and the environment.*

Discharge of hazardous substances contained in industrial wastewater (emission) into a receiving water body will cause an immission, the size of which may be estimated as concentrations of substance in time and space in the receiving waters. The nature and extent of this immission is determined by the interaction between the emission and the receiving environment - the immision then determines the nature and the extent of the impact on the natural ecosystems, *fig 1.*

Risk estimation of industrial effluents seeks a measure of the likelihood of adverse effects on biological systems, owing to the discharge of substances from the effluent, using a methodology of sideways extrapolation. Experimental basis for such risk estimation for sideways extrapolation may be gained by
   - hazard assessment,
   - *in situ* experiments and studies of the receiving waters
in order to
   - predict environmental fate of discharged substances,
   - measure the impact on exposed ecosystems.

The ultimate goal of risk estimation of industrial effluents is an evaluation of the environmental consequences of the situation. Normally, discharged amounts of substance are so small that risk estimation should be conducted solely towards *direct, environmental exposure* [7], *i.e.* excluding the general, environmental exposure, owing to estimated negligible effects of the individual effluents on world environment. However, there might be situations in which there are so many similar industrial effluents countrywide or worldwide that significant general environmental exposure will be recognized. In that case an integrated environmental impact assessment of the particular type of industrial production leading to such general exposure would have to be conducted, *e.g.* as has been done with bleached wood pulp production in Sweden[8,9]. Principles of impact assessment of discharges leading to a general environmental exposure are very similar to the ones applied to chemicals[7].

3. PREDICTED ENVIRONMENTAL FATE OF SUBSTANCES.

Predicting the environmental fate of substances discharged by an industrial effluent requires solving of mass- balance equations about specified volumes, V of interest[10]:

$$V(dc_i/dt) = J_i + \Sigma R_i + \Sigma T_i + \Sigma W, \text{ where}$$

c_i = concentration of the chemical in compartment *i*
J = transport through the system
R = reactions within the system
T = transfer from one phase to another
W = inputs

Environmental compartments to be considered in connection with discharge through industrial effluent are mainly water and bed[11]. Fundamental processes that have to be considered are[12]:
- transport,
  - hydrodynamic transport,
  - sorption,
  - sediment transport,
  - benthic, boundary layer transport,
  - volatilization,
- transformation and degradation,
  - direct photolysis,
  - indirect photolysis,
  - hydrolysis,
  - microbial transformation/ degradation.

Although significant in predicting adverse biological effects, bioaccumulation in biota, and biotransformation/degradation reactions other than the microbial play a negligible role in the total mass balance of substances and dynamics in the receiving waters, owing to the small mass fraction of biota to the total mass of the system[12].

Facing the modelling of fate of substances of complex industrial effluents, one has to cope with the fact that[13]
- all hazardous substances may not have been identified during hazard assessment,
- possibly, the fate of a number of substances are affected by the presence of other substances (*e.g.* synergism).

Therefore, three approaches may be used, alone or in combination[13]
- model general properties of the effluent,
- model chemical classes by choosing representative compounds
- model one compound at a time.

Physical models of microecosystems that may have been used during hazard assessment or *in situ* receiving waters monitoring of selected single substances or chemical classes of compounds may give valuable input to the modelling of fate, *fig. 2* [13].

Hazardous, chemical class of substances discharged by industrial effluents may be grouped according to the principal fates in the near-discharge-point environment,
- easily biodegradable liquid and suspended solids,
- slowly biodegradable, hydrophilic liquid and suspended solids,
- solid phase substance.

The following items of basic information concerning the industrial effluent are either available, mainly from hazard assessment, or assessible[14],

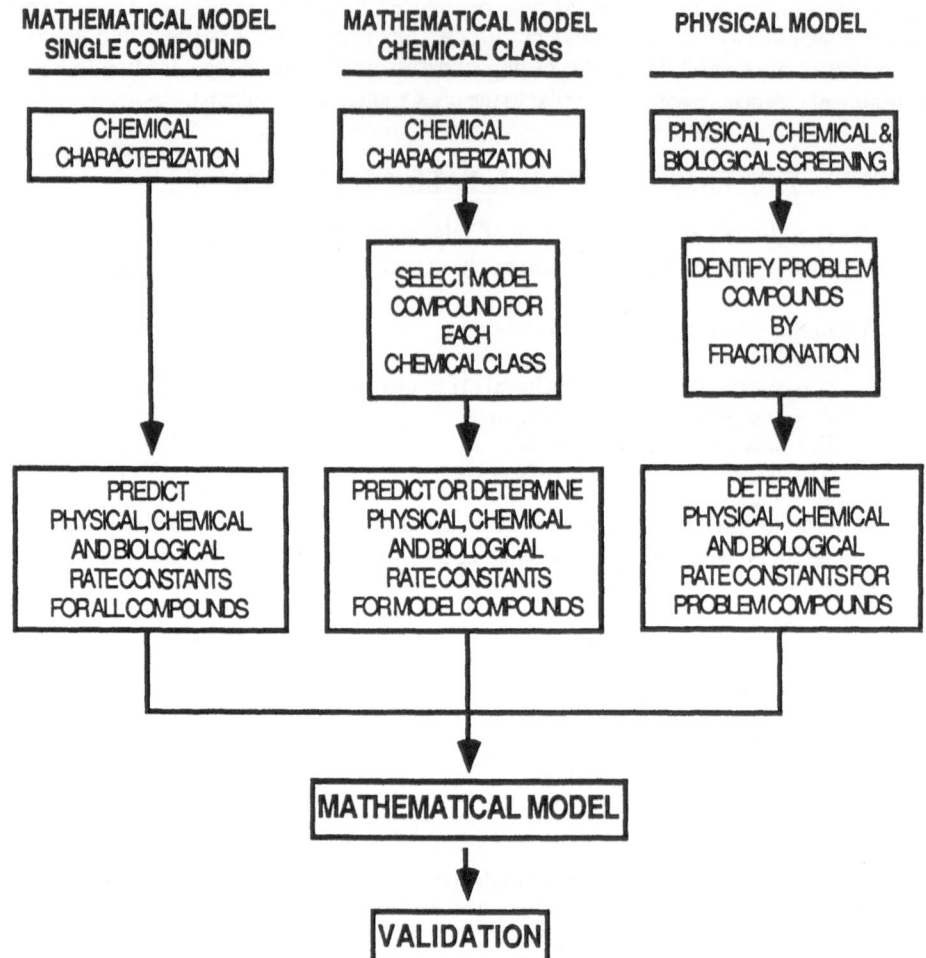

| MATHEMATICAL MODEL SINGLE COMPOUND | MATHEMATICAL MODEL CHEMICAL CLASS | PHYSICAL MODEL |

Fig. 2. *Approaches for the modelling of fate of discharged substances[13].*

- discharged rates of substances,
- possible, environmental fate of discharged substances,
- geographical discharge point of effluent,
- hydraulic conditions (currents *etc.*),
- natural physico-chemical conditions (salinity *etc.*),
- potential vulnerability of the target communities of exposure.

On that basis mathematical modelling should be used to describe concentration-time profiles of the various liquid and suspended solids, and solid phase substances should be calculated using hydrodynamic models in such a way that maximum accuracy will be obtained for the most sensitive compartments or sectors of the ecosystem[14], identifying sedimentation areas of the receiving waters etc.

Mathematical models are based on the principle of conservation of mass, and for a volume segment of receiving waters the models have to include the mecanisms depicted on, *fig. 3* [15]. The loading rate from the effluent W_T is the primary input to the water column, and the relevant transformation (degradation) reactions (summerized in the rate constants k_1 and k_2) are: hydrolysis, oxidation, biodegradation, photolysis and volatilization. The adsorption/desorption reaction is assumed to be at an equilibrium, and the fraction of the

Mechanisms

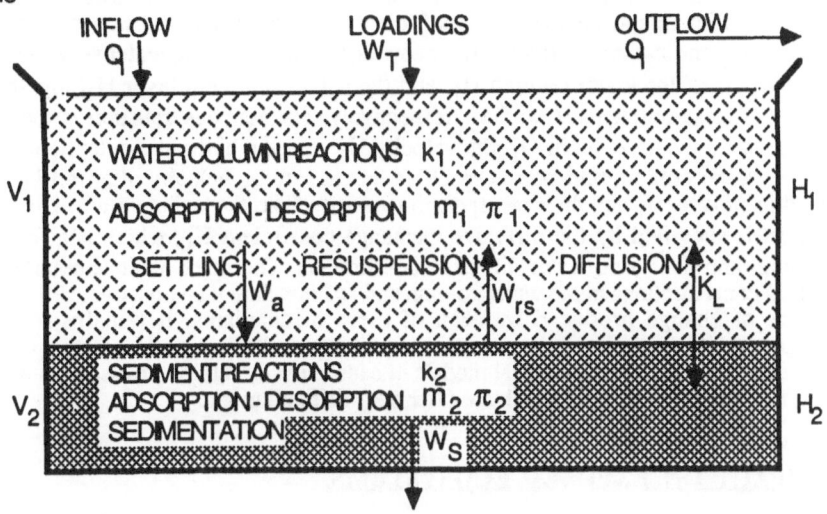

Definitions

Parameter	Water Column	Sediment
<u>Chemical/Biological</u>		
Loading rate (kg/day)	W_T	
Sum of hydrolysis, oxidation, biodegradation, photolysis and volatilization rates (/day)	k_1	k_2
Partition coefficients (liter/kg)	π_1	π_2
<u>Physical</u>		
Solids concentration (mg/l)	m_1	m_2
Depth	H_1	H_2
Volumes (m³)	V_1	V_2
Flowrate (m³/day)	Q_1	
Velocity (m/day)	$U_1=Q_1H_1/V_1$	
Detention time	$t_{01}=V_1/Q_1$	
Settling velocity	w_a	
Resuspension velocity		w_{rs}
Diffusion exchange coefficient (cm/day)		K_L
Sedimentation		
Velocity (mm/yr)		w_s
Rate coefficient (/day)		$K_s=w_s/H_2$
<u>Concentrations</u>		
Total dissolved + particulate (µg/l)	c_{T1}	c_{T2}
Particulate $f_p=m\pi{\cdot}10^{-6}/(1+m\pi{\cdot}10^{-6})$	f_{p1}	f_{p2}
Dissolved $f_d=1/(1+m\pi{\cdot}10^{-6})$	f_{d1}	f_{d2}

Fig. 3. Mathematical modelling the fate of discharged substances in a receiving water volume segment[15].

chemical mass that is either dissolved or absorbed to particu-lates is determined by the mass of adsorbing solids m and the partition coefficients π in the water column (m_1 and π_1) and the sediment (m_2 and π_2), respectively. The geometry of the receiving water segment is specified by the depths of the water column H_1, and the active sediment layer H_2, together with the volumes of these segments V_1 and V_2, respectively. The aqueous transport is specified by the flowrate of the water column Q_1, the velocity U_1, and the hydraulic detention time t_{01}. The sediment/water column transport is specified by the settling velocity of adsorbing particles to the sediment layer w_a and their resuspension velocity from the sediment w_{rs}. Dissolved-phase transport is specified by the diffusive exchange coefficient K_L, which is a function of the interstitial and overlying water diffusion coefficients. The mass-balance equations for total chemical concentrations in the water column and sediment are then constructed from the sum of all rates of change produced by each of these mechanisms. It remains to solve these complicated differential equations and to interpret the results, e.g. steady-state approximations. The precision of such mathematical models of cause heavily rely on precise input data from the actual receiving waters, data which are often very difficult to get.

4. IMPACT ON EXPOSED ECOSYSTEMS.

The impact on the marine ecosystem should be concerned with both pelagic and benthic ecosystems and evaluated on basis of
- extrapolation of results of effect analysis during hazard assessment using advection-dispersion mathematical models,
- extrapolation of results obtained from physical models of microecosystems,
- extrapolation of experimental results of *in situ* studies of bio-accumulation and ecotoxicity,
- ecoepidemiological studies and biological surveys.

There are many difficulties in assessing the ecological significance of effects observed in the laboratory. Stress tolerance and stability and recovery potential of exposed ecosystems are particularly important parametres. Ecologists would have to identify key properties of ecosystems of the receiving waters estimating those parametres[14].

Easily biodegradable substances may cause oxygen depletion and death from suffocation. However, toxic effects from those substances are expected to be only short-term or within a limited area. Slowly biodegradable hydropholic substances could turn out to be quite persistent in the environment, but still, toxic effects would be seen only in a limited area owing to advection-dispersion in the receiving waters. Often, hydrophilic substances are having quite pronounced acute toxic effects on biotic systems, owing to a high bioavailability. Slowly biodegradable hydrophobic substances have the propensity of being both persistent and bioaccumulative. Biomagnification and human exposure to the substances may thus occur, and long-term ecological effects will be possible. A solid phase substance has the propensity to geoaccumulate, thus possibly creating anaerobic conditions in sedimentation areas and, also, possibly releasing hazardous low molecular mass substances on degradation[14].

Transplantation or cage experiments have frequently been used for *in situ* studies. Fish or mussels have been put in cages, and growth rate, mortality, bioconcentration of discharged substances, organoleptic tests of taste and smell, or *in situ*-preference/avoidance studies have been performed[1,16].

Ecoepidemiological studies are a retrospective method, aiming at establishing a connection between an effluent and the health of a receiving ecosystem. The impact of one or more effluents after their interaction with various surrounding factors may be studied through community composition and diversity of communities of organisms, and contents of specific substances in seawater, biota and sediments. Evaluation of such obtained results is quite complicated, and today the nature of these complex interactions is not fully understood[16], e.g.[17,18]. Several methods are available for describing statistically a community of organisms, e.g. by indices and diagrams as the Jaccard and Czekanowski methods, cf.[2,16].

5. SUMMARY OF RISK ESTIMATION.

Risk estimation of industrial effluents by evaluating the environmental consequences of the situation concerning the discharge of substances may be summarized in a flow scheme, table I.

The experimental program of risk estimation has hardly been standardized, but is rather suited for the purpose. Generally, it is of major importance to differenciate between fresh-waters, brakish waters, and marine waters when extrapolating from hazard assessment data. Possibly, the modelling of environmental fate, impact and effects on biotic systems of the receiving waters by discharged substances could, in most instances, be evaluated by

- identifying primary, affected areas of the receiving waters where biological effects can be directly recognized owing to the discharge, *e.g.* by biological survey,
- identifying secondary, affected areas where life conditions are expected to have deteriorated owing to the discharge, *e.g.* by extrapolations of laboratory experiments using advection-dispersion modelling,
- identifying areas of the receiving waters exposed to substances of the effluent, *e.g.* by *in situ* chemical determinations in biota and sediment, and/or fate modelling of effluent substances.

Table I. Flow scheme of risk estimation of industrial effluents

1. Evaluation of available information usable for extrapolation of fate and effect of discharged substances directly exposing ecosystems of the receiving waters,
   - hazard assessment report,
   - geology, hydraulogy, and ecology of the receiving waters,
   - possible impact from other sources.

2. Planning and execution of risk estimation experimental program,
   - predicting primo and ultimate fate of discharged substances,
     - mathematical modelling the fate on the basis of hazard assessment data,
     - *in situ* determination of substances,
       - ambient water,
       - sediment,
       - biota,

   - predicting biological impact,
     - mathematical modelling the effects on the basis of hazard assessment data,
     - ecoepidemiological surveyance of benthic and pelagic flora and fauna,

   - *in situ* ecochemicological and ecotoxicological experiments on transplanted species,
     - bioaccumulation,
     - growth rate,
     - mortality,
     - histopathalogical effects.

3. Preparation of a risk estimation report evaluating
   - experimental program,
   - modelling the fate and impact of discharged substances,
   - needs for further risk estimation program,
   - environmental consequences.

Unfortunately, it may in many instances turn out, that our scientific knowledge on functioning and vulnerability of natural ecosystems is too poor to predict long-term, environmental consequences of many industrial effluents. Consequently, it has often been seen, that politicians are making decisions based on an inclomplete, scientific knowledge.

6. ACKNOWLEDGEMENTS.

The author gratefully acknowledges many valuable discussions and inputs from friends and colleagues, particularly Palle Lindgaard-Jørgensen and Kim Christiansen. This work is part of a project study in progress conducted for the Commission of the European Communities, DG XII.

REFERENCES

1. Folke, J., Toxicol. Environ. Chem., **10**(3), 201-224 (1985).
2. Folke, J., Environmental Impact Assessment for Marine Ecosystems of Wheat and Rye Straw Pulp Bleaching and Combined Mill Effluents, (Chalmers University of Technology & University of Gothenburg, 1985).
3. Folke, J., Behaviour and Transformation of Organic Micropollutants in Water Treatment Processes (J. Folke & J, Rivera Aranda (eds.), Concerted action, COST 641, WP4, OMP/57/85, Commission of the European Communities, XII/ENV/7/85, Bruxelles, 1985), pp1-9.
4. Folke, J., The Role of the Determinations of Physico-Chemical Behaviour and Transformation Reactions of Organic Micro-pollutants in Environmental Hazard Assessment of Industrial Effluents (W. Giger & J. Zeyer (eds.), Concerted Action, Cost 641, WP3, OMP/59/85, Commission of the European Communities, XII/ENV/37/85, Bruxelles 1985, pp 13-26.
5. Folke, J., Hazard Identification of Industrial Effluents (Workshop organized within the framework of the Concerted Action, Cost 641, Working Party 4, held in Copenhagen, J. Folke (ed.), Denmark, 20-22 May, 1985), proceedings in press.
6. Kates, R.W., Risk Assessment of Environmental Hazard. SCOPE 8. (John Wiley & Sons, New York, 1978).
7. Schmidt-Bleek, F., A.W. Klein & W. Haberland, Chemicals in the Environment. Chemicals Testing and Hazard Ranking - the Interaction between Science and Administration (Christiansen, Kock & Bro-Rasmussen, Symposium proceedings, 18-20 October 1982, Technical University of Denmark, 1983), pp 11-48.
8. Kringstad. K.P. & L. Strömberg, Environmentally Harmonized Production of Bleached Pulp. Final Report (in Swedish). (Swedish Forest Product Research Laboratory, Stockholm, 1982).
9. Södergren, A., P. Jonsson, K. Kringstad, S. Lagergren & M. Olsson, Research program for the field environment/cellulose: biological effects of forest industry discharges (in Swedish). (National Swedish Environment Protection Board, Bulletin No. PM 1673, Stockholm, 1983)
10. O'Conner, D.J. & J.P. St.John, Modeling the Fate of Chemicals In the Aquatic Environment (K.L. Dickson, A.W. Maki & J. Cairns, Jr. (eds.). Ann Arbor Science, Ann Arbor, 1982), pp 13-34.
11. Klein, A.W. & F. Schmidt-Bleek, Modeling the Fate of Chemicals In the Aquatic Environment (K.L. Dickson, A.W. Maki & J. Cairns, Jr. (eds.). Ann Arbor Science, Ann Arbor, 1982), pp 73-92.
12. Burns, L.A., Modeling the Fate of Chemicals In the Aquatic Environment (K.L. Dickson, A.W. Maki & J. Cairns, Jr. (eds.). Ann Arbor Science, Ann Arbor, 1982), pp 101-126
13. Bergman, H.L., & J.S. Meyer, Modeling the Fate of Chemicals In the Aquatic Environment (K.L. Dickson, A.W. Maki & J. Cairns, Jr. (eds.). Ann Arbor Science, Ann Arbor, 1982), pp 247-267.
14. Landner, L., Ecotoxicological Testing for the Marine Environment (Persoone, Jaspers & Claus, State University of Ghent and Institute of Marine Scientific Research, Bredane, Belgium, Vol. 1, 1984), pp 657-687.

15. Di Toro, D.M., D.J. O'Conner, R.V. Thomann & J.P. St. John, Modeling the Fate of Chemicals In the Aquatic Environment (K.L. Dickson, A.W. Maki & J. Cairns, Jr. (eds.). Ann Arbor Science, Ann Arbor, 1982), pp 165-190.
16. Lindeström, L., H. Hultberg, K. Beijer & L. Landner, Traditional and new methods for investigation of impact in aquatic ecosystems by effluents from pulp and paper mills (IVL-B602, Swedish Water and Air Pollution Research Institute, Stockholm, 1981).
17. Folke, J. J. Birklund, A. K. Sørensen & U. Lund, Chemosphere, **12**,1169-1181 (1983).
18. Folke, J. & J. Birklund, Toxicol. Environ. Chem.,**10**, 41-65 (1985).

Das NTA-Problem

S. Kanowski

Umweltbundesamt D 1000 Berlin 33

Zusammenfassung

Die dringende Notwendigkeit, den Phosphoreintrag
in von Eutrophie bedrohte Gewässer zu verringern und die darauf zurückzuführenden gesetzlichen Beschränkungen haben die Industrie veranlaßt,
Phosphatsubstitute zu entwickeln. NTA erwies sich
als die am besten geeignete Substanz. Um als Phos-
phatersatzstoff geeignet zu sein, ist eine gute
Gebrauchstauglichkeit nicht hinreichend; die Substanz darf nach ihrem Gebrauch keine Schäden in
der aquatischen Umwelt hervorrufen. Das Verhalten
und Schicksal von NTA in der Umwelt, insbesondere die Abbaufähigkeit in biologischen Kläranlagen,
der Einfluß auf die Eliminierbarkeit von Schwermetallen in Kläranlagen und die Remobilisierung von
Schwermetallen aus Gewässersedimenten wurden inten-
siv erforscht. Es ist nachgewiesen, daß NTA nach
Adaption normalerweise biologisch gut abgebaut wird,
daß der Abbau aber schwankt in Abhängigkeit vom An-
lagentyp, der Wassertemperatur, der NTA-Konzentration,
der Wasserhärte und den Betriebsparametern. Falls NTA
uneingeschränkt eingesetzt wird, können sich in den
Oberflächengewässern Konzentrationen einstellen, die
zur Remobilisierung von Schwermetallen und zur Ver-
unreinigung von Trinkwasser führen. Dies trifft insbesondere für stark mit Abwasser belastete Gewässer
zu. Aus diesem Grund sollte der Einsatz von NTA als
Gerüststoff in Waschmitteln entsprechend den örtli-
chen Gegebenheiten begrenzt und das Verhalten der NTA
in der Umwelt sorfältig beobachtet werden.

Summary

Because of the evident need to reduce the input of Phosphorus to
surface waters and the resultant restrictions on the use of phosphates as builders in detergent formulations, the detergent
industry turned its attention to substitutes; NTA was considered
the best substance. To be acceptable as a substitute builder the
substance should have an adequate performance and no
undesirable effects should be produced in the environment when
the substances are used in large quantities associated with the
present scale of detergent use. The behaviour and fate of NTA
in the environment with particular reference to the removal of

NTA during waste water treatment and the effects of NTA on heavy metal solubility both during treatment and in the receiving waters have been widely studied. It is concluded that NTA removal during secondary biological treatment is after adaption normally high but is subject to considerable variation both temporal and between works as a result of changes in NTA load, temperature, water hardness and treatment process parameter. If NTA is excessively used the resulting concentrations in surface waters may be sufficient to remobilise heavy metals and contaminate potable waters particularly in areas of low effluent dilution and high water re-use. Therefore the use of NTA as a detergent builder should be limited and the behaviour in the environment carefully monitored.

1. Einleitung

Nitrilitriessigsäure (Nitrilotriacetat, NTA, $N(CH_2CO_2H)_3$ ist eine Aminopolycarbonsäure, die unter Namen wie Trilon A, Dissolvine A, Versene oder Komplexon vermarktet wird. Im analytischen Zweig der Komplexometrie wird sie seit vielen Jahrzehnten zur Metallionentitration benutzt. Seit längerer Zeit wird sie neben der verwandten Ethylendiamintetraessigsäure (EDTA) u. a. zur Wasserenthärtung, zur Maskierung von Schwermetallionen in der Galvanotechnik und zur äußeren Reinigung des Rohrsystems von großen Kesselanlagen verwendet. Etwa 1970 gewann der Einsatz von NTA eine neue Dimension, als große Textilwaschmittelhersteller in Nordamerika begannen, NTA als Ersatzstoff für Natriumtripolyphosphat (NTPP,STP) einzusetzen. Es war und ist zu erwarten, daß eine schärfere Gesetzgebung gegen den Einsatz von Phosphaten in Wasch- und Reinigungsmitteln zu einer großen Zunahme des NTA-Einsatzes führen wird, sofern keine gesetzlichen Regelungen erfolgen. Deshalb ist das NTA-Problem eigentlich ein Phosphor-Problem.

2. Das Phosphor-Problem (1)

2.1 Phosphateinträge in Gewässer und Maßnahmen zu ihrer Reduzierung

Die starke Eutrophierung von Seen, Talsperren und staugeregelten Flüssen und die damit einhergehende Verschlechterung der Gewässerqualität hat in vielen Ländern Anlaß zur Erforschung der Ursachen und für Maßnahmen zu ihrer Eindämmung gegeben. Sorge bereitet, daß inzwischen Eutrophierungserscheinungen auch in der Nordsee beobachtet werden. Als Ursache ist in den meisten Fällen der zivilisatorisch bedingte große Eintrag von Phosphaten in die Gewässer auszumachen (2). Bild 1 zeigt die dramatische Entwicklung der Phosphorkonzentration im Bodensee in den letzten 40 Jahren. Die Zunahme der Phosphorkonzentration ist insbesondere bedingt
- durch die Kanalisierung der Einzugsgebiete, aber auch durch
- die Einführung der Phosphate in die Textilwaschmittel,
- erhöhte Einträge aus der landwirtschaftlichen Düngung (Erosion)
- erhöhte Einträge aus der Massentierhaltung (Jauche, Gülle),
- atmosphärischem Eintrag von Staub aus der Verbrennung fossiler Brennstoffe.

Eine hinreichende Verringerung des Phosphateintrags in durch Eutrophierung gefährdete Gewässer ist i.a. nur durch Maßnahmen an __allen__ Quellen erreichbar. Beispielsweise verringert die wichtige Maßnahme der Phosphatfällung in kommunalen Kläranlagen nicht die Phosphate, die bei starken Regenfällen mit dem Abwasser über die Kanalüberläufe direkt in die Gewässer gelangen (Mischkanalisation). Ebensowenig wird dem Abwasser Phosphat entzogen, das nicht an eine Kläranlage angeschlossen ist. Der bisher weniger beachtete Eintrag aus der Luft ist durch die Diskussion über die Gewässerversauerung durch den sauren Regen bekannter geworden und als Problem erkannt worden. Vom japanischen Biwa-See wird vermutet, daß nach weitgehender Sanierung des Einzugsgebietes nunmehr über 50 % des Phosphats aus der Atmosphäre eingetragen wird und dies den See nach wie vor bedroht.

Eine gesetzliche Begrenzung bzw. ein Verbot des Phosphatgehalts in Wasch- und Reinigungsmitteln ist eine in vielen Ländern praktizierte Maßnahme. Sie hat den Vorteil, an stark mit Abwasser belasteten Gewässern relativ schnell und auch relativ billig den Eintrag von Phosphat deutlich zu verringern.

2.2 Phosphate und Waschmittel

Textilwaschmittel, die für alle Temperaturen geeignet sind (Vollwaschmittel, Alltemperaturwaschmittel) bestehen aus 10-15 Produkten bzw. Produktgruppen, von denen die folgenden vier die Hauptmenge stellen:

		Masse-%
Gerüststoffe (builder)	(z.B. Phosphate)	20
Waschaktive Substanzen	(anion. und nichtion. Tenside)	15
Bleichmittel	(Perborat)	25
Alkalien	(Soda, Silikat)	15

Das Waschen von Textilien besteht im wesentlichen aus drei Vorgängen,
- der Enthärtung des Waschwassers, d.h. der Entfernung der Calciumionen durch die alternativen Vorgänge Ausfällung, Komplexierung, Ionenaustausch bzw. Dispergierung des Calciumcarbonats,
- Schmutzablösung und Dispergierung (primäre Waschwirkung)
- Verhinderung der Wiederverschmutzung aus der Waschflotte, Verhinderung von Ablagerungen auf Textilien und Waschmaschine (sekundäre Waschwirkung).

Die Gerüststoffe haben an allen drei Vorgängen entscheidenden Anteil und das Natriumtripolyphosphat ($Na_5P_3O_{10}$) hat sich waschtechnisch als besonders vorteilhaft erwiesen.

Die mit den angekündigten oder vollzogenen Phosphatbegrenzungen einhergehende Suche nach Phosphatersatzstoffen hat in den letzten 20 Jahren einen großen Umfang angenommen. Im deutschen Patentamt sind in den letzten 20 Jahren über 200 Patente angemeldet worden. Ein allseits (waschtechnisch und ökologisch) befriedigender Ersatzstoff konnte bisher nicht gefunden werden. Dies ist die Ursache dafür, daß in vielen Ländern zwar Phosphatbegrenzungen bestehen, aber nur in wenigen Fällen der Schritt zum Phosphatverbot beschritten worden ist. Tabelle I gibt einen Überblick über Phosphatbegrenzungen.

Tabelle I:

Länder mit Begrenzungen des Phosphatgehalts in Waschmitteln

	% P im Waschmittel	Zeitpunkt
D	5	1.1.1984
CH	0,5	1.7.1986
NL	6	1983
AU	7	1.1.1985
	5	1.1.1987
USA*	0,5	
Can	2,2	1983
Nor	3	1.1.1986
Swe		
Finnland		
Italien		
Japan	z.T. 0	

* Great Lakes Staaten

In der Bundesrepublik Deutschland wurde auf der Grundlage des Waschmittelgesetzes verordnet, daß der Phosphatgehalt* in Textilwaschmitteln ab 1.1.1984 nur noch ca. 5 Masseprozent betragen darf. Dies entspricht einer Abnahme des Gehalts um ca. 50 % im Vergleich mit 1975. Diese Maßnahme hat dazu geführt, daß der durch Wasch- und Reinigungsmittel verursachte Phosphateintrag in die Gewässer von ca. 40 % auf ca. 25 % abgenommen hat. Eine weitere Reduzierung des Phosphatgehalts in Waschmitteln auf gesetzlichem Wege ist z.Z. nicht vorgesehen.

2.3 Phosphatersatzstoffe

Unter den Phosphatersatzstoffen sind insbesondere das Zeolith A (Kapazität 250 000 jato) und das NTA (westliche Welt ca. 100 000 jato) daneben in geringerem Umfang das Citrat zu nennen. In jüngster Zeit wurden Produkte entwickelt, die aufgrund eines ausgeprägten threshold-Effects und guter Dispergierwirkung in geringen Konzentrationen (0-1 bzw. 0-3 %) einen entscheidenden Beitrag zum Wascherfolg liefern können.

Als threshold-Effekt bezeichnet man die Eigenschaft einer Substanz, im unterstöchiometrischen Einsatz die Ausfällung von schwerlöslichen Erdalkalisalzen zu verhindern bzw. zu verzögern. Der Verzögerungseffekt gilt als ausreichend, da eine Verhinderung von Ablagerungen auf Gewebe und/oder Heizstab (Inkrustierung) nur für den Zeitraum des Wasch- und Spülvorgangs erforderlich ist. Bei den Verbindungen handelt es sich um Copolymere verschiedener Polycarbonsäuren und um organische Phosphonate.

Unter den geprüften Ersatzstoffen hat sich das NTA als besonders gebrauchstauglich und vollwertiger Ersatzstoff erwiesen. Im Unterschied zum STP hat es außerdem den Vorteil hydrolysebeständig zu sein. Vom STP im Waschmittel werden bei der Waschpulverproduktion im Sprühturm, durch Lagerung und während des Waschvorganes i.a. über 20 % hydrolysiert und damit inaktiv. Neben NTA ist vor allem das Zeolith A, ein ökologisch völlig unbedenklicher Stoff, von Interesse, da es in größerem Umfang eingesetzt wird. Zeolith enthärtet das Waschwasser durch Ionenaustausch. Da dieser Vorgang langsamer erfolgt als die üblichen Io-

nenreaktionen und daher Carbonatausfall nicht völlig vermieden
werden kann, erfordert die Benutzung von Zeolith die parallele
Verwendung einer geringen Menge eines weiteren Gerüststoffes
bzw. eines Additivs mit threshold effekt.

Citrat erfüllt die Anforderungen eines Gerüststoffes auch
in Kombination mit anderen nur unzulänglich. Phosphonate haben
die ökologisch unangenehme Eigenschaft, biologisch nicht abbaubar zu sein. Polycarboxylate sind ebenfalls nicht abbaubar, werden aber bei der biologischen Abwasserreinigung zum größten Teil
mit dem Klärschlamm entfernt. In welchem Umfang sie als Wasch-
mittelbestandteil akzeptabel sind, bedarf der näheren Untersuchung.

Nachfolgende Tabelle gibt einen Überblick über die Änderun-
gen der Zusammensetzung von Vollwaschmitteln in der Bundesrepublik in den letzten 10 Jahren unter dem Einfluß der Phosphatge-
setzgebung.

---

* Begrenzt ist der Gehalt an Phosphor pro Liter Waschflotte in
Abhängigkeit von der Wasserhärte

Tabelle II:

Änderung der Waschmittelzusammensetzung in den letzten Jahren*

	1975 von/bis	1984 von/bis
Waschaktive Substanzen		
Anionische Tenside	4/10	7/10
Nichtionische Tenside	2/6	3/7
Seife	(1/6)	2/4
Gerüststoffe (Builder)		
Phosphat (STP)	27/40	18/26
NTA	-	2/5 (7)
Zeolith	-	10/23
Natriumsilikat	3/7	4/8
Additive		
Polycarboxylat	-	1/3
Phosphonat	-	2
EDTA	1	2

Außerdem: Perborat, Bleichmittelaktivator, Bleichmittelstabili-
sator, Soda, Magneiumsilikat, Carboxymethylcellulose, Natrium-
sulfat, optische Aufheller, Enzyme, Parfüm, kationische Tenside, Wasser   *Mittelwerte deutscher Vollwaschmittel

3. NTA als Phosphatersatzstoff

3.1 Allgemeines

Die bedeutendste chemische Eigenschaft der NTA ist ihre Fähig-
keit mit vielen mehrwertigen Metallionen wasserlösliche Komplexe zu bilden (Bild 2). Es bilden sich Chelatkomplexe hoher Sta-
bilität, im wesentlichen im stöchiometrischen Verhältnis NTA/
Metallion von 1:1. Diese Komplexe entstehen in Konkurrenz zu
schwerlöslichen Niederschlägen der Metallionen mit Anionen wie
Hydroxid, Carbonat oder Sulfat. Bild 3 zeigt Stabilitätskonstanten einiger Metall-Gerüststoff-Komplexe. Die Konstanten für
NTA liegen deutlich über den für Natriumtripolyphosphat.

Herstellung von NTA:

NTA wird durch Cyanmethylierung synthetisiert (7)

$$3\ CH_2O + 3\ NaCN + 3\ H_2O \xrightarrow{100^\circ\ C,\ 20\ bar} N\ (CH_2CO_2Na)_3 + 2\ NH_3$$

Nebenprodukte: Glykokoll, Iminodiessigsäure, Hexamethylen-
tetramin, Glykolsäure

Physikalische Eigenschaften:

Molekulargewicht H_3NTA = 191; Na_3 NTA = 257
Löslichkeit bei 20° C 640 g/l Wasser

3.2 Ökologische Eigenschaften von NTA

3.2.1 Der Abbau von NTA und ihr Verhalten in Kläranlagen (8)

Das in Wasch- und Reinigungsmitteln eingesetzte NTA gelangt
vollständig in das Abwasser. Der Verbleib der NTA ist in ent-
scheidendem Maße vom Abbau- bzw. Eliminationsverhalten in der
wäßrigen Phase abhängig, und damit vor allem vom Vorhandensein,
der Art und Effizienz einer Abwasserreinigungsanlage.
      NTA und die Metall-NTA-Komplexe sind grundsätzlich unter
aeroben Bedingungen biologisch abbaubar. Abbaugeschwindigkeit
und Abbaugrad hängen jedoch stark von den Milieufaktoren ab.
Voraussetzung für den Abbau ist eine ausgeprägte Adaptation der
Mikroorganismen, wofür in Kläranlagen erfahrungsgemäß 10 - 100
Tage erforderlich sind. Der Abbauweg wird durch folgendes Sche-
ma beschrieben; es findet keine Anreicherung von Zwischenpro-
dukten statt (4,9).

Verlauf des aeroben Abbaus von NTA

Der Abbau ist bei höheren NTA-Konzentrationen (≥ 5 mg/l)
konzentrationsunabhängig (Zeitgesetz 0.Ordnung) (10). Solche
Konzentrationen sind realistisch. Vergleichsweise wurden in ca-
nadischen Kläranlagenzuflüssen bis zu 20 mg/l NTA gemessen; unter deutschen Verhältnissen wäre bei NTA als einzigem Gerüststoff in Waschmitteln mit 10-34 mg/l zu rechnen. Bei niedrigen
Konzentrationen (≤ 2 mg/l) kann der Abbau durch ein Zeitgesetz
1. Ordnung beschrieben werden.
      In Abwesenheit abbauhemmender Stoffe kann davon ausgegan-
gen werden, daß bei Temperaturen des Rohabwassers von 15° C NTA
im Mittel zu 80 % und bei niedrigen Temperaturen um 5° C zu etwa 50 % abgebaut wird.
      Der Abbau der NTA-Alkali- und Erdalkalimetall-Komplexe so-
wie der NTA-Komplexe von Aluminium, dreiwertigem Eisen, Chrom,
Mangan, Blei und Kobalt verläuft etwa gleich schnell wie der
von NTA selbst. Als schwer abbaubar gelten die NTA-Komplexe von
Nickel und Quecksilber, z.T. auch die von Cadmium, Kupfer und
Zink.

Höhere Gehalte von Erdalkalien fördern den biologischen Abbau von schwerabbaubaren NTA-Schwermetallkomplexen durch Umkomplexierung.

Der Abbau von NTA ist deutlicheren Streuungen unterworfen als der Abbau der durch den Summenparameter BSB_5 erfaßten organischen Abwasserinhaltsstoffe. Die Eliminierung von Zink im mechanisch-biologischen Klärprozeß wird durch NTA signifikant herabgesetzt. Dies wird in mehreren Arbeiten auch für Blei und Nickel vermutet. Dieses Ergebnis ist vor allem im Zusammenhang mit der vorstehend berichteten Tatsache der größeren Störanfälligkeit des NTA-Abbaus von Bedeutung. Falls der NTA-Abbau für mehrere Stunden unterbrochen wird, könnte die üblicherweise erfolgende 50-70 %ige Adsorption der Schwermetalle an den Klärschlamm in dieser Zeitspanne aufgehoben werden. Dieses Problem gehört zu denen, die in der deutschen NTA-Studie unter den "Offenen Fragen" aufgelistet ist (siehe Anhang).

Die Fällung von Phosphaten mit Eisen- und Aluminiumsalzen im Rahmen der weitergehenden chemischen Abwasserbehandlung wird in bisher nicht näher ermitteltem Ausmaß gestört. Dies kann von Bedeutung sein, falls eine Fällung auf Rest-Phosphorgehalte wesentlich unter 1 mg/l vorgenommen werden soll.

Der anaerobe Abbau von NTA in Faultürmen findet nach einer Adaptationsphase von 1 bis 2 Wochen statt, sofern ca. 30 % Schlamm zugeführt wird, der unter aeroben Bedingungen an NTA adaptiert worden ist. Bei der anaeroben Abwasserreinigung in Faulgruben, Faulteichen und geschlossenen Fermentern wird kein NTA-Abbau beobachtet.

Mittlere und hohe Salzkonzentrationen (über 15 g/l) im Wasser hemmmen bzw. unterbinden den biologischen Abbau. Im Ästuarbereich konnten keine Bakterien gefunden werden, die NTA dort abbauen, wohl aber Bakterien, die unter Süßwasserbedingungen NTA abbauen. Offenbar ist unter den salinen Bedingungen der Metabolismus inhibiert.

3.2.2 NTA-Komplexe im Gewässer (11)

NTA ist unter Gewässerbedingungen zu fast 100 % an Kationen in der Lösungsphase gebunden. Dabei werden aufgrund der Größe der Stabilitätskonstanten zunächst überwiegend Schwermetall-NTA-Komplexe gebildet, bis diese Kationen durch Bindung an NTA "verbraucht" sind (12). Für die in Bild 4 aufgeführte Rheinwasseranalyse zeigt Bild 5 die Chemische Spezifikation von NTA im Konzentrationsbereich von 10-10.000 µg/l NTA aufgrund komplexchemischer Berechnungen von Hennes (13). Im unteren Konzentrationsbereich liegt NTA hauptsächlich als Kupfer-, Nickel- und Zink-Komplex vor, im oberen mangels Schwermetalle als Calcium-Komplex.

Bild 6 zeigt, daß sich die Spezifikation des Kupfers im Rheinwasser bei 100 µg/l NTA mit steigendem pH-Wert nur unwesentlich ändert. Die Berechnungen von Hennes wurden unter Berücksichtigung von Eisenhydroxid durchgeführt. Trotz der großen Stabilitätskonstante des Fe III-NTA Komplexes wird die chemische Spezifikation des NTA durch Fe nur unwesentlich beeinflußt, da aufgrund des Löslichkeitsprodukts von $Fe(OH)_3$ weit über 99 % der Gesamteisenmenge (hier 1,3 mg/l) ungelöst vorliegen. Bild 7 zeigt den Anteil von gelöstem Fe(III) als Funktion der NTA-Konzentration.

Wenn NTA in ein Gewässer gelangt, kann es zu einer Remobilisierung von Schwermetallen kommen, die im Sediment festgelegt

sind. Das Ausmaß der Remobilisierung hängt sowohl von der Lage
der chemischen Gleichgewichte als auch von den kinetischen Fak-
toren (Kontaktzeit und Mischungsintensität) ab. Laboruntersu-
chungen mit Sediment und Schwebstoffen haben gezeigt, daß aus
Hamburger Hafenschlamm bereits bei 0,1 mg/l H_3NTA Kupfer und
Nickel in meßbarem Umfang remobilisiert wurden und aus Neckar-
sediment Nickel oberhalb von 0,1 mg/l H_3NTA. Dietz führte am
Ruhrsediment mit 1 mg/l NTA Remobilisierungsversuche durch.
Bild 8 zeigt, daß nach 45 h NTA komplett als Zn-, Ni- und Cu-
Komplex vorlag (14).

Das Komplexierungsvermögen von NTA für Schwermetalle ka n
neben der Remobilisierung auch eine Abnahme der Sorption zur
Folge haben, d.h. eine Verminderung der Schwermetallentfernung
aus dem Wasserkörper durch Bindung und Ablagerung im Sediment.
In Laborversuchen wurden solche Effekte ab 0,2 mg/l H_3NTA für
Nickel, Cadmium und Zink beobachtet.

Das Problem der Remobilierung von Schwermetallen ist im Zusammenhang mit der Trinkwassergewinnung durch Uferfiltration und
künstlicher Grundwasseranreicherung von Bedeutung. Einerseits
zeigten Modellversuche, daß ab einer Konzentration von 1 mg/l
NTA Schwermetalle aus Filtermaterial mobilisiert werden, andererseits zeigen niederländische Erfahrungen, daß 1-2 mg/l NTA
die Schwermetallbilanz nicht verändern und das NTA nach 6-8 Wochen Adaption im Boden vollständig biologisch abgebaut wird.

3.2.3 Wirkungen von NTA auf die Gewässerbiozönose (15)

Bei den denkbaren NTA-Konzentrationen von bis zu 1 mg/l in Oberflächengewässern sind toxische Schädigungen von Einzelorganis-
men und Biozönosen praktisch auszuschließen. Eine Akkumulation
von Schwermetallen in aquatischen Organismen erscheint möglich,
da sie vom Angebot an Schwermetallen im Gewässer abhängt, das
von der NTA-Konzentration im Gewässer beeinflußt werden kann.
Es wurden sowohl durch Förderungen als auch Hemmungen des Algenwachstums beobachtet. Eine Stimulierung des Algenwachstums könnte sowohl durch Aufhebung der Hemmung toxischer Schwermetalle,
durch Chelatisierung als auch durch Erhöhung der Verfügbarkeit
essentieller Spurenelemente erfolgen. Beispielsweise könnte die
Blaualgenbildung im Gewässer zunehmen, da diese Art besonders
empfindlich auf Kupfer reagiert. In der Literatur wird auch ein
möglicher Einfluß des Eintrags von Stickstoff durch NTA auf
stickstofflimitierte, phosphorreiche Seen diskutiert.

Wegen der ungelösten Fragen in diesem Zusammenhang wird
der NTA-Einsatz an verschiedenen Stellen durch Monitoringpro-
gramme begleitet.

3.2.4 Verhalten von NTA bei der Trinkwassergewinnung (16)

In welchem Umfang NTA bei der Trinkwassergewinnung aus
Oberflächenwasser entfernt wird, hängt in hohem Maße von der
Art der Aufbereitung ab.

Filtration und Sedimentation haben praktisch keinen Elimi-
nierungsgrad für NTA-Komplexe. Flockung und Fällung mit Filtration erreichen eine 10-20 %ige Eliminierung, wobei der Nickel-
NTA-Komplex nicht entfernt wird. Andererseits werden Flockung
und Fällung mit Eisen- und Aluminiumsalzen durch NTA oberhalb
von 0,5 mg/l H_3NTA empfindlich gestört.

H_3NTA wird relativ gut an Aktivkohle adsorbiert; dies gilt
nicht für die Erdalkali- und Schwermetall-Komplexe. Da NTA nur

komplexiert vorliegt, ist keine wesentliche Adsorption zu erwarten. Andererseits kann ein gewisser NTA-Abbau in der Aktivkohle stattfinden.

Durch Ozon werden bei pH 7 die NTA-Komplexe gleich gut abgebaut wie Phenol. Huminsäuren werden bei pH 7 schneller oxidiert, so daß bei Ozonmangel mit dem Passieren von NTA gerechnet werden muß. Unter deutschen Verhältnissen (Ozoneinsatz von 1-2 mg/l) kann mit einer NTA-Eliminierung von 50-70 % gerechnet werden.

Durch Chlor und Chlordioxid wird NTA unter Wasserwerksbedingungen nicht eliminiert.

Durch Bodenpassage mit Verweilzeiten von über 20 Tagen kann eine Eliminierung von 90 % erreicht werden.

Aus diesen Ergebnissen ist zu schlußfolgern, daß NTA durch ausgefeilte Aufbereitungsverfahren nahezu vollständig, durch einfache Aufbereitungsverfahren nur in unbedeutendem Ausmaß entfernt wird (Bild 9).

3.3 Toxikologische Eigenschaften von NTA

Ein 1970 bekanntgewordenes Ergebnis, daß NTA die teratogene Eigenschaft von Methylquecksilber erhöht, wurde in späteren Untersuchungen nicht bestätigt. Allerdings war das erstgenannte Ergebnis in den USA Anlaß, auf den NTA-Einsatz in Waschmitteln für längere Zeit zu verzichten.

NTA hat nur eine geringe akute und chronische Toxizität, die für die im Trinkwasser zu erwartenden Konzentrationen ohne Bedeutung ist. Es ist nicht mutagen und nicht teratogen. Bei hohen Konzentrationen verursacht NTA Tumorbildung in den Nieren und im Harntrakt von Ratten und Mäusen. Eine konservative Risikoabschätzung der EPA gelangt zu einer durch NTA im Trinkwasser verursachbare Krebshäufigkeit der Bevölkerung von 1:500 000 (17,4).

Neuerdings wird über eine tumorpromovierende Wirkung der NTA berichtet. Nach gemeinsamer Verabreichung eines Nitrosamins mit großen Mengen NTA an Ratten stieg die Nierentumorrate je nach NTA-Menge von 33 bis auf 100 % (18-20). Untersuchungen über den Einfluß von größeren NTA-Aufnahmen auf das Verhalten von Schwermetallen und physiologisch wichtiger Kationen bei Ratten führten zu keinen einheitlichen Ergebnissen (21). Neben Einflüssen auf die Resorption eissentieller Spurenelemente bzw. Mineralstoffe sind auch für Schwermetalle veränderte Aufnahmebedingungen durch NTA-Zufuhr bei Mensch und Tier denkbar. Zusammenfassende Darstellungen liegen in (22,23) vor.

3.4 Umweltanalytik von NTA

Zur Bestimmung von NTA im µg-Konzentrationsbereich, wie er in Oberflächengewässern und im Trinkwasser vorliegt, ist die Gaschromatographie geeignet.

In der Bundesrepublik ist eine Norm nach DIN/DEV in Vorbereitung (24).

Prinzip:

Das Nitrilotriacetat wird aus den mit Formaldehyd konservierten Wasserproben auf einem Anionenaustauscher angereichert. Die Elution erfolgt mit konzentrierter Ameisensäure, so daß eine optimale Wiederfindungsrate des NTA gewährleistet ist. Nach

Überführung in den n-Tributylester und Extraktion mit n-Hexan
wird die gaschromatographische Bestimmung mit Hilfe eines stick-
stoffselektiven Detektors durchgeführt. Als innerer Standard
dient Hepta- und/oder Oxtadecansäurenitril. Zur quantitativen
Auswertung verwendet man Bezugskurven, die mit Lösungen anstei-
gender NTA-Konzentrationen und konstanter Konzentration des in-
neren Standards hergestellt werden.

Das Verfahren ist für den Konzentrationsbereich 1 µg bis
200 µg H_3NTA pro Liter Wasser bei Anwendung von 50 ml Probe geeignet.

Für höhere Konzentrationen von 1 bis 10 mg/l, wie sie im
Abwasser auftreten, ist die polarographische Methode zweckmäßig.
Hierbei wird die im Wasser enthaltene NTA durch Zugabe von Bis-
mutionen in ihren Bismut-Komplex überführt, der polarographisch
bestimmt wird. Eine DIN/DEV-Norm ist in Vorbereitung.

3.5 Gesetzliche Regelungen und Umfang des NTA-Einsatzes

NTA ist zur Zeit in keinem Land verboten; folgende Regelungen sind bekannt:

Bundesrepublik Deutschland:

Im Ergebnis der umfangreichen Literatur- und Fall-Studie zu NTA
hat die Fachgruppe Wasserchemie in der Gesellschaft Deutscher
Chemiker der Bundesregierung 1983 empfohlen, den NTA-Einsatz zu
bogrenzen (siehe Anhang). Der NTA-Gehalt in Waschmitteln sollte
3,4 Masse-% im Mittel nicht überschreiten, und in Oberflächengewässern sollten nicht mehr als 200 µg/l als 99-Perzentil erreicht werden. Außerdem sollte der Einsatz von NTA von einem
großflächigen Monitoringprogramm und einem umfangreichen Forschungsprogramm begleitet werden. Monitoring und Forschung werden insbesondere empfohlen, um sicherzustellen, daß der NTA-
Einsatz nicht zu einer Remobilisierung von Schwermetallen aus
den hochbelasteten Gewässersedimenten führt.

USA:

NTA wurde Ende der sechziger Jahre in einigen Waschmitteln verwendet. Nach Auftauchen des Verdachts im Jahre 1970, NTA ver-
stärke die teratogene Kapazität von Methylquecksilber, hat die
Industrie auf Bitten des Surgeon General auf NTA verzichtet.
Nach Wegfallen dieses Grundes, führte die Feststellung eines
krebserzeugenden Potentials Mitte der siebziger Jahre zum Fort-
bestehen des Verzichts. Nachdem die EPA 1980 die Bedenken, NTA
stelle ein gesundheitliches Risiko dar, weitgehend zurückge-
stellt hat, wird in Indiana (totales Phosphatverbot) NTA eingesetzt. Der Wunsch der Waschmittelindustrie, NTA auch in anderen
Bundesstaaten in Waschmitteln zu verwenden, hat im Staat New
York nach Abschluß eines 1984 durchgeführten hearings nunmehr
zu einer Gesetzesinitiative geführt, die ein Verbot bezweckt.
Dies geschah, obwohl von Seiten der Industrie angeboten wurde,
NTA zurückzunehmen, falls in den Trinkwasserbrunnen (insbesondere auf Long Island) 3 µg/l NTA überschritten werden sollte.

Canada

Seit den frühen siebziger Jahren wird NTA in Waschmitteln eingesetzt. Der mittlere Massegehalt liegt bei 15 %. Wegen der
sehr günstigen hydrologischen Verhältnisse sind die NTA-Konzentrationen in den Gewässern gering; im Trinkwasser werden im
Mittel 3 µg/l, maximal 30 µg/l gefunden.

Schweiz (26)

Die Schweiz hat im Juli 1985 ein Phosphatverbot für Textil-
waschmittel zum 1. Juli 1986 ausgesprochen (- 0,5 % Phosphor).
Sie gestattet gleichzeitig den Einsatz von Na_3 NTA bis zu 5
Masse-%. Es wird damit gerechnet, daß die NTA-Konzentration im
Trinkwasser 3 µg/l nicht übersteigen wird.

Niederlande:

In den Niederlanden bestehen Vereinbarungen zwischen Industrie
und Behörden über eine Phosphatbegrenzung. Gegen den Einsatz
von 6500 t NTA pro Jahr (im Mittel 5 % in den Waschmitteln) und
gegen NTA-Konzentrationen bis zu 50 µg/l im Trinkwasser bestehen keine Bedenken.

Die folgende Tabelle gibt einen Überblick über den NTA-Ver-
brauch (27,28).

TABELLE

Einsatz von NTA in Waschmitteln

Land	% NTA im Waschmittel	t/a 1983	Richtwert im Trinkwasser µg/l
Canada	10	17 000	50
USA (Indiana)	15	1 500	
Finnland, Schweden, Norwegen	10	200	
Schweiz	5	500	(3)
Italien	3	500	
Niederlande	12	500	50
BR Deutschland	0-7	2 900	(3)

Aufteilung des NTA-Einsatzes in der Bundesrepublik

	t/a 1985 1. Hj. x 2
Waschmittel	2 200
Reinigungsmittel, Enthärter	500
Chemische Industrie	40
Papier-Industrie	15
Textil-Industrie	100
Sonstige Abnehmer	80

4. Die NTA-Situation

NTA wird insbesondere in Ländern mit strengen Auflagen zur
Phosphatbegrenzung in Waschmitteln von den Herstellern als an-
gemessener Ersatzstoff betrachtet.
NTA wird von keiner Seite als ökologisch und gesundheit-
lich völlig unbedenklich eingestuft, andererseits ist die Verwendung zur Zeit in keinem Land verboten.
Die staatlicherseits eingenommene Haltung gegenüber NTA
reicht von völliger Freigabe verbunden mit Monitoringauflagen
(Canada) bis zu einem angestrebten Verbot der Verwendung in
Waschmitteln (Staat New York).
Die unterschiedliche Haltung zum NTA in den einzelnen Ländern resultiert sowohl aus unterschiedlichen hydrologischen und
umweltpolitischen Gegebenheiten als auch aus der unterschiedli-

chen Bewertung der vorliegenden Forschungsergebnisse über NTA.
   Bei der Beurteilung von NTA werden insbesondere folgende
Gegebenheiten berücksichtigt:
- der Grad der Erfassung und das Ausmaß der biologischen Reinigung des kommunalen Abwassers,
- die Gefahr der Gewässereutrophierung durch Waschmittelphospha-
  te,
- die aktuelle Belastung der kommunalen Abwässer und die histo-
  rische Belastung der Gewässersedimente mit toxischen Schwermetallen,
- die hydrologischen und klimatischen Verhältnisse unter Berücksichtigung der Bevölkerungsdichte sowie
- das Ausmaß und die Art der Trinkwassergewinnung aus Oberflächengewässern und aus Einzelbrunnen.
   Aus waschtechnischer Sicht erscheint der Einsatz von NTA
wünschenswert, aber selbst im Falle eines totalen Verbots von
Phosphaten nicht zwingend.
   Es sind Waschmittelformulierungen auf dem Markt, die unter
Hartwasserbedingungen ohne Phosphat, Phosphonat und NTA ein hin-
reichendes bis gutes Waschergebnis ermöglichen.
   Andererseits muß betont werden, daß gegen einen geringen
Einsatz von NTA mit der Folge des Auftretens einiger Mikrogramm
NTA pro Liter im Trinkwasser keine wissenschaftlich fundierten
Argumente vorgebracht werden können; dies gilt insbesondere,
wenn man die zulässigen Höchstkonzentrationen toxischer Stoffe
in den nationalen und supranationalen Trinkwasserrichtlinien
zum Vergleich heranzieht.

5. Schrifttum

(1) "Phosphor, Wege und Verbleib in der Bundesrepublik
     Deutschland"
     Verlag Chemie, Weinheim/New York (1978)
(2) Vollenweider, R.A. "Scientific fundamentals of the
     eutrophication of lakes and flowing waters ..."
     OECD Paris, Techn.Rep. DAS/CSI 68.27 (1968)
(3) "Ecological effects of non-phosphate detergent builders:
     Final report on NTA"
     Internat. Joint Commission, Great Lakes Research Advisory
     Board (1978)
(4) US-EPA Final Report - NTA - April 22, (1980)
(5) "Die aquatische Umweltverträglichkeit von Nitrilotriace-
     tat (NTA)"
     Fachgruppe Wasserchemie, Ges.Deutscher Chemiker
     Verlag Hans Richarz, St. Augustin (1984)
(6) R. Perry et.al. "Environmental Aspects of the use of NTA
     as a detergent builder"
     (Review Paper) Water Res. Vol.18, 255-276, (1984)
(7) Ullmanns Encyklopädie der techn. Chemie, Bd. 17 339-341,
     4. Aufl. (1979) Verlag Chemie Weinheim
(8) in (5) Schrifttum S. 157, 178, 249
(9) Warren, C.B. "Biodegradation of NTA and NTA-metal ion com-
     plexes: Survival in Toxic Environments" Academic Press,
     N.Y. 473-496 (1974)
(10) Gudernatsch, H. "Verhalten von NTA im Klärprozeß und im
     Abwasser", Gas-Wasserfach, Wasser-Abwasser 111, 511-516
     (1970)

(11)   in (5) Schrifttum S. 204
(12)   Rubin, M. Biol. Trace Elem.Res. 2/1, 1-19 (1980)
(13)   Hennes, E.Ch. "Berechnung der Komplexierung der Schwerme-
       talle im Rheinwasser durch NTA"
       Ber.d.Inst. für Radiochemie vom 5.5.1983; Auszug in (5)
       293-305293-305
(14)   Dietz, F. "Zur Frage der Remobilisierung von Schwermetal-
       len durch NTA"
       Korrespondenz Abwasser 29, 692-693 (1982)
(15   in (5) Schrifttum S. 232
(16)   in (5) Schrifttum S. 266
(17)   Hiasa, Y. et.al "NTA: Promoting effects on the developm.
       of renal tubular cell tumors in rats treated with N-ethyl-
       N-hydroxyethylnitrosamine"
       J. Nat. Cacer Inst. 72, 483-489 (1984)
(18)   Anderson, R.L. "A review of the environm. and mammalian
       toxic. of NTA"
       The Procter and Gamble Comp., Cincinnati, Ohio, April 9,
       1984
(19)   Cantilena, Jr. et.al. "The Effect of chelating agents on
       the excretion of endogenous metals"
       Toxicol. appl. Pharmacol. 63, 344-350 (1982)
(20)   Keen, C.L. et.al. "Effect of dietary iron, copper, and
       zinc chelates of NTA on trace metal conc. in rat milk and
       maternal and pup tissues"
       J. Nutrit. 110, 897-906 (1980)
(21)   Michael, W.R. et.al. "Effect of Na_3NTA on the metabolism
       of selected metalions"
       Toxicol. appl. Pharmacol. 24, 519-529 (1973)
(22)   "Testimony pertiment to the proposed regulation banning
       the use of NTA in N.Y.state"
       Bureau of Toxic Substances Assessment N.Y. State Dept. of
       Health; Jan. 25, 1985
(23)   "Does the use of NTA in detergent products pose a poten-
       tial risk to human health?"
       Univ.Associated for Res. and Educ. in Pathology Inc., Be-
       thesda, Maryland, USA. Report to the Procter and Gamble
       Comp., (1984)
(24)   in (5) S. 269
(25)   Matheson, D.H. "NTA in the Canadian Environment"
       Inland Water Directorate, Ottawa, Canada (1977)
(26)   "Änderung der Schweizer Waschmittel-Verordnung vom
       3. Juli 1985"
(27)   Berth, P. "Der Einsatz von NTA in Ländern außerhalb der
       Bundesrepublik Deutschland"
       in (5) 407-412
(28)   BASF AG Bericht an den Bundesminister des Innern, Bonn
       (1985)

Hinweis:

Im Kapitel 3.2 "Ökologische Eigenschaften von NTA" sind weitgehend die Ergebnisse der deutschen NTA-Studie (5) übernommen
worden.

Anhang

Folgerungen und Empfehlungen der deutschen NTA-Studie (5)

Die von der Fachgruppe Wasserchemie der Gesellschaft Deutscher Chemiker eingesetzte Koordinierungsgruppe hat aus den Ergebnissen der Studie nachfolgende Schlußfolgerungen gezogen:

1. Aufgrund der in der Bundesrepublik Deutschland vorliegenden Gewässersituation und der möglichen Auswirkungen soll NTA nicht unbegrenzt eingesetzt werden.

2. Die Menge des eingesetzten NTA soll regelmäßig erfaßt werden.

3. Die NTA-Konzentration in den Gewässern ist zu überwachen mit dem Ziel, evtl. Auswirkungen rechtzeitig erkennen zu können. Die Entwicklung der NTA-Konzentrationen im Trinkwasser sollte parallel dazu beobachtet werden.

4. Der Einsatz von NTA ist mit der Durchführung eines Monitoring-Programmes zur Ermittlung von NTA-Konzentrationen und evtl. Einflüssen auf die aquatische Umwelt und auf das Schwermetallverhalten zu verknüpfen.

5. Auf der Grundlage des gegenwärtigen Wissensstandes darf der NTA-Gehalt an gewässergütewirtschaftlich repräsentativen Stellen in den Oberflächengewässern der Bundesrepublik Deutschland $0,2$ mg/l H_3NTA (99-Perzentilwert) nicht überschreiten. Die repräsentativen Stellen sind so auszuwählen, daß die hier durchgeführten Untersuchungen Auskunft über folgende Gesichtspunkte ergeben:
   - Entwicklung der NTA-Konzentration in den Gewässern der Bundesrepublik Deutschland
   - Beeinflussung des Rohwassers für die Trinkwassergewinnung
   - Auswirkungen auf die aquatische Umwelt.

   Insbesondere wird als eine repräsentative Stelle der Meßpunkt "Zornige Ameise" an der Ruhr vorgeschlagen.
   Weitere repräsentative Meßstellen sollten für andere Oberflächengewässer im Rahmen des Monitoring-Programmes an gut durchmischten Gewässerbereichen festgelegt werden.

6. Aus der o.g. Konzentration für NTA in den Gewässern errechnet sich auf der Basis der Fallstudie Ruhr nach dem gegenwärtigen Wissensstand eine jährliche Einsatzmenge von höchstens 25.000 t Na_3NTA in der Bundesrepublik Deutschland. Dies entspricht rechnerisch einem mittleren Gehalt an Na_3NTA von $3,4$ % in allen Waschmitteln.

7. Ergibt sich eine auf die Verwendung von NTA zurückzuführende signifikante Erhöhung des Schwermetall-Konzentrationspegels in den Kläranlagenabläufen, so ist die NTA-Einsatzmenge zu vermindern. Dies gilt auch für den Fall, daß die Abwasserreinigung bzw. die Trinkwasserversorgung beeinträchtigt oder ungünstige Auswirkungen auf die aquatische Ökologie festgestellt werden.
   Haben sich bei Erreichen der NTA-Einsatzhöchstmenge keine signifikant meßbaren Erhöhungen der Schwermetall-Konzentrationen abgezeichnet und sind keine ungünstigen Auswirkungen auf die Trinkwasserversorgung, Abwasserreinigung und die aquatische Ökologie festgestellt worden, so ist eine weitere begrenzte Erhöhung des Einsatzes von NTA möglich.

8. Aufgrund der exemplarischen Bedeutung der Ruhr für die Trinkwasserversorgung und ihrer schwermetallbedingten Emp-findlichkeit gegenüber Komplexbildnern soll als erste Stufe des Monitoring-Programmes möglichst bald mit stichprobenar-tigen Kontrollen des NTA-Gehaltes in der Ruhr am Meßpunkt "Zornige Ameise" begonnen werden.
Um vorerst einen unnötig großen Analysenaufwand zu vermei-den, kann mit dem Beginn der nächsten Stufe des noch zu for-mulierenden NTA-Monitoring-Programmes in der Bundesrepublik Deutschland gewartet werden, bis der NTA-Konzentrationslevel in der Ruhr (Zornige Ameise) 0,01 mg/l H_3NTA erreicht hat. Darüber hinaus sollten bereits laufende Schwermetall-Unter-suchungsprogramme in der Bundesrepublik Deutschland zur Er-mittlung des gegenwärtigen Schwermetallpegels in unseren Ge-wässern zur Auswertung herangezogen bzw. erforderlichenfalls intensiviert werden.
9. Ergeben die Untersuchungen an der Ruhr (Zornige Ameise), daß 0,1 mg/l H_3NTA als 99-Perzentilwert erreicht werden, so ist eine Zwischenbilanz zu ziehen, insbesondere bezüglich des Abbauverhaltens von NTA. Erforderlichenfalls ist die NTA-Einsatzmenge neu festzusetzen.

Die Koordinierungsgruppe hat aus den Ergebnissen der Stu-die außerdem die Schlußfolgerung gezogen, daß zu einigen Pro-blemen des Einsatzes von NTA noch offene Fragen bestehen, de-ren Klärung unbedingt erforderlich ist. Die Bearbeitung der nachfolgend aufgeführten offenen Fragen sollte nach Auffassung der Koordinierungsgruppe so schnell wie möglich beginnen, spä-testens jedoch mit der zweiten Stufe des Monitoring-Programm-mes.
1. Einfluß von NTA auf die Elimination von Schwermetallen im technischen Maßstab bei der kommunalen Abwasserreinigung (Anschlußwert > 5.000 E, Summe der Konzentrationen der toxi-schen Schwermetalle > 2 mg/l).
2. Präzisierung des prozentualen NTA-Abbaugrades in Kläranlagen unter technischen Bedingungen.
3. Einfluß von kurzfristigen Konzentrationsschwankungen anorga-nischer und organischer Stoffe sowie von NTA selbst im Roh-abwasser auf den aeroben NTA-Abbau.
4. Mögliche Hemmung des NTA-Abbaus durch leicht abbaubare Kon-kurrenznährstoffe sowie durch nicht abbaubare Substanzen.
5. NTA-Abbau unter anaeroben Bedingungen, insbesondere in Faul-gruben und Abwasserteichen.
6. Einfluß von NTA in Konzentrationen < 1 mg/l auf die Unter-grundpassage unter besonderer Berücksichtigung unterschied-lich zusammengesetzter Untergrundmaterialien.
7. Verhalten von NTA in realistischen Konzentrationen in Fil-tersystemen mit vorausgegangener Schwermetall-Kontamination und stabilen anaeroben Verhältnissen unter naturnahen biolo-gischen, chemischen und hydraulischen Randbedingungen.
8. Remobilisierung von Schwermetallen aus Sedimenten und Schwe-bestoffen im Konzentrationsbereich unterhalb 0,5 mg/l H_3NTA.
9. Wirkung von NTA auf das Algenwachstum in komplexen Ökosyste-men im Rahmen des Monitoring-Programmes an gestauten Flüssen und in experimentellen in-situ-Einrichtungen.
10. Geschwindigkeit des Abbaus unter den Bedingungen deutscher Ästuarien in Abhängigkeit vom Salzgehalt, der Temperatur und

und der Biomassedichte.
11. Eliminierung von NTA bzw. NTA-Metall-Komplexen durch die
    Ozonung sowie bei der A-Kohle-Filtration mit und ohne bio-
    logisch aktiver Stufe (steril, insteril, Adaptationszeit
    nach Reaktivierung), vor allem bei niedrigen Temperaturen.
12. Erhöhung der Aufeisenungsrate (Braunwasserproblem) in Was-
    serleitungen durch NTA in geringen Konzentrationen und Be-
    einflussung der Schutzschichtbildung.

Bild 1 : Mittlere Phosphorkonzentration im Bodensee-Obersee

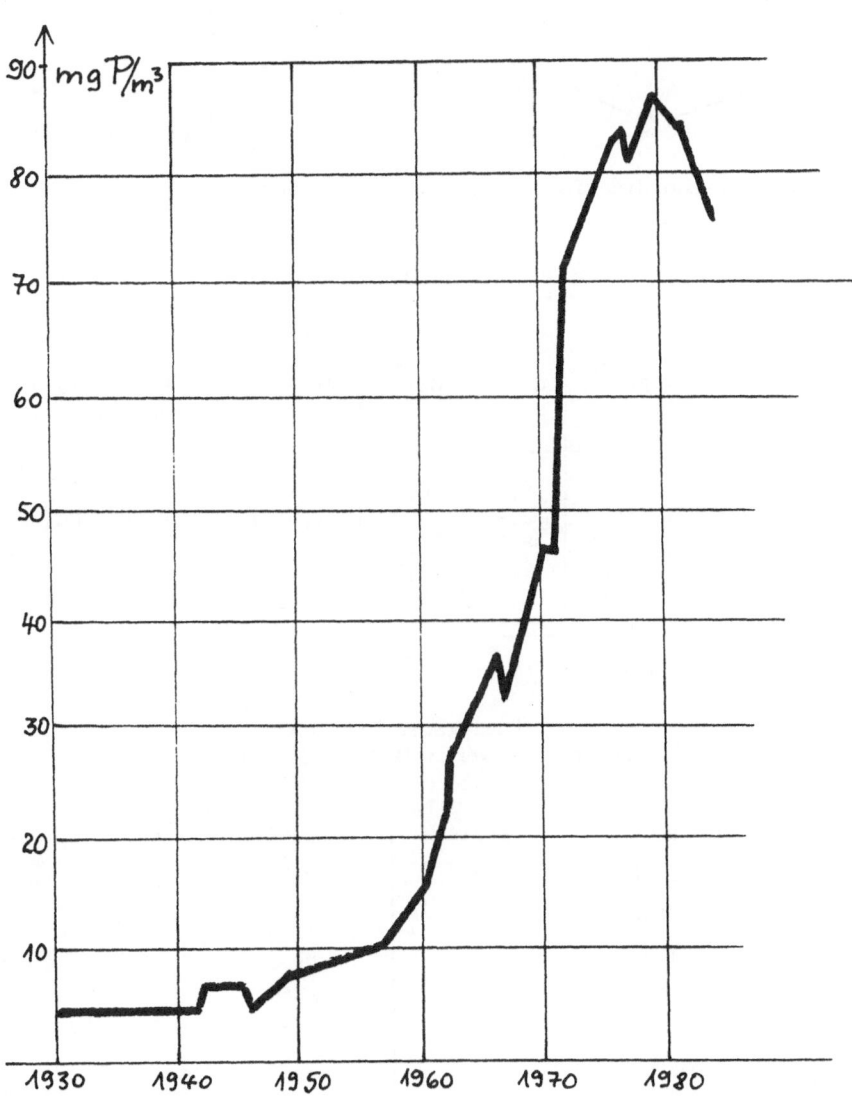

Bild 2 : Metallchelate mit Nitrilotriessigsäure

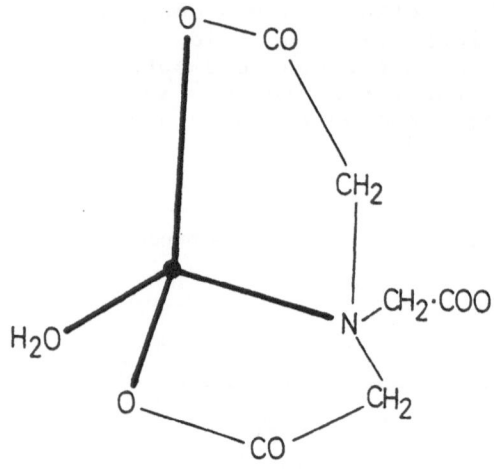

4er-Koordination

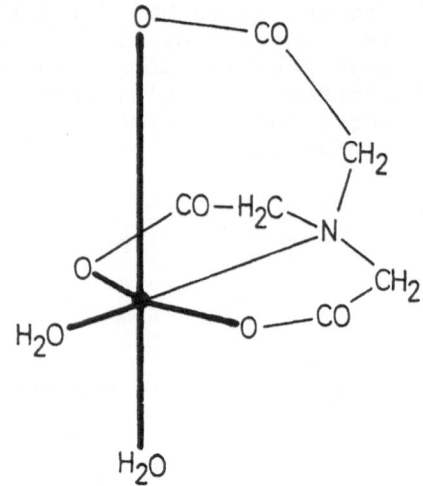

6er-Koordination

Bild 3 : Stabilitätskonstanten

lg Kml*

	Fe^{3+}	Cu^{2+}	Cd^{2+}	Ni^{2+}	Beobachtete Remobilisierung
NTPP		8,3	6,6	6,7	
NTA	15,9	12,9	9,8	11,3	ab 0,2 mg/l Cu,Zn, Ni
EDTA	16,0	18,7	16,4	18,5	
HEDP	16,2	11,8	9,1	9,2	ab 3 mg/l Fe
Polycarboxylat	(29)	2,7	1,9		

* $K_{ml} = [ML] / [M] \cdot [L]$

Bild 4 : Rheinwasser (Strom - km 906)
(Jahresmittelwerte 1979, filtriert (13))

Ca^{2+}	80	mg/l	Cu^{2+}	5	µg/l
Mg^{2+}	11,8	"	Zn^{2+}	20	"
NH_4^+	0,89	"	Cd^{2+}	0,2	"
Cl^-	167	"	Pb^{2+}	3	"
HCO_3^-	153	"	Hg^{2+}	0,3	"
SO_4^{2-}	78	"	Ni^{2+}	9	"
PO_4^{3-}	1,27	"	Fe^{3+}	1,3*	mg/l

* unfiltriert

Bild 5 : Chemische Spezifikation von NTA im
Rheinwasser + 10 µg/l - 10 mg/l H$_3$NTA (13)

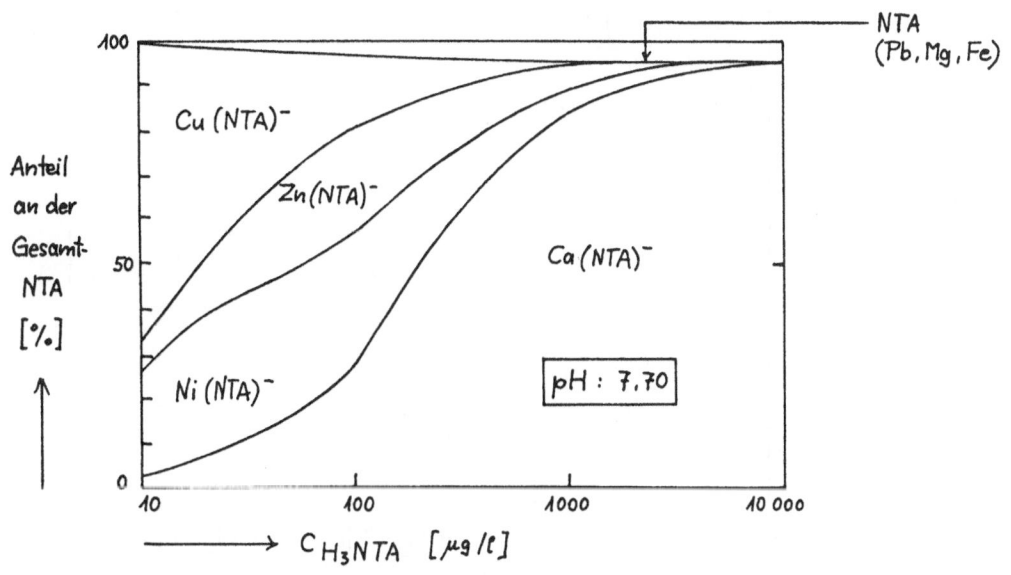

Bild 6 : Chemische Spezifikation des Kupfers im Rheinwasser
+ 100 µg/l H$_3$NTA (13)

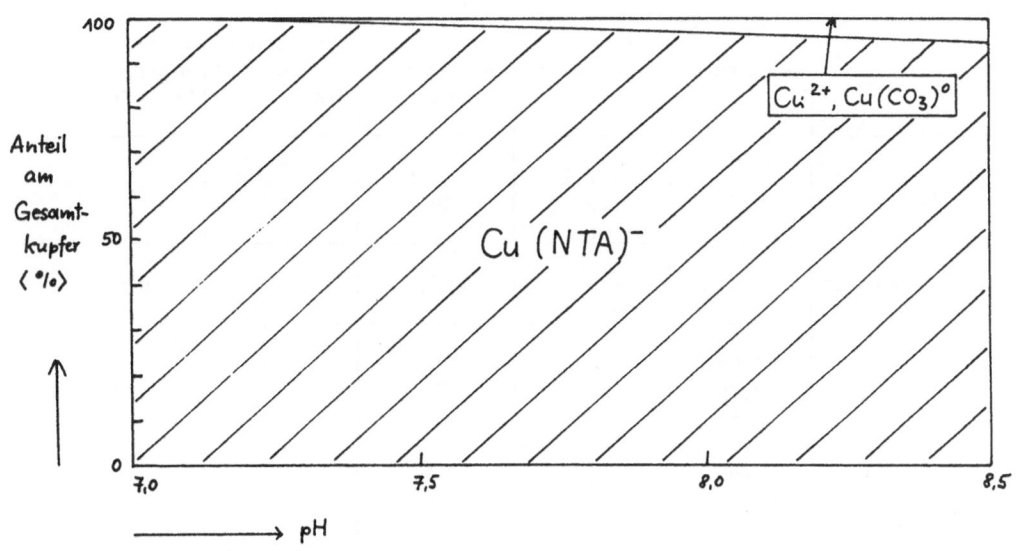

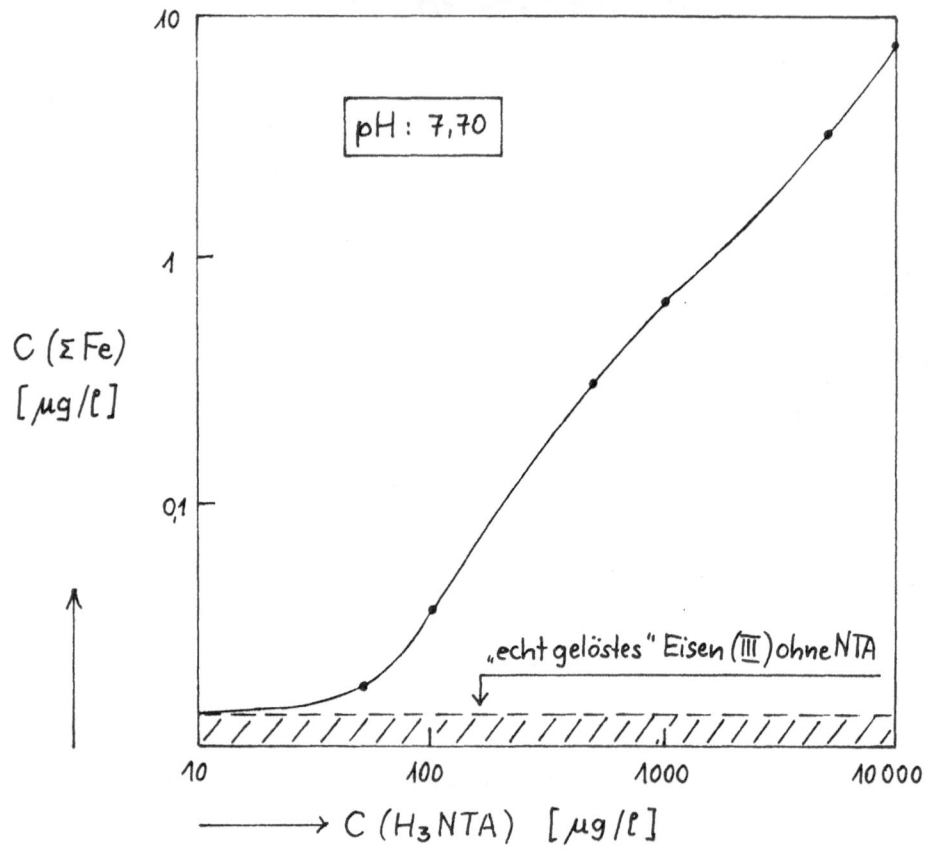

Bild 7 : Konzentration an Fe (III) "gelöst" im Rheinwasser
+ 10 μg/l – 10 mg/l H_3NTA (13)

pH: 7,70

$C (\Sigma Fe)$ $[\mu g/\ell]$

„echt gelöstes" Eisen (III) ohne NTA

$\longrightarrow C (H_3NTA)$ $[\mu g/\ell]$

Bild 8 : Remobilisierung von Schwermetallen durch NTA (14)

	Sediment	Ruhrwasser	Kontroll-versuch	nach 45 h remobilisiert Versuch mit NTA	entspricht NTA (1:1 Komplex)
pH		7,1	7,5	7,6	
Ca mg/l		50	49	48	
Schwermetalle	mg/g TR	μg/l	μg/l	μg/l	μg/l
Zn	3,04	70	75	353	813
Ni	0,26	22	25	78	173
Cu	0,61	22	12	24	36
Cd	0,025	<1	<1	1	2
Pb	0,46	< 10	< 10	< 10	–
Fe	46,0	25	30	34	–

Umsatz an NTA :   1024 μg/l

Bild 9 : Trinkwassergewinnung

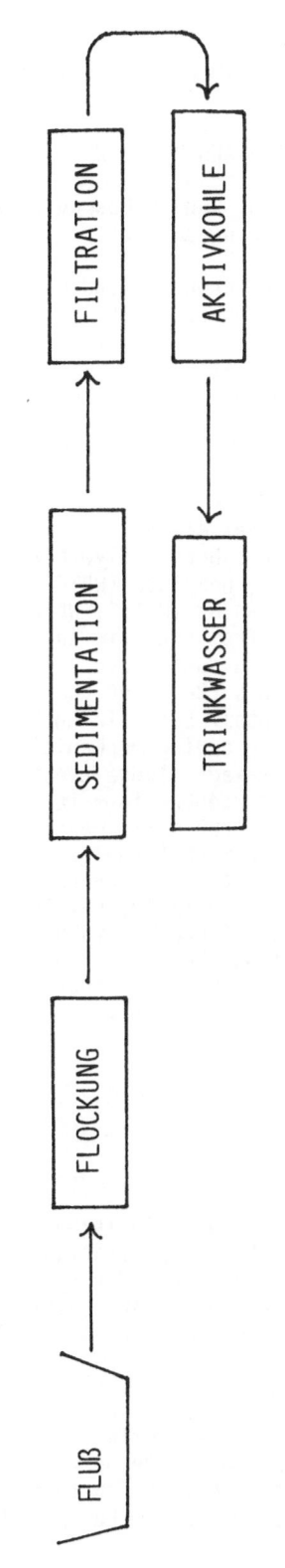

EINFACHE AUFBEREITUNG :

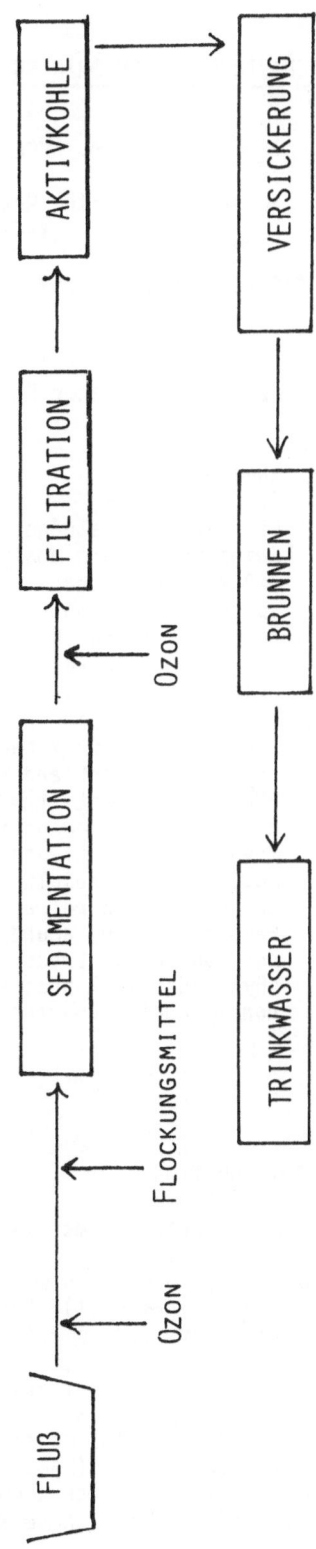

CHEMISCH – PHYSIKALISCHE AUFBEREITUNG :

BEHAVIOUR OF NONIONIC SURFACTANTS IN BIOLOGICAL WASTE WATER TREATMENT

M. AHEL*, W. GIGER** AND M. KOCH***

*Center for Marine Research Zagreb, Rudjer Boskovic Institute,
Y-41001 Zagreb, Yugoslavia

**Swiss Federal Institute for Water Resources and Water Pollution
Control (EAWAG), CH-8600 Dübendorf, Switzerland

***Cantonal Laboratory,
Department for Water Pollution Control, CH-8005 Zürich, Switzerland

Summary

Effluents and sludges from municipal sewage treatment plants in
Switzerland were analyzed for nonylphenol polyethoxylates (NPnEO,
n=3-20), nonylphenol mono- and diethoxylate (NP1EO, NP2EO), corresponding nonylphenoxy carboxylic acids (NP1EC, NP2EC) and nonylphenol (NP). These chemicals derive from nonionic surfactants of the
alkylphenol polyethoxylate type and contribute significantly to the
organic loading in sewage effluents (2-13% of the total dissolved
organic carbon in primary effluents). Specific analytical methods
were used to study their behaviour during mechanical-biological
sewage treatment and subsequent sewage sludge treatment. The parent
NPnEO surfactants (n=3-20) with concentrations in raw and mechanically treated sewage from 400 to 2'200 mg/m^3 were efficiently removed by the activated sludge treatment (usually more than 90%).
However, this resulted in the formation of several metabolic products, and the refractory nature of NP1/2EO, NP1/2EC and NP was
recognized. The abundance of the different metabolites varied depending on the treatment conditions. Both biotransformations and
physicochemical processes determine the behaviour and fate of the
nonylphenolic substances in sewage treatment.

1. INTRODUCTION

Detergents are manufactured and consumed in large quantities for
personal care, household use and industrial applications. In 1982, approximately 30 million metric tons of synthetic detergents and soaps were
produced worldwide (1). Surfactants are the major organic components of
the detergents and comprise several weight percent of the detergent formulations. About 30% of all surfactants used are nonionic surfactants,
and almost all of them are produced by the addition of ethylene oxide to
various hydrophobic substances such as fatty alcohols, alkylphenols and
fatty acids (2). Two major types of nonionic surfactants are linear alcohol polyethoxylates (LAEO) and alkylphenol polyethoxylates (APnEO,
Fig. 1) which comprise about 80% of all nonionic surfactants used. The
end use markets for these surfactants in Western Europe, USA and Japan in

1982 were 500'000 and 310'000 metric tons for LAEO and APnEO respectively (3,4).

It should be pointed out that both LAEO and APnEO are very complex mixtures of oligomers, isomers and homologues. The major difference between these two surfactant types lies in the hydrophobic part of the molecule which determines their different behaviour in waste water treatment and in the aquatic environment. LAEO are readily biodegradable to CO_2 and H_2O (Fig. 2) both in waste water treatment (5) and at trace concentrations in natural waters (6). In contrast, biodegradation of APnEO is much slower (7) leading to the formation of different refractory metabolites under aerobic and anaerobic conditions (Fig. 3, ref. 8-12). In addition, further chemical treatment of effluents can cause the formation of chlorinated and brominated derivatives of APnEO metabolic products which are suspected of having mutagenic properties (11).

To assess their environmental acceptability it is important to evaluate whether the biotransformation of the parent molecules can change the aquatic toxicity of the nonionic surfactant residues. Thus, it was found that the toxicity of LAEO disappears more rapidly than it would be expected from analytical measurements of LAEO residual concentration (13). In contrast, the toxicities of NP, NP1EO and NP2EO formed by the biotransformation of NPnEO are much higher than the toxicity of the corresponding parent compounds (8,14,15).

Based on the above mentioned environmental characteristics of LAEO and APnEO it follows that APnEO represent the critical group of nonionic surfactants. Therefore the behaviour of APnEO in waste water and sewage sludge treatment as well as their environmental fate should be examined with special care. In this paper we report on investigations into the behaviour of APnEO in several full scale mechanical-biological sewage treatment plants in the area of Zürich, Switzerland.

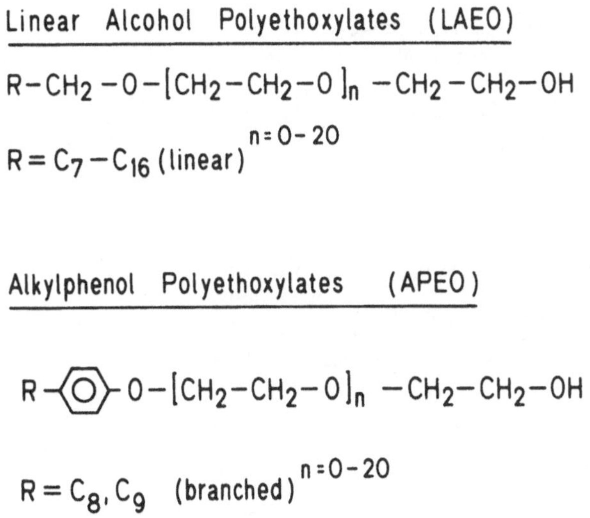

Linear Alcohol Polyethoxylates (LAEO)

$$R-CH_2-O-[CH_2-CH_2-O]_n-CH_2-CH_2-OH$$

$$n = 0-20$$

$$R = C_7-C_{16} \text{ (linear)}$$

Alkylphenol Polyethoxylates (APEO)

$$R-\underset{}{\bigcirc}-O-[CH_2-CH_2-O]_n-CH_2-CH_2-OH$$

$$n = 0-20$$

$$R = C_8, C_9 \text{ (branched)}$$

Fig. 1: Molecular formulae of linear alcohol polyethoxylates and alkylphenol polyethoxylates.

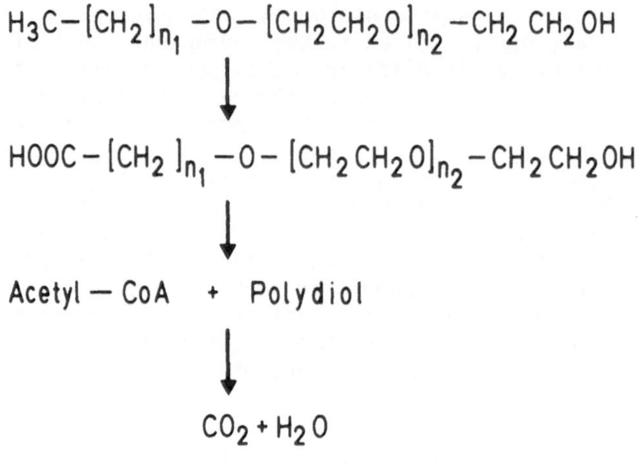

H3C−[CH2]n1 −O−[CH2CH2O]n2 −CH2CH2OH
HOOC−[CH2]n1 −O−[CH2CH2O]n2 −CH2CH2OH
Acetyl − CoA + Polydiol
CO2 + H2O

Fig. 2: Biodegradation pathway of linear alcohol polyethoxylates.

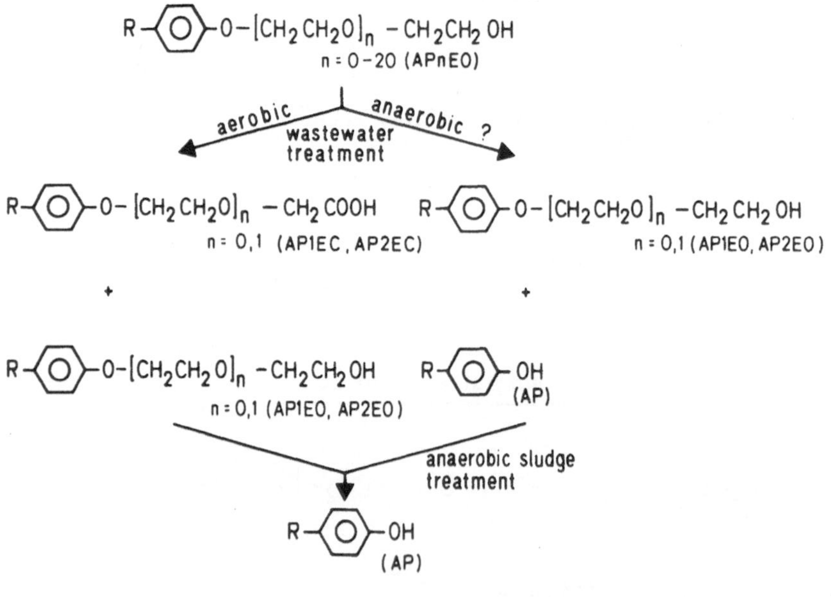

R−O−[CH2CH2O]n −CH2CH2OH
n = 0−20 (APnEO)
aerobic
wastewater
treatment
anaerobic ?
R−O−[CH2CH2O]n −CH2COOH
n = 0,1 (AP1EC, AP2EC)
R−O−[CH2CH2O]n −CH2CH2OH
n = 0,1 (AP1EO, AP2EO)
+
+
R−O−[CH2CH2O]n −CH2CH2OH
n = 0,1 (AP1EO, AP2EO)
R−OH
(AP)
anaerobic sludge
treatment
R−OH
(AP)

Fig. 3: Aerobic and anaerobic biotransformation of alkylphenol polyethoxylates.

2. EXPERIMENTAL

2.1. Sewage treatment plants (STP)

Several STP with population equivalent capacities varying from 4'000 to 240'000 were studied. The scheme of a typical STP investigated in this work is presented in Fig. 4. After mechanical treatment of the raw waste water, mostly of municipal origin, the primary effluents were biological-ly treated by the activated sludge process. The effluents from the se-condary clarifiers were discharged into the receiving waters, and in some STP a tertiary treatment for phosphorous removal was included.
Generally, primary and secondary sludges were treated by anaerobic digestion.

2.2. Sampling

24h- and 2h-composite samples were collected with automatic sampling devices operating either flow or time proportional. During collection the samples were kept at 4°C. All composite samplings were performed at dry weather conditions. In addition, grab samples from effluents and sludges were taken. Samples for the determination of alkylphenolic compounds were preserved by the addition of formaldehyde (1% v/v) and stored at 4°C.

2.3. Analytical methods

The analytical methods used for the specific determination of alkyl-phenolic compounds are described in details elsewhere. An overview and references are given in Table I. Dissolved organic carbon (DOC) and ni-trogen parameters (NH_3/NH_4^+) were determined by standard methods within 24 h.

Table I: Analytical Methods

Compound	Enrichment	Separation	Detection	Reference
APnEO n=3-20	gaseous stripping into ethyl acetate	normal-phase HPLC	UV-absorption	16
AP AP1EO AP2EO	steam-distillation/ extraction	normal-phase HPLC	UV-absorption	17
AP1EC AP2EC	solvent-extraction	normal-phase HPLC	UV-absorption and/or UV-fluorescence	18

HPLC: high-performance liquid chromatography
UV: ultraviolet

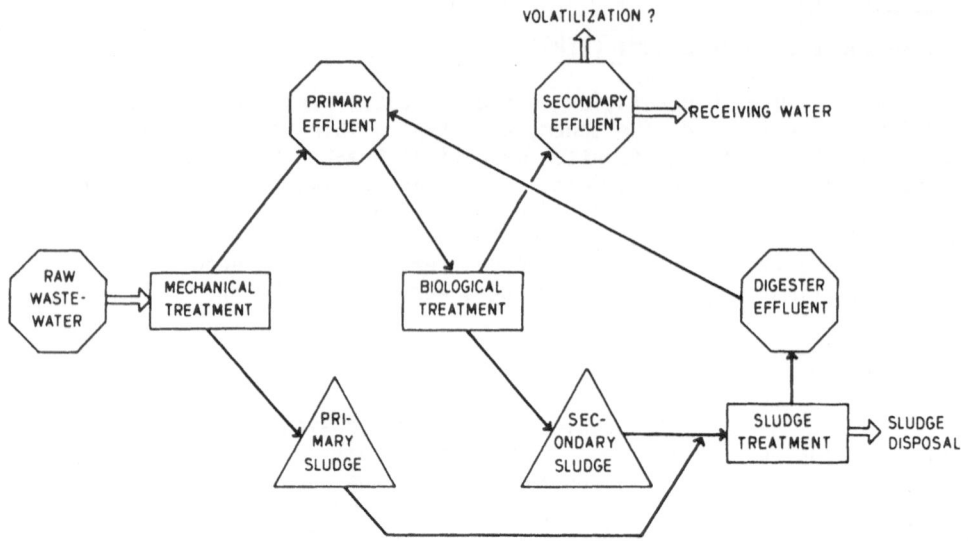

<u>Fig. 4</u> Scheme of a mechanical-biological sewage treatment plant.

3. RESULTS AND DISCUSSION

3.1. Occurrence of alkylphenol polyethoxylates in raw waste waters and effluents

Raw and mechanically treated municipal waste waters contained alkylphenol polyethoxylates (APnEO, n=3-20) in the range of 400 to 2'200 mg/m³ (Fig. 5) with no significant difference between the two types of samples. The most frequent values were measured in the range from 800-1000 mg/m³. This is, to our knowledge, the first report on phenolic nonionic surfactants in waste waters which were determined by specific measurements rather than by analytical methods yielding collective parameters. Concentrations of individual oligomers are added up to give the total APEO concentrations in mg/m³ (16). More detailed analyses showed that the bulk of the APEO were 4-nonylphenol polyethoxylates (NPnEO) with smaller amounts of 2-nonyl, 4-decyl- and 4-octylphenol polyethoxylates. NPnEO contributed substantially to the total organic load and represented 2.8-13% of the DOC content in the primary effluents of the investigated STP.

Loading dynamics of the studied STP showed typical diurnal and daily variations. Diurnal variations of NPnEO and boron concentrations in primary and secondary effluents of the STP Zürich Glatt were determined by analyses of 2h-composite samples. A maximum concentration in the primary effluent was observed in late afternoon followed by a minimum at night (Fig. 6). A similar variation was found for the concentrations of boron which derives mainly from the detergent bleaching agent sodium perborate. The concentrations of NPnEO in secondary effluents were always significantly lower, showing an efficient removal of these compounds by activated sludge treatment (more than 95%), while boron was only slightly affected.

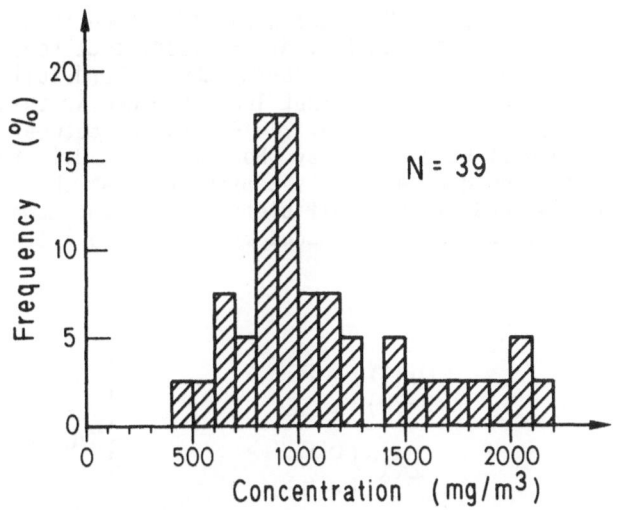

Fig. 5: Frequency distribution of the total concentrations of nonylphenol polyethoxylates (NPnEO, n=3-20) in raw sewage and primary sewage effluents.

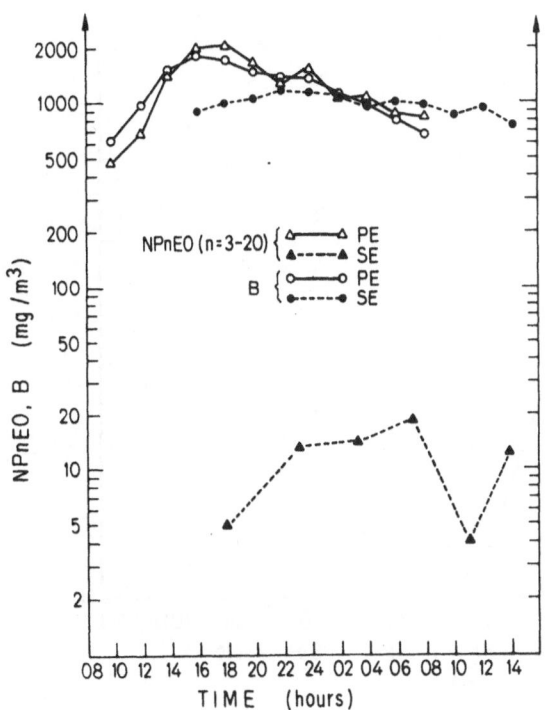

Fig. 6: Diurnal variation of nonylphenol polyethoxylates (n=3-20) and boron in primary (PE) and secondary effluents (SE) of the STP Zürich-Glatt.

Daily variations of NPEO and DOC load were determined in the same STP by analyses of 24h-composite samples. As expected, a decrease of the load for both NPnEO and DOC was observed during the weekend followed by an increase on Monday (Fig. 7). Total input load by NPnEO in this STP with a population equivalent capactity of 240'000 varied between 25.6 kg/d on Sunday to 45.1 kg/d on Friday. However, only 0.13 to 1.15 kg/d were found in the secondary effluents confirming that the elimination of NPnEO (n=3-20) is very efficient. During the same period the DOC load of 780 to 1300 kg/d was reduced by only 67 to 71%.

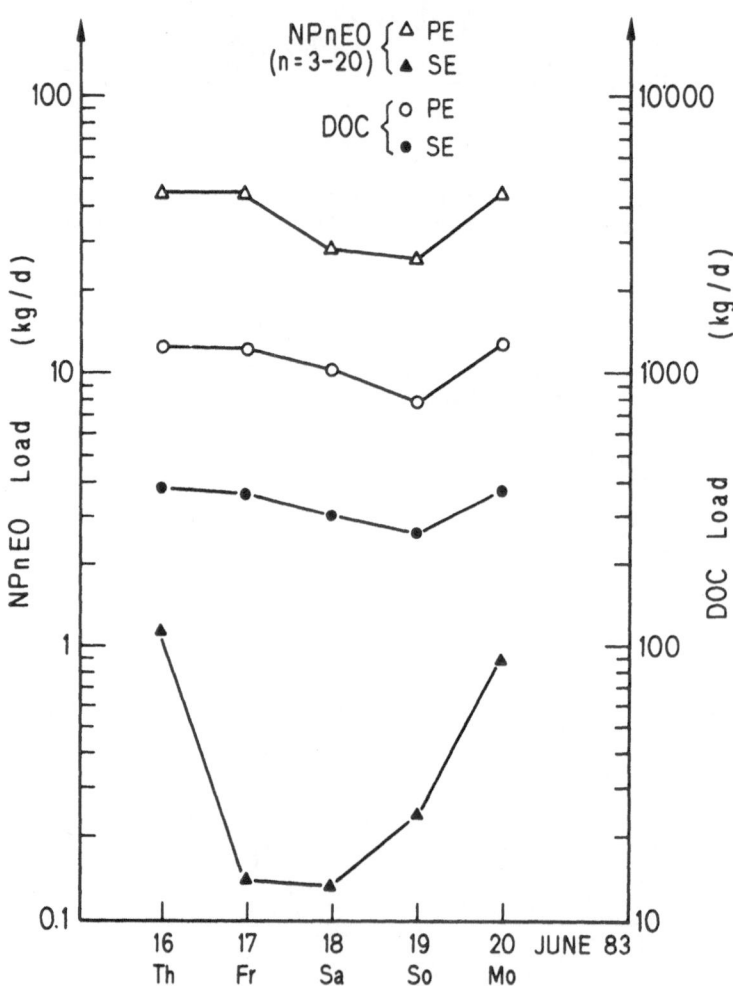

Fig. 7: Daily variation of nonylphenol polyethoxylates (n=3-20) and DOC in primary (PE) and secondary effluents (SE) of the STP Zürich-Glatt.

3.2. Occurrence and behaviour of biotransformation products of APnEO during biological waste water treatment

It might be concluded that the elimination of parent APnEO shows that no serious problems are associated with the use of alkylphenolic surfactants. However, in order to obtain a more complete insight into the behaviour and fate of these compounds in waste water treatment it is necessary to monitor not only the parent compounds but also other nonylphenolic compounds which can be formed by biotransformation processes. A comparison of the full range oligomer distributions (n=0-20) in the most commonly used commercial APEO surfactants (Fig. 8) and in waste waters (Fig. 9) illustrates clearly the importance of nonylphenolic metabolic products. Poisson distributions maximising at AP9/10EO are generally found in the surfactants which are used in laundry detergents (Fig. 8). Alkylphenol ethoxylates with only one or two ethoxy units per molecule are very minor constituents (less than 1%), and nonethoxylated alkylphenols are absent. Raw waste water and primary effluents, however, typically contained bimodal NPEO oligomer distributions with one maximum around NP7EO and the second at NP1EO and NP2EO (Fig. 9). Nonylphenol was also present in substantial concentrations. Aerobic and anaerobic biotransformation processes proceeding both in the sewers and during the mechanical treatment are thought to cause these occurrences. The results of the analyses of secondary effluents of the STP Bassersdorf and Opfikon (Fig. 9) represent typical NPnEO distributions observed in STP which are operated at low-loading, nitrifying and at high-loading, non-nitrifying conditions, respectively. A bimodal NPnEO distribution with dominating NP1EO (400 $\mu mol/m^3$) remained in the secondary effluent at Opfikon although 75% of the higher NPnEO (n=3-20) were removed by the biological treatment. In the secondary effluent at Bassersdorf, the intermediate NPnEO (n=5-10) were minor constituents and the higher NPnEO (n=10-20) were not detectable. Since the input concentrations of NPnEO were almost identical for both mentioned STP it follows that the operating conditions of a STP drastically influence the behaviour of NPnEO in the activated sludge treatment, as it was previously suggested (8).

However, NPnEO (n=3-20), NP1EO, NP2EO and NP are not the only important alkylphenolic compounds which can be found in sewage effluents. Carboxylated derivatives, mainly nonylphenoxy acetic acid (NP1EC) and nonylphenoxyethoxy acetic acid (NP2EC) (see Fig. 3) must also be considered. In fact, according to their behaviour in sewage treatment the nonylphenolic substances can be divided into four types: (a) NPnEO (n=3-20), (b) NP1EO and NP2EO, (c) NP and (d) NP1EC and NP2EC (see Fig. 3). Fig. 10 shows the concentration levels of these four compound types in primary and secondary effluents of four sewage treatment plants. In order to allow direct comparisons between the different classes of nonylphenolic compounds concentrations are expressed in molar units. The distributions of the various nonylphenolic compounds in primary and secondary effluents differed considerably. Among them the NPnEO (n=3-20) were far the most abundant types in primary effluents with concentrations of 1.3-3.8 mmol/m³. They were accompanied by irregularly high concentrations of NP1EO, NP2EO and NP. As mentioned, this clearly indicated that the biodegradation of NPnEO started before the activated sludge treatment. The comparatively low concentrations of nonylphenoxy carboxylic acids suggested that these biotransformations proceeded in oxygen depleted and/or anaerobic environments which are very probable in sewers. In contrast to the primary effluents, the most abundant nonylphenolic compounds in the secondary effluents were NP2EC and NP1EC (0.3-1.1 mmol/m³). It is obvious that NPEC are favourably formed under aerobic conditions resulting in a net production in the secondary treatment. Under the same conditions NPnEO (n=3-20) as well as NP were eliminated to various extents in all STP. NP1EO/NP2EO represented the intermediate types of compounds. The relationship between their formation and biodegradation rates during the activated sludge treatment can vary and leads either to a net elimination or to a net formation of these compounds (e.g., in the STP Dübendorf).

Results showing the precise data on the elimination of each type of nonylphenolic compound as well as the overall elimination by the secondary treatment (based on 24h-composite samples) are given in Table II. The eliminations of NPnEO (n=3-20) in the STP Fällanden, Zürich-Glatt and Niederglatt were high (more than 90%) while in the STP Dübendorf a substantially lower elimination of only 76% was observed. This difference becomes enhanced when the elimination rates for NP1EO, NP2EO and NP are compared. The elimination of these nonylphenolic compounds in the STP Dübendorf were only 29 and 37% compared with 79-80% and 84-93% for the other three STP. NPEC were produced from nonylphenol ethoxylate compounds by the activated sludge treatment and their concentrations in secondary effluents increased by a factor of 3.2-6.0. Thus, the overall elimination of nonylphenolic compounds was only 26% for the STP Dübendorf compared with 69-78% for the other three STP, which is considerably lower than the recommended required elimination of 80% (19).

TABLE II Elimination of different nonylphenolic compounds by activated sludge treatment in four STP.

| | Elimination (%) | | | |
Compounds	Fällanden	Dübendorf	Zürich-Glatt	Niederglatt
1. NPnEO, n=3-20	93	76	92	92
2. NP1EO+NP2EO	79	29	80	79
3. NP1EC+NP2EC	-230	-499	-221	-232
4. NP	93	37	84	84
5. Sum 1-4	69	26	71	78
6. NH_3/NH_4^+	98	6	0	99

$$\text{Elimination (\%)} = \frac{C_{PE}-C_{SE}}{C_{PE}} \times 100$$

C_{PE}: Concentration on primary effluent

C_{SE}: Concentration in secondary effluent

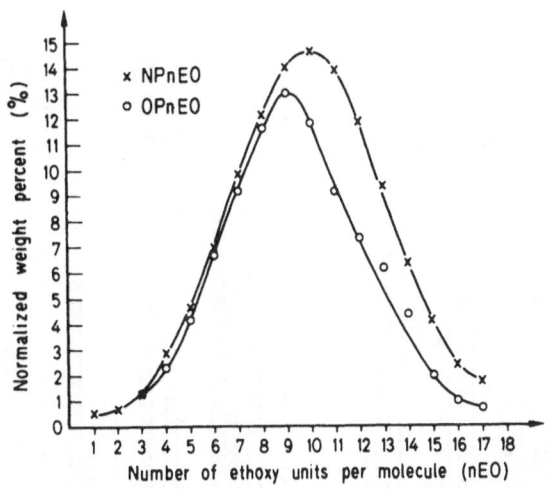

Fig. 8: Weight distribution of APEO oligomers in commercial nonionic surfactants. NPnEO: Marlophen 810; OPnEO: Synperonic OP10.

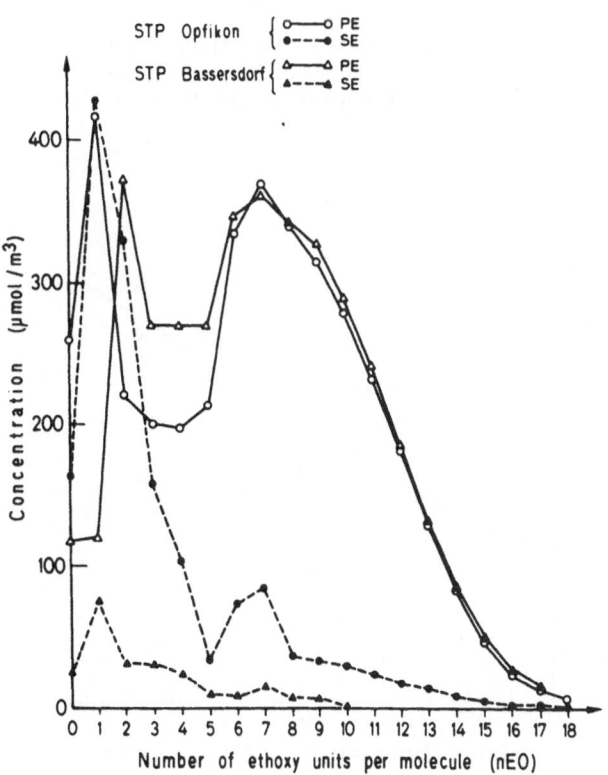

Fig. 9: Concentrations of individual NPnEO oligomers in primary (PE) and secondary effluents (SE) of the STP Bassersdorf and Opfikon.

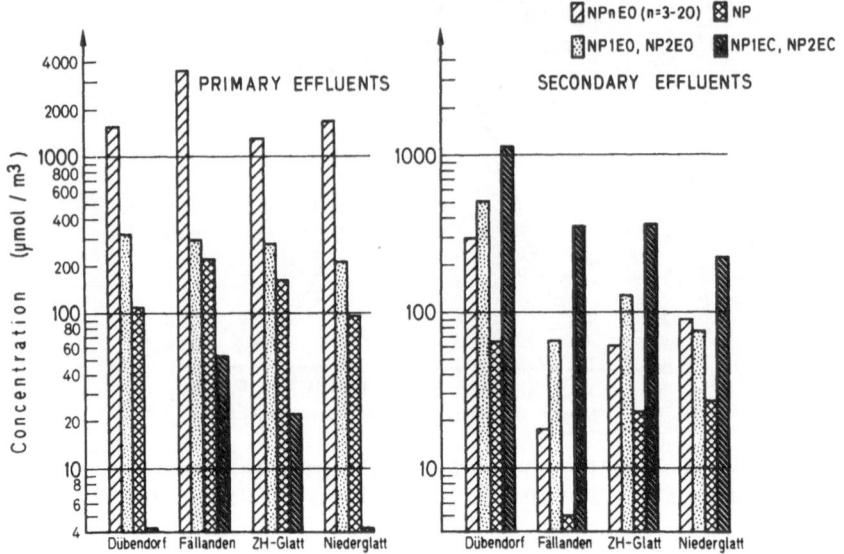

Fig. 10: Distribution of nonylphenol polyethoxylates and their metabolites in primary and secondary effluents of the STP Dübendorf, Fällanden, Zürich-Glatt, and Niederglatt.

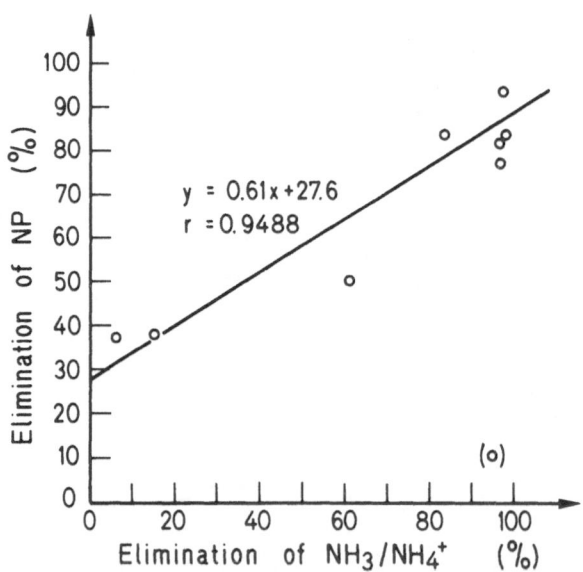

Fig. 11: Correlation of ammonia and nonylphenol (NP) elimination by activated sludge treatment.

The elimination efficiency of nonylphenolic compounds by the activated sludge process is well correlated with the removal of NH_3/NH_4^+ showing that nitrifying conditions are essential for the efficient biotransformation of these substances. This is shown in Fig. 11 for nine STP. Nonylphenol is chosen because it seems to be a typical metabolic product the formation of which is favoured under anaerobic conditions whilst its biodegradation is enhanced aerobically. A linear regression was observed which is described by the equation

$$E_{NP}(\%) = 0.61 \ E_{NH_3/NH_4^+}(\%) + 27.6(\%)$$

and a relatively high correlation coefficient $r = 0.9488$. The only point which did not fit well on the obtained regression line represents the result of the STP Nänikon which receives a great portion of waste water from a detergent manufacturing chemical plant.

3.3. The role of physicochemical processes in the distribution of APnEO and their metabolites in sewage treatment plants

The strong dependence of elimination of nonylphenolic compounds on nitrifying conditions in the STP indicated that biological processes played the dominant role. However, the effect of physicochemical processes should also be addressed, in particular the partitioning of NPnEO and their metabolites between the aqueous and solid phases. This is shown in Fig. 12 which gives the distributions of four nonylphenolic compound types between effluents and sludges of the STP Uster. The influence of partitioning between primary sludge and primary effluent on the elimination of lipophilic NP, NP1EO and NP2EO from the aqueous phase is obvious by comparison with their concentrations in raw sewage and primary effluent. No similar behaviour was observed for higher NPnEO (n=3-20). Similarly, the tertiary treatment caused a decrease of the lipophilic NP, NP1EO and NP2EO but had little or no effect on the more hydrophilic NPEC. The role of the partitioning on the distribution of the nonylphenolic compounds in the aeration tank and in the secondary clarifier is apparent from the concentration ratios between activated sludge and secondary effluent. This ratio varied from 10'500 for NP to 520 for NPEC. The strong accumulation of NP in the anaerobically digested sludge (2.6 mmol/kg) can be explained both by lipophilic partitioning of NP and its precursors and by biotransformation of NP1EO, NP2EO and NPEC under anaerobic conditions (9).

Data on the physicochemical properties of NPEO, NPEC and NP which govern their distributions between aqueous and solid phase are relatively scarce. The aqueous solubilities of p-NP, p-NP1EO, and p-NP2EO at 20°C were determined in our laboratory as 5.2 mg/l, 2.9 mg/l and 3.1 mg/l respectively . A log K_{OW} value of 4.2 for NP determined by HPLC was reported (15). The sorption constants for different activated sludges were by a factor of 2 to 8 higher for NP (10'000 - 26'000 ml/g) than those for NP1EO (3'900 - 11'000 ml/g) and for NP2EO (1'300 - 6'900 ml/g).

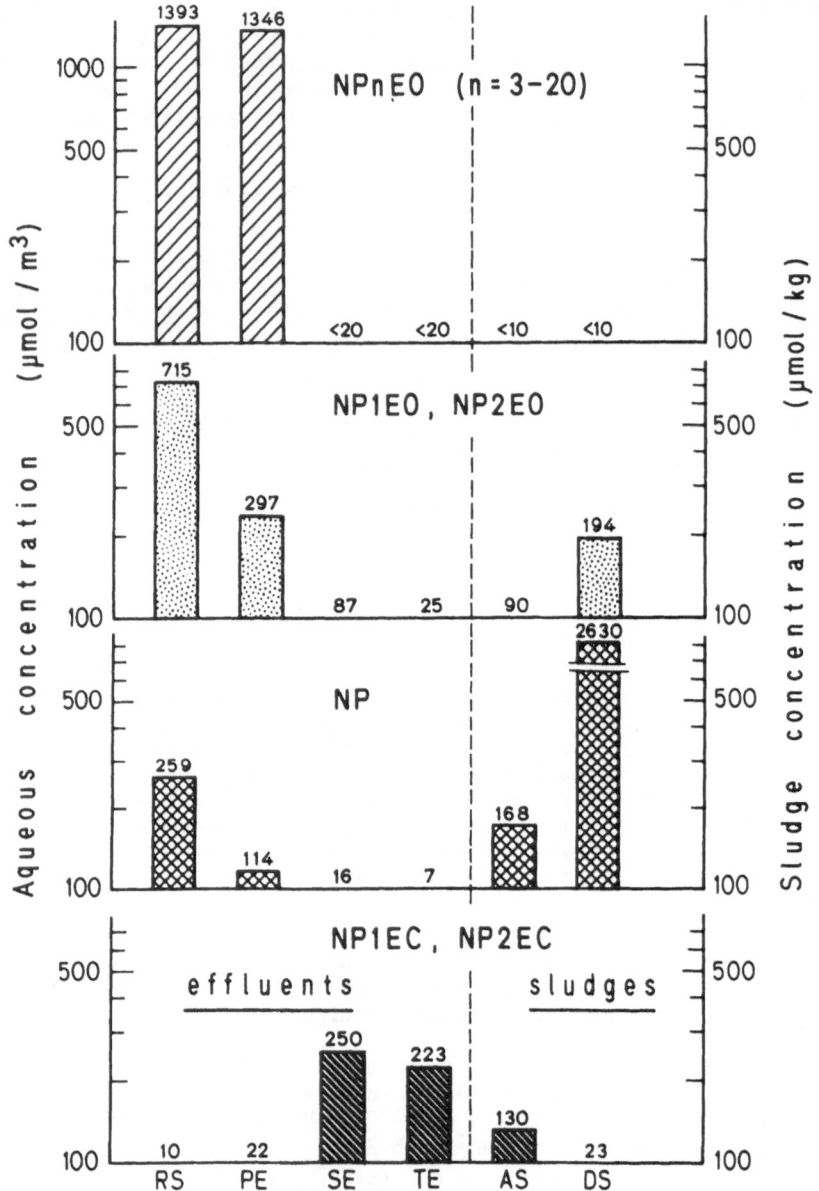

Fig. 12: Distribution of nonylphenolic compounds in effluents and sludges of the STP Uster. RS: raw sewage; PE: primary effluent; SE: secondary effluent; TE: tertiary effluent; AS: activated sludge; DS: digested sludge.

4. CONCLUSIONS

The nonionic surfactants of alkylphenol polyethoxylate type are important organic constituents of municipal waste waters and contribute to the DOC by 2.5-13%. The elimination of the parent NPnEO oligomers (n=3-20) by biological transformations in the STP is very efficient (more than 90%). However, the relatively complex biodegradation pattern of these compounds is poorly understood. The most important feature is the formation of several refractory metabolic products (NP1EO, NP2EO, NP, NP1EC, and NP2EC) which contribute considerably to the overall loads of nonylphenolic compounds especially in secondary effluents and sludges. The formation of certain types of metabolic products depends on the waste water treatment conditions. Generally, NPEC are formed preferentially under aerobic conditions whereas NP is produced under anaerobic or oxygen poor conditions. NP1EO and NP2EO represent the intermediate types of compounds. Nitrifying conditions in the STP are important for the elimination of NPEO and NP but are not sufficient for the elimination of NPEC. Thus, even in nitrifying, low-loaded STP, the overall elimination of alkylphenolic substances is lower than 80%.

Taking into account the high toxicity for aquatic organisms of the lipophilic persistent metabolites (14,15) it must be concluded that APnEO are not environmentally acceptable surfactants.

ACKNOWLEDGMENT

This work was supported in part by the Swiss Department of Commerce (Project COST 641) and by the Swiss National Science Foundation (Nationales Forschungsprogramm 7D, research project on "Organic Contaminants in Sewage Sludge"). We thank J. Zobrist and C. Jaques for boron determinations, H. Bolliger for drafting the figures and B. Schneider for typing the manuscript.

REFERENCES

1. Schneider, R., Tenside Deterg., 21, 212-215 (1984).

2. Schick, M.S., "Nonionic Surfactants", Marcel Dekker Inc., New York, 1967.

3. Werdelmann, B.W., In: "Proceedings of the World Surfactants Congress", Vol. 1, Kürle Verlag, Gelnhausen, F.R. Germany., 1984, pp 3-21.

4. Haupt, D.E., Tenside Deterg., 20, 332-337 (1983).

5. Schöberl, P., Kunkel, E., Espeter, K., Tenside Deterg., 18, 64-72 (1981).

6. Vashon, R.D., Schwab, B.S., Environ. Sci. Technol., 16, 433-436 (1982).

7. Kravetz, L., J. Am. Oil Chem. Soc., 58, 58A-65A (1981).

8. Stephanou, E., Giger, W., Environ. Sci. Technol., 16, 800-805 (1982)

9. Giger, W., Brunner, P.H., Schaffner, C., Science, 225, 623-625 (1984).

10. Giger, W., Ahel, M., Schaffner, C., In: "Analysis of Organic Water Pollutants", Angeletti, G., Bjørseth, A., Eds., Reidel Publishing Company, Dordrecht, Holland, 1984, pp 280-288.

11. Reinhard, M., Goodman, N., Mortelmans, K.E., Environ. Sci. Technol.,16, 351-362 (1982).

12. Brüschweiler, H., Gämperle, H., Schwager, F., Tenside Deterg., 20, 317-324 (1983).

13. Turner, A.H., Abram, F.S., Brown, V.M., Painter, H.A., Water Res., 19, 45-51 (1985).

14. Janicke, W., Bringmann, G:, Kühn, R., Gesundheits-Ingenieur, 90, 133-138 (1969).

15. McLeese, D.W., Zitko, V., Sergant, D.B., Burridge, L., Metcalfe, C.D., Chemosphere, 10, 723-730 (1981).

16. Ahel, M., Giger, W., Anal. Chem., 57, 2584-2590 (1985).

17. Ahel, M., Giger, W., Anal. Chem., 57, 1577-1583 (1985).

18. Ahel, M., Giger, W., Determination of Surfactant Derived Alkylphenoxy Carboxylic Acids in Environmental Samples By High-Performance Liquid Chromatography. Manuscript in Preparation.

19. OECD, Pollution by Detergents, Paris (1971).

A DATA BASE ON BEHAVIOUR AND EFFECTS OF ORGANIC POLLUTANTS IN WASTE WATER TREATMENT PROCESSES

P. Lindgaard-Jørgensen and B. Neergaard Jacobsen
Water Quality Institute
11, Agern Alle - DK 2970 Hørsholm
DENMARK

Summary

A data base including data on physical/chemical properties of organic chemicals, biodegradability data, levels of toxic effects on microorganisms, and analysed in- and outlet concentration from Waste Water Treatment Plants (WWTP) in Denmark and other countries has been constructed.

The data base consists of data obtained by reviewing the scientific litterature on 126 organic chemicals and 17 compound groups - primarily from the EEC-lists - both black and grey - the US EPA Priority Pollutant List, and some other chemicals known to be used in Denmark.

In the data base it is possible to sort the chemicals with respect to such parameters as for instance log P_{ow}, solubility in water, boiling point, degradability, toxicity etc.

The litterature shows that chemicals with boiling points below 150°C or with Henry's law constant higher than 10^{-3} atm m^3/mol should be expected to be stripped off to some extent in the WWTP, and that compounds with log P_{ow} exceeding 3-4 or solubilities below 0,2-2/g/l should be expected to adsorb to the sludge in the WWTP.

Organic compounds expected to pass the WWTP unchanged are those which are hydrophilic (log P_{ow} < 2-3) with boiling points > 150°C and showing poor degradability.

As a result, chemicals which are expected to be stripped off, to adsorp to the sludge, or to pass the WWTP unchanged, have been listed.

1. INTRODUCTION

A huge amount of chemicals are on the market today (perhaps 50,000-70,000). Some are used in households and industries, and a number of these chemicals are discharged to municipal waste water treatment plants.

In many cases the inlet concentration to the treatment plant of these chemicals discharged from non-point sources is low, often at the ppb-level. Exceptions are the cases where one point source, for instance an industry, is the dominant source of the chemicals discharged to the treatment plant. Although the inlet concentration of a chemical might be low, it is of importance to know the fate of the chemical in the treatment plant, with special emphasis to the elimentation processes.

Also determination of toxic levels of the chemicals on microorganisms from the treatment plant compared to inlet concentrations might elucidate whether toxic effects should be expected under normal circumstances (non-accidential).

A limited number - 126 specific compounds and 17 compound groups - out of the huge amount of chemicals known to be marketed, have beeh evaluated in this study with respect to behaviour and effect in a waste water treatment plant. These chemicals are present on a Danish list of potential hazardous chemicals and include chemicals from the EEC-lists, US-EPA Priority Pollutant List and some other chemicals known to be used in Denmark.

Collection of data and organization in the data base is described elsewhere (1).

2. BEHAVIOUR

A review of the scientific literature on pilot and full scale experiments with waste water treatment plants (WWTP's) revealed that the behaviour of chemicals depends on their physical/chemical properties and structure (2, 3, 4, 5).

Figure 1 shows in principle the behaviour of organic chemicals in WWTP's.

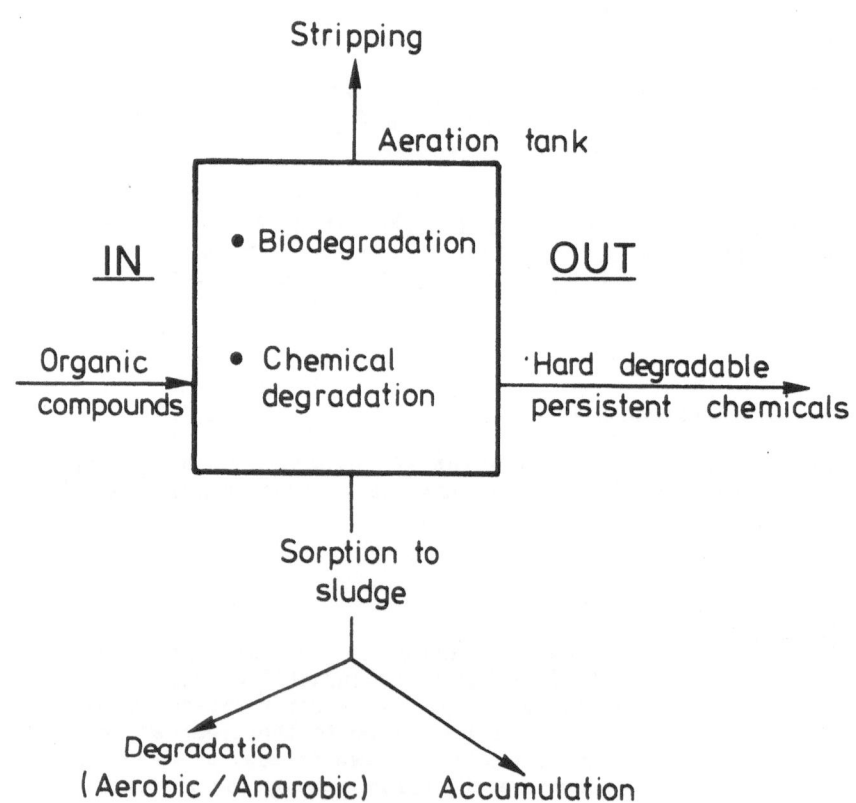

Figure 1:   Principles of the behaviour of organic chemicals in WWTP's

The four major processes of elimination of chemicals which can be of importance are:

- sorption to sludge (removal with excess sludge)

- stripping to the atmosphere

- chemical degradation (hydrolysis, oxidation, photochemical degradation

- biodegradation

For a given chemical the process or processes of importance for elimination from the WWTP depends among others on the physical/chemical properties of the chemical, the inlet concentration, but also on the construction, operation, and chemical load of the plant.

3. SORPTION

Pilot plant and full scale experiments with spiked chemicals showed the sorption to sludge to depend on the lipophilicity of the chemical, and the sludge concentration measured as organic carbon in the plant (6, 7, 8).

Lipophilicity can be described by the octanol/water partition coefficient or the water solubility, and the above mentioned experiments showed chemicals with log P_{ow} > 3 to be present in the sludge (9, 10, 11, 12). Significant elimination on the other hand was only shown for chemicals having log P_{ow} ≥ 4-6 (11, 12, 13). Table I shows some of the chemicals which to some extent must be expected to be sorbed to the sludge in a WWTP.

Table I: Chemicals susceptible of being sorbed to the sludge in WWTP's
(log P_{ow} ≥ 3, solubility < 2 g/l)

Chemical	log P_{ow}	Solubility in water g/l
Antracen	4.45	7.5E-5
Benzidin		0.40
Carbonhexachlorid, Perchlorethan	3.93	0.05
4-Chlor-m-cresol	3.10	3.9
4-Chlor-o-cresol	3.10	
O-chlortoluen	3.42	
Dichlobenil, Casoron		0.018
1,2-Dichlorbenzen	3.38	0.1
1,3-Dichlorbenzen	3.38	0.1
1,4-Dichlorbenzen	3.39	0.05
2,4-Dichlorphenoxy acetic acid (2,4-D)		0.54
4,6-Dinitro-o-cresol		0.13
EDTA		0.5
Ethylbenzen	3.15	0.14
Hexachlor-1,3-butadien		0.002
Hexachlorbenzen	6.18	0.0
Lindan	3.72	7.3E-3

(cont.)

Mecarbam		<1
Mesithylen		0.02
Naftalen	3.3	0.03
Parathion-ethyl	3.81	0.02
Pentachlorphenol	5.01	0.014
Phenylether	4.20	0.02
Styren		0.28
Sulfotepp		0.025
1,2,4-Trichlorbenzen		0.02
2,4,5-Trichlorphenol	3.39	1.2
2,4,6-Trichlorphenol		0.8
Xylen	3.15	0.2

4. STRIPPING

Elimination of a chemical by stripping from the water phase to the gas phase depends primarily on the vapour pressure and the water solubility of the chemical and the aeration efficiency in the plant (8, 14, 15, 16, 17).

The physical/chemical property describing the tendency to evaporate is Henry's law constant (11), and it has been shown that chemicals with a Henry's law constant > 10^{-3} atm m^3/mol must be expected to be stripped from a WWTP (15, 16, 17).

As a simple rule-of-thumb, substances with boiling points less than 150°C must be expected to evaporate to some extent (13).

Table II shows the chemicals among the 126, that have been evaluated, which must be expected to be able to be stripped from WWTP's.

Table II: Chemicals susceptible of being stripped off from the aeration plant in a WWTP. (H $\geq$ 10^{-3} atm $\cdot$ m^3/mol or boling point $\leq$ 15°C)

Chemical	Henry's law const, Atm*m^3/mol	Boiling point $^{\circ}$C
Acetonitril		82.0
Antracen	1.4E-3	340
Benzen	6.0E-3	80.1
1-Butylacetat		125
2-Butylacetat		112
Butylamin, aminobutan		78
Carbonhexachlorid, Perchlorethan	5.0E-3	187
Carbontetrachlorid	2.3E-2	76.7
Chlorbenzen	3.8E-3	132
Chloroform	3.1E-3	62
1,2-Dichlorbenzen	1.9E-3	179
1,3-Dichlorbenzen	2.6E-3	172
1,4-Dichlorbenzen	2.1E-3	173.4
1,1-Dichlorethan	5.8E-3	57.3

(cont.)

1,2-Dichlorethan	1.1E-3	83.5
Diethylamin		56
Diisopropylether		69
Dioxan, Glycolethylenether		101
Epichlorhydrin	1.2E-3	116
Ethylacetat		77
Ethylbenzen	8.4E-3	136.2
Ethylchlorid	1E-2	12.4
Ethylether, Diethylether		35
Formaldehyd		-20
Hydrazin		113
Mesityloxid		130
MethylCellosolve		135
Methylchlorid	2.4E-2	-24
Methylenchlorid	3.2E-3	41
Oxiran		11
N-propanol		97.4
Propylenchlorid	2.8E-3	96.8
Pyridin		115
Styren	4.0E-3	145.2
Tetrachlorethylen, Perchlorethylen	8.3E-3	121.4
Tetrahydrofuran		65.5
Toluen	6.3E-3	110.8
1,1,1-Trichlorethan, Methylchloroform	4.9E-3	76
Trichlorethylen	1E-2	86.7
Vinylchlorid	2.4	-13.9
Xylen	5.1E-3	138.4

5. CHEMICAL DEGRADATION

The most significant process of chemical degradation is hydrolysis. Some chemical structures are known to hydrolyze in water, although the rate of hydrolysis shows great variations (18). (Table III).

Table III: Types or organic functional groups that are potentially susceptible to hydrolysis.

Alkyl halides	Nitriles
Amides	Phosphonic acid esters
Amines	Phosphonic acid esters
Carbamates	Sulfonic acid esters
Carboxylic acid esters	Sulfonic acid esters
Epoxides	Sulfuric acid esters

Photochemical degradation is not believed to be a significant elimination process, due to the low transmission efficiency of light in polluted
water (18).

6. BIODEGRADATION

A quantitative description of biodegradability in WWTP's based on physical/chemical properties of chemical structures is not possible with today's knowledge. Also, results from full scale and pilot plant experiments are very limited (8, 19, 20, 21).

On the other hand, some ranking is possible, based on results from simple laboratory tests as for instance the OECD Screening Test or the Zahn-Wellens Test.

A ranking in four groups is proposed as follows:

Group 1: Chemicals expected to degrade are those showing a high degradability in a screening test providing poor conditions for degradability a.o. (The OECD's Screening Test and the Closed Bottle Test).

Group 4: Chemicals expected to persist are those which are not degraded even under optimal test conditions allowing a.o. good possibilities for adaptation.

Group 2 Chemicals expected to show variable degradability (decreasing from
and 3: group 2 to 3) are those which show different results depending on the source of inoculum used in the test, adaptation status of inoculum, and other factors. It is believed, that the degradability of these chemicals in group 2 and 3 depends very much on how the plant is run.

As a result of what has been mentioned previously, chemicals that have Henry's law constant less than 10^{-3} atm m^3/mol or boiling point higher than 150°C, and are hydrophilic (log P_{ow} < 2-3) showing poor degradability (group 4) should be expected to pass the WWTP unchanged (Table IV).

Table IV: Chemicals expected to pass a WWTP unchanged (log P_{ow} ≤ 4, or water solubility > 0,1 g/l and H ≤ 10^{-3} atm, or B_p ≥ 75°C and biodegradability Group 4).

2,4,5-Trichlorphenol
Bis-(2-chlorethyl)ether
Dioxan
4-Nitrochlorbenzen
1,1,2,2-Tetrachlorethan
1,1,2-Trichlorethan

7. EFFECTS

Effects of the chemicals on WWTP's can be: inhibition on microbial processes, effects on flocculation, or other processes.

This project only included effects on the microbial processes of heterotrophic microorganisms, nitrifiers and anaerobic bacteria.

Table V shows a summary of the toxic levels of some chemicals on heterotrophic microorganisms (22, 23, 24).

Table V: Toxic levels of chemicals on heterotrophic microorganisms (PSEUD = Pseudomonas putida, MIT = Microtox system, AS = Activated sludge). Results obtained from 22 or own experiments.

Chemical	Effect level	Test	Effect concentration
Acrolein	EC10	PSEUD	0.4
Acrylonitril	EC10	PSEUD	53
Anilin	EC10	PSEUD	130
Benzen	EC10	PSEUD	92
1-Butylacetat	EC10	PSEUD	115
2-Butylacetat	EC10	PSEUD	78
Butylamin, aminobutan	EC10	PSEUD	800
Carbontetrachlorid	EC10	PSEUD	30
Cetyl-trimethyl-ammoniumchlorid	EC10	PSEUD	20
4-Chlor-o-cresol	EC20	AS	2.5
Chlormethylphenoxy acetic acid (MCPA)	EC20	AS	173
Chlormethylphenoxypropan acid (MCPP)	EC20	AS	134
Chloroform	EC10	PSEUD	125
2-Chlorphenol	EC10	PSEUD	30
4-Chlorphenol	EC10	PSEUD	20
O-Chlortoluen	EC10	PSEUD	15
M-cresol (3-Methylphenol)	EC10	PSEUD	53
O-cresol (2-Methylphenol)	EC10	PSEUD	33
P-cresol (4-Methylphenol)	EC10	PSEUD	30
1,2-Dichlorbenzen	EC10	PSEUD	15
1,3-Dichlorbenzen	EC50	MIT	97
1,2-Dichlorethan	EC10	PSEUD	135
2,4-Dichlorphenol	EC10	PSEUD	6
Dichlorphenoxypropan acid (2,4-DP)	EC20	AS	153
Diethylamin	EC10	PSEUD	10000
Diisopropylether	LC 0	PSEUD	125
2,4-Dimethylphenol (2,4-Xylenol)	LC 0	E.COLI	500
4,6-Dinitro-o-cresol	EC10	PSEUD	16
2,4-Dinitrophenol	EC10	PSEUD	115
Dioxan, Glycolethylenether	EC10	PSEUD	2700
EDTA	EC10	PSEUD	105
Epichlorhydrin	EC10	PSEUD	55
Ethylacetat	EC10	PSEUD	270
Ethylbenzen	EC10	PSEUD	12
Formaldehyd	EC10	PSEUD	14
Hydroquinon, quinol	EC10	PSEUD	58
Methylenchlorid	LC 0	PSEUD	1000
Methylethylketon	EC10	PSEUD	1150
	LC 0	PSEUD	
Nitrobenzen	EC10	PSEUD	7

(cont.)

4-Nitrophenol	EC10	PSEUD	4
Pentachlorphenol	EC50	MIT	0.08
Phenol	EC10	PSEUD	64
N-propanol	EC10	PSEUD	2700
Pyridin	EC10	PSEUD	340
Sexton, anon	EC10	PSEUD	180
Styren	EC10	PSEUD	72
Tetrahydrofuran	EC10	PSEUD	580
Toluen	EC10	PSEUD	29
1,2,4-Trichlorbenzen	EC10	BOD	5
1,1,2-Trichlorethan	EC10	PSEUD	93
Trichlorethylen	EC10	PSEUD	65
2,4,5-Trichlorphenol	EC20	AS	10
Triethanolamin (TEA)	EC10	PSEUD	10000
Vinylchlorid	EC10	PSEUD	120

Table VI shows ratios between inlet concentrations and significant toxic levels (1).

Table VI: Ratios between expected mean and max. concentration in municipal wastewater versus toxicity on heterotrophic microorganisms.

Chemical	mid. conc. pseud. tox	max. conc. pseud. tox
Toluen	0.00744	0.44827
1,2,4-Trichlorbenzen	0.00620	0.86000
Pentachlorphenol	0.00375	0.00875
Ethylbenzen	0.00183	0.06083
Trichlorethylen	0.00116	0.02769
Methylenchlorid	0.00050	0.04900
Benzen	0.00021	0.01695
Chloroform	0.00013	0.00344
1,2-Dichlorbenzen	0.00006	0.02933
O-cresol (2-methylphenol)	0.00005	0.00033
Phenol	0.00004	0.00017
4-Chlor-o-cresol	0.00004	
Carbontetrachlorid	0.00003	0.06333
1,3-Dichlorbenzen	0.00002	0.00278
1,1,2-Trichlorethan	0.00001	0.00145
2,4-Dichlorphenol	0.00001	0.00005
1,2-Dichlorethan	0.00000	0.56296
4-Chlorphenol	0.00000	
2-Chlorphenol	0.00000	
Vinylchlorid		0.03250

As can be seen, no single chemical has been measured in concentrations having a significant effect. On the other hand WWTP's receive discharges of many chemicals which in combination might affect the activity of the microorganisms in the plant.

8. DISCUSSION

The data base provides a useful tool for handling the huge amount of data collected in this study.

The sorting algorithms in the data base also provides a tool in the hazard evaluation of chemicals with respect to waste water treatment plant processes and receiving waters. Because of the limited data (physical/chemical data, degradability data and data on toxic effects on microorganisms) which is needed to perform the hazard evaluation with respect to WWTP's, it is believed, that many other than the 126 chemicals included in this study can be evaluated with only a limited effort.

Some international cooperation on this field would be of great importance in collecting data and evaluating other chemicals, than those included in this project.

9. ACKNOWLEDGEMENTS

The study has been conducted by contract to the National Agency of Environmental Protection, Denmark. The active participation from the Agency in collecting physical/chemical data is gratefully acknowledged.

10. REFERENCES

(1) Neergaard Jabobsen, B., Lindgaard-Jørgensen, P., Settergren Sørensen, P., Collection, organization and utilization of data on toxic organic pollutants in relation to secondary waste water treatment. Proceedings of a workshop organized within the framework of the concerted action "Behaviour and transformation of organic micropollutants in water treatment processes. COST 641 held in Copenhagen, Denmark, May 20-22, 1985 (to be published).

(2) Feiler, H., Fate of priority pollutants in publicly owned treatment works, Interim Report. EPA-440/1-80/301 (1980).

(3) DeWalle, F.B., et al., Presence of priority pollutants and their removal in sewage treatment plants. Draft final report, Grant R 806102, U.S. EPA, Cincinnati, Ohio.

(4) Interim methods for the measurement of organic priority pollutants in sludge. U.S. EPA, Cincinnati, Ohio (Sept. 1979).

(5) Cohen, J.M., Rossman, L. and Hannok, National survey of municipal waste waters for toxic chemicals. Presented at the 8th U.S.-Japan conference on sewage treatment technology, Cincinnati, Ohio, October (1981).

(6) McCarty, P.L., and Reinhard, M., Statistical evaluation of trace organics removal by advanced waste water treatment. J. Water Pollut. Control Fed., 52, 1907 (1980).

(7) Strier, M.P., and Gallup, J.D., Removal pathways and fate of organic priority pollutants in treatment systems - chemical considerations. Proc. 37th Ind. waste conf., West Lafayette, Indiana (1982).

(8) Kincannon, D.F., et al., Removal mechanisms for biodegradable and non-biodegradable toxic priority pollutants in industrial waste waters. Proc. Ind. wastes symposia, 54th annual WPCF conf., Detroit, Mich. (Oct. 1981).

(9) Kincannon, D.F. and Stover, E.L., Fate of organic compounds during biological engineering (Environmental Engineering Division) annual meeting, Atlanta, GA, July, 1981.

(10) Petrasek, A.C., Austern, B.M., Pressly, T.A., Winslow, L.A. and Wise, R.H., Behaviour of selected organic priority pollutants in waste water collection and treatment systems, JWPCF 55 (10), 1286-1296, (1983).

(11) KinCannon, D.F., Stover, E.L., Nichols, V. & Medley, P., Removal mechanisms for toxic priority pollutants. JWPCF, 55(2), 157-163, (1983).

(12) Matter-Müller, C., Gujer, W., Giger, W. & Stum, W., Non-biological elimination mechanisms in a biological sewage treatment plant. Prog. wat. tech. 12, 299-314, (1980).

(13) Patterson, J.W. & Kodukala, P.S., Biodegradation of hazardous organic pollutants. Chemical engineering progress, 77(4), 48-55, (1981).

(14) Thibodeaux, L.J., Air stripping of organics from waste water. A compendium. Proc. of the second natl. conf. on complete water. Reuse, AIChE, 358-378, Chicago, Ill., (1975).

(15) Kincannon, D.F., and Stover, E.L., Stripping characteristics of priority pollutants during biological treatment. Paper presented at the secondary emissions session of the 74th annual AIChE meeting, New Orleans, La. (1981).

(16) Thibodeaux, L.J., and Millican, J.D., Quantity and relative desorptionrates of air strippable organics in industrial waste water. Environ. sci. technol. 11, 9, 879-883 (Sept. 1977).

(17) Roberts, P.V., et al., Volatilization of organic pollutants in waste water treatment - model studies. Draft final report EPA no. EPA-R-806631. Munic. Environ. Res. Lab., Cincinnati, Ohio (Jan. 1982).

(18) Lyman, W.J., Reehl, W.F., Rosenblatt, D.H. (EDS), Handbook of chemical property estimation methods. Environmental behaviour of organic compounds, Mc-Graw-Hill, (1982).

(19) Stover, E.L. and Kincannon, D.F., Biological treatability of specific organic compounds found in chemical industry waste waters. JWPCF, 55, 1, 97-109, (1983).

(20) Tabak, H.H., Quave, S.A., Mashni, C.I. and Barth, E.F., Biodegradability studies with organic priority pollutant compounds. JWPCF, 53(10), 1503, (1981).

(21) Wukasch, R.F., Grady, C.P.L. and Kirsch, E.J., Prediction of the fate of organic compounds in biological waste water treatment systems. AIChE symposium series (water 1980), 137, no. 209, (1981).

(22) Beckman Instruments. Microtox EC-50 values. Microtox^{tm} application notes, (1983).

(23) Bringmann, G. and Kühn, R., Comparison of the toxicity tresholds of water pollutants to bacteria, algae and protozoa in the cell multiplication inhibition test. Water res. 14, 231-241, (1980).

(24) Verschueren, Karel, Handbook of environmental data on organic chemicals. Second edition, Van Nostrand Reinholdt Company 1983. ISBN: 0-442-28802-6.

S. JOHNSEN[*], J. KRANE[**], G.E. CARLBERG[*], E. AAMOT[**], L. SCHOU[***]

[*]   Center for Industrial Research
        P.b. 350 Blindern, 0314 OSLO 3, Norway

[**]   Department of Chemistry, University of Trondheim
        7055 DRAGVOLL, Norway

[***]   IKU, P.O.Box 1883
        7001 TRONDHEIM, Norway

Summary

A complex mixture of PAH was added to tap water and aquatic humus
respectively. The samples were stored for variable time periods from
4 to 56 days, and PAH was recovered by solvent extraction. 4 main
series of samples were analysed. These were PAH in tap water, in
humus water, in chlorinted tap water and in chlorinated aquatic humus.
The last two series were chlorinated 3 days before solvent extraction.
9 of the components in the PAH-mixture were quantitatively analysed by
GC and GC/MS.

The recovered amount of PAH from aquatic humus was found to be less
than from tap water, due to adsorbtion of PAH to aquatic humus.
Aquatic humus sampled at different times of the year showed variable
behaviour towards adsorbtion of PAH. For chlorinated tap water, a
selective decrease in PAH-recovery was observed compared to the tap
water samples. The reason for this was assumed to be reaction
(oxidation) of PAH by active chlorine. This effect was even more
pronounced for the chlorinated aquatic humus samples. For some
samples a significant decrease in recovery for all the analysed
PAH-components was observed. It was also found that growth of microorganisms in the sample resulted in a recovery decrease for some of
the analysed PAH components.

1. INTRODUCTION

Aquatic humus, recognized by the characteristic yellow-brown colour,
is the water extractable fraction of soil humus. The composition and
structure of humus has been studied for many years and many papers have
been published on this task (1,2,3,4). So far there has been no agreement
of a final model for "an average humus molecule". However, it has been
established that aquatic humus includes a large number of different organic
functional groups. Aquatic humus' ability to adsorb or complex inorganic
and organic micropollutants is well known, and has been shown by many
different authors (5,6,7,8,9,10,11,12,13). Gjessing and Berglind (7) and
Carlberg and Martinsen (6) showed that aquatic humus could adsorb polycyclic aromatic hydrocarbons (PAH). Carlberg and Martinsens work stated that
the adsorbtion of different PAH-components at low concentrations were

selective and not dependent upon the time of contact between PAH and the aquatic humus.

In Norway most of the drinking water sources are surface waters, often containing relatively high amounts of aquatic humus. Because Norway receives a substantial amount of long transported pollutants via the atmosphere in addition to those from local sources, the aquatic humus can be expected to contain complexed organic micropollutants. Much of the humus and thereby the complexed pollutants are removed from those waters using advanced water treatment processes. However, even these treated waters will contain some organic humus. The normal disinfection procedure in Norway is chlorination.

It is well known from the literature that treatment of humus water by chemical oxidants such as ozone and chlorine affects the colour and molecular size variation of the aquatic humus (14,15,16,17,18,19). It is also known that chlorination of PAH in water systems results in reaction between PAH and chlorine (20,21,22).

The purpose of this work was to find out whether chlorination of aquatic humus containing a complex mixture of PAH affected the recovery of PAH (by solvent extraction) compared to recovery from non chlorinated humus water.

2. EXPERIMENTAL SECTION
2.1 Polycyclic Aromatic Hydrocarbons

The mixture of PAH used in the work was isolated from creosote, a tar distillate including a neutral fraction which contains around 60 different PAH components. From the complex mixture of PAH, 9 components were selected and quantified in order to simplify the calculations. These were naphthalene, biphenyl, acenaphthene, fluorene, phenanthrene, fluoranthene, pyrene, chrysene/triphenylene and benzo(a)pyrene. Deuterium labelled biphenyl, d_{10}-biphenyl, 3,6-dimethylphenanthrene and β,β'binaphthyl were used as internal standards for the quantitative analyses.

2.2. Tap water and humus water samples

The tap water samples were treated water from Jonsvannet, a drinking water reservoar for the city of Trondheim, Norway. This water contains low levels of aquatic humus. (Colour below 15 mg Pt/l). Humus water was collected from a marsh (Heimdalsmyra) south of Trondheim, at different times of the year. The humus water was kept in 10 L glass bottles in a fridge at 4 ^{0}C in the dark before use. In order to remove particles, the humus water was filtrated by 0.45 μm filters.

The humus water was diluted with filtrated tap water to a colour of 90 mg Pt/l before PAH was added. The diluted and filtrated humus water was transferred to dark glass flasks (1 L), and PAH was added using the procedure described below.

To the 1 L diluted humus samples a solution of the PAH-mixture in acetone (10 μl) was injected into the water and stirred for 2 hrs. The amount of PAH added to the water samples was well below the solubility for the different components. The total concentration of PAH was approx. 250 μg/l. In the same way PAH was added to 1 L tap water samples. The tap water samples served as blanks to measure the efficiency of the extraction procedure. The humus- and tap water samples were stored at room temperature in darkness for different time periods. Table 1 shows the series of humus samples and tap water samples and the time of storage.

TABLE 1
Non-chlorinated humus- and tap water series

Sample	Storage period (days)	Time of humus water sampling
Sp 4	4	Spring May
Sp 23	23	" "
A 14	14	Autumn Sept/Oct
A 21	21	" "
A 56	56	" "
W 21	21	Winter Jan/Feb
W 4	4	" April
W 7	7	" "
W 14	14	" "
W 21 B	21	" "
T 4	4	Tap water
T 7	7	"
T 14	14	"
T 21	21	"
T 56	56	"

2.3 Chlorinated tap water and humus water samples

Some of the humus and tap water samples were chlorinated 3 days before
the termination of the storage period. These samples were prepared as
described above with two additions. In order to keep the pH at a constant
pH 7.2, a buffer was added to each sample (10 % phosphate buffer, consisting of 62 vol. % 1/15 M Na_2HPO_4 and 38 vol. % KH_2PO_4). To avoid growth of
microorganisms, 0.02 weight % NaN_3 was added to the samples.

Chlorination was performed by adding a relatively concentrated solution of NaOCl (~ 1.0 g/l) to the samples. The amount of NaOCl added was
large enough to give a rest concentration of about 2.0 mg/l free active
chlorine in the samples, measured by neocomparator. Table 2 shows the
series of chlorinated humus- and tap water samples and the time of storage.

TABLE 2
Chlorinated humus- and tap water series

Sample	Storage time days	Time of humus water sampling
A 4 Cl	4	Autumn Sept./Oct.
A 14 Cl	14	"
W 14 Cl	14	Winter Jan./Feb.
W 21 Cl	21	"
T 4 Cl	4	Tap water
T 14 Cl	14	"
T 21 Cl	21	"

To one tap water sample and one humus water sample containing phosphate buffer, NaN_3 was not added. Microorganisms were allowed to grow in
the samples for 14 days, and extraction was then carried out.

2.4 Extraction

The PAH was recovered from the samples by cyclohexane extraction. Each sample was extracted twice with 50 ml solvent under magnetic stirring. The two extracts were mixed, internal standards added and the sample was dried with anhydrous sodiumsulphate. The samples were concentrated to 1.0 ml under a gentle stream of highly purified nitrogen gas.

2.5 Analyses

The extracts from the non chlorinated samples were analysed by a HP 5880A gas chromatograph using a 25 m SE-54 fused silica capillary column with i.d. = 0.33 mm and film thickness 0.17 µm. On column injection was used. A HP-ISTD-computer program was used for quantification.

The extracts from the chlorinated samples gave very complex gas chromatograms, and problems with overlapping peaks often occurred. In addition to the GC-analyses these samples were analysed by GC/MS, using selected ion monitoring (SIM). The analyses were carried out by a HP 5840/5985 GC/MS system with a HP 21 MX-E computer system. The column used was a 25 m S6E.BP-5 fused silica capillary column with i.d. = 0.32 mm and film thickness 0.50 µm. The ionization mode was electron impact of 70 eV at 200 °C ion source temperature. Splitless injection at 265 °C was used.

3. RESULTS

3.1 Precision and accuracy of the results

The results given for each humus- and tap water sample is the average of three parallels. To picture the precision of the results presented, it is important to know the uncertainty in the recovery values. Two factors are of importance for the precision of the results. One is the precision of the analytical method used, the other is the deviation between the sample parallels. To establish the precision of the GC and GC/MS analytical methods, a standard solution of the 9 components and the 3 internal standards was analysed 5 times. Relative response factors for the 9 PAH components were calculated and standard deviation for 5 analyses was found. It was found that the precision for the GC-analyses was extremely good. The relative standard deviation for the relative response factors of the 9 components was between 0 and 0.8 per cent. This is probably due to the injection technique. It is known from the literature that on column injection gives very high precision (23,24).

The precision of the GC/MS SIM analyses was not quite as good as for GC. Relative standard deviation was found to be between 1.3 and 5.6 per cent for the 9 components. The deviation between parallels was usually found to be around 5 per cent, However, for samples giving a low recovery of PAH the precision was worse, even as bad as 100 % for one component (naphthalene) between two parallels in sample A 14 Cl.

The linearity of FID and MS as detectors was also tested. Both were found to give linear response inside the actual concentration ranges for this work.

3.2 Tap water samples

The recovery of the nine PAH components from tap water was on average 90-100 %, for most samples between 95 and 100 %. The recovery was not dependent upon the storage time. However, one sample, T 56, gave lower recoveries for the three components biphenyl, acenaphthene and fluorene. The recovery for these three components were 67, 82 and 75 % respectively. The reason for this is not known.

The recovery values obtained for the other sample types have been adjusted according to the loss of PAH from tap water samples.

3.3 Humus water

Figure 1 shows per cent recovery for 8 of the 9 analysed PAH components from the aquatic humus spring sample (May). It can be seen from the figure that the recovery is dependent upon the storage time and to a certain degreee also on the specific PAH component.

For the autumn- and winter-humus samples the picture was quite different. The recovery of the PAH-components was with very few exceptions, around 90-95 % for all these samples.

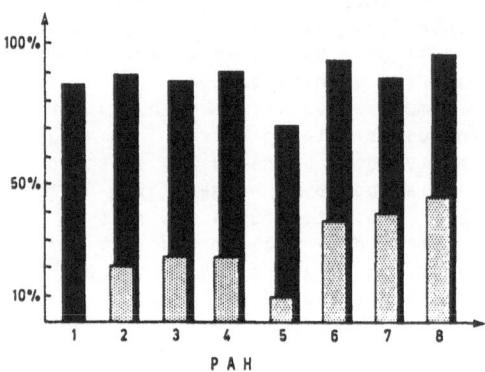

■ 4 days storage

☐ 23 days storage

1. Naphthalene	5. Fenanthrene
2. Biphenyl .	6. Fluoranthene
3. Acenaphthene	7. Pyrene
4. Fluorene	8. Chrysene

Fig. 1 Per cent recovery of PAH compounds from humus water after 4 and 23 days storage. Absence of naphthalene after 23 days is due to analytical problems

3.4 Chlorinated tap water and humus water

The recoveries of 8 PAH components from chlorinated tap water and humus water after 14 days storage is shown in figure 2. The humus quality used was taken in autumn (Sept./Oct.). The recovery of PAH from chlorinated tap and humus water was found to be independent of storage time. The recovery of PAH from chlorinated humus water taken in winter (Jan/Feb) showed a similar pattern.

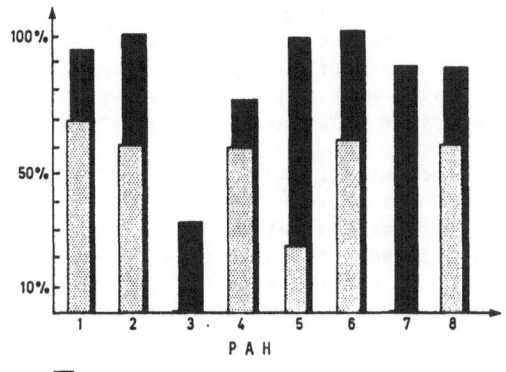

Tap water

Humus water

Fig. 2 Per cent recovery of PAH compounds
from chlorinted tap water and humus
water

For the samples in which microorganisms were allowed to grow, a decrease in recovery of the components naphthalene, biphenyl, acenaphthalene
and fluorene was observed. The recovery of the four components was 5 %,
45 %, 85 % and 75 % respectively. The results from tap and humus water
samples were similar.

4. DISCUSSION

The PAH recoveries from the analyses of the cyclohexane extracts of
the non chlorinated tap water samples, show that this method is satisfactory for PAH analysis in these water samples. Our goal with the experiments
from storage of PAH in non chlorinated humus water was to find out whether
the adsorbtion/complexation of PAH to aquatic humus was dependent upon the
contact time. We also wanted to control the capacity of adsorbtion/com-
plexation of PAH to aquatic humus. However, it was soon understood that an
extremely important parameter for adsorbtion/complexation of PAH to aquatic
humus, was the time of the year when the humus was sampled from the natural
source. Unfortunately, this was realized to late in the project period to
make a complete investigation of this effect.

The results from the analyses of spring humus samples, Fig. 1, show
that this aquatic humus quality was able to adsorb/complex relatively large
amounts of PAH. The adsorbtion/complexation of PAH in these samples was to
a certain degree selective, with phenanathrene giving less recovery than
the other analysed components. This is in accordance with the results from
earlier work by Carlberg and Martinsen (6), where no phenanthrene was
recovered after a short contact time with aquatic humus. For the humus
water taken at other times of the year, the recovery of all the analysed
PAH components was much higher than for the spring samples, usually above
90 %. Anyway, one should have in mind that the concentration of PAH in the
samples are relatively high compared to earlier work. Even a slight reduction in recovery compared to the tap water samples could mean that PAH is
adsorbed/complexed to the aquatic humus. However, this has to be investigated further using lower PAH concentrations.

Chlorinating PAH in tap water result in decreased recovery for some of
the analysed components. This is due to reaction between PAH and active

chlorine. The reaction was found to be selective. Acenaphthene and fluorene showed the lowest recovery from tap water, while biphenyl, phenanthrene and fluoranthene were recovered almost qauntitatively. The components naphthalene, pyrene and chrysene also showed decreased recovery compared to the non chlorinated tap water samples. Later work done in our laboratories indicates that the reaction of PAH with active chlorine may be dependent upon the amount of NaOCl added.

When the same experiments were carried out for PAH in humus water, a significant decrease in recovery for all the analysed components was observed. Figure 3 illustrates this effect for the autumn humus. This humus' quality was shown to give very little adsorbtion/complexatin of PAH, but recovery from the chlorinated samples was generally quite low. The decrease in recovery of the analysed components for the chlorinated humus water samples was also found to be selective. This is most pronounced for the two components acenaphthene and pyrene. Acenaphthene was detected, but the signal to noise ratio was below the limit of quantification in all samples. Pyrene also gave a very low recovery in all chlorinated humus water samples, but its signal to noise ratio was well above the limit of quantification for most of the samples. The recovery of fluorene decreased also for all the chlorinated humus water samples, but the effect was not as strong as for acenaphthene and pyrene.

The recovery of PAH from molecular weight fractions of aquatic humus is now being studied. The actual fractions are MW < 10000, 10000 < MW < 30000 and MW > 30000. Both chlorinated and non chlorinated samples are studied.

Growth of microorganisms seem to influence the recovery of some of the PAH components. It is therefore necessary to use for example NaN_3 to avoid such growth.

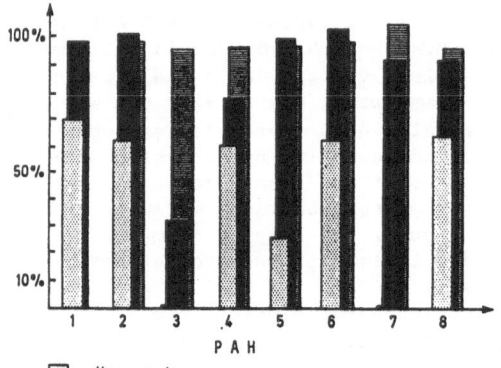

☐ Humus water

■ Chlorinated tap water

▦ Chlorinated humus water

Fig. 3 Per cent recovery of PAH compounds
from aquatic humus, chlorinated
tap water, and chlorinated aquatic
water

5. CONCLUSION
The results from the nonchlorinated samples show that aquatic humus can adsorb/complex PAH. This ability is probably dependent upon the sampling time of humus water from the natural source.

Chlorination of PAH in tap water and humus water results in decreased recovery of PAH from the samples.

There are, as far as the authors can see, two possible explanations for the larger decrease in recovery of PAH from chlorinated humus water compared to chlorinated tap water. One is that aquatic humus act as a catalyst for the reaction between PAH and active chlorine. The other possibility is that chlorination of aquatic humus influence the ability of adsorbtion/complexation of PAH. It is not possible from the existing results to give a full explanation of the observed effects.

6. REFERENCES

(1) GJESSING, E.T. Physical and Chemical Characteristics of Aquatic Humus. Ann Arbor Sci. Mich. USA 1976.
(2) THURKMAN, E.M., WERSHAW, R.L., MALCOLM, R.L., PICKNEY, A.J. Molecular Size of Aquatic Humic Substances. Org. Geochem. Vol. 4 1982, 27-25.
(3) SCHNITZER, M., KAHN, S.U. Humic Substances in the Environment. Dekker New York 1972.
(4) LLAO, W., RUSSEL, F., JOHNSEN, J.D., MILLINGTON, D.S. Structural Characterisation of Aquatic Humic Material. Environm. Sci. Technol. 16 1982, 403-410.
(5) CARTER, C.W., SUFFET, I.H. Binding of DDT to Dissolved Humic Material. J. Am. Chem. Soc. Vol. 16 No 11 1982, 735-740.
(6) CARLBERG, G.E., MARTINSEN, K. Complexation of Organic Micropollutions to Aquatic Humus. Sci. Tot. Env. 25 1982, 245-254.
(7) GJESSING, E.T., BERGLIND, L. Adsorption of PAH to Aquatic Humus. Arch. Hycrobiol. 1 92 1981, 24-30.
(8) TRYLAND, O. Humic Water dissolves metallic iron. Vatten 3 1976 271-273.
(9) BENES, P., GJESSING, E.T., STEINES, E. Interaction between humus and trace materials in fresh water. Water Res. 10 1976, 711-716.
(10) SAAR, R,A., WEBER, J.H. Complexation of Cadmium (II) with water and soil derived fulvic acid: Effect of pH and fulvic acid concentration. Can. J. Chem: 57 1979, 1263-1268.
(11) BOEHM, P.D., QUINN, J.G. Solubilisation of hydrocarbons by the dissolved organic matter in sea water. Geochem. & Cosmochem. Acta 37 1973, 2459-2477.
(12) OGNER, G., SCHNITZER, M. The occurence of alkanes in fulvic acid. A soil humic fraction. Geochem. & Cosmochem. Acta 34 1970, 921-928.
(13) OGNER, G., SCHNITZER, M. Humic substances: Fulvic acid - dietylphtalat complexes and the role in pollution. Science 170 1970, 317-319.
(14) CHRISTMAN, R.T., LLAO, W.T., MILLINGTON, D.S., JOHNSON, J.D. Oxidative degradation of Aquatic Humic material.
(15) NORWOOD, D.L., JOHNSON, J.D., CHRISTMAN, R.F. Reactions of Chlorine with selected Aromatic models of Aquatic Humic material. J. Am. Chem. Soc. 14 2 1980, 187-190.
(16) BECHER, G., CARLBERG, G.E., HONGSLO, J.K., MONARCA, S. High Performance Size Exclusion Chromatography of Chlorinated Humic Water and Mutagenicity studies using the Microscole Fluctuation Assay. Environm. Sci. Technol. In press.
(17) CHRISTMAN, R.F., JOHNSON, J.D., HASS, J.R., PFAENDER, F.K., LLAO, W.T., NORWOOD, D.L., ALEXANDER, H.J. Natural and Model Aquatic Humics: Reactions with Chlorine. Water Chlorination Env. Imp. and Health Effects. Ann Arbor Sci. Vol. 2 1978, 15-28.

(18) NORWOOD, D.L., JOHNSON, J.D., CHRISTMAN, R.F., MILLINGTON, D.S. Chlorination Products from Aquatic Humic Material at Neutral pH. Water Chlorination Env. Imp. and Health Effects. Ann Arbor Sci. Vol. 4 Book 1 1983, 191-200.

(19) CHRISTMAN, R.F., JOHNSON, J.D., PHAENDER, F.K., NORWOOD, D.L., WEBB, M.R. Chemical Identification of Aquatic Humic Chlorination Products. Water Chlorination Env. Imp. and Health Effects. Ann Arbor Sci. Vol. 3 1980, 75-84.

(20) TAYMAZ, K., WILLIAMS, D.T., BENOIT, F.M. Chlorine Dioxide oxidation of Aromatic Hydrocarbons commonly found in Water. Bull. Environm. Contam. Toxicol. 23 1979, 398-404.

(21) OYLER, A.R. Implications of treating Water containing PAH with Chlorine. A GC-MS study. Environm. Health Persp. 46 1982, 73-86.

(22) HARRISON, R.I., PERRY, R., WILLINGS, R.A. PAH in Raw, Potable and Waste Waters. Water Res. 9 1975, 331-346.

(23) SCHOMBURG, G., HUSMANN, H., RITMANN, R. Comparative investigations on Split, Splitless and On Column sampling. J. Chromatog. 204 1981, 85-96.

(24) KNAUS, K., FULLEMANN, J., TURNER, M.P. On Column injection - a precise and acurate injection technique. Technical paper no 94. HP company 1982.

PROPERTIES OF MUTAGENIC COMPOUNDS FORMED DURING CHLORINATION OF HUMIC-WATER

L. KRONBERG[1], B. HOLMBOM[2] and L. TIKKANEN[3]

[1]Department of Organic Chemistry, Åbo Akademi, SF-20500 Turku, Finland.
[2]Laboratory of Forest Products Chemistry, Åbo Akademi, SF-20500 Turku, Finland.
[3]Technical Research Centre of Finland, Food Research Laboratory, SF-02150 Espoo, Finland.

Summary

Mutagenic compounds were formed during treatment of a natural humic water with chlorine. Mutagenic activity increased with increased chlorine dosage and acid chlorination conditions. No reduction in activity was observed during storage of water at pH 2.1 and 20°C for 12 days. Lability of activity was noted in water stored at pH≥4. Mutagens were not oxidized by air and were considered as non-volatile compounds. Extraction with diethyl ether and sorption/desorption with XAD resins were the most efficient of the methods studied for extracting mutagenicity from water. Ether-water distribution of mutagenicity at different pH values indicates that the main active compounds are organic acids with dissociation constants between 4 and 6. Known mutagens from pulp mill effluents can account only for a minor part of the observed mutagenicity.

1. INTRODUCTION

The presence of Ames-mutagenic compounds in drinking water, treated with chlorine, has been noticed in several studies (1). It is generally assumed that the mutagenic compounds are products of reactions of chlorine with humic substances (2). In a recent study (3) we found good correlations between mutagenic activity of a natural humic water treated with chlorine and of drinking waters prepared from surface water containing relative high amounts of humic material. Furthermore reversed phase high performance liquid chromatographic fractionation of extracts of chlorine treated humic water and drinking water revealed that the main mutagens in both extracts could be collected within the same retention times (4). Chlorine treated humic water offers, due to its pronounced activity, a suitable material for studies of mutagens most likely present in chlorinated drinking waters.

Determination of the structure of the main mutagens would be an important step towards a correct assessment of their effects on human health (1,5). One approach to structural elucidation of mutagens is purification and isolation of active compounds by various chromatographic techniques followed by mass spectrometric analyses of pure fractions (6,7). Knowledge of some chemical and physical properties of the main mutagens would facilitate the development of fractionation methods and the handling of active fractions.

In this work we have studied the recovery of mutagenic activity from chlorine treated humic water by liquid-liquid extraction and by sorption/desorption with XAD resins, the diethyl ether - water distribution of activity at different pH values, changes in mutagenicity during storage of water at different pH values, stability against air oxidation and volatility of mutagenic compounds. In addition the influence of chlorine dosage and chlorination pH on mutagenicity of humic water were determined.

2. EXPERIMENTAL

Water sample. Humic water with a total organic carbon (TOC)content of 29 mg/l was collected from a lake situated in a marsh region in the southwestern part of Finland.

Chlorination procedure. Chlorination was performed at pH 7.0 by the addition of a freshly prepared solution of chlorine in distilled water. The amount of chlorine added was 29 mg/l thus achieving the same Cl_2/TOC ratio (1:1) as commonly used during disinfection of drinking waters in Finland. Chlorination was carried out for 60 h to a chlorine residual of <0.1 mg/l (colorimetric determination with N,N-diethyl-p-phenylenediamine). Chlorinated water was examined either immediately or after storage for a few days at pH 2.1 and 4°C.

The effect of chlorine dosage and chlorination pH on mutagenicity was studied by the application of chlorine dosages of 14.5, 29.0 and 43.5 mg/l at pH 3.0, 7.0 and 9.0.

Liquid-liquid extraction procedure. Volumes of 200 ml chlorinated water were acidified to pH 2.1 with 4M HCl and extracted with one 50 ml portion and two 25 ml portions of solvent. Solvents used for determination of extractability of mutagenic activity were diethyl ether (freshly distilled), ethyl acetate, benzene, dichloromethane and hexane. The ether-water distribution of mutagenicity was determined at different pH values of the water phase. Extractions were performed immediately after pH adjustment. Moreover ether-water extractions were carried out when the influence on mutagenicity of chlorination conditions, storage time and boiling was studied.

XAD extraction procedure. A detailed description of the XAD extraction procedure used has been given previously (3). A volume of 2 l water was passed over columns of XAD-4/8. Adsorbed organics were eluted in reverse direction with 100 ml of ethyl acetate.XAD extracts were used to determine the volatility and the sensitivity to air oxidation of mutagenicity.

Influence of storage time and pH on mutagenicity. Samples of chlorine treated humic water were stored at pH 2.1, 4.0, 7.0, 8.5 and 10.0 at 20°C. After storage times of 1, 5 and 12 days the samples were extracted with ether.

Stability against air oxidation and volatility of mutagens. A portion of the XAD extract was evaporated to dryness and stored in an open vessel on a laboratory bench. After 22 h the extract was redissolved in DMSO (dimethyl sulfoxide). Volatility of mutagens were studied by subjecting XAD extracts to rotary evaporation at 10 mmHg for 10 min at 60°C and 100°C, respectively. The residues were dissolved in DMSO.

Mutagenicity tests. The Ames test method used for determination of mutagenicity has been described in detail previously (3). Throughout the work mutagenicity was monitored by Salmonella typhimurium tester strain TA100 without metabolic activation. Prior to Ames test extracts were evaporated just to dryness under a stream of nitrogen and redissolved in DMSO. Results of tests are given as mean value of revertants (rev) from three plates. The standard deviation (S.D.) for revertant counts was typically <10%.

3. RESULTS AND DISCUSSION

Formation of mutagenicity was promoted by high chlorine dosages and acid reaction conditions (figure 1). A linear relation was found for mutagenicity of water and applied chlorine dosages at chlorination pH 3.0. However, at chlorination pH 7.0 and 9.0 an increase of the Cl_2/TOC ratio from 1.0 to 1.5 did not result in an increased activity. During chlorination at pH 7.0 and 9.0 a decline in pH was noted and counteracted by the addition of phosphate buffer and NaOH.

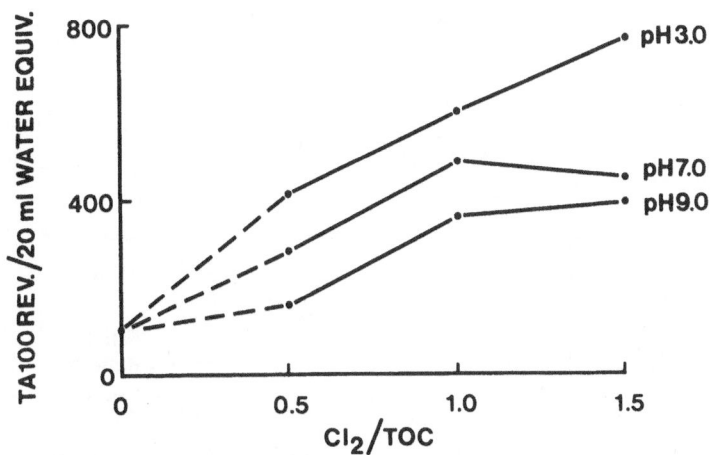

FIGURE 1. Mutagenicity of chlorine treated humic water as a function of chlorine dosage and chlorination pH.
TOC of humic water = 29 mg/l.
Chlorine dosages = 14.5, 29.0and 43.5 mg/l.

These results agree with those of Meier et al (2) who studied mutagenicity of a water solution of commercial soil humic acid treated with chlorine. As stated by Meier and co-workers lower activity in samples chlorinated at neutral and basic conditions can be due to instability of mutagens and/or unfavourable formation conditions.

It is not known whether mutagenic compounds produced at different chlorination pH are identical. For this reason we continued our work with samples of humic water chlorinated at pH 7.0 and a Cl_2/TOC ratio of 1/1.

Chlorine treated humic water could be stored at pH 2.1 for 12 days without any significant reduction in activity (table I). Water stored at pH 4.0, 7.0 and 8.5 for 1 day showed a 10% decrease in activity and in water stored at pH 10.0 about 40% of activity was lost. After storage times of 5-12 days these samples had lost between 60-80% of the mutagenicity. Samples stored at pH≥7 consumed hydroxyl ions and pH had to be readjusted daily. Decline in pH during storage and the alkali lability of mutagenic activity has also been noted by Meier et al (4).

TABLE I. Influence of water pH and storage time on mutagenic activity

| pH | TA100 rev. for 20 ml water equiv. ±S.D.[a] | | | |
	0 day[b]	1 day	5 days	12 days
2.1	565 ± 39	598 ± 15	552 ± 11	565 ± 19
4.0		529 ± 28	247 ± 10	158 ± 15
7.0		537 ± 24	199 ± 12	244 ± 7
8.5		536 ± 30	267 ± 14	264 ± 14
10.0		375 ± 34	183 ± 16	-

- = not measured
a) mean value of three plates ±S.D., spontaneous rev = 113 ±11.
b) humic water treated with chlorine for 60 h. After chlorination pH adjustment was carried out (= 0 day). Samples were stored at 20°C.

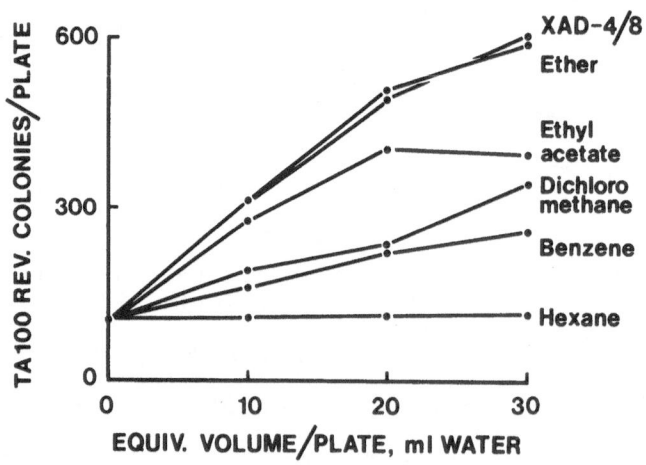

FIGURE 2. Recovery of mutagenic activity from water (pH 2.1) by extraction with different solvents and by sorption/desorption with XAD resins.

Instability of mutagens at neutral pH were also demonstrated in an experiment where chlorine treated humic water was boiled at pH 7 for 10 min. A 70% reduction in activity was noted (table II). This reduction could not be attributed to

volatilization of mutagens as no loss in activity occured when
water was boiled at pH 2.1.

Extraction with ether and adsorption to XAD resins were
the most efficient of the methods studied for extracting
mutagenicity from water (figure 2). The ethyl acetate extract
exhibited a somewhat lower mutagenicity. Less than half of the
activity in the ether extract was found in dichloromethane and
benzene extract.The hexane extract was nonmutagenic.

The lower activity of the ethyl acetate extract compared
to ether and XAD extracts can be due to the high amount of
organic material extractable with ethyl acetate and co-extrac-
tion of substances with inhibiting effects in Ames test.
The same difference in mutagenicity yield of ether and
dichloromethane extractions has been reported by Coleman et
al (8). The negative response of the hexane extract shows that
the responsible mutagens are fairly polar compounds.

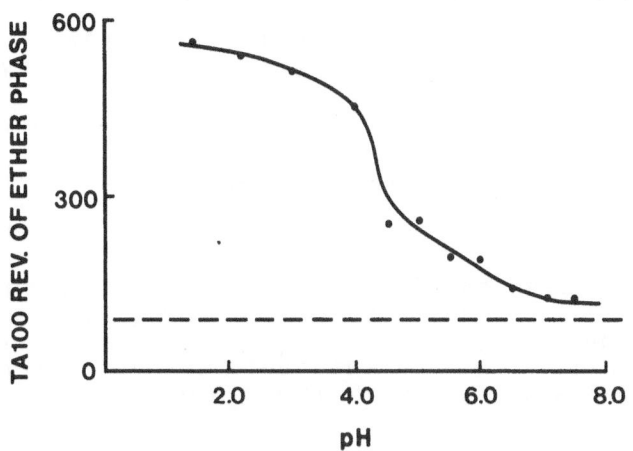

FIGURE 3. Distribution of mutagenicity between diethyl ether
and water at different pH values. The test volume was 20 ml
water equivalents. ---- = spontaneous revertants.

Extraction of activity at different pH values of the
water phase revealed that the highest yield of activity was
obtained at pH≤3 (fig. 3). A 50% decrease in mutagenicity was
observed when extractions were performed at pH 5.0 and 6.0.
Above pH 6.5 hardly any activity could be extracted. Previously we found (3) that the activity of XAD extracts of acidified
water exceeds the activity of neutral by a factor of ten.

The observed distribution of mutagenicity between ether
and water at different pH values suggests that about 50% of
the activity of chlorine treated humic water is caused of
organic acids with dissociation constants of about 4. The
remaining amount of activity could be attributed to organic
acids with pKa values of 5-6.

As the observed distribution of mutagenicity between
organic phase and water does not agree with the distribution
observed for the strong mutagen 3-chloro-4-(dichloromethyl)-5-
hydroxy-2(5H)-furanone (6) or for 2-chloropropenal (9), both

compounds isolated from kraft chlorination effluents, these compounds can be only minor mutagens in extracts of chlorine treated humic water.

Mutagenicity of the XAD extract was not oxidized by air and was not lost during rotary evaporation at 10 mmHg or during boiling at pH 2.1 (table II).

TABLE II. Stability of mutagenic activity against rotary evaporation, boiling and air oxidation.

TA100 revertants for 20 ml water equivalents ±S.D.					
Evaporation[a]			Boiling[b]		Air[c]
N$_2$	60°C	100°C	pH 7	pH 2.1	oxidation
471 ± 11	452 ± 35	436 ± 35	205 ± 26	458 ± 48	473 ± 28
154 ± 20[d]			113 ± 10[d]	±	129 ± 7[d]

a) evaporation was carried out by a gentle stream of nitrogen and by rotary evaporation at 10 mmHg for 10 min at 60°C and 100°C, respectively.
b) chlorine treated humic water was boiled at pH 7 and 2.1 for 10 min.
c) a dried portion of the XAD-4/8 extract was stored in an open vessel for 22 h on a laboratory bench.
d) spontaneous revertants.

The information obtained in this study of chemical and physical properties of mutagens formed during chlorination of humic water should facilitate the development of fractionation and purification methods of mutagenic extracts, aiming at a final structural elucidation of main mutagen(s).

4. REFERENCES
1. Lober J.T., Mutat. Res. 76 (1980) 242-268.
2. Meier J.R., Lingg R.D. and Bull R.J., Mutat. Res. 118 (1983) 25-41
3. Kronberg L., Holmbom B. and Tikkanen L., Vatten 41 (1985) 106-109
4. Kronberg L., Suomen Akatemian julkaisuja 3 (1985) 62-69.
5. Kool H.J., van Kreijl C.F. and and Zoeteman B.C.J., Crit. Rev. Environ. Control 12 (1982) 307-357.
6. Holmbom B., Voss R.H., Mortimer R.D. and Wong A., Environ. Sci. Technol. 18 (1984) 333-337.
7. Kronberg L., Holmbom B. and Tikkanen L., Sci. Total Environ., in press.
8. Coleman W.E., Munch J.W., Kaylor W.H., Streicher R. P., Ringhand H.P. and Meier J.R., Environ. Sci. Technol. 18 (1984) 674-681.
9. Kringstad K.P., Ljungquist P.O., de Sousa F. and Strömberg L.M., Environ. Sci. Technol. 15 (1981) 562-566.

Behaviour of Organic Micropollutants during Infiltration of
River Water into Ground Water : Results of a Field Study in the
Glatt Valley, Switzerland

C. SCHAFFNER, M. AHEL AND W. GIGER

Swiss Federal institute for Water Resources and
Water Pollution Control (EAWAG)
CH - 8600 Dübendorf, Switzerland

Summary

Nitrilotriacetic acid (NTA), pentachlorophenol (PCP), nonylphenol
(NP), nonylphenol monoethoxylate (NP1EO) and nonylphenol diethoxylate
(NP2EO) have been studied at a field site of river water infiltrating
into the ground water. During one year samples were collected at
approximately monthly intervals from the river and from the ground
water using observation wells in different distances from the river
bed. NTA, occurring in the river at concentrations of 8 to 83 mg/m^3,
was rapidly eliminated within the first two meters of infiltration by
biological degradation. The PCP concentrations of 0.10 to 0.25 mg/m^3
in the river were lowered to 0.02 to 0.07 mg/m^3 in the ground water
after an infiltration distance of 7 m. NP, NP1EO and NP2EO showed a
behaviour corresponding to an intermediate persistance in comparison
to the two other micropollutants.

Introduction

The fate of organic micropollutants during ground water infiltration is of
great interest since many water works use bank filtration as a first step
in the treatment of river water for public water supplies. Field and laboratory studies are necessary to enhance our knowledge on the behaviour
of organic chemicals during infiltration of river water to ground water.
In this paper we report on recent field work studying pentachlorophenol
(PCP), nonylphenol (NP), nonylphenol monoethoxylate (NP1EO), nonylphenol
diethoxylate (NP2EO) and nitrilotriacetate (NTA). The field site for our
investigation was in the lower Glatt Valley, Switzerland, where the Glatt
River infiltrates into a quarternary fluvioglacial valley fill aquifer.
The Glatt River is a small, rather heavily polluted perialpine river
which receives effluents from ten mechanical-biological treatment plants
of municipal waste water (1). At the field site the average discharge of
the river is approximately 8 m^3/sec and permanent infiltration of the
river through a saturated zone can be assumed. Observation wells allowed
the sampling of freshly infiltrated water at various distances (2.5-13 m)
from the river. During one year seventeen sample series were collected at
approximately monthly intervals including samples from the river and from
four ground water observation wells.

Analytical Methods

	Enrichment	Derivatization	Separation	Detection	Ref.
PCP	C-18-silica	acetylation	HRGC	ECD	(2)
NTAH$_3$	anion-exchange	esterification to tri-n-butylester	HRGC	NPD	(3)
NP NP1EO NP2EO	steam- distillation/ extraction	---	normal-phase HPLC	UV	(4,5)

Results and Discussion

The observed averages and ranges of concentrations are given in Table 1. It was concluded that NTA is eliminated rapidly during ground water infiltration. Starting from a range of 8 to 83 mg/m^3 and an average of 27 mg/m^3 in the river after 7 m of infiltration only 0.5 mg/m^3 are left corresponding to an elimination of 98% (Fig. 1). Low temperatures in winter (4-6°C) and reduced oxygen contents in summer had no effect on the efficient elimination of NTA. This result is highly important in addressing the question as to what extent NTA might reach bank filtrated waters from polluted rivers (Fig. 2). The phenolic pollutants were eliminated according to the sequence:

$$NP1EO \approx NP2EO > NP > PCP$$

This is based on the decrease of the average concentrations over the first seven meters of infiltration. In particular, PCP turned out to be rather persistent in the ground water but not to such a degree as tetrachloroethylene and other chlorinated solvents which had been studied earlier (6). (Fig. 3,4)

References

(1) Ahel, M., Giger, W., Molnar-Kubica, E. and Schaffner, C. In "Analysis of Organic Water Pollutants", Angeletti, G., Bjørseth, A., Eds., Reidel: Dordrecht, Holland, 1984, pp. 280-288.

(2) Renberg, L. and Lindström, K., J. Chromatogr. 214, 327-334 (1981).

(3) Schaffner, C. and Giger, W., J. Chromatogr. 312, 413-421 (1984).

(4) Ahel, M. and Giger, W., Anal. Chem. 57, 1577-1583 (1985).

(5) Giger, W., Ahel, M. and Schaffner, C. In "Analysis of Organic Water Pollutants", Angeletti, G., Bjørseth, A., Eds., Reidel: Dordrecht, Holland, 1984, pp 91-109.

(6) Schwarzenbach, R.P., Giger, W., Hoehn, E. and Schneider, J.K., Environ. Sci. Technol. 17, 472-479 (1983).

Sample Compound	Glatt River	Observation well			
		P 1	P 2	P 3	P 4
Pentachlorophenol (PCP)	0.17 (0.08 - 0.35)	0.12 (0.08 - 0.16)	0.09 (0.05 - 0.14)	0.05 (<0.02 - 0.08)	0.03 (<0.02 - 0.09)
Nitrilotriacetic acid (NTA H_3)	27 (8.4 - 83)	1.3 (0.4 - 3.6)	1.2 (0.4 - 3.2)	0.5 (0.25 - 1.1)	0.3 (<0.2 - 1.1)
Nonylphenol (NP)	4.1 (0.7 - 26)	1.0 (0.14 - 3.1)	0.5 (<0.1 - 1.4)	0.5 (<0.1 - 1.5)	0.3 (<0.1 - 1.1)
Nonylphenol monoethoxylate (NP1EO)	7.5 (2.0 - 20)	1.0 (<0.1 - 4.8)	0.3 (<0.1 - 1.5)	0.2 (<0.1 - 0.3)	0.1 (<0.1 - 0.2)
Nonylphenol diethoxylate (NP2EO)	8.2 (0.8 - 21)	0.4 (<0.1 - 1.6)	0.3 (<0.1 - 1.1)	0.2 (<0.1 - 0.4)	0.1 (<0.1 - 0.2)
Distance from river bed (m)	0	2.5	5	7	13

Table 1: Concentrations of selected organic micropollutants in the Glatt River and ground water observation wells. Average concentrations in mg/m^3. ( ) Observed ranges of 17 determinations.

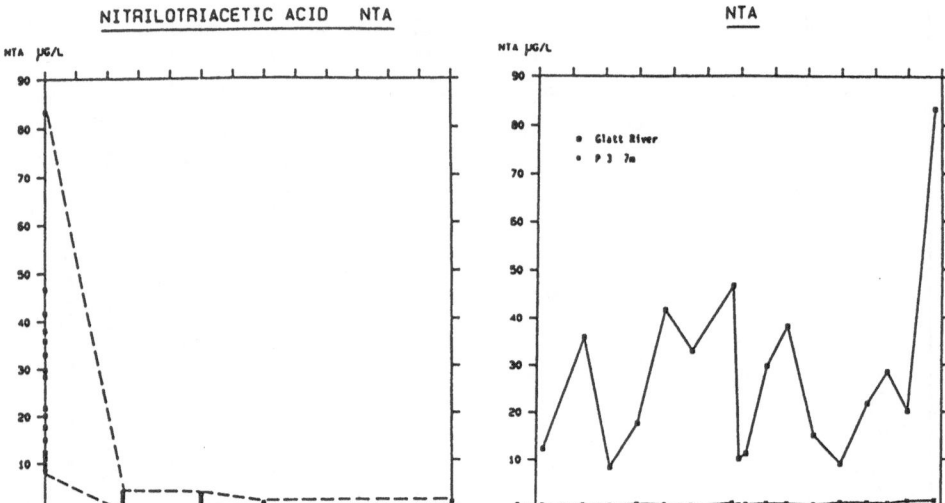

Fig 1. Concentration profile of NTA in the Glatt River and ground water observation wells.

Fig 2. NTA concentrations in the Glatt River and observation well P3 as a function of time.

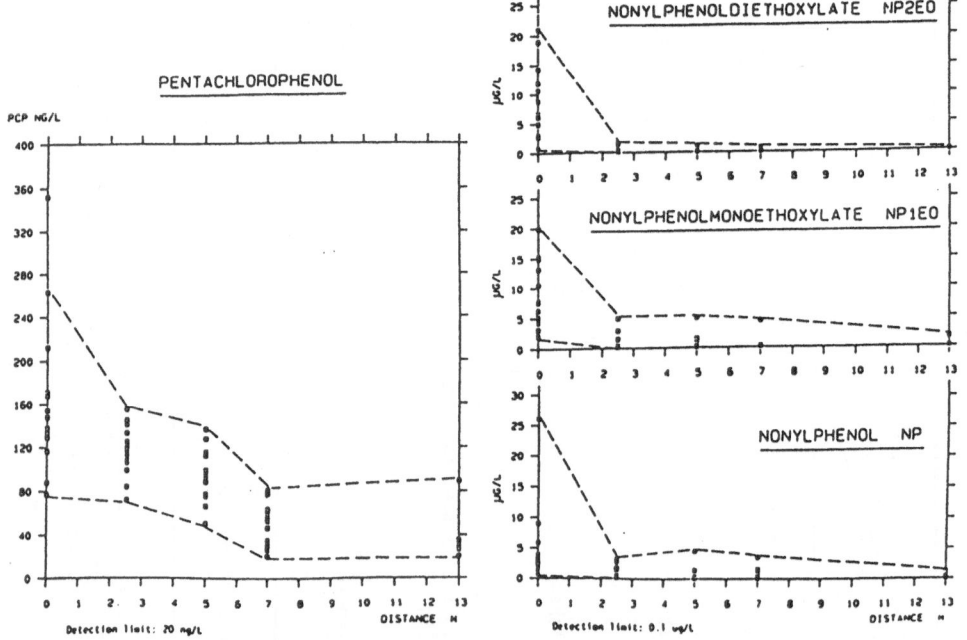

Fig 3 and 4. Concentration profiles of pentachlorophenol, nonylphenol and nonylphenolethoxylates in the Glatt River and ground water observation wells.

THEORETICAL ASPECTS AND FUTURE ACTIVITIES

Mathematical modelling of the behaviour of organic micropollutants in the aquatic environment

Quantitative structure-activity relationships in aquatic toxicology

Indices for the prediction of environmental properties of hetero-atomic polycyclic aromatic pollutants

Integration and use of the list of organic micropollutants in the aquatic environment in the ECDIN data bank

Environmental research of the European Communities in the future

MATHEMATICAL MODELLING OF THE BEHAVIOUR OF
ORGANIC MICROPOLLUTANTS IN THE AQUATIC ENVIRONMENT

D. M. IMBODEN
Swiss Federal Institute for Water Resources and Water Pollution Control
(EAWAG), Dübendorf (Switzerland)

Summary

Modelling of the behaviour of micropollutants in the aquatic environment serves to generalize to other systems the knowledge gained from
measurements in a specific situation, provided that certain rules are
obeyed. In order to use a model as a tool for predicting concentrations in natural systems, calibration of the model is necessary. The
model should be as simple as possible and only contain quantities
which can be actually measured. The proper choice of model dimensionality, a key factor of model complexity, can be made by comparing reaction and mixing rates.

1. THE OBJECTIVE OF MODELLING

The behaviour of micropollutants in the aquatic environment is extremely complex, since it simultaneously reflects the interplay between the
specific properties of the chemical constituent and the environmental
systems. The physico-chemical characteristics of the constituent, such as
in situ reactivity, sorption, and volatilization may be quite complicated
already, especially if the chemical transformations are mediated by microbial processes. If they are combined with the complexity of the mixing and
transport pattern of a natural aquatic system, the resulting behaviour is
even more puzzling to an observer who tries to understand the faith of the
pollutant in the environment.

In fact, the ability to measure the distribution of a specific pollutant in a given natural system is still a significant step behind the ability to understand and interprete the measured concentrations. While the
former yields information only on one specific situation (that is on something which has happened already), the later provides the base for genera-
lizing the knowledge acquired for one situation to different systems and
times, that is for making predictions.

Models are tools for surmounting the barrier between measuring and
understanding; in fact, the more complicated the situation, the more important is the role of the models, since modelling means an intentional
simplification of a complex and thus intractable system. The model is not
meant to reflect all the detailled structure of the real system, but to
select and enhance the important features.

This article gives a brief overview of some principles which are
important for the mathematical modelling of environmental systems. It is
neither intended to present one specific model nor to summarize the state-
of-the-art of aquatic modelling (see (1)).

2. REACTION VERSUS TRANSPORT RATES

The concentration C of a chemical constituent at some specific location in an aquatic system (a lake, for instance) may undergo a temporal change due to two kinds of processes: On the one hand, the chemical may be transported from or to the location due to mixing processes such as diffusion or water currents; on the other hand, the chemical may be locally produced or destroyed by some chemical or biological reaction.

The best way to compare the two types of phenomena is by introducing the concept of rates of concentration change. For an in situ reaction, such as hydrolysis, photolysis, biodegradation etc., the temporal change $(dC/dt)_{reaction}$ may be expressed as

$$(dC/dt)_{reaction} = - k_r C \tag{1}$$

The reaction rate is positive if the chemical is consumed, negative if it is produced. Note that eq. (1) is valid whether the reaction is first-order or of another kind ! For a first-order reaction (such as radioactive decay), k_r is independent of C. However, if the reaction is not first-order (but second order, for instance), the rate k_r depends on the concentration C (e.g. $k_r = k \cdot C$, for a second order reaction). Thus, the rate k_r is nothing else than the actual relative concentration decrease due to the reaction:

$$k_r = - \frac{1}{C} \left(\frac{dC}{dt} \right)_{reaction} = - \frac{d}{dt} (\ln C)_{reaction} \tag{2}$$

k_r has the dimension $[time]^{-1}$. The inverse,

$$\tau_r = k_r^{-1} \qquad [time] \tag{3}$$

is the mean reaction-time of the chemical. If the reaction is first-order, the reaction time is constant and equal to the mean life time of the chemical with respect to the reaction. Otherwise, τ_r varies with the concentration C, and the mean life time of the chemical is more complicated to evaluate.

In a similar way as done for reactions we can introduce a mixing rate k_{mix} or mixing time $\tau_{mix} = k_{mix}^{-1}$. Let us select a "test volume" (for instance a cube of volume L^3) and ask the question: What is the time τ_{mix} needed to mix the water (or the chemical) within this cube ? If the mixing occurs by water currents (advection), this time is of the order

$$\tau_{mix} = \frac{L}{v} \text{ or } k_{mix} = \frac{v}{L} \tag{4}$$

(v: current velocity);
if the mixing is by (molecular or turbulent) diffusion, the time is of the order

$$\tau_{mix} = \frac{L^2}{D} \text{ or } k_{mix} = \frac{D}{L^2} \tag{5}$$

where D (dimension [length^2 time^{-1}]) is the diffusion coefficient. Note that we have used the expression "of the order" to indicate that the above equations are order of magnitude estimations, only.

Comparing τ_r and τ_{mix} yields a first and simple answer to the question whether reaction or mixing would be dominant within the test volume (2). If $\tau_r \gg \tau_{mix}$, the mixing processes would immediately destroy all concentation inhomogeneities which may result from a spatial variation of the reaction rate within the volume. Yet, if $\tau_{mix} \gg \tau_r$, the concentration distribution may become nonconstant if the reaction rate varies, since the mixing is too slow to equalize different consumption or production rates. Finally, if τ_r and τ_{mix} are of the same order, both kinds of processes are important for the concentration distribution.

The analysis of time scales (or rates) is a simple but extremely powerfull tool for the interpretation of field data and for the selection of the proper model. The next chapter shall illustrate this point.

3. MODEL DIMENSION

Organic micropollutants usually enter the aquatic system at a boundary, either as a point source (e.g. input of waste water and polluted rivers into lakes or ground water systems) or as surface source (input from atmosphere by gas exchange or precipitation; diffusive runoff into rivers and lakes, infiltration of surface waters into groundwater). - An exception is given, for instance, by the case of a pollutant which is derived by chemical transformation from another chemical distributed over the whole system (volumetric source); we are not discussing such a situation any further. - To the expert who wants to assess the faith and pathway of the pollutant, the following questions are important:

(1) How does the spatial concentration pattern of the pollutant look like if the boundary sources are known ?
(2) What can be deduced from the measured concentration distribution with respect to the chemical transformations of the pollutant within the aquatic system ?

The first question is important for designing the optimal sampling strategy, the second to generalize the information on the chemical in such a way in order to make predictions for other systems. Of course, the questions are not independent of each other: If the concentration field is known from measurements, information can be deduced with respect to the reactivity of the pollutant; on the other hand, the reactivity gives information on the spatial distribution of the chemical (see section 2). The mathematical model suitable for describing the two aspects of the pollutant, transport and reactivity, should have the simplest possible form but still depict the important features of the situation. The spatial dimensionality of the model is a key factor determining model complexity. It extends from the simplest completely mixed system ("zero" dimension), the one-dimensional model (river model, or vertical lake/ocean model) to the fully three-dimensional transport model. The principle of rate analysis serves as a base for chosing the proper model dimension.

As an example, let us consider the case of nitrilotriacetic acid (NTA) in surface waters. NTA is used in detergents and other consumer products; it enters the aquatic environment in (treated or untreated) waste water. Biodegradation is considered the only significant removal process for NTA. laboratory experiments yield reaction rates of about 2d^{-1} at concentrations around 50 mg m^{-3}, but only about 0.1 d^{-1} at 5 mg m^{-3}. These values embrace the concentrations measured in Swiss rivers near urban areas. At concentrations which are typical for Swiss lakes (1 mg/m^3 or less), the reaction rate are between 0.02 d^{-1} and about 0.005 d^{-1} (3).

In most rivers, typical vertical and lateral mixing times lie between less than a minute and less than an hour. Thus, NTA can be considered to be homogeneously distributed within a river cross-section, with exception of the river section just down-stream of a waste water input. A one-dimensional model is the proper choice for describing NTA in such rivers.

In lakes, with areas not larger than a few hundreds square kilometers, the relevant horizontal mixing rates are 0.1 d^{-1} or more (4). Thus, no significant horizontal NTA concentration gradients are expected (again, with exception of the region close to waste water inlets). In contrast, vertical mixing may be as small as 10^{-5} d^{-1} during the summer (lake stratification), but much larger (0.001 to 1 d^{-1}) during the winter (lake circulation) (4). Thus, we expect to find significant vertical concentration gradients for NTA during the summer. Measurements from several Swiss lakes confirm this conclusions (3). Application of a one-dimensional vertical lake model is the appropriate way for the interpretation of NTA concentrations in these lakes (4).

4. THE GOLDEN RULES OF MODELLING

As long as models are simply used to "mimic" measured data, any model which yields a satisfying agreement between field data and calculation seems to be appropriate. Yet often, very different models lead to similar results. Therefore, if the model should also serve to make predictions (for the same or another system), it is vital to know which model is correct, especially since the different models giving near identical results for one system may lead to very different results for another one.

There is no absolutly safe way to find the correct model, since we can never test all possible models. Yet, there are a few rules which make the search for a model with predictive power more secure:
(1) Use the simplest model compatible with the measured data set (simple with respect to dimensionality, number of chemical species, number of parameters).
(2) Do not introduce variables (concentrations of chemical species) which are not measured. "Hidden" variables often make the model arbitrary.
(3) Every model should contain a list of conditions which have to be checked for the specific situation to which the model should be applied.
(4) Validate a model by applying it to a set of data which was not used when the model was constructed and calibrated. If the model describes the new data set without change of calibration parameters, its predictive credibility is greatly enhanced.

Obviously, the information available for a certain pollutant in a certain aquatic environment is often too scarce in order to fulfil the above requirements, especially the one on calibration. The process of modelling is an iterative one, beginning with a vague idea and hopefully ending with a validated concept. As always in science, success not only depends on the application of rules, but also on intuition and good luck.

REFERENCES

(1) JORGENSEN, S.E. (1984). Modelling the fate and effect of toxic sub-
     stances in the environment. Ecolog. Modelling, vol. 22, special
     issue.
(2) IMBODEN, D.M. and LERMAN, A. (1978). Chemical models of lakes. In A.
     LERMAN [ed.], Lakes: Chemistry, Geology, Physics, Springer, New York.
(3) GIGER, W., SCHNEIDER, J., IMBODEN, D.M., SCHWARZENBACH, R.P., MOLNAR-
     KUBICA, E. and SCHAFFNER, C. (1985). Transport and transformation
     processes of organic pollutants in Swiss lakes. Abstract Div.
     Environm. Chemisty, Amer. Chem. Soc., Chicago, Illinois, September
     1985.
(4) IMBODEN, D.M. and SCHWARZENBACH, R.P. (1984). Spatial and temporal
     distribution of chemical substances in lakes: Modelling concepts. In
     W. STUMM [ed.], Chemical processsis in lakes, Wiley, New York.

QUANTITATIVE STRUCTURE-ACTIVITY RELATIONSHIPS IN AQUATIC TOXICOLOGY

HANS KÖNEMANN
Directorate-General for Environmental Protection,
Chemicals Division, Leidschendam,.
The Netherlands

Summary
An overview is given of the state-of-the art in quantitative
structure-activity relationships (QSARs) in aquatic toxicology.
Three entrances are used to discuss the literature: the types of
toxicity tests used in QSAR analysis, the nature of chemicals for
which QSARs have been determined succesfully and the physical and
chemical parameters used.

The reasons for succes of the QSAR approach in aquatic
toxicology are discussed and it is concluded that calculation of
QSARs helps in some cases to predict the toxicity of chemicals
and is in all cases useful for a better understanding of the
factors which determine the toxicity of a chemical.

1. INTRODUCTION

This paper is not meant to give a complete scientific review of
the use of quantitative structure activity relationships. It is
intended to give a· short introduction on its background and some
impression of its possibilities and limitations. A rather extensive
list of references is included to help readers with a more deep-going
interest.

The study of the relationship between biological activity and
physico-chemical parameters has a long history, starting in the end of
the last century with a rapid development in the last 20 years, with
the work of Corwin Hansch as the catalysing factor (1,2). This
development took mainly place in the field of pharmacology, where
structure-activity relationships proved to be useful in drug design.
The calculation of quantitative structure-activity relationships
(QSARs) for aquatic toxicity studies started 10 years later. One of
the earliest applications of the Hansch approach has been presented by
Kopperman et al (1974). They calculated a QSAR for the toxicity of a
series of phenols to Daphnia magna:

$$\log 1/c = 0.500 \, \pi + 0.4537 \, F + 0.636 \, R + 3.731$$
$$n = 14 \quad r = 0.978$$

c is the 48h EC 50
π is the hydrophobic substituent constant
F and R are electronic parameters

This equation shows the main characteristics of a QSAR in the Hansch
tradition: a multiple regression equation in logarithmic form is
calculated, in which toxicity is considered as a function of physical
and chemical properties of molecules or molecular fragments. A
thorough discussion of the Hansch approach can be found e.g. in Martin
(4), Seydel and Schaper (5) and Topliss (6). Some other examples of
early applications of QSAR on aquatic toxicity (and bioconcentration)
can be found in Veith & Konasewich (7). In particular the last few
years the number of publications in this field is rapidly increasing.

2. Analysis of the literature

In stead of giving a detailed literature review, we prefer to use a few entrances to characterise and analyse the available literature along these lines.

2.1. Types of tests in QSAR studies
species used

Many fish species are used, clams, crustaceans, algae and bacteria. The quality of the calculated QSARs does not seem to be dependent on the actual choice of the species. This choice therefore seems to reflect what is generally used in aquatic toxicity. These species have one thing in common: for an efficient gas (O_2, CO_2) exchange with water, they have a high surface/content ratio, which also allows an efficient exchange of chemicals with water. We are only aware of two systematic comparisons of QSARs of the same group of chemicals to different aquatic species (8, 9). In both cases QSARs have been calculated for the toxicity of chemicals, resp. to 4 and 19 species. The groups of chemicals are rather small and not optimal for QSAR calculation. The great number of species, however, makes it possible to draw a tentative conclusion: for chemicals with a non-specific mode of action, the applicability of a QSAR is not very species dependent. This is probably more than a circular argument since the specificity of the toxicity of chemicals depends on the mode of action as will as on the kinetics of the chemicals. The latter variable is less species dependent for aquatic animals than for terrestrial animals.

test duration

Most studies concern acute lethality tests with aqueous exposure of a duration of a few days. Only recently some results have been published of QSAR calculations for on sublethal effects, such as reproduction and growth of Daphnia magna (8, 10) and the development of early life stages of fish (11, 12) as toxicity parameter. The reason for this is clear: many quantitative toxicity data are needed to calculate a useful QSAR. Producing (sub)chronic data is rather laborious and often they are less precise than acute data, which affects the quality of a calculated QSAR. Therefore, in its analyses and conclusions, this paper is mainly limited to acute tests.

2.2. Types of compounds in QSAR studies

Most QSARs have been calculated for series of homologues, in particular for aromatic compounds. Phenol-derivatives have probably most widely been studied (3, 13-18). Some studies have been reported on aniline-derivatives (15, 19) and on chlorobenzenes (18, 20) and aliphatic amines (21). Schultz et al have published QSAR calculations for nitrogeneous heterocyclic compounds (22-24). Also organometal compounds have been studied: several relationships for organotin compounds have been published (25-27). All these QSARs have in common that for a series of rather close homologues high correlation coefficients are found, frequently in the range of 0.95 - 0.99. In a few cases it appeared to be possible to aggregate series of close homologues to larger series, without affecting the quality of the QSAR very much. More aggregations are probably possible when additional parameters are used.

There is one noticeable exception on the apparent rule of
limitation of a QSAR to a narrower or broader series of homologues.
The QSAR calculated by Könemann for chlorobenzenes (20) proved to be
also applicable to several other types of non-reactive,
non-electrolyte chemicals, such as (chloro)toluenes, cloroalkanes and
₋alkenes, alcohols and ethers. For this relationship the range of
toxicity (LC50-fish) covers more than 6 orders of magnitude (fig. 1).

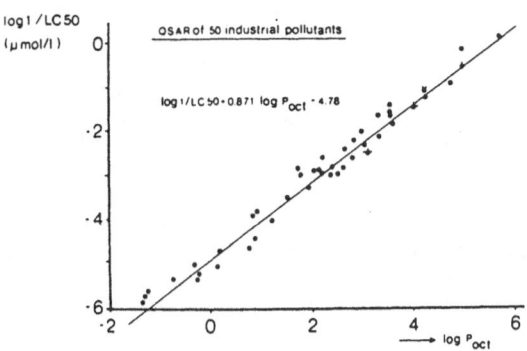

Fig. 1 QSAR for 50 industrial pollutants (from ref. 43)

As a matter of fact, this finding was not very new, something similar
was already known since the end of the last century (28-31): the
toxicity of narcotic substances is not very structure-specific and
mainly depends on its physico-chemical characteristics (aqueous
solubility, fat/water partition coefficient). The validity and the
broad applicability of this QSAR has been confirmed by other authors
(32-34). All these authors stress that, with some exceptions for
different reasons, this QSAR gives the minimum toxicity of a chemical:
each non-electrolytic organic substance is at least as toxic as
predicted by this QSAR. If a chemical exerts a more specific type of
action or when it is more reactive, it will be more toxic than
predicted.
      This indication of the types of compounds for which QSARs have
been calculated is far from complete. Only the most frequently studied
groups of chemicals have been mentioned. The literature shows in this
respect at the moment an ongoing diversification.

2.3. Types of parameters
hydrophobicity:
      In nearly all QSARs a parameter for hydrophobicity is used, most
frequently the octanol/water partition coefficient (P) or the water
solubility. These two are highly correlated, in particular for
liquids. The advantage of P is that it can be calculated easily in
many cases on the bases of the work of Hansch and Leo (35) or Rekker
(36). Other parameters which are sometimes used (particularly in older
studies) are moleculair weight, number of substituents etc. As a
matter of fact, for homologuous series many parameters give satisfying

results, since they are all correlated with P and water solubility. Our advise would be to use the most generally available parameter - log P - in all cases to represent hydrophobicity, in order to facilitate the comparability of QSARs, unless strong arguments against its use can be given.

It is necessary to stress that the quality of a QSAR can be impaired by the inaccuracy of the log P values used. The experimental determination of log P is somewhat tricky for high values: phase separation is usually incomplete, leading to apparantly lower P-values. The highest experimental value found in literature is the best, in general (for water solubility the opposite holds: the lower the value, the better). Because of these experimental problems, calculated log P values may give better results than experimental values (20). As an alternative for the shake-flask determination, HPLC or TLC techniques are sometimes use, based on the correlation between P and liquid chromatographic retention times (37-39). It seems to be necessary, however, to develop experimental P-values with other techniques than flask-shaking, avoiding phase separation problems, for series of highly hydrophobic chemicals.

The reason for the succes of hydrophobicity in predicting toxicity is obvious. The bioconcentration of a chemical depends largely on this factor, as is generally known (40).

electronic and reactivity parameters

The most generally applied electronic parameters is the Hammett constant ( ) (originally developed to predict reactivity in organic synthesis (52) or Δ pK$_a$ from which the Hammett constant is deduced. Some authors reached some improvement by splitting up the Hammett constant in a field and resonance parameter, according to Swain and Lupton (3, 41). The use of electronic parameters is desirable where chemicals have strong polar groups, such as phenols and anilines and aliphatic amines. Such chemicals can be ionized to different degrees dependent on their pK$_a$ and the environmental pH.

Since it is well-known that ionisation affects the uptake of chemicals, it is not surprising that QSARs for this type of chemicals are pH dependant. This has been extensively discussed by Könemann (16) and Saarikoski (17). The latter one gives one pH-independent QSAR with a separate correction formula.

A relatively new parameter, which can partly be mentioned electronic in nature, is the rate constant for nucleophilic substitution reactions with n-nitrobenzylpyridine. This parameter had been used as QSAR parameter for in vitro experiments measuring mutagenicity. Hermens et al (42) showed that it could explain the difference in 14-day LC50s (fish) of relatively unstable organohalides. A drawback of this parameter is that no calculation method for rate constants is available nor an extensive data base.

steric parameters

In aquatic toxicology steric parameters are less frequently used in QSAR analysis than in pharmacology. The only parameter used more or less frequently seems to be the molar refractivity (MR), an indicator of Van der Waals interactions and therefore electronic in nature but often strongly correlated with molar volume (44, 24). For this reason MR is used to represent steric influences.

molecular connectivity indices

Molecular connectivity indices (MC) are topological parameters, calculated from numerical values assigned to the atoms in a molecule depending on their valencies introduced in its present form by Hall and Kier (45). An advantage of MC is the fact that they can be calculated without any need for experimental determinations. Several MC parameter are used, but most frequently the first order MC index is used. In early publications it was suggested that MC was a good predictor for water solubility, boiling point and partition coefficient. In a more theoretical paper Kier and Hall (46) stressed the electronic information contained in MC. So many virtues seem to be more than any type of parameter can cope with. Thus it is not surprising that it could be shown that MC could replace log P to predict fish toxicity for a limited homologues series of chemicals, but that it failed for a non-homologues series, where log P was still succesful (20).

In several recent publications MC is used as QSAR parameter (47-49), either alone or in combination with other parameters. It seems to be too early, however, to predict a bright future for MC.

other parameters

In QSAR analysis outside the field of aquatic toxicology several other parameters are used frequently. A clear and concise overview of parameters used in QSAR analysis is given in Hansch and Leo (35), which also contains an extensive list of partition coefficients and electronic and steric constants.

statistical treatment and selection of parameters

In most cases multiple linear regression analysis is applied. Often many regression equations are calculated and the preferred equation is selected on the basis of the significance of the different parameters, the correlation coefficient and the standard deviation. In the ideal situation this standard deviation is only determined by the reproducibility of the toxicity tests. The more parameters are used, the more toxicity data are necessary to perform a good analysis. The ready availability of computerized data processing techniques in every laboratory sometimes seems to hinder a critical assessment of the importance of QSAR parameters. The final choice of the QSAR equation should, in our opinion, not only depend on the statistical characteristics of the equation but also depend on the possibility to understand the influence of the parameters in a QSAR equation. This does not only apply to the selection of the parameters, but also to the form in which they appear in the equation. This is not an easy task in multiple regression analysis. In some cases, however, it is obvious that simple linear regression analysis is not the most appropriate. Well known are parabolic and bilinear relationships of biological activity with log P (50, 51). Hermens showed that the reactivity parameter he used also had a bilinear log influence on toxicity (42).

A bilinear influence can also be expected for pK_a if chemicals with op proper pK_a-range are selected, since for weak acids the role of pK_a can be attributed to the dissociation (and consequently lower uptake) and this influence will be absent for high pK_a values. Until so far no author has investigated this possibility. In some cases (41, 52) the rejection of pK_a as a parameter in QSAR studies with phenols can probably be explained in this way.

3. Reasons for success of QSAR approach in aquatic toxicology.

The application of QSARs in this field has been rather successful in contrast e.g. with the QSAR analysis of mammalian LD50s. There is a simple explanation for this fact. This succes of QSARs in aquatic toxicology has probably mainly a kinetic background. Several factors make the kinetics in aquatic toxicity studies simpler than in oral mammalian toxicity studies, such as a continuous exposure and the relative low rates of metabolic processes in aquatic organisms. We do not suggest that the toxicity of a chemical to fish, and therefore its correlation with physical and chemical parameters can be explained on the basis of kinetics only. Rather we suggest that the more complex kinetics in mammals are a serious obstacle in QSAR-analysis.

Not only the hydrophobic influence in QSARs has a kinetic background. This is also the case for the electronic parameters (Hammett, pK_a) as has been discussed above. This does not mean that occurrence of these parameters can be explained on the basis of kinetics only. They also play a part in the toxicant-receptor interaction.

4. Usefulness and applicability of QSARs

Apart from the possibility to predict the toxicity of chemicals on the basis of QSARs, this approach is useful to bring some line in toxicity data. It helps to identify what structural features of molecules are important determinants of their toxicity. Furthermore, if for a series of chemicals a QSAR with a high correlation coefficient can be calculated, this indicates that these chemicals have something in common: a similar mode of action and similar kinetics (although it can not be considered as a proof of this). This fact has been used by Könemann (53) and Hermens (10, 19) in their study of the toxicity of mixtures of chemicals. In this light it is particularly interesting to see what the limitations of a QSAR are, which chemicals fall (unexpectedly or not) outside its boundaries.

In absence of toxicity data for many existing chemicals the possibility of predicting a chemical's toxicity on the basis of a QSAR is important, e.g. for priority setting purposes. This seems to be possible for several series of homologues. For most of these cases, however, the number of chemicals which have been used for QSAR calculation is rather small and the boundaries of its validity have not yet been fully determined. This may change rapidly, since the number of scientists working in this field is extending, but on the basis of the present information many QSARs are primarily suited for comparison purposes and less for prediction.

There are at least two classes of chemicals for which QSARs are available which can be used for prediction purposes. The first one consists of phenol derivatives. It has already been mentioned that several authors have calculated QSARs for phenols. These QSARs should only be used with good professional judgement, to single out those chemicals which can be expected to be outliers. The relatively large number of calculated QSARs for phenols is very helpful in this respect.

The second class of chemicals which should be mentioned is the one of non-reactive, non-electrolyte chemicals. The importance of this QSAR has already been discussed: it predicts the minimum toxicity of a vast range of chemicals. The difficult question is: which ones are more toxic than predicted by this QSAR? It is probably possible to develop a list of structural characteristics of chemicals which make them likely to deviate from this QSAR, e.g. since they are "reactive" and therefore more toxic than predicted. Such a list, however, does not yet exist and therefore also this QSAR can only be used with professional judgement.

It can be concluded that, if QSARs are used by experts who are aware of the limitations of these relationships, they can be a useful tool for predicting the toxicity of certain types of chemicals. Such an estimate may at least indicate the order of magnitude of a chemical's toxicity. In several other cases QSARs may assist in interpreting toxicity data. The use of the increasing number of QSARs can certainly be of help in preparing priority lists of chemicals. Calculated toxicities can, however, never replace experimental data and are not reliable enough to be used in final hazard assessments. QSAR studies can be very valuable to bring some order in the great number of environmental chemicals by organising them into a smaller number of groups of chemicals. Since it is likely that chemicals within one QSAR have a common mode of action, this may also facilitate the study of the mechanism of toxic action.

5. Future developments
Further progress in the study of structure-activity relationship can be expected in the next few years. Two developments seem to be particularly promising. The first one is the QSAR-analysis of larger numbers of data. Such an exercise would be of great importance to be able to find the significance of more than the two most frequently used QSAR parameters (P and pK_a). The second development is the introduction of one particular additional QSAR parameter, chemical reactivity, already mentioned under 2.3.

Acknowledgements
I am grateful to prof.dr. D. Calamari (University of Milano) and dr. J. Hermens (University of Utrecht) for useful suggestions and comments on the manuscript.

7. REFERENCES

(1) HANSCH, C., R.M. MUIR, T. FUJITA, P. PEYTON, P.P. MALONEY F. GEIGER and M. STREICH (1962). The correlation of biological activity of plant growth regulators and chloromycetin derivatives with Hammett constans and partition coefficients. J.Am. Chem. Soc. 85, 2817
(2) HANSCH, C. and T. FUJITA (1964). P-σ-π Analysis. A method for the correlation of biological activity and chemical structure. J.Am. Chem. Soc. 86, 1616

(3) KOPPERMAN, H.L, R.M. CARLSON and R. CAPEL (1974). Aqueous chlorination and ozonization studies. 1. Structure—toxicity correlations of phenolic compounds to Dapnia magna. Chem. Biol. Interactions 9, 245.

(4) MARTIN, Y. (1978). Quantitative Drug Design, Dekker, New York.

(5) SEYDEL, J.K. and K.J. SCHAPER (1982). Quantitative structure—pharmacokinetic relationships and drug design. Pharmac. Ther. 15, 131.

(6) TOPLISS, J.A., Ed. (1983). Quantitative structure—activity relationships of drugs. Academic Press, New York.

(7) VEITH, G.D. and D.E. KONASEWICH (eds) (1975). Great Lakes Advisory Board, Windsor, Ontario. Proc. Symp. on Structure—activity correlations in studies of toxicity and bioconcentration with aquatic organisms.

(8) CALAMARI, D., S.GALASSI, F.SETTI and M.VIGHI (1983). Toxicity of selected chlorobenzenes to aquatic organisms. Chemosphere 12, 253.

(9) SLOOFF, W., J.H. CANTON and J.L.M. HERMENS (1983). Comparison of the susceptibility of 22 fresh water species to 15 chemical compounds. Part I. (sub)acute toxicity tests. Aquatic Toxicology 4, 113.

(10) HERMENS, J., E. BROEKHUIJZEN, H. CANTON and R. WEGMAN (1985). Quantitative structure activity relationships and mixture toxicity studies of alcohols and chlorohydrocarbons: effects on growth of Daphnia magna. Aquatic Toxicology 6, 209.

(11) BIRGE, W.J. and R.A. CASSIDY (1953). Structure—activity relationships in aquatic toxicology. Fundam. Appl. Toxicol. 3, 359.

(12) CALL, D.J., L.T. BROOKE, M.L. KNUTH, S.H. POIRIER and M.D. HOOGLAND (1985). Fish subchronic toxicity prediction model for industrial organic chemicals that produce narcosis. Env. Tox. & Chem. 4, 335.

(13) ZITKO, V., D.W. McLEESE, W.G. CARSON and H.E. WELCH (1975). Toxicity of alkyldinitrophenols to some aquatic organisms. Bull. Environm. Contam. Toxicol. 16, 508.

(14) DURKIN, P.R. (1978). Biological impact of various chlorinated phenolics and related compounds on Daphnia magna. Tappy Environm. Conf. Proc., p. 165.

(15) McLEESE, R.W., V. ZITKO, D.B. SERGEANT, L. BURRIDGE and C.D. METCALFE (1979). Structure—lethality relationships for phenols, anilines and other aromatic compound in shrimps and clams. Chemosphere 8, 53.

(16) KÖNEMANN, H. and A. MUSCH (1981). QSARs in fish toxicity studies. Part 2: the influence of pH on the QSAR of chlorophenols. Toxicology, 19, 223.

(17) SAARIKOSKI, J. and M. VILUKSELA (1982). Relation between physicochemical properties of phenols and their toxicity and accumulation in fish. Toxicol. Environm. Safety 6, 501.

(18) LIU, D., K. THOMSON and K.L.E. KAISER (1982). Quantitative structure—toxicity relationship of halogenated phenols on bacteria. Bull. Environm. Contam. Toxicol. 29, 130.

(19) HERMENS, J., P. LEEUWANGH and A. MUSCH (1984). Quantitative structure—activity relationships and mixture toxicity studies of chloro- and alkylanilines at acute lethal levels to the guppy. Ecotoxicol. Environ. Safety 8, 388.

(20) KÖNEMANN, H. (1981). Fish toxicity tests with mixtures of more than two chemicals: a proposal for a quantitative approach and experimental results. Toxicology, 19, 229.

(21) CALAMARI, D., R. DA GASSO, S.GALASSI, A.PROVINI and M.VIGHI (1980). Estimating the hazard of eight amines on aquatic life. Chemosphere 9, 753.

(22) SCHULTZ, T.W. and M. CAJINA-QUEZADA (1980). Structure-toxicity relationships of selected nitrogenous heterocyclic compounds. Arch. Environm. Contam. Toxicol. 11, 353.

(23) SCHULTZ, T.W. and M. CAJINA-QUEZADA (1982). Structure-toxicity relationships of selected nitrogenous heterocyclic compounds II. Dinitrogen molecules. Arch. Environm. Contam. Toxicol. 11, 353.

(24) SCHULTZ, T.W. and B.A. MOULTON (1985). Structure-activity relationships of selected pyridines. I Substituent constant analyses. Ecotox. and Env. Safety 10, 97.

(25) WONG, P.T.S., Y.K. CHAM, O. KRAMAR and G.A. BENGERT (1982). Structure-toxicity relationships of tin compounds on algal. Can. J. Fish. Aquat. Sci. 39, 483.

(26) LAUGHLIN, R.B., R.B. JOHANNESSEN, W.J. FRENCH, H.E. GUARD and F.E. BRINCKMAN (1985). Structure-activity relationships for organotin compounds. Environ. Tox. & Chem. 4, 343.

(27) VIGHI, M. and D. CALAMARI (1985). QSARs for organotin compounds on Daphnia magna. To be published.

(28) RICHET, C. (1893). Sur le rapport entre la toxicité et les propriétés physiques des corps. C.R. Soc. Biol. 54, 775.

(29) OVERTON, E. (1897). Ueber die osmotischen Eigenschaften der Zelle und ihre Bedeutung für die Toxikologie und Pharmakologie. Z.Phys. Chem., 22, 189.

(30) MEYER, H. (1899). Zur Theorie der Alkoholnarkose. 1. Welche Eigenschaft der Anästhetica bedingt ihre narkotische Wirkung?. Arch. exp. Pathol. Pharmakol. 42, 109.

(31) FERGUSON, J. (1939). The use of chemical potential as indices of toxicity. Proc. R. Soc. London Ser. B. 127, 387.

(32) LIPNICK, R.L. and W.J. DUNN (1983). An MLAB-study of aquatic structure toxicity relationships. Proc. 4th European Symp. Chemical Structure-Biological Activity, quantitative approaches, Bath, England, September 6-9, 1982. Elsevier, Amsterdam.

(33) VEITH, G.D., D.J. CALL and L.T. BROOKE (1983). Structure-toxicity relationships for the fathead minnow, Pimephales promelas: Narcotic industrial chemicals. Can.J. Fish. Aquat. Sci 40, 743.

(34) LIPNICK, R.L., D.E. JOHNSON, J.H. GILFORD, C.K. BICKINGS and L.D. NEWSOME (1985). Comparison of fish toxicity screening data for 55 alcohols with QSAR predictions of minimum toxicity for nonreactive nonelectrolyte organic compounds. Env. Tox. & Chem. 4, 281.

(35) HANSCH C. and A.J. LEO (1979). Substituent constants for correlation of biological activity and chemical structure. J. Wiley, New York.

(36) REKKER, R.F. (1977). The hydrophobic fragmental constant. Elsevier, Amsterdam.

(37) MCCALL, J.M. (1975). Liquid-liquid partition coefficients by HPLC. J. Med. Chem. 18, 549.

(38) KÖNEMANN, H., R. ZELLE, F. BUSSER and W.E. HAMMERS (1979). Determination of log P_{oct} values of chloro-substituted benzenes toluenes and anilines by high-performance liquid chromatography on ODS-silica. J.Chromatogr. 178, 559.

(39) RENBERG, L., G. SUNDSTRÖM and K. SUNDH-NYGÅRD (1980). Partition coefficients of organic chemicals derived from reversed phase thin layer chromatography. Chemosphere 9, 683.

(40) VEITH, G.D., K.J. MACEK, S.R. PETROCELLI and J. CARROL (1980). An evaluation of using partition coefficients and water solubility to estimate bioconcentration factors for organic chemicals in fish. in: Aquatic Toxicology, Proceedings of 3rd annual symposium on aquatic toxicology. ASTM STP 707.

(41) HALL, L.H., L.B. KIER and G. PHIPPS (1984). Structure-activity relationships studies on the toxicities of benzene derivatives: I An additivity model. Env. Tox. & Chem. 3, 355.

(42) HERMENS, J., F. BUSSER, P. LEEUWANGH and A. MUSCH (1985). Quantitative correlation studies between the acute letual toxicity of 15 organic halides to the guppy and chemical reactivity towards n-nitrobenzylpyridine. Toxicol. Environ. Chem. 9, 219.

(43) KÖNEMANN, H. (1980). Structure-activity relationships and additivity in fish toxicity studies of environmental pollutants. Ecotoxicol. Environ. Safety 4, 415.

(44) KAISER, K.L.E., D.G. DIXON and P.V. HODSON (1984). QSAR studies on chlorophenols, chlorobenzenes and para-substituted phenols. In: QSAR in Environmental Toxicology, p. 189 K.L.E. Kaiser (ed.), D. Reidel Publ. Comp., Dordrecht.

(45) HALL, L.H. and L.B. KIER (1976). Molecular connectivity in chemistry and drug research. Academic Press, New York.

(47) HALL, L.H. and KIER L.B. (1984). Molecular connectivity of phenols and their toxicity to fish. Bull. Environ. Contam. Toxicol., 32, 354.

(48) BASAK, S.C., D.P. Gieschen and V.R. Magnuson (1984). A quantitative correlation of the LC50 values of esters in Pimephales promelas using physicochemical and topological parametes. Environ. Tox. & Chem. 3, 191.

(49) GOVERS, H., C. RUEPERT and H. AIKING (1984). Quantitative structure-activity relationships for polycyclic aromatic hydrocarbons, correlation between molecular connectivity, physico-chemical properties, bioconcentration and toxicity in Daphnia pulex. Chemosphere 13, 227-236.

(50) HANSCH, C. (1971). Quantitative structure-activity relationships in drug design. In: Drug Design Vol. I (p. 296), E.J. Ariëns, ed., Acad. Press, New York.

(51) KUBINYI, H., (1977). Non-linear dependence of biological activity on hydrophobic character: the bilinear model. In: Biological activity and chemical structure (p. 239), J.A. Keverling Buisman, ed., Elsevier, Amsterdam.

(52) BENOIT-GUYOD, J.L., C. ANDRÉ, G. TAILLANDER, J. RODRAT and A. BOUCHERLE (1984). Toxicity and QSAR of chlorophenols on Lebistes reticulates. Ecotoxicoly Environ. Safety 8, 227.

(53) KÖNEMANN, H. (1981) Quantative structurs-activity relationships in fish toxicity studies. Part 1: A relationship for 50 industrial pollutants Toxicology 19, 209.

(54) JAFFE, H.H. (1953). A re-examination of the Hammett equation. Chem. Rev. 53, 191.

INDICES FOR THE PREDICTION OF ENVIRONMENTAL PROPERTIES OF HETERO-ATOMIC POLYCYCLIC AROMATIC POLLUTANTS

H. GOVERS and P. de VOOGT
Institute for Environmental Studies
Free University, P.O. Box 7161,
1007 MC AMSTERDAM, THE NETHERLANDS

Summary

It is shown in this study, that the partition coefficient n-octanol/ water of nonpolar polycyclic aromatics can be predicted rather accurately via Rekker fragmental values ($r = 0.903-0.985$; $s = 0.24-0.32$) and via RPHPLC capacity factors ($r = 0.976$; $s = 0.18$). The possibilities of Molecular Connectivity Indices for this purpose are not yet known. However, for the subgroup of aromatics with pyridine-like rings these indices turn out to be equal or better predictors than Rekker fragmental values ($r = 0.996$; $s = 0.11$ and $r = 0.995$; $s = 0.13$, respectively). Biological activity, c.q. growth inhibition of Tetrahymena pyriformis, of aromatics with pyridine- or pyrazine-like rings can be predicted more accurately via Molecular Connectivity Indices ($r = 0.992$; $s = 0.19$) than via experimental values of the partition coefficient ($r = 0.970$; $s = 0.36$). The (dis-) advantages of structural, chromatographic and partition indices with respect to the prediction of transport properties and biological activity are discussed briefly.

1. INTRODUCTION

The study of the relationships between molecular structure and physical properties on one side and transport properties and biological activity on the other side (i.e. QSAR and PAR, see Fig.1) can reduce substantially the need for complex, extensive and expensive experimental determinations (1). Moreover, it can give more insight in kinetic and toxic mechanisms for groups of congeneric compounds.

The experimental determination, via the shake flask method, of partition coefficients n-octanol/water (c.q. Log K_{ow}) has some clear disadvantages with respect to time, costs, applicability to compounds having a low aqueous solubility, interferences by impurities, acidity conditions , formation of complexes and emulsions and finally, the need for new analytical methods (2, 3). As an alternative the use of cheap, relatively fast and reproducible chromatographic systems has been described recently (2, 4, 5, 6, 7, 8). However, drawbacks remain with respect to polar compounds, stability of the stationary phase, Log K_{ow} range and accuracy for heterogeneous groups of compounds. A second alternative, therefore, may be the prediction of Log K_{ow} from structural properties of molecules. Among these are Rekker fragmental values (9) and molecular connectivity indices , MC, (4, 10, 8). Rekker fragmental values have to be derived from extensive sets of experimental Log K_{ow} values. Molecular connectivity indices (11, 12) encode information about size, branching, cyclization, unsaturation and hetero-atom content. They can be derived directly from molecular structure.

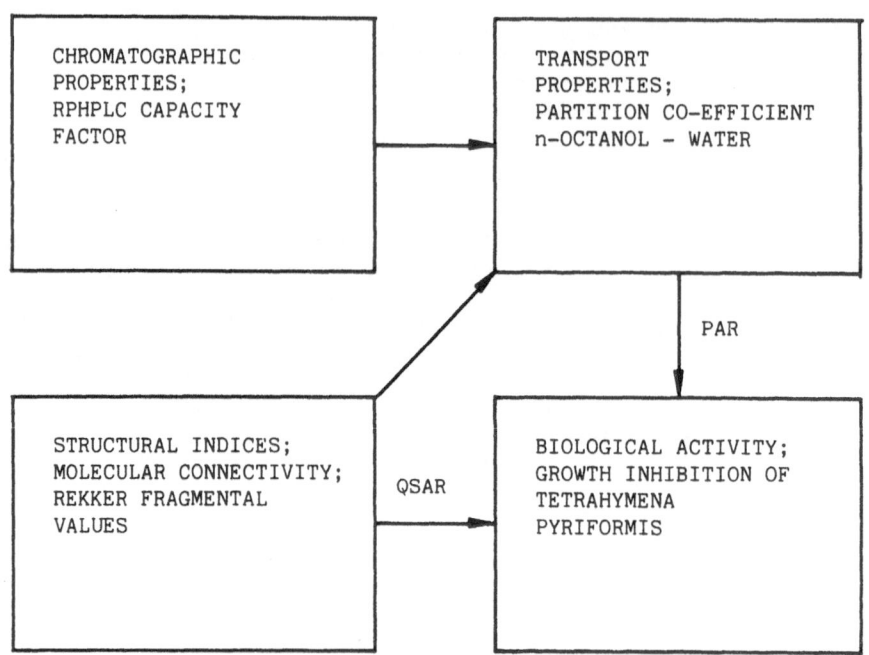

Figure 1. Relationships between Structural Properties, Chromatograph
Properties, Transport and Biological Activity

Many structural and physical properties have been used for the
prediction of biological activities within more or less congeneric groups of
compounds (13, 14, 15, 16, 17, 18, 19, 20). Among these are Log K$_{ow}$ and
MC (21, 22). Here we shall focus on the use of these two indices for the
prediction of growth inhibition of the freshwater ciliate Tetrahymena
pyriformis as measured by Schultz et al (22). This should be considered
as an example of the method only. The long term purpose of our work implies
the prediction of lethality and reproducibility of Daphnia and bacteria
and of Ames test mutagenicity.
      Recently we have shown (10, 8) that MC and/or chromatographic
indices (e.g. the logarithm of the RPHPLC capacity factor, Log k') are
very promising with respect to the prediction of transport properties and
biological activity of the highly congeneric series of unsubstituted
polycyclic aromatic hydrocarbons (PAH). Here we extend our work to
polycyclic aromatics (NH-PAH, N-PAH, O-PAH, S-PAH and Cl-PAH) with
pyrrole-, pyridine- and pyrazine-like rings, with oxygen or sulphur
containing heterocyclic rings and rings with chlorine substituents (see
Fig.2). In this way a vast number of polycyclic aromatics is considered.
Prediction from molecular structures or physical properties requires the
congenericity of the compounds under investigation and the similarity of
interactions in which they are involved. Therefore, this study extends to

the relatively nonpolar compounds to verify the applicability of the concept to aromatics containing at most two identical hetero-atoms. In the near future, the concept will be extended to both more complex and more polar compounds.

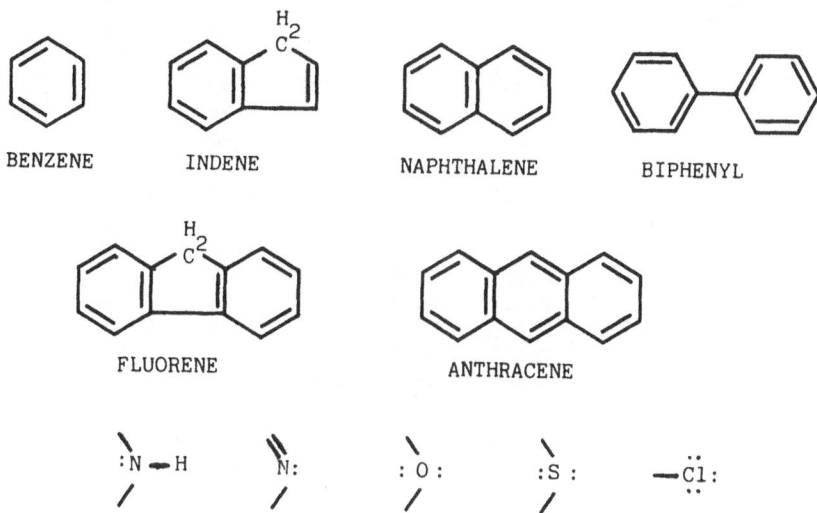

BENZENE   INDENE   NAPHTHALENE   BIPHENYL

FLUORENE   ANTHRACENE

Figure 2. Structures of Parent aromatic compounds and
Functional groups (substituents)

2. COLLECTION OF DATA

Relevant data for the derivation of correlation (linear regression) equations to be used for the prediction of Log K_{ow} and Log BR (biological activity, i.e. growth inhibition) are summarized in Table I. These data were obtained in the following way.

Experimental values of Log K_{ow} and the growth inhibition parameter Log BR were taken from miscellaneous literature sources denoted in the table.

Reversed phase HPLC capacity factors (Log k') were measured in a system consisting of a Kontron LC 410 pump and a Pye Unicam LC/UV absorbance detector containing a 0.8 microliter flowcell with 1 mV output, operating at a wavelength of 254 nm. A stripchart recorder (Kipp BD8) was used. Injection was carried out using a Rheodyne model 7125 valve with a 20 microliter loop. A Lichrosorb 5 RP 18 (Merck, 150 x 4.6 mm) column was used as a stationary phase and the mobile phase consisted of methanol (Baker, Analyzed Reagent) and deionized water (60:40, v/v). The solvent mixtures were degassed by an ultrasonic bath and vacuum pumping. The flow rate was 1.0 ml/min. The acidity of the mobile phase was buffered at pH = 7 with NaH_2PO_4/NaOH. The system was thermostatited at 27°C by a Kratos URA 200 system. The retention volume of an injection of an aqueous $NaNO_3$ (Baker, Analyzed Reagent) solution was taken as the void volume (^0V_o) of the column. The polycyclic aromatics were obtained from Janssen Chimica and used without further purification. These compounds were dissolved in methanol or toluene (Baker, Analyzed Reagent, redistilled) and stored in the dark at room temperature. The retention volumes (V_r) of the compounds were measured at least three times, individually. Log k' was calculated from Log k' = Log $[(V_r - V_o)/V_o]$.

Table I. Partition coefficients (Log K_{ow}), Molecular Connectivity Indices ($^1\chi^v$, $\Delta^1\chi^v$) and RPHPLC capacity factors (Log k') of aromatics.

Compound	Log K_{ow}^{exp}	Log BR[i]	Log k'[j]	Log K_{ow}^{k}	$^1\chi^{v,1}$	$\Delta^1\chi^{v,1}$
1. Benzene	2.11[a]			2.02	2.000	0.000
2. Indene	2.92[b]		0.972	3.14	3.211	
3. Naphthalene	3.36[a]		1.094	3.30	3.405	0.000
4. Biphenyl	3.95[c]			3.97	4.071	0.000
5. Fluorene	4.18[c]			4.12	4.612	
6. Anthracene	4.54[a]		1.674	4.57	4.809	0.000
7. Pyrrole	0.75[c]			0.79	1.577	
8. Indole	2.13[d]		0.340	2.07	2.988	
9. Carbazole	3.29[b]		0.912	3.34	4.405	
10. Pyridine	0.62[c]	-1.19		0.70	1.850	-0.150
11. Quinoline	2.02[c]	0.01		1.98	3.264	-0.140
12. 4-Phenyl-pyridine	2.55[e]			2.65	3.921	-0.150
13. 2-Phenyl-pyridine	2.63[f]			2.65	3.931	-0.140
14. Acridine	3.40[c]	1.40	0.944	3.25	4.679	-0.130
15. Pyrazine	-0.22[f]	-1.82		-0.51	1.699	
16. Quinoxaline	0.84[b]	-0.30		0.76	3.124	
17. 4,4'-Bipyridyl	1.24[g]			1.33	3.771	-0.300
18. 2,4'-Bipyridyl	1.48[g]			1.33	3.781	-0.290
19. 2,3'-Bipyridyl	1.43[g]			1.33	3.781	-0.290
20. 2,2'-Bipyridyl	1.67[g]			1.33	3.791	-0.280
21. Phenazine	2.84[b]	1.40	0.803	2.04	4.549	
22. Benzofuran	2.67[b]		0.722	2.18	2.889	
23. Dibenzofuran	4.12[b]		1.420	3.46	4.313	
24. Thiophene	1.81[b]			1.46	2.411	
25. Benzothiophene	3.11[d]		0.937	2.73	3.765	
26. Monochlorobenzene	2.84[b]			2.76	2.477	
27. 4-Chlorobiphenyl	4.61[h]			4.71	4.548	
28. p-Dichlorobenzene	3.39[b]			3.51	2.954	
29. 4,4'-Dichlorobenzene	5.36[h]			5.45	5.025	

a. See ref.(3)
b. See ref.(23)
c. See ref.(9)
d. Average values of ref.(23)
e. Average values of ref.(23) and ref.(24)
f. See ref.(24)
g. See ref.(29)
h. See ref.(25)
i. See ref.(22)
j. This study, see text
k. Calculated after ref.(9), see text
l. This study, see text

Log K_{ow} values were calculated after Rekker en de Kort (9) using fragmental values of hydrogen (f_H = 0.182), carbon (f_C = 0.155), pyrrolyl (f = 0.61), nitrogen in rings with 2 nitrogen atoms (f_N = -0.929), pyridinyl in rings with 1 nitrogen atom (f = 0.52), chlorine (f_{Cl} = 0.924), oxygen (f_O = -0.439), sulphur (f_S = 0.11) and C_M (f = 0.289). It has to be stressed that the f_O- and f_S- values used are derived from reference compounds rather dissimilar to our S- and O-compounds.

The molecular connectivities $^1\chi^v$ ($= \sum_{ij} 1/\sqrt{\delta_i^v \delta_j^v}$, δ_i^v = number of non hydrogen valence electrons on atom i) were calculated according to Kier and Hall (11) using δ_i^v-values as defined by Kier and Hall (26). $\Delta^1\chi^v$ (= $^1\chi^v$ of the compound minus the $^1\chi^v$ of its parent PAH) is a MC index used for a 2 parameter linear correlation equation for the prediction of Log K_{ow} of aromatics with pyridine-like rings (See next section).

All calculations were carried out via a simple desk top calculator.

3. CORRELATIONS

In order to investigate the predictive power of structural plus chromatographic indices and partition coefficients linear one parameter correlation equations ($y = A + Bx_1$) were derived from the data of Table I. In one case a second parameter (i.e. the equation $y = A + Bx_1 + Cx_2$) was applied. The results are summarized in Table II.

Table II. Correlation equations for the prediction of Log K_{ow} and Log BR (y) from partition coefficients, RPHPLC capacity factors and structural indices (x). For the definition of symbols and compounds involved, see Table I and the text. Given are the coefficients A, B and C, the number of data (N), the correlation coefficient (r) and the standard error (s).

y	A	B	x_1	C	x_2	N	r	s	Eqn.nr.
Log K_{ow}	0.218	0.957	Log K_{ow}^{Rek}			29	0.985	0.24	(1)
	−0.914	1.005	$^1\chi^v$			29	0.736	0.92	(2)
	0.822	0.803	Log K_{ow}^{Rek}			10	0.903	0.32	(3)
	1.419	1.853	Log k'			10	0.976	0.18	(4)
	0.096	0.978	Log K_{ow}^{Rek}			13	0.995	0.13	(5)
	0.385	0.869	$^1\chi^v$	8.139	$\Delta^1\chi^v$	13	0.996	0.11	(6)
Log BR	−1.530	0.914	Log K_{ow}^{exp}			6	0.970	0.36	(7)
	−3.367	1.028	$^1\chi^v$			6	0.992	0.19	(8)

From the eqns.(1)-(4), holding for the broad group of aromatics with C, H, N, O, S (and Cl in eqns.(1) and (2)) atoms it can be inferred that Rekker fragmental values and RPHPLC capacity factors can predict Log K_{ow} with a standard error of about 0.2 - 0.3 Log K_{ow} units. In particular Log k' (eqn.(4)) seems to be a valuable predictor. The Rekker method also works rather accurately (eqn.(1)).

One single MC index ($^1\chi^v$) does not work well for the broad group of compounds (eqn.(2), s = 0.92). This is illustrated also in Fig.3, showing a broad band of dots around the solid curve. However, the additional drawing of separate dotted curves for subgroups of aromatics strongly suggests the possibility of substantial improvement of the predictive power via the introduction of additional MC indices, one at a maximum per type of hetero-atom or substituent. Only compounds 17 (4,4'-bipyridyl) and 21 (phenazine) have an experimental Log K_{ow} value which differs by more than 0.3 log unit from the predicted value.

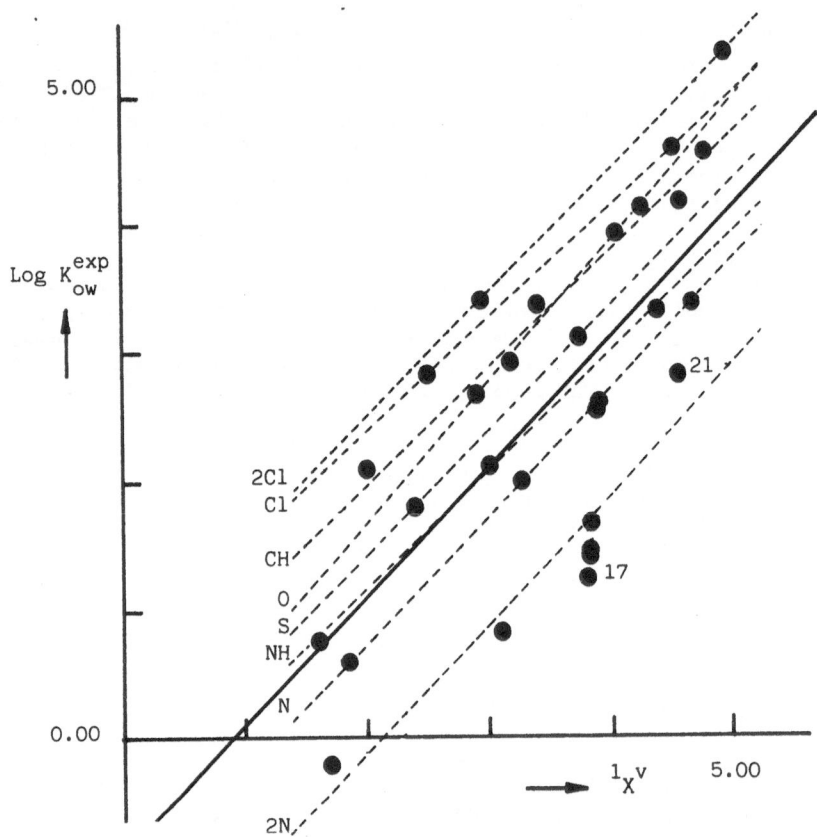

Figure 3. Plot of Log K_{ow} vs. $^1X^V$ for Eqn.(2) (solid curve)
and for subgroups of aromatics with one type of
substituent (dotted curves).
Compound figures refer to Table I.

Therefore we selected the subgroup of (four) parent aromatics (nrs.
1, 3, 4 and 6 of Table I) plus (nine) aromatics with pyridine-like
rings (nrs. 10-14 and 17-20 of Table I) and applied a second MC parameter,
which is related directly to the nitrogen substituent and its next atomic
neighbours, $\Delta X^V = {}^1X^V$ minus the $^1X^V$ of the parent compound. In this way
the correlation improved markedly (eqn.(6)); a standard error of s = 0.11
was obtained, close to the accuracy of experimental values of Log K_{ow} and
at least as good as the Rekker method prediction (eqn.(5)). The linear
regression eqn. (6) is drawn in Fig.4. Only compounds 10 (pyridine) and
20 (2,2'-bipyridyl) deviate by more than 0.1 log unit from the predicted
value. Moreover an interesting improvement is obtained with respect to the
Rekker method for isomeric compounds (nrs. 12, 13 and 17-20), which are
almost resolved in the MC method but obtain identical values in the Rekker
prediction.

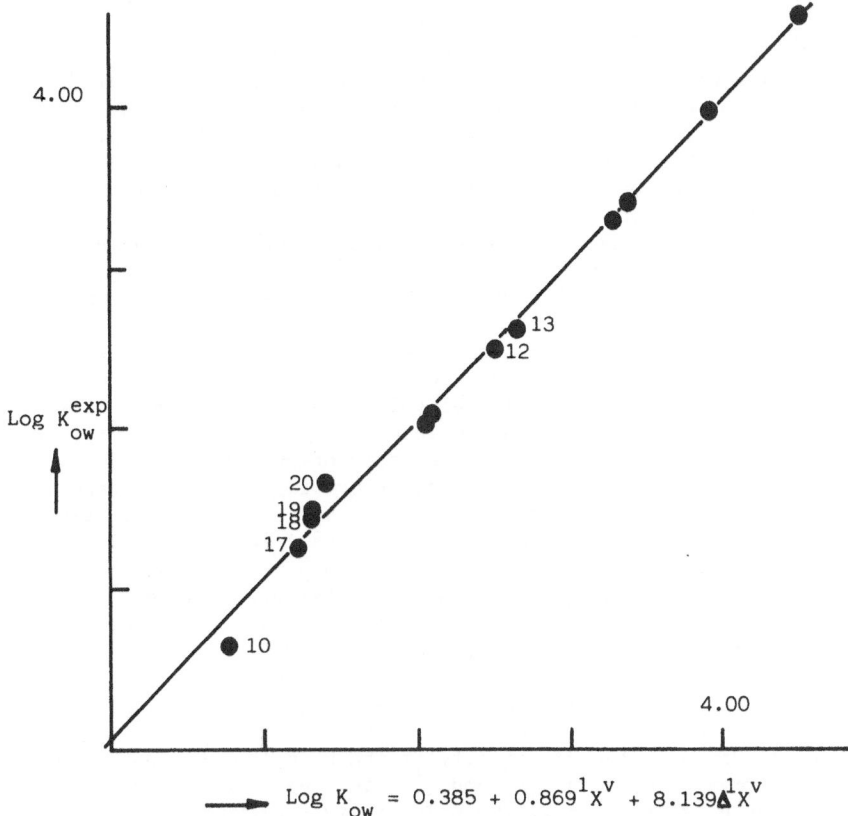

Figure 4. Plot of Eqn. (6), see text; Log K_{ow} as a function of $^1X^V$ and Δ^1X^V for parent and pyridine-like aromatics. Compound figures refer to Table I.

Finally biological activity, c.q. Log BR, is predicted even better via $^1X^V$ (eqn. (8); s = 0.19) than via experimental Log K_{ow} values (eqn. (7); s = 0.36) as concerns pyridine (nrs. 10, 11, 14)- and pyrazine (nrs. 15, 16, 21)- like ring systems. Again the standard error is close to experimental accuracy (22) for the $^1X^V$ predictor.

4. DISCUSSION AND CONCLUSIONS

No brief, complete or otherwise satisfactory answer can be given to the question why the indices used above work so well. However, a chromatographic system can mimic, at least in principle, the partition of a compound between n-octanol and water to a high degree by the choice of appropriate stationary and mobile phases. Rekker fragmental values are closely related to substituent constants as defined by Leo et al (23). As a consequence they are incorporated relatively easy into linear free energy (extra-)thermodynamic theories for partition coefficients. The validity of partition coefficients and structural properties for the prediction of biological activity has found some theoretical basis in e.g. the early theory of Hansch and Fujita (13), which clearly explains that factors other than Log K_{ow} can be important (cf. our eqns. (7) and (8)).

MC indices are not yet put into a sound theory which relates them to partition coefficients or biological activity. However, it is known (12),

that the $1/\sqrt{\delta_i^v\ \delta_j^v}$ – values of bonded atom pairs can be expressed as a
function of volumes and electronegativities of the participating atoms i
and j. Therefore, one theoretical line might be the study of the dependence
of molecular interaction (e.g. between active- and receptor-molecules)
from $^1X^v$, which is a sum of atomic pair contributions. The other line
should be, to our opinion, the incorporation of MC indices into an (extra-)
thermodynamic model.

Apart from the need for both further theoretical development and more
experimental values for partition coefficients and biological activities
of polycyclic aromatic compounds, some (dis-)advantages of the indices
used are already clear.

Rekker fragmental values have to be derived from experimental Log K_{ow}
values. In a similar way coefficients A, B, C .. etc of MC indices have
to be derived from experimental values. Also, the number of times a
certain fragment is met in a molecule can be counted directly as can be the
MC indices. However, there are some important differences.

In the Rekker method a steadily growing number of fragmental values
for each new substituent has to be derived in order to obtain accurate
predictions of Log K_{ow}. As an example the f_O and f_S fragments used in our
calculations are based on non furan- and non thiophene- like reference
compounds and, consequently, cause the relatively large standard error
(s = 0.32) of eqn. (3). In order to give an (excellent) prediction the
MC method appears to need also additional MC indices, like ΔX_i^v, for each
hetero-atom. However, in the case of pyridine-like rings the Rekker method
needs four fragmental values (for H, C, C_M and pyridinyl), whereas the
MC method only needs three coefficients (A, B, C) (cf. eqns. (5) and
(6)). Moreover the versatility of the MC method has not been applied
exhaustively; only the $^1X^v$ and ΔX^v indices were tested in this study and
no higher order indices. The possibility remains that for Cl-, O-, S- and
NH- aromatics (see Fig.3), the Log K_{ow} vs. $^1X^v$ linear regression curves
of which are rather close to that of the parent compounds (CH), one
single MC index will be sufficient instead of four. As a further
advantage of the MC method we have pointed already at the successful
resolution of isomeric compounds, which is not possible in the Rekker
method.

The log of the capacity factor, Log k', turned out to be a useful
predictor of Log K_{ow} for a broad group of aromatics (see eqn. (4)). A
standard error as low as s = 0.18 could be obtained,which is in between
the errors for unsubstituted PAH (8) and for a combined group of nonpolar
and polar compounds (5). Our chromatographic system will be optimized in
the next future for stationary and mobile phases or via a combined use of
Log k' values from different chromatographic systems (cf. additional
parameters in the Rekker and MC methods). In addition it is interesting
to note that even a relatively simple method as RPTLC can predict
experimental Log K_{ow} values rather well (28, 8, 27).

Finally, we would like to mention the interesting possibility of the
prediction of biological activity directly from MC indices, without the
help of partition coefficients, as expressed by our example on growth
inhibition (eqns. (7) and (8)).

REFERENCES

(1) KOCH, R. (1982). Chemosphere 11, 511-520.
(2) RAPAPORT, R.A. and EISENREICH, S.J. (1984). Environ. Sci. Technol.
     18, 163-170.

(3) KARICKHOFF, S.W., BROWN, D.S. and SCOTT, T.A. (1979). Water Research
     13, 241-248.
(4) WELLS, M.J.M., CLARK, C.R. and PATTERSON, R.M. (1981).
     J. Chromatogr. Sci. 19, 573-582.
(5) EADSFORTH, C.V. and MOSER, P. (1983). Chemosphere 12, 1459-1475.
(6) HAMMERS, W.E., MEURS, G.J. and LIGNY, C.L. de (1982).
     J. Chromatography 247, 1-13.
(7) XIE, T.M., HULTHE, B. and FOLESTAD, S. (1984). Chemosphere 13,445-459.
(8) RUEPERT, C., GRINWIS, A. and GOVERS, H. (1985). Chemosphere 14,
     279-291.
(9) REKKER, R.F. and KORT, H.M. de (1979). Eur. J. Med. Chem. 14, 479-488.
(10) GOVERS, H., RUEPERT, C. and AIKING, H. (1984). Chemosphere 13, 227-
     236.
(11) KIER, L.B. and HALL, L.H. (1977). Eur. J. Med. Chem. 12, 307-312.
(12) KIER, L.B. and HALL, L.H. (1981). J. Pharmac. Sci. 70, 583-589.
(13) HANSCH, C. and FUJITA, T. (1964). J. Amer. Chem. Soc. 86, 1616-1626.
(14) HANSCH, C., LEO, A., UNGER, S.H., KIM, K.H., NIKAITANI, D. and
     LIEN, E.J. (1973). J. Med. Chem. 16, 1207-1216.
(15) SMITH, I.A., BERGER, G.D., SEYBOLD, P.G. and SERVE, M.P. (1978).
     Cancer Research 38, 2968-2977.
(16) LIEN, E.J. (1982). J. Clinic. and Hospital Pharmacy 7, 101-106.
(17) YUAN, M. and JURS, P.C. (1980). Toxicology and Applied Pharmacology
     52, 294-312.
(18) HODES, L. (1981). J. Chem. Inf. Comput. Sci. 21, 132-136.
(19) KOCH, R. (1982). Chemosphere 11, 497-509.
(20) EGOROVA, N.A., KRASOVSKIJ, G.N. and KOCH, R. (1982). Chemosphere 11,
     915-919.
(21) HALL, L.H. and KIER, L.B. (1981). Eur. J. Med. Chem.- Chem. Therap.
     16, 399-407.
(22) SCHULTZ, T.W., KIER, L.B. and HALL, H.L. (1982). Bull. Environm.
     Contam. Toxicol. 28, 373-378.
(23) LEO, A., HANSCH, C. and ELKINS, D. (1971). Chem. Rev. 71, 525-616.
(24) HANSCH, C. and LEO, A. (1979). Substituent Constants For Correlation
     Analysis in Chemistry and Biology. John Wiley & Sons, New York.
(25) BRUGGEMAN, W.A. (1983). Thesis, University of Amsterdam.
(26) KIER, L.B. and HALL, L.H. (1983). J. Pharmac. Sci. 72, 1170-1173.
(27) REKKER, R. (1984). J. Chromatogr. 300, 109-125.
(28) VOOGT, P. de, and GOVERS, H. (1985).Structural and Chromatographic
     Predictors of n-Octanol/Water partition coefficients. International
     Conference: Dioxin '85; Bayreuth, FRG, september 16-19.
(29) PIJPER, P.J. (1980). Thesis, Free University Amsterdam.

INTEGRATION AND USE OF THE LIST OF ORGANIC MICROPOLLUTANTS IN THE AQUATIC ENVIRONMENT IN THE ECDIN DATA BANK

W. PENNING, G. AINA and B. HELM

Commission of the European Communities, Directorate-General XII,
Joint Research Centre, Ispra Establishment, I-21020 ISPRA (VA), Italy

Summary

In 1973 the Council of Ministers of the European Communities decided to include a project for a data bank on environmental chemicals in the Environmental Research Programme of the European Communities. The data bank, which was originally conceived as a feasability study, has since been 1981 present on the information market and has been well accepted by external users. The COST 641 list of organic micropollutants has been converted to the ADABAS format and made available on line to the users of ECDIN.

1. INTRODUCTION

The ECDIN project, which started in 1973, ended its pilot phase recently and a fully operational version is proposed. ECDIN will enable all persons engaged in environmental management and research to obtain on-line reliable information on chemical substances with environmental impact.

Previous publications on ECDIN (1-10) describe the conceptual framework for the project, the development of the toxicological files and the software needs for ECDIN. This paper deals with the description of the final on-line version, the results of the pilot operational phase.

2. SCOPE OF ECDIN

ECDIN, in the operational phase, enable its users engaged in environmental management and research to obtain detailed scientific data, which ideally have already been evaluated by a scientist or a scientific pannel. The data base has also been designed to give a rapid response to questions regarding the potential danger of chemical substances, and their impact on the environment.

The basic principle of the data bank is to store relevant information on chemical substances produced in sizeable quantities regardless the form in which they are used, their intended functions or their presumed degree of harmlessness. Moreover chemicals having a high toxicity, selected natural toxic products are also included.

Applying these criteria the ECDIN data bank intends to store information on 62000 single chemical substances.

3. DATA COVERAGE

Bearing in mind the aim of the ECDIN data bank, it is necessary to store information on different subjects, chemical identification, synonyms in different languages, properties, use patterns, disposal, toxicity and behaviour in the environment. At present for the 62000 chemicals ECDIN has data in one or more of the 23 possible subject files which are on-line available.

For its major files such as toxicity and environmental sciences ECDIN stores two types of data:

1. data from scientific literature which have been validated by an expert or data evaluated by an expert panel.

2. data, not necessarily validated or evaluated, extracted from larger data files produced by other organisations or working groups, e.g. the ciclops file on micropollutants.

Numerical data, controlled words and free text fields are used in a way which allows searching and formatting with ADABAS Data Base Management System capabilities.

The data files which are offered to the on-line users together with the number of chemicals covered and the number of records in the file are listed below:

DATA FILE	RECORDS	COMPOUNDS
1 Chemical Synonyms	331069	62304
2 Physicochemical Properties	31053	4130
3 Transport and Handling		
4 Chemical Structure Diagram	50696	50696
5 Chemical Producers	13154	5477
6 Chemical Processes	531	611
7 Uses	734	734
8 Production and Consumption Statistics	3978	719
9 Export Statistics	1657	88
10 Import Statistics	634	62
11 Occupational Health and Safety	3996	573
12 Occupational Exposure Limits	1673	738
13 Directive 67/548/EEC	932	762
14 Classical Toxicity	41655	18365
15 Aquatic Toxicity	18219	678
16 Mutagenicity		
17 Carcinogenicity	1784	134
18 Odour and Taste Threshold Limits	1073	574
19 Effects of Soil Microorganisms	1922	72
20 Concentrations in environmental matrices	16157	1308

DATA FILE	RECORDS	COMPOUNDS
21 Analytical Methods	1318	995
22 Legislation	9318	1253
23 Teratogenicity		
24 Concentration in Human Media	5814	617
25 Concentration in Animal Media	2714	120
26 Metabolism in the Aquatic Environment		
27 Metabolism in Soil	1752	102
28 Bioaccumulation in the Aquatic Environment	1368	160

4. USAGE OF THE ECDIN DATA BANK

After having functioned as an in-house research project for several years, in 1981 an experimental user study was launched with a limited number of selected outside users. Having obtained relatively positive results, in 1983, a proper marketing study was performed by Datacentralen, Copenhagen. The aim of this study was to verify the operational stage of ECDIN, its acceptability of different users and to test the opportunity of making ECDIN publicly available.

Therefore, ECDIN exists in two configurations:
1. "Scientific" ECDIN at the Commission services at the Joint Research Centre, which is responsible for data collection, validation and evaluation, software development and file design.
2. Publicly available ECDIN, at present available on Datacentralen's host computer in Copenhague, responsible for the on-line distribution against payment to the customers world-wide. Publicly available ECDIN is a subset of "scientific" ECDIN.

The marketing study has identified the different user categories which have a potential interest in the ECDIN data bank:
- government or local authorities . . . . 13 %
- public research organizations . . . . 23 %
- library, information services . . . . 15 %
- chemical industry . . . . 32 %
- pharmaceutical industry . . . . 6 %
- other industries . . . . 11 %

The results of the marketing study were satisfactory although an increase in data coverage in the various fields is requested by the on-line users.

5. FUTURE OF ECDIN

Having completed the marketing study in July 1985, the Commission of the European Communities will take still this year a decision on the final destination of the project. It is foreseen that the two configuration model: "scientific" ECDIN and the publicly available ECDIN will be maintained in the future.

6. INTEGRATION OF LIST OF MICROPOLLUTANTS IN ECDIN

As described in detail in a previous publication (11) ECDIN intended to include data compiled on behalf of the COST 641 management committee by the Water Research Centre (WRC), Stevenage. This has now been completed and available for use by the on-line users.

The computerized data were obtained from WRC and have been converted to a special ADABAS ECDIN format. This operation has involved various steps:

- identification of the chemical substances
  this step presented the most difficult in the conversion process. About 79% of the substances present in the original file were properly defined. The unidentified substances were mainly substances without environmental interest, biological materials, mixtures of chemicals which were determined as mixtures in the chemical analysis, e.g. benzopyrenes etc.
- development of the conversion programs
  according the general rules of the ECDIN data base all abbreviations used in the original printed version (12) for the analytical technique, country, laboratory where the analysis has been performed, etc, have been fully expanded. The bibliography on data can be obtained on-line if the user requires so.
- development of the display of the data file.

The data on concentrations in surface water for each single chemical can be accessed in two different ways:

1. subdivided by sample type
2. subdivided by country in which the chemical has been identified

An example shows the organisation of data for DDT:

```
|  (ECDIN)      Environmental Concentrations      Super Menu  |
|          (COST 641 Water Research Centre - Stevenage)       |
|                                                             |
|   Substance 0001381 DDT                                     |
|               ***** Available Menus *****                   |
|                                                             |
|      A              By Sample Type                          |
|      B              By Country                              |
|_____|
```

```
| (ECDIN)          Environmental Concentrations          Menu |
|          (COST 641 Water Research Centre - Stevenage        |
|                                                             |
| Substance 0001381 DDT                                       |
|                                                             |
|       Available Data                     Count              |
|                                                             |
|    A   Fauna                              18                |
|    B   Finished Water                     17                |
|    C   Flora                               2                |
|    D   Industrial Effluent                 1                |
|    E   Plankton                            6                |
|    F   Rainfall                            3                |
|    G   Sediment                           11                |
|    H   Sewage                             16                |
|    I   Solid Sewage Waste                  3                |
|    J   Subterranean Water                  1                |
|    K   Surface Water                     102                |
```

```
| (ECDIN)          Environmental Concentrations          Report |
|          (COST 641 Water Research Centre - Stevenage)         |
|                                                               |
| Substance 0001381  DDT                                        |
|                                                               |
| Location of Sampling: THE NETHERLANDS                         |
| Sample Type: FAUNA: PROSOBRANCHES/BIVALVES                    |
| Sampling Period: 1976                                         |
| Analytical Technique: GAS LIQUID CHROMATOGRAPHY               |
| Concentration: 0.003-0.005 MG/KG                              |
|                                                               |
| Comment: IN COASTAL WATERS                                    |
|                       BIBLIOGRAPHY                            |
| J.K. QUIRIJNS ET AL., SURVEY OF THE CONTAMINATION OF DUTCH    |
| COASTAL WATERS BY CHLORINATED HYDROCARBONS, INCLUDING THE     |
| OCCURENCE OF METHYLTHIOPENTACHLOROBENZENE AND DI(METHYL-      |
| THIO)TETRACHLOROBENZENE. THE SCI. OF THE TOTAL ENVIR., 13,    |
| 225, (1979).                                                  |
```

7. USE OF THE CICLOPS DATA FILE IN ECDIN

The data file is fully incorporated with the other data files in the ECDIN system and can therefore also be used linked with other data files. This enables the ECDIN system to reply to various questions which may be of some interest:
- concentrations of environmental compounds in river Rhine?
- which chemicals have been found in surface waters in concentrations higher than the LC50 for rainbow trout?
- concentration of propoxur in drinking water?

8. CONCLUSIONS

There is a growing interest in the ECDIN data bank, both on the European and extra-European information market. The computerisation and the on-line distribution of the list of micropollutants in water, has contributed to this.

It is nevertheless necessary that the collaborators to the COST 641 action maintain the data file continuously updated with recent data from monitoring programs.

REFERENCES

(1) GEISS F. and BOURDEAU Ph. (1976). In "Environmental Quality and Safety", F. Coulston and F. Korte (Eds), Thieme (Stuttgart), Academic Press (New York), vol. 5, p. 15, "ECDIN, An EC Data Bank for Environmental Chemicals", paper presented at the Third International Symposium on Chemical and Toxicological Aspects of Environmental Quality, Tokyo, Japan, November 19-22 (1973). .

(2) GEISS F., PETRIE J.H., BOURDEAU Ph., and OTT H. (1975). The Environmental Chemical Data and Information Network (ECDIN), paper presented at the UNEP Workshop on an International Register of Potentially Toxic Chemicals, Bilthoven, the Netherlands, January 6-11 (1975).

(3) BONI M., GEISS F., PETRIE J.H., and TOWN W.G. (1976). The development of a Data Network on Chemicals and their Effects on the Environment. Paper presented at the EURIM 2 Conference, Amsterdam, March 22-25, 1976. Proceedings published by ASLIB.

(4) NØRAGER O., TOWN W.G., and PETRIE J.H. (1978). Analysis of the Registry of Toxic Effects of Chemicals Substances (RTECS) Files and Conversion of the Data in these Files for Input to the Environmental Chemicals Data and Information Network (ECDIN). J. Chem. Information Computer Science, vol. 18, n. 3.

(5) PROCTOR D., ROBSON A., VEAL M.A., PETRIE J.H. and TOWN W.G. (1978). Development of an Exchange Format for the European Environmental Chemicals Data and Information Network (ECDIN). Information Processing & Management, vol. 14, 429-443.

(6) PETRIE J.H., POWELL J., and TOWN W.G. (1978). Computerized Data Handling in the Environmental Chemicals Data and Information Network, in Generalized Data Management Systems and Scientific Information. OECD Nuclear Energy Agency, Paris.

(7) OTT H., GEISS F., and TOWN W.G. The Environmental Chemicals Data and Information Network (ECDIN) and Related Activities of the European Communities. Proceedings of the Second International Symposium on Aquatic Pollutants, Noordwijkerhout, the Netherlands, September 26-28 (1977). Published in "Aquatic Pollutants - Transformation and Biological Effects, O. Hutzinger, van Lelyveld L.H., and B.C.J. Zoeteman (Eds.), Pergamon Press.

(8) TOWN W.G., POWELL J., and HUITSON P.R. Recent Experience of the Application of a Commercial Data Base Management System (ADABAS) to a Scientific Data Bank (ECDIN). Paper presented at the Seventh Cranfield International Conference on Mechanised Storage and Retrieval of Information, July 17-20 , 1979.

(9) COST Project 64b Management Committee. A Comprehensive List of Polluting Substances which have been Identified in Various Fresh Waters, Effluent Discharges, Aquatic Animals and Plants, and Bottom Sediments. EUCO/MDU/40/74 (October 1974): Second Edition EUCO/MDU 73/76 (1976).

(10) PENNING W., and TOWN W.G. (1981). Development of the Toxicological Files in Environmental Chemicals Data and Information Network (ECDIN). Clinical Toxicology, vol. 18, n. 10: 1169-1181.

(11) TOWN W.G., and PENNING W. (1982). Integration of the COST 64B List of organic micropollutants in the ECDIN Data Base. COST Project 64b BIS. Concerted Action Analysis of Organic Micropollutants in Water. Activity Report of the Community-Cost Concertation Committee covering the period October 1978-December 1981, Commission of the European Communities, XII/ENV/17/82, OMP/29/82, 387-399.

(12) Cost Project 64b BIS (1984). Concerted Action Analysis of Organic Micropollutants in Water. An inventory of polluting substances which have been identified in various fresh waters, effluent discharges, aquatic animals and plants, and bottom sediments. Volume III. Commission of the European Communities, OMP/51/84.

ENVIRONMENTAL RESEARCH OF THE EUROPEAN COMMUNITIES IN THE FUTURE

G. ANGELETTI and H. OTT
Directorate-General for Science, Research and Development
Commission of the European Communities, Brussels

Summary

A general overview of the environmental research programmes of the European Communities is given, summarizing important achievements of the past programmes and outlining future activities. The programme proposal for contract research and co-ordination for the period 1986 - 1990, submitted by the Commission to the Council, is discussed with particular emphasis on research on organic micropollutants in the aquatic environment.

1. INTRODUCTION

The participants of this conference may wish to obtain some information on the role of the Concerted Action "Organic Micropollutants in the Aquatic Environment" within the overall research activities of the Communities in the environmental field. The scope of this paper is to review these research activities, to summarize the achievements, and to describe the orientation for the future, with particular emphasis on the Concerted Action "Organic Micropollutants in the Aquatic Environment", subject of this symposium.

Main aim of the programme is to provide the scientific basis for the implementation of the Environment policy of the Communities which is directed to protect the human health and the environmental resources.

A further objective of the research programme is to promote or improve the European cooperation of the research activities executed in the various countries, and to launch common projects involving different laboratories in the Member States.

After more than ten years of Community environmental research, a careful review of the programme was undertaken. Earlier this year, the Commission proposed a new (fourth) programme, to be effective from 1 January 1986 (1) for five years. It should be therefore worthwhile to outline the main achievements obtained so far, and review the orientation of Community research.

2. THE COMMUNITY ENVIRONMENTAL RESEARCH PROGRAMME

Community environmental research has evolved in parallel with the environmental policy of the European Communities in order to take into account its objectives, and the allocations provided by the Council have always increased since 1973, proving the importance given by the Member States to the environmental problems.

During the 1970's, several COST projects were incorporated in the programme, with the participation of European Non-Member States; environmental research was also incorporated in the programme of the Joint Research Centre and gradually increased.

With the approval of the first Framework Programme for Community Scientific and Technical Activities in 1983 (reviewed two years ago in Oslo (2)), the Research Action Programme (RAP) concept and a closer connection between contract research, concerted action and JRC research was introduced. It is clear that such a programme can contribute to improve not only the living and working conditions, but also the industrial competitivity, the rational use of energy and agricultural productivity. The scientific content of the present programme 1981 – 1985 includes five research areas:

1)   Sources, pathways and effects of pollutants

      implemented by contract research and four concerted actions:

      -    Organic Micropollutants in the Aquatic Environment (COST 641);
      -    Physico-Chemical Behaviour of Atmospheric Pollutants (COST 611);
      -    Air Pollution Effects on Terrestrial and Aquatic Ecosystems (COST 612);
      -    Benthic Coastal Ecology (COST 647);

2)   Reduction and prevention of pollution and nuisances

      implemented by contract research and a concerted action:

      -    Treatment and use of organic sludges and liquid agricultural wastes (COST 681)

3)   Protection, conservation and management of the natural environment

      implemented by contract research

4)   Environment information management

      implemented by contract research

5)   Complex interactive systems: man-environment interactions

The first sector covers all phenomena related to pollution origin, fate and transformation of pollutants in the environment, the effects on human beings and on ecosystems, with particular emphasis on air quality and water quality, encompassing problems related to acid deposition and their effects, and to the fate of organic pollutants in the aquatic environment, subject of this symposium.

   Research area 2 deals with the development of new and advanced techniques for the reduction and/or the prevention of the pollution from industrial processes and other sources. Particular attention is given to the clean technologies, in order not only to reduce or avoid pollution and wastes, but also to significantly save energy and raw materials in some selected industrial sectors: chemical industry, metal finishing and coating, food and feed, pulp and paper, textile and tanning industry. Emphasis also is given to the problems related to the management of the waste, including the treatment and recycling of toxic and dangerous waste.

Research executed in area 3 is directed to improve the understanding of the structure and function of the ecosystems and the qualitative and quantitative aspects of biogeochemical cycles of nitrogen, sulphur, carbon, phosphorus, in order to predict negative consequences caused by man and to suggest preventive actions.

Within research area 4 a data bank on environmental chemicals (ECDIN - Environmental Chemicals Data and Information Network) has been developed at the JRC Ispra in order to collect and make available data on chemicals which constitute a potential risk for health and/or environment.

The 5th area covers interdisciplinary research to study the global interactions between man and his environment.

3. ACHIEVEMENTS OF COMMUNITY ENVIRONMENTAL RESEARCH

It is widely recognized that the Community environmental research has contributed to improve the knowledge and cooperation of European laboratories in environmental sciences and many results have served as basis for environmental policy.

The results of the environmental research programme have been evaluated by an independent group of qualified experts which noted the amount and quality of the research going on in the Community, funded or co-ordinated by the Commission, and acknowledged with high satisfaction the achievements of the Community's Environmental Research Programme. The report of this group is available (3).

The achievements are well documented and a large number of publications has been produced giving access via extensive references to specific research items funded by the Commission.

Due to the wide range of the areas covered by the programme, it is impossible here to give details of the results.

For certain areas like air pollution, organic micropollutants in water, sewage sludge, most research in Europe is now incorporated in Concerted Actions.

The programme has also contributed to the development of abatement technologies and of processes for the treatment and removal of wastes.

The research has improved the knowledge on the natural environment and functioning of ecosystems.

For contract research, summary reports have been published regularly by the Commission, and contractors have been encouraged to publish the results in the open literature. For Concerted Actions a special effort has been consecrated to collate the results in activity reports giving the main achievements and to edit the proceedings of numerous European symposia and workshops on selected topics, making them available to the scientific community and to regulatory bodies.

As far as regarding the use of the results for regulation, some examples can be given:
-       many data on the effects on health and the environment of various pollutants (heavy metals, organic) have been used by the Commission to propose quality objectives for several substances (e.g.with regard to 129 dangerous substances related to the Council Directive 76/464/EEC, (4);
-       the results from sewage sludge research provide the scientific basis of a proposed directive on the use of sewage sludge in agriculture;

-   the elucidation of the conversion mechanisms of the
    atmospheric pollutants constitute the scientific basis for various
    regulations for the control of air pollution.

4. ORIENTATIONS FOR THE FUTURE

In preparing the proposal for a new 5-year programme, mentioned above,
the Commission took into account in particular:

a. the updating of the Environmental Policy of the Communities, laid
    down in the 3rd Environment Action Programme (5) and the
    priorities defined in this document;
b. the Research Policy of the Communities, as defined in the
    Framework Programme mentioned above.

Furthermore, a number of specific reports, issued by international
bodies or prepared on behalf of the Commission, have been considered.

The Commission took also account of the advice of the formal advisory
bodies, such as the Management and Co-ordination Advisory Committee
"Environment and Climatology" (CGC) and of the Concertation Committees of
various COST Actions.

In the following, the content of the programme covering 11 research
areas, is summarized. For a more detailed description, the already
mentioned proposal (1) may be consulted.

1. HEALTH EFFECTS OF POLLUTANTS

    -   chronic and late effects at low exposure levels and early
        indicators of health effects;
    -   epidemiology and exposure trends;

2. ECOLOGICAL EFFECTS OF POLLUTANTS

    -   effects on sensitive key species;
    -   effects on ecosystems;

3. ASSESSMENT OF CHEMICALS

    -   development and assessment of testing procedures;
    -   replacement of vertebrates used for toxicity testing;
    -   structure/activity relationships;
    -   evaluation of chemicals;

4. AIR QUALITY

    -   analysis, sources, transport, transformation of pollutants
        (COST project 611 with complementary contract research);
    -   effects of air pollution on the natural environment (COST
        project 612 with complementary contract research);
    -   effects of air pollution on materials;
    -   stratospheric chemistry;
    -   remote sensing techniques;
    -   indoor air quality (by concerted action);

5. WATER QUALITY

- analytical methods (COST Project 641);
- biotic and abiotic degradation of pollutants (COST
  project 641);
- eutrophication;
- remote sensing techniques;

6. SOIL QUALITY (incorporating part of COST project 681)

- analytical methods;
- behaviour of pollutants on soil;
- effects of pollutants on soil;
- effects of agricultural and forestry practice on soil
  quality;

7. NOISE RESEARCH

8. ECOSYSTEM RESEARCH

- basic research on the functioning of ecosystems
  (incorporating also COST project 647);
- effects of agricultural practice and urbanisation on
  ecosystems, loss of genetic diversity;
- environmental oceanography;
- biogeochemical cycles;
- conservation of flora and fauna (new concerted action);

9. WASTE RESEARCH

- waste management;
- organic wastes (incorporating part of COST project 681);
- toxic and dangerous waste;
- abandoned disposal sites;

10. REDUCTION OF POLLUTION

- advanced abatement technologies;
- clean technologies;

11. SCIENTIFIC BASIS OF ENVIRONMENTAL LEGISLATION AND MANAGEMENT

The auditory of this conference may be particularly interested in
the future of COST project 641. This project has been
incorporated essentially unchanged with regard to the
redefinition of its scientific content in 1983. It covers the
following areas:

a. Analytical Methodologies and Data Treatment

- Basic analytical techniques, including sampling and sample treatment, gas chromatography, high pressure liquid chromatography, mass spectrometry;

- Specific analytical problems, in particular analysis of selected classes of compounds, such as those likely to be regulated by Council Directive 76/464/EEC, chlorinated paraffins, tensides, optical brightners, metal-organic and organo-phosphorous compounds;

- Collection and treatment of analytical data

b. Physical/Chemical Behaviour of Organic Micropollutants in the Aquatic Environment

- distribution and transport mechanisms
- structure/activity relationships
- bioavailability and bioaccumulation

c. Transformation Reactions in the Aquatic Environment

- chemical and photochemical reactions
- biological transformations

d. Behaviour and Transformation of Organic Micropollutants in Water Treatment Processes

- infiltration
- waste water treatment
- drinking water treatment (including haloform formation).

The Concertation Committee will certainly in the near future review the programme in the light of the results of this symposium.

REFERENCES

1. Commission of the European Communities (1985).
Proposal for a Council Decision adopting three multiannual Research and Development programmes in the field of Environment, 1986 - 1990, COM (85) 391 final of 24 July 1985.

2. A. KLOSE, G. ANGELETTI, C. WHITE; Integration of the Environment Research Action Programme in the Framework Programme for Community Scientific and Technical Activities 1984 - 1987.
Proceedings of the Third European Symposium on "Analysis of Organic Micropollutants in Water", held in Oslo, 19 - 21 September 1983 (EUR 8518); D. Reidel Publishing Company.

3. Commission of the European Communities (1985).
   Evaluation of the Community's Environmental Research
   Programmes (1976 - 1983); Research Evaluation - Report N° 14,
   XII/720/84-EN.

4. Council of the European Communities (1982).
   List of substances which could belong to List I of Council
   Directive 76/464/EEC.
   O.J. N° C 176 of 14 July 1982.

5. Council of the European Communities (1983);
   Council Resolution of 7 February 1983 on the Third
   Environmental Action Programme 1982 - 1986.
   O.J. N° C 46 of 17 February 1983.

LIST OF PARTICIPANTS

AUSTRIA

P.B. CZEDIK-EYSENBERG
Fachverband der Chemischen
Industrie Österreich
Ketzergasse 471
A-1238 Wien-Rodaun

D. DWORSKY
Raasdorferstr. 60
A-2301 Gr. Enzersdorf

G. FRAUENWIESER
Österr. Bundesinstitut für
Gesundheitswesen
Stubenring 6
A-1010 Wien

P. HANKE
Univer GmbH -Werk
Simmering
7. Haidequerstr. 4
A-1110 Wien

C. LEOPOLTER
Bundesanstalt f. Lebensmittel-
untersuchung
Kinderspitalgasse 15
A-1090 Wien

F. OESTERREICHER
Bundesanstalt f. Lebensmittel-
untersuchung
Kinderspitalgasse 15
A-1090 Wien

R. MECL
Bundesanstalt f. Lebensmittel-
untersuchung
Kinderspitalgasse 15
A-1090 Wien

K. STANGEL
Bundesministerium für
Gesundheit und Umweltschutz
Stubenring 1
A-1010 Wien

I. VOVK
International Atomic Energy
Agency
Wagramerstrasse 5
P.O. Box 100
A-1400 Wien

R. WEISS
Österr. Wasserwirtschafts-
verband
An der Hülben 4
A-1010 Wien

BELGIUM

S. BONOTTO
SCK/CEN
Centre d'Etude de l'Energie
Nucléaire
Boerentang 200
B-2400 Mol

H. DE HENAU
Procter & Gamble
100 Tenselaan
B-1820 Strombeek-Bever

C. DEWAELE
Laboratory of Organic Chemistry
State University of Ghent
Krijgslaan 281 S4
B-9000 Gent

D. QUAGHEBEUR
Instituut voor Hygiëne
en Epidemiologie
J. Wytsmanstraat 14
B-1050 Brussel

I. TEMMERMAN
Instituut voor Hygiëne
en Epidemiologie
J. Wytsmanstraat 14
B-1050 Brussel

F. VAN HOOF
Study Syndicate for Water
Research
C/o Antwerp Waterworks
Mechelsesteenweg 64
B-2018 Antwerpen

CANADA

F.W. KARASEK
Chemistry Department
University of Waterloo
Waterloo, Ontario
Canada, N2L-3E1

DENMARK

A. BÜCHERT
National Food Institute
19 Morkhoj Bygade
DK-2860 Soborg

A. DAMBORG
Water Quality Institute
11, Agern Allé
DK-2970 Horsholm

J. FOLKE
Water Quality Institute
11, Agern Allé
DK-2970 Horsholm

C. GRON
Water Quality Institute
11, Agern Allé
DK-2970 Horsholm

N. HANSEN
Levnedsmiddelkontrollen I/S
1 Dyregaardsvej
DK-2740 Skovlunde

P. LINDGAARD-JORGENSEN
Water Quality Institute
11, Agern Alle
DK-2970 Horsholm

FEDERAL REPUBLIC OF
GERMANY

P. HENSCHEL
Umweltbundesamt
Bismarckplatz 1
D-1000 Berlin 33

O. HÜTZINGER
University of Bayreuth
Postfach 3008
D-8580 Bayreuth

S. KANOWSKI
Umweltbundesamt
Bismarckplatz 1
D-1000 Berlin 33

P.G. LAUBEREAU
Hessische Landesanstalt
für Umwelt
Aarstrasse 1
D-6200 Wiesbaden

K. LEVSEN
Universität Bonn
Abteilung Physikalische Chemie
Wegelerstr. 12
D-5300 Bonn

B. SCHINK
Universität Konstanz
Fakultät für Biologie
Postfach 5560
D-7750 Konstanz

FINLAND

L. KRONBERG
Department of Organic
Chemistry
Abo Akademi
Akademigatan 1
SF-20500 Abo

I. KUNINGAS
Helsinki City Water and
Sewage Work
Kyläsaarenkatu 10
SF-00550 Helsinki

M. MÄKELÄ
National Board of Waters
P.O. Box 250
SF-00101 Helsinki

M. SALKINOJA-SALONEN
University of Helsinki
Dept. of General Microbiology
Mannerheimintie 172
SF-00280 Helsinki

S. REKOLAINEN
Water Research Institute
P.B. 250
SF-00101 Helsinki

FRANCE

P. ARPINO
Ecole Polytechnique
Department of Chemistry
Route de Saclay
F-91128 Palaiseau

C. HENNEQUIN
I.R.C.H.A.
F-91710 Vert-le-Petit

M. MARCHAND
IFREMER
Centre de Brest
BP. 337
F-29273 Brest

R. MASSOT
Commissariat à l'Energie
Atomique
CEA. CENG.
B.P. 85 X
F-38041 Grenoble Cedex

GREECE

N. MIMICOS
NRC "DEMOCRITOS"
Aghia Paraskevi Attikis
GR-15210 Athens

E. STEPHANOU
Dept. of Environmental Chemistry
Chemical Technology University
of Crete
P.O. Box 1470
GR-Iraklion, Crete

HUNGARY   P. PASZTO
                                    Anstalt für Wasserwirtschaft
                                    Alkotmany u. 29
                                    H-1054 Budapest

IRELAND   P.J. FLANAGAN
                                    An Foras Forbartha
                                    St. Martin's House
                                    Waterloo Road
                                    IRL-Dublin 4

                                    B. CALLAN
                                    An Foras Forbartha
                                    St. Martin's House
                                    Waterloo Road
                                    IRL-Dublin 4

                                    C. O'DONNELL
                                    An Foras Forbartha
                                    St. Martin's House
                                    Waterloo Road
                                    IRL-Dublin 4

ITALY   D. BOTTA
                                    Politecnico
                                    Dipartimento di Chimica
                                    Industriale e Ingegneria
                                    Chimica
                                    Piazza Leonardo da Vinci 32
                                    I-20133 Milano

                                    S. GALASSI
                                    Istituto di Ricerca sulle
                                    Acque - CNR
                                    Via Occhiate
                                    I-Brugherio (Milano)

                                    T. LA NOCE
                                    Istituto di Ricerca sulle
                                    Acque - CNR
                                    Via Reno 1
                                    I-00198 Roma

A. LIBERATORI
Istituto di Ricerca sulle
Acque - CNR
Via Reno 1
I-00198 Roma

E. MANTICA
Politecnico - Dipartimento
Chimica Industriale e Ingegneria
Chimica
Piazza Leonardo da Vinci 32
I-20133 Milano

NORWAY   **A. BJØRSETH**
SCATEC
P.O. Box 147
N-1312 Slependen

E. GJESSING
Norwegian Institute for
Water Research
P.O. Box 333
Blindern
N-Oslo 3

S. JOHNSEN
Centre for Industrial Research
P.O. Box 350
Blindern
N-0314 Oslo 3

SPAIN   **E. BAONZA DEL PRADO**
Centro de Estudios Hidrograficos
Paseo Bajo Virgen del Puerto 3
E-28006 Madrid

J. CAIXACH GAMISANS
Laboratoria Espectrometria de
Masses - IQBO - CSIC
J. Girona Salgado 18-26
E-08034 Barcelona

A. CHICOTE AYUSO
Centro de Estudios Hidrograficos
Paseo Bajo Virgen del Puerto 3
E-28006 Madrid

J. GONŻALES-NICOLAS
Centro de Estudios Hidrograficos
Paseo Bajo Virgen del Puerto 3
E-28006 Madrid

J. RIVERA ARANDA
Consejo Superior
Investigaciones Sientificas
Jorge Girona Salgado S/N
E-Barcelona 3

F. VENTURA
Laboratoria Espectrometria de
Masses - IQBO - CSIC
J. Girona Salgado 18-26
E-08034 Barcelona

SWEDEN

H.A. GRANMO
National Swedish Environment
Protection Board
Kristineberg Marine Biological
Station P1 2130
S-45034 Fiskebäckskil

B. JOSEFSSON
Dept. Analytical and Marine
Chemistry
Chalmers University of
Technology
S-41296 Gothenburg

L. RENBERG
National Swedish Environment
Protection Board
Box 1902
S-17125 Solna

G. SUNDSTRÖM
National Swedish Environment
Protection Board
Box 1902
S-17125 Solna

SWITZERLAND

M. AHEL
EAWAG
Swiss Federal Institute for
Water Resources and Water
Pollution Control
CH-8600 Dübendorf

D. IMBODEN
EAWAG
Swiss Federal Institute for
Water Resources and Water
Pollution Control
CH-8600 Dübendorf

E. KUHN
EAWAG
Seenforschungslaboratorium
CH-6047 Kastanienbaum

M. MÜLLER
FAW
Swiss Federal Research
Station Wädenswil
CH-8820 Wädenswil

C. SCHAFFNER
EAWAG
Swiss Federal Institute for
Water Resources and Water
Pollution Control
CH-8600 Dübendorf

R. SCHWARZENBACH
EAWAG
Swiss Federal Institute for
Water Resources and Water
Pollution Control
CH-8600 Dübendorf

J. ZEYER
EAWAG
Seenforschungslaboratium
CH-6047 Kastanienbaum

THE NETHERLANDS

J.P. BOON
DGW - Rijkswaterstaat
P.O. Box 8039
NL-4330 MA Middelburg

H. GOVERS
Institute for Environmental
Studies
Free University
P.O. Box 7161
NL-1007 MC Amsterdam

H. HEIDA
Environmental Research
Laboratory
City of Amsterdam
Amstelveenseweg 88-90
NL-1075 XY Amsterdam

H. KÖENEMANN
Ministry of Housing, Physical
Planning, and Environment
P.O. Box 450
NL-2260 MB Leidschendam

A. OPPERHUIZEN
Laboratory of Environmental
and Toxicological Chemistry
Nieuwe Achtergracht 166
NL-1018 WV Amsterdam

J.R. PARSONS
Laboratory of Environmental
and Toxicological Chemistry
University of Amsterdam
Nieuwe Achtergracht 166
NL-1018 WV Amsterdam

C. WOLFF
Shell Internationale
Petroleum Maatschappij B.V.
P.O. Box 162
NL-2501 AN The Hague

G. SCHRAA
Department of Microbiology
Agricultural University
H. Van Suchtelenweg 4
NL-6703 CT Wageningen

S.M. SCHRAP
Laboratory of Environmental
and Toxicological Chemistry
University of Amsterdam
Nieuwe Achtergracht 166
NL-1018 WV Amsterdam

T.H.M. SIJM
Laboratory of Environmental
and Toxicological Chemistry
University of Amsterdam
Nieuwe Achtergracht 166
NL-1018 WV Amsterdam

R. VAN DEN BERG
Netherlands Institute for
Fishery Investigations
Haringkade 1
NL-1976 CP Ijmuiden

P.C.M. VAN NOORT
Rijksinstituut voor Volks-
gezondheid en milieuhygiëne
A. van Leeuwenhoeklaan 9
NL-3720 BA Bilthoven

R.C.C. WEGMAN
National Institute of
Public Health and Environmental
Hygiene
P.O. Box 1
NL-3720 BA Bilthoven

UNITED KINGDOM

B. CRATHORNE
Water Research Centre
Environment
Henley Road
UK-Marlow, Bucks. SL7 2HD

A. DOBBS
Water Research Centre
Medmenham Laboratory
P.O. Box 16
UK-Marlow, Bucks. SL7 2HD

B.J. HARLAND
Imperial Chemical Industries PLC
Brixham Laboratory
Freshwater Quarry
UK-Brixham, Devon TQ5 8BA

H.A. PAINTER
Water Research Centre
Medmenham Laboratory
P.O. Box 16
UK-Marlow, Bucks. SL7 2HD

C.B. WALLER
Thames Water
Technical Investigation Service
Nugent House
Ground Floor South, RBH
Vastern Road
UK-Reading, Berks. RG1 8DB

YUGOSLAVIA

V. DREVENKAR
Institute for Medical Research
and Occupational Health
M. Pijgade 158
YU-41000 Zagreb

Z. FRÖBE
Institute for Medical Research
and Occupational Health
M. Pijgade 158
YU-41000 Zagreb

COMMISSION OF THE EUROPEAN
COMMUNITIES

G. ANGELETTI
DG XII/G - 1
Rue de la Loi 200
B-1049 Brussels

H. OTT
DG XII/G - 1
Rue de la Loi 200
B-1049 Brussels

D. NICOLAY
DG XIII/A - 2
Bâtiment Jean Monnet
Plateau du Kirchberg
L-2920 Luxembourg

K.K. JOHANNSON
DG XII/G - 1
Rue de la Loi 200
B-1049 Brussels

W. PENNING
Joint Research Centre
CEC - ISPRA
I-21020 Ispra (Varese)

INDEX OF AUTHORS